W9-CRX-682

WITHD...
CALTECH LIBRARY...

PLEASE STAMP DATE DUE, BOTH BELOW AND...

DATE DUE | DATE DUE | DATE DUE

Solar Phenomena in Stars and Stellar Systems

NATO ADVANCED STUDY INSTITUTES SERIES

*Proceedings of the Advanced Study Institute Programme, which aims
at the dissemination of advanced knowledge and
the formation of contacts among scientists from different countries*

The series is published by an international board of publishers in conjunction
with NATO Scientific Affairs Division

A	Life Sciences	Plenum Publishing Corporation
B	Physics	London and New York
C	Mathematical and Physical Sciences	D. Reidel Publishing Company
		Dordrecht, Boston and London
D	Behavioural and Social Sciences	Sijthoff & Noordhoff International
		Publishers
E	Applied Sciences	Alphen aan den Rijn and Germantown
		U.S.A.

Series C – Mathematical and Physical Sciences

Volume 68 – Solar Phenomena in Stars and Stellar Systems

QB
520
N37
1980
cop. 1

ASTR

Solar Phenomena in Stars and Stellar Systems

*Proceedings of the NATO Advanced Study Institute
held at Bonas, France, August 25-September 5, 1980*

edited by

ROGER M. BONNET
*Centre National de la Recherche Scientifique,
Verrieres-le-Buisson, France*

and

ANDREA K. DUPREE
*Harvard Smithsonian Center for Astrophysics,
Cambridge, Mass., U.S.A.*

D. Reidel Publishing Company

Dordrecht : Holland / Boston : U.S.A. / London : England

Published in cooperation with NATO Scientific Affairs Division

Library of Congress Cataloging in Publication Data

NATO Advanced Study Institute (1980: Bonas, France)
    Solar phenomena in stars and stellar systems.

    (NATO advanced study institutes series. Series C, Mathematical and
physical sciences ; v. 68)
    'Published in cooperation with NATO Scientific Affairs Division.'
    Includes index.
    1. Sun–Congresses. 2. Stars–Congresses. I. Bonnet, R. M.
(Roger M.) II. Dupree, Andrea K. III. North Atlantic Treaty
Organization. Division of Scientific Affairs. IV. Title. V. Series.
QV520.N37   1980   523.8   81–5065
ISBN 90–277–1275–1   AACR2

Published by D. Reidel Publishing Company
P.O. Box 17, 3300 AA Dordrecht, Holland

Sold and distributed in the U.S.A. and Canada
by Kluwer Boston Inc.,
190 Old Derby Street, Hingham, MA 02043, U.S.A.

In all other countries, sold and distributed
by Kluwer Academic Publishers Group,
P.O. Box 322, 3300 AH Dordrecht, Holland

D. Reidel Publishing Company is a member of the Kluwer Group

All Rights Reserved
Copyright © 1981 by D. Reidel Publishing Company, Dordrecht, Holland
No part of the material protected by this copyright notice may be reproduced or utilized
in any form or by any means, electronic or mechanical, including photocopying,
recording or by any informational storage and retrieval system,
without written permission from the copyright owner

Printed in The Netherlands

TABLE OF CONTENTS

INTRODUCTION

This book represents the proceedings of a NATO Advanced Study Institute which was held at Bonas from August 25 till September 5, 1980 and was devoted to the study of "Solar Phenomena in Stars and Stellar Systems". It is intended for a broad audience. Students and post-doctoral scientists for example can discover new aspects of astrophysics. The general spirit of the ASI was aimed at presenting a unified aspect of astrophysical phenomena which can be studied intensively on the Sun although they are of a much more general nature. On the other hand, specialists in solar or stellar physics will find here the latest theoretical developments and/or the most recent observations made in their own field of research. An extensive bibliography will be found throughout the various sections, to which the reader may refer, for more detailed developments in various specific areas.

In the past, stellar and solar astrophysics have more or less followed their own independent tracks. However, with the rapid development of modern techniques, in particular artificial satellites like the International Ultraviolet Explorer and the Einstein Observatory, which provide a new wealth of data, it appears that chromospheres, coronae, magnetic fields, mass loss and stellar winds, etc.. . ., are found not only in the Sun but occur also in other stars. Frequently these other stars represent quite different conditions of gravity, luminosity, and other parameters from those occurring in the Sun.

The Sun is no longer an isolated astrophysical object but serves the role of representing the basic element of comparison to a large class of objects. The book reviews these phenomena as exhaustively as possible and a generalization is constantly attempted. When necessary, as for instance in the case of solar flares, the problems are also studied from the basic physics point of view.

The volume is separated into 4 sections. An overview of the problem is given in the first section. The physics of Stellar Interiors is reviewed in Section 2. Section 3 deals with the crucial aspect of existence and physics of chromospheres and coronae and how they relate to the presence of convective phenomena in stellar subsurface layers. Section 4 addresses the

R. M. Bonnet and A. K. Dupree (eds.), Solar Phenomena in Stars and Stellar Systems, ix–x.
Copyright © 1981 by D. Reidel Publishing Company.

question of variability which might be of interest   not   only   to
the astrophysicist but also to those who want to know whether our
star will present long term luminosity variations in relation   to
the climate on the earth.   These four sections reflect the
overall breakdown of the sessions which kept the <u>86</u> scientists of
the ASI busy for nearly two complete weeks.

The organization of this ASI and the edition   of   this   book
would   not   have been possible without the generous assistance of
NATO, the U.S. National Science Foundation, The   Centre   National
de la Recherche Scientifique (CNRS), the Centre National d'Etudes
Spatiales (CNES), the Delegation Generale a la   Recherche   Scientifique et Technique (DGRST),   the   French Ministry of Foreign
Affairs and the personnel of the Laboratoire   de   Physique   Stel-
laire et Planetaire (LPSP) and of the hospitality and ambience to
be found at the Chateau de Bonas.   The   Editors   wish   to   acknowledge them all here for their total and generous devotedness.

THE SUN AS A STAR:
SOLAR PHENOMENA AND STELLAR APPLICATIONS

Robert W. Noyes

Harvard-Smithsonian Center for Astrophysics
Cambridge, Massachusetts 02138 USA

1.   INTRODUCTION.

Our Sun is a run-of-the-mill star, having no obvious
extremes of stellar properties. For this reason it is perhaps
more, rather than less, interesting as an astrophysical object,
for its sameness to other stars suggests that in studying the
Sun, we are studying at close hand common, rather than unusual
stellar phenomena. Conversely, comparative study of the Sun and
other solar-type stars is an invaluable tool for solar physics,
for two reasons: First, it allows us to explore how solar
properties and phenomena depend on parameters we cannot vary on
the Sun--most fundamentally, rotation rate and mass. Second,
study of solar-like stars of different ages allows us to see how
stellar and solar phenomena depend on age; study of other stars
may be one of the best ways to infer the earlier history of the
Sun, as well as its future history. In this review we shall
concentrate on phenomena common to the Sun and solar-type (main
sequence) stars with different fundamental properties such as
mass, age, and rotation.

The Sun's age may be estimated from terrestrial geologic
evidence as about 4.5-5 x 10^9 years. It is impossible to
determine the ages of other main sequence stars in the solar
neighborhood to anything like the accuracy of our knowledge of
the Sun's age. Detailed calculations of luminosity evolution for
a few stars whose evolution has carried them noticeably above the
main sequence (Perrin et al. 1977) indicate that some nearby
solar-like stars have ages considerably in excess of the Sun.
Younger stars which have not yet begun to evolve off the main
sequence are much harder to date; and the age of most other field

1

R. M. Bonnet and A. K. Dupree (eds.), Solar Phenomena in Stars and Stellar Systems, 1–31.
Copyright © 1981 by D. Reidel Publishing Company.

stars relative to the Sun is poorly known. However, the ages of
younger clusters (e.g. the Hyades and Pleiades) are much easier
to date due to the rapid evolution off the main sequence of their
more massive members, so that under the assumption of synchronous
birth of stars in clusters, it is possible to identify samples of
solar-type stars with reasonably well-defined ages younger, as
well as older, than the Sun.

Originally, as expressed in the Russell-Vogt theorem, it was
thought that knowledge of the mass and the chemical abundance
distribution throughout the Sun is sufficient to determine its
present character, and its past and future evolution, uniquely.
If this theorem were strictly correct, solar and stellar physics
by now would be largely a closed book. However, the theorem did
not include another set of parameters that give rise to some of
the most interesting aspects of solar-stellar physics, and the
main subject matter of this Institute. These parameters are
stellar rotation and, in close relation, stellar magnetic fields.

2. THE SUN AND THE ROTATION-ACTIVITY-MAGNETIC FIELD-AGE
CONNECTION.

'In the past 15 years, as a result of pioneering work of
Wilson (1966), Kraft (1967), Skumanich (1972) and others, a very
satisfactory empirical scenario has been developed, whereby many
aspects of stellar behavior come together in a plausible physical
fashion. The basis of this scenario is the fact that stars later
than about F5 have well-developed convection zones. A small
part of the non-radiative energy carried upward in convective
motions creates a high-temperature corona, and the corona in turn
gives rise to a thermally-driven stellar wind (as we discuss
below, the details of creation of the corona and wind have
recently changed markedly with the realization of the importance
of magnetic fields in that process, but the overall scenario is
not seriously changed). Due to the presence of magnetic fields
on late-type stars (originally simply inferred from stellar Ca II
observations on the basis of the observed Ca II-magnetic field
relation in the Sun) the stellar wind is expected to co-rotate
out to the Alfven radius r_A (Weber and Davis 1967), where the
magnetic energy density $B_A^2/8\pi$ falls below the kinetic energy
density $1/2 \rho_A v_A^2$ of the stellar wind flow. B_A and ρ_A are the
magnetic field strength and gas density at the Alfven radius. r_A
is the "lever arm" for angular momentum loss by the stellar wind,
and is many times the stellar radius R_*; (in the Sun $r_A/R_\odot \sim 20$).
Therefore the angular momentum loss $2/3 r_A^2 \dot{M} \Omega$, (where $\dot{M}$ is the mass
loss rate, and Ω the angular velocity of the solar wind at the
Alfven radius) is greatly enhanced due to the presence of the
magnetic field. Thus stars with convective zones (and surface
fields) rapidly spin down to the low angular velocities

characteristic of most stars later than F5.

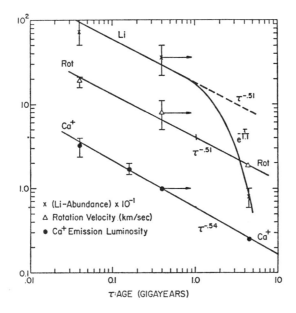

Figure 1.   Ca$^+$ emission, rotation, and lithium abundance _versus_ stellar age. From Skumanich (1972).

Skumanich (1972) showed that, empirically, the spindown of
surface rotation speed v_{rot} for these stars seems to obey the law
$v_{rot} \sim T^{-1/2}$ where T is the age of the star (Figure 1).   In
addition, he found that the Ca II K-line flux of main sequence
stars also declines with age according to the   same   law (see
Figure 1). This relation is of particular significance because,
as pointed out by Babcock and Babcock (1955) and Leighton (1959),
there is one-to-one corespondance between areas on the Sun with
strong surface fields and areas of strong overlying chromospheric
Ca II emission. This correlation has been put on a quantitative
basis by Frazier (1970) and Skumanich _et al._ (1975); the relation
between field strength and Ca II emission in the Sun is roughly
monotonic from weak-field regions in the quiet Sun all the way to
bright plages. While considerable scatter in the data shows that
more than simply field _strength_ determines chromospheric heating,
the observed chromospheric emission level is a useful indicator
of the mean field strength per unit area. To the extent that the
rather tight relation for the Sun may be extended to solar-like
stars, we may infer from the data in Figure 1 that in stars the
surface magnetic field strength decreases with age, and at the
same rate as the decrease of rotation.   Direct   measurements of

surface fields on other stars would of course be very welcome in
putting this inference on a more solid footing.

Skumanich's findings were the first evidence of a rotation-
activity correlation in stars, where activity here refers both to
surface magnetic fields and the closely related chromospheric and
coronal emission enhancements. Throughout these proceedings
there will be considerable discussion of more modern evidence for
the rotation-activity connection.

The Sun, indicated in Figure 1 by the symbols at 4.5 billion
years age, plays an important role in determining the $T^{-1/2}$
relation for both the rotation rate and the Ca II emission. More
recently, questions have been raised whether the Sun does in fact
lie nicely along the $T^{-1/2}$ relation defined by cluster and field
stars, or whether the Sun's Ca II emission and rotation rate are
anomalously low. Thus, Blanco et al. (1974) in comparing Ca II
emission of the quiet Sun and solar plages with integrated Ca II
emission from field stars, deduce that the Ca II emission from
the Sun (including active regions) lies about a factor 3 below
that of solar-like field stars of the same age.   In addition,
Smith (1979), by using a new technique to measure rotational
broadening of spectral lines, has concluded that solar type
stars, while following a $T^{-1/2}$ rotation-age law, are on the
average rotating some 2.5 times faster than found by Skumanich.
This would mean that the Sun, if it were "average," would have an
equatorial rotation speed of 5 km/sec rather than its actual
value of 1.9 km/sec.

If both Smith's and Blanco et al.'s results are correct, the
Sun still shows roughly the appropriate rotation-activity
relation, but is anomalously deficient in both quantities for its
age. However, both Blanco et al.'s results and Smith's results
should be treated with some caution. Blanco et al. selected some
of their stars for comparison from Wilson's (1968) list of
candidates for a search for Ca II activity cycles, and this
selection effect may bias the stellar fluxes upward. Smith's
spin-down relation for solar-type stars shows considerable
scatter when individual stars, rather than binned averages, are
considered, at least partly because age determinations for
individual stars are too uncertain. In short, it appears that
more data on rotation and activity of solar-like field stars is
needed before it can be concluded whether the Sun is anomalous.

In principle, one could measure the solar spin-down$_2$ torque
directly and thus infer whether it has followed the $T^{-1/2}$ law of
Figure 1. As Zirker (1981) shows in these proceedings, a crude
estimate of the angular momentum loss from the observed solar
wind flux and an estimate of the Alfven radius r_A yields a
spindown rate $\dot{L}/L \sim (10^{10}$ years$)^{-1}$. This estimate simply shows

that the solar wind can in principle have played its postulated role in the scenario discussed above, but it lacks the accuracy required for detailed spindown calculations. Detailed spindown calculations would require knowledge of a) the three-dimensional structure of the solar wind, b) the photospheric field and its extrapolation up to the Alfven radius, c) the history of variability of both of these quantities over the lifetime of the Sun, and d) the coupling of surface layers suffering spindown with deeper internal layers that may contain much of the angular momentum resident in the Sun. It is not surprising that detailed estimates of solar spindown vary considerably. Recently, for example, Mochnacki (1980) has concluded that the interplanetary magnetic field is significantly lower than previously estimated, from which he concludes that the Alfven radius lies at only $12R_{\odot}$, and the spindown time is $\sim 5 \times 10^{10}$ years. Such a spindown rate, if constant over the lifetime of the Sun, would imply an initial rotation rate barely larger than the present one, leading to the conclusion that because of some accident of birth solar rotation was, initially, anomalously low. No firm conclusion can be drawn at the present time since solar wind data cannot yet provide definitive clues on spindown, so all possibilities from a relatively rapidly rotating initial Sun with large spindown to a slowly rotating initial Sun with small spindown should be considered viable.

A recent survey of late-type stars in the solar neighborhood (Vaughan and Preston 1980) suggests that the rotation and/or activity history of the Sun could have been more complex, in a potentially very interesting way. Figure 2 (Vaughan and Preston 1980) shows the relation between Ca II emission and B-V color i.e. for 396 solar neighborhood field stars. In this figure the Sun appears as a vertical line at B - V = 0.66, extending over a range spanning Ca II emission levels at minimum and maximum solar activity. It is interesting to note that even at activity minimum the Sun shows more Ca II emission than many other stars of the same spectral type. These less active stars are inferred by Vaughan and Preston to be older than the Sun (from kinematical and evolutionary arguments); yet some of these, whose activity variation was surveyed by Wilson (1978, see also Vaughan 1980), still showed long-term variability. Thus even at solar minimum the Sun has a Ca II emission level above some stars with variable chromospheres.

More significant in Figure 2 is the presence of an apparent gap in the distribution of Ca II emission, dividing the stars with $.5 \lesssim (B - V) \lesssim 1.0$ into two groups, of relatively higher and lower Ca II emission level. This gap appears to be a real effect and not a selection effect in the data. Vaughan and Preston note that stars above the gap appear to be younger than stars below the gap, from kinematical arguments; this of course would also be implied by the age-activity relation discussed earlier.

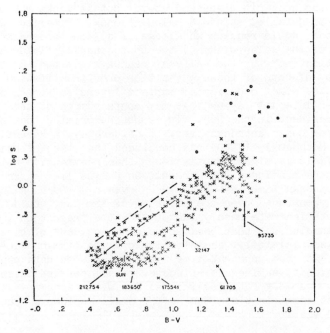

Figure 2. The flux index S, measuring Ca II H and K
emission relative to the continuum, for 396 solar
neighborhood main sequence stars, plotted versus broad-
band color, B-V. The dashed lines indicate position of
Hyades stars. The "gap" discussed in the text lies
below and parallel to the dashed lines. From Vaughan
and Preston 1980.

The apparent reality of the gap in Figure 2 raises some
interesting questions. Using the Skumanich scaling law and
working back from the age of the Sun and its location in Figure
2, one may infer that solar-type stars (B - V ~ .66) should have
traversed the gap at age perhaps ~ 10^9 years. (Of course this
conclusion depends on whether the Sun did in fact follow a $T^{-1/2}$
relation in its activity and rotation, and we have already seen
the possibility cannot be ruled out that the Sun has always been
a slow rotator.) If solar-like stars did indeed cross such a gap
at age ~ 10^9 years, this implies that their Ca II emission
rapidly decreased at that time. Such a more-or-less
discontinuous decrease in level of (magnetic) activity at a
certain epoch in a star's evolution would have obvious importance
to dynamo theory--a fact to which we shall return in the next
section.

3. SOLAR AND STELLAR ACTIVITY CYCLES

The most fundamental aspect of the Sun's magnetism and activity is its quasi-periodic activity cycle. The question of how such an activity cycle may be manifested on other stars is one of the major topics for discussion at this Institute. Here we shall briefly review the phenomenology of the solar activity cycle, with particular attention to aspects that may hold clues to its origins in the magnetic dynamo. Later we shall try to place the observed phenomena of solar activity in a broader stellar context.

Sunspots are dark regions with strong magnetic fields that appear on the solar surface, last for a few days to several months, and then decay and disappear. Their surface brightness at 6000A is only about 10% of the photosphere, due to their cool temperature ($T_{eff} \sim$ 4200K in a spot, about 1600K less than in the photosphere). Magnetic fields within spots are typically about 3000 gauss, giving a magnetic pressure $B^2/8\pi$ that makes up the pressure deficit of the cool gas within the spot relative to the outside photosphere. Thus spots appear to be approximately in magnetohydrostatic equilibrium.

Sunspots often occur in groups, sometimes of great complexity. Often a group has a predominant spot located in the western (or leading, in the sense of rotation) part of the group, and one or more following spots trailing after. With few exceptions the leading and following spots are of opposite magnetic polarities, while all leading spots in a given hemisphere have the same polarity and all leading spots in the other hemisphere have the opposite polarity.

Spots wax and wane in numbers with an approximate 11-year period (Figure 3), defining the activity cycle. A cycle is defined to begin when spots first appear at relatively high north and south latitudes. Subsequent spot appearances are at ever-decreasing latitudes until the spots produced late in the cycle occur rather close to the equator.

During this entire episode of spot emergence the spot groups obey the magnetic polarity laws described above. However, after the disappearance of the last near-equatorial spots a new episode of spot emergence begins, again at high latitudes. The time between successive initiations of high-latitude spot emergence is about 11 years. However, in each new cycle the polarity relations are the reverse of those of the previous cycle. Thus the magnetic activity cycle is approximately 22 years.

It has been recently shown that another component of solar magnetic flux emergence may play an important role in the solar

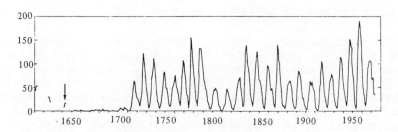

Figure 3. Annual mean sunspot numbers, AD 1610-1974, from Eddy, Gilman, and Trotter (1976). The "Maunder minimum" occurred from 1645 to 1715.

magnetic activity cycle. Golub, Davis, and Krieger (1979) report that X-ray bright points, which overlie small ephemeral bipolar magnetic regions, also appear to have a solar-cycle dependence of their emergence. The ephemeral active regions and bright points are like miniature active regions in magnetic structure and coronal heating, but differ in that their lifetime is only a few hours. Perhaps a more fundamental difference is that to first order, bright point and ephemeral active region emergence appears approximately to be in anti-phase with the sunspot cycle; i.e. maximum bright point emergence rate occurs near spot minimum, and vice versa. This was interpreted by Golub, Davis, and Krieger as a modulation of the spatial scale of magnetic flux emergence, with the result that the variation of the total flux on all scales is much less than the variation of sunspot flux alone. In fact, the data do not rule out the possibility that the total flux emergence rate is approximately constant over time.

The picture is much more complex than outlined above, due to the fact that bright points and ephemeral active regions seem to emerge more or less uniformly over the Sun, rather than solely in the preferred latitudes of active region emergence. Thus, bright points occur also in the quiet Sun, remoted from active regions, and at all latitudes including the poles. They appear prominently in coronal holes (see Section 4) because of their enhanced contrast there.

A third important type of photospheric magnetic flux concentration, in addition to sunspots and ephemeral active regions, is photospheric flux knots. These are localized regions of "quiet Sun" magnetic field, which are now recognized to be concentrated into small regions of size a few hundred km or less and field strength 1-2 kilogauss (Stenflo 1976). As a result of increased spatial resolution of modern solar instrumentation, there is today no solid evidence for non-zero field concentrations substantially less than 1 kilogauss; this curious and potentially very important result is discussed in these proceedings by Parker (1981).

Surrounding sunspots are <u>photospheric</u> <u>plages</u>, regions of photospheric field concentration averaging a few hundred gauss. High-resolution magnetic data suggest, however, that their magnetic field strength is also in the 1-2 kilogauss range of other photospheric flux concentrations. The difference between plages and quiet-Sun magnetic field concentrations could be one mainly of total flux per unit area rather than field strength, the area-averaged flux being much higher in plages than in the quiet Sun.

It is natural to ask whether evidence of magnetic activity can be detected directly on other solar-type stars. The most obvious signs of solar magnetic activity--sunspots--might be expected to produce a diminution of stellar brightness when spots cross the face of a star. There are two apparent difficulties with such expectations, one conceptual and one practical. The conceptual difficulty is that there is no <u>a priori</u> reason why the radiative flux missing in sunspots should not be simply compensated by a minute increase in flux in the much larger photospheric areas surrounding the spot. Such a "bright ring" effect has been long sought in solar observations, although it has not yet been found.

Recently, Foukal and Vernazza (1979) have carried out a cross-correlation study of Abbot's extensive set of data on the solar constant as monitored at the surface of the Earth, and have found both (a) a positive correlation with the presence of photospheric faculae on the disk, and (b) a negative correlation with the presence of sunspots on the disk. Since faculae, when observed near the limb, show a slight enhancement of continuum emission, both correlations are in the direction expected from the relative brightness of the respective surface features, and in addition the amount of the inferred brightness change ($\sim 7 \times 10^{-4}$) is approximately the expected value, if the surrounding photosphere were unaffected.

The fact that a signal is seen in the solar data implies that the energy "blocked" by spots does not appear instantaneously, and therefore must be temporarily "stored" in the convection zone to appear at a later time (perhaps spread over a large area or long time so that it could easily escape detection). Similarly the excess energy radiated by faculae must come from some subsurface storage agent (presumably the magnetic field) rather than by tapping the convective flux that is emerging concurrently. Foukal and Vernazza estimated that there is a phase lag of about one day between the diminution of brightness and the appearance of spots, suggesting that the convective energy flow is blocked at a depth such that the difference between the times for convective flow to the surface and buoyant rise of the magnetic flux was about a day. This is not inconsistent with the depth of $\sim 1.5 \times 10^4$ km often associated with supergranulation flow; however, in view of the noisiness of the solar constant data and the lack of precision in calculating the rate of flow of convective energy, and the dynamic effects of magnetic buoyancy, quantitative estimates are not yet possible. The important point is that spots and faculae do indeed seem to have a signature in the total brightness of the Sun as a star. It is likely that in the near future more precise data from solar constant monitors in space will pin down the important question of phase lags.

The second difficulty alluded to above in detecting solar-type spots on other stars is a practical one. Sunspots, because of their relatively small size relative to the solar radius, cover an aggregate of no more than about 10^{-3} of the solar disk even at sunspot maximum, so the diminution of brightness should at most be only 10^{-3} magnitude; hence very precise photometry would be required to detect starspots covering the same practical area on another star.

An obvious question is what determines the size of sunspots, and how that size might scale to other stars. Little is known about the first question. Spots generally fall in the size range 20,000 to 50,000 km, not far different from the characteristic size (~ 3000 km) of the large-scale convective cells known as supergranulation. It has been speculated (Bumba and Howard 1965; Mullan 1973) that the two phenomena are related, the size of both depending on the structure and depth of the convection zone. Rucinski (1979) assumed such a relation and, using convective envelope models, predicted that both supergranulation and spot sizes should decrease toward later-type main sequence stars.

The facts of the matter, as discussed in detail elsewhere in this Institute (Hartmann 1981a, Hall 1981), are that there is good evidence for large dark features on other stars, especially BY Dra and RS CVn stars, with many characteristics which suggest a close similarity to sunspots. The dark features could be either enormous single spots or large groups of relatively small spots. In any case, if these features are physically similar to sunspots, i.e. if they are formed as a consequence of strong concentrations of magnetic fields in the photosphere, their detailed physics may well be markedly different; the surface area covered is a significant fraction of the total stellar surface, and they can be expected to alter the star's surface properties or its subsurface convective flow in a way quite beyond the capabilities of sunspots.

A strong proof of a physical similarity between dark features in stellar photospheres and sunspots would be the direct detection of magnetic fields in spots themselves, or at least in the surrounding active regions. Unfortunately, detection of fields in spots themselves is very difficult, just by virtue of the fact that spots are dark. To get around this it might be useful to make observations of candidate spot stars in the infrared, where sunspots appear brighter relative to the photosphere, and to record spectral lines which are preferentially formed in cooler regions (cf. Ramsey and Nations 1980) and which have large Zeeman sensitivity.

Recently Robinson, Worden, and Harvey (1980) have reported the detection of magnetic fields associated with plages on the active-chromosphere stars χ Boo A and 70 Oph A. Their observing method does not measure the net magnetic field from the star, as detected by traditional Babcock-type polarization observations, for the net field may well be very small: leading and following polarities would tend to cancel, leaving a very small net residual. (On the Sun this residual amounts to only about one gauss, well below the $\sim$ 100 gauss sensitivity of Babcock-type analyses; see e.g. Severny et al. 1970). However, by recording the differential broadening of Zeeman-sensitive and Zeeman-insensitive lines, without regard to polarization, it is possible to measure both the average value of the plage fields (from the amount of broadening) and the fractional area of the star covered with magnetic fields (from the relative strength of the broadened component of the Zeeman-sensitive line relative to its un-broadened component). Values of about 2 kilogauss were obtained, not unlike solar values, and fractional coverage of 10–40% was found, somewhat larger than for the Sun. This new technique, if verified and extended to other stars, will be an extremely promising tool for direct measurement of the basic parameter of stellar activity.

We have already discussed an indirect, but far more readily-observable measure of stellar magnetic activity: emission in the Ca II H and K lines. As we have already mentioned, on the Sun this emission is produced wherever there is a magnetic field, independent of its sign; hence the H and K lines measure the summed magnetic flux of both polarities, rather than the algebraic sum of the flux. Most of the chromospheric Ca II emission arises above photospheric plages; these chromospheric plages are also called active regions for they are spatially associated with all the varied phenomena of solar activity in the chromosphere and corona (Section 4).

The fact that much of the disk-integrated Ca II emission arises in active regions led Sheeley (1967) to study, from Mt. Wilson Ca II spectroheliogram archives, whether the sunspot cycle could be detected in disk-integrated Ca II K line emission. He measured a 40% modulation in this emission from the spectroheliograms, a result which has been recently reproduced by direct observations of the Sun as a star (White and Livingston 1980).

The possibility of detecting activity cycles on solar-like stars led O.C. Wilson (1968) to undertake an intensive synoptic study of K-line emission for 91 stars; this work has extended through the present. The first results (Wilson 1978) showed that cycles are indeed detectable on many late-type stars, and that they exhibit many similarities to the Sun, as well as many interesting differences. Figure 4 illustrates a few of the stars reported by Wilson (1978). Of the total sample of F, G, and K stars surveyed, about 40% showed evidence of cycles. Another 40% experienced short-term variability of K-line emission so large (see Figure 4d) that even if a cycle were present, it might go undetected. Significantly, those stars showing large short-term variability also have the largest mean level of K-line emission.

The period of those activity "cycles" that are seen in Wilson's (1978) data seems to vary from as short as about 7 years to as long as 10 or 11 years--i.e. the length of the observation period. Longer periods would not be distinguished as cycles in the data, and periods as short as a year or two could easily be lost in the aliasing noise due to the annual cycle of observations of most stars. However it is perhaps noteworthy that no clear periods in the range 2-5 years are seen in the data.

The amplitude of clearly defined cycles varies from less than 10% to about 35% . These amplitudes, which refer to the variation of the total intensity within 1A bands centered on the Ca II H and K lines, are best compared with the amplitude of solar modulation over the cycle by referring to measurements made by Wilson (1978), using the same instrument, of the solar Ca II emission reflected from the moon. The amplitude of the solar cycle found in this way was about 10%. Thus the amplitude of solar cycle variation is near, but not quite at, the minimum level observed for solar-type stars.

A recent analysis (Vaughan 1980) of Wilson's data in the light of the "gap" uncovered in Vaughan and Preston's (1980) "neighborhood survey" (Figure 2) shows that cycles are characteristically found in "old" stars (i.e. those below the gap in Figure 2) while "young" stars (i.e. those above the gap) characteristically exhibit large short-term variability with no clear evidence of cyclical behavior. (Cyclical behavior could, however, be present in the latter class if it were small enough in amplitude to be masked by the large short-term variations.)

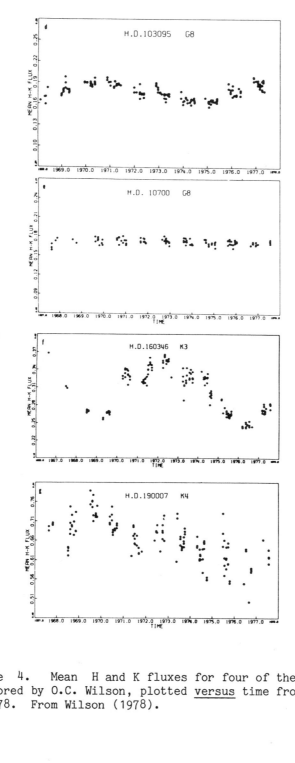

Figure 4. Mean H and K fluxes for four of the stars
monitored by O.C. Wilson, plotted <u>versus</u> time from 1967
to 1978. From Wilson (1978).

If the Sun's $Ca_{1/2}$II emission has decreased with time according to the $T^{-1/2}$ relation of Skumanich (1972), such a conclusion would suggest that perhaps the Sun did not always have an activity cycle. The Sun would initially have exhibited large short-term activity variations until it crossed the "gap" at an age of about 10^9 years, after which the large short-term variability would have decreased and the quasi-periodic activity cycle began. Alternatively, if the Sun were always a slow rotator, it could always have had an activity cycle more or less like its present one. Discussions of this sort, however, must be considered as pure speculation until much more evidence is obtained on the nature of the gap in Figure 2 and of the presence or absence of stellar activity cycles above and below the gap.

One other important aspect of solar variability deserves mention in the context of possible related stellar properties. The period 1645 to 1715 in Figure 3, known as the Maunder minimum, is one of very much reduced solar activity (Eddy 1976). During the time of the Maunder minimum sunspots were very rare, even at times of extrapolated activity maximum. For a period of 32 years not a single sunspot was observed on the northern hemisphere of the Sun. The total number of spots reported during the Maunder minimum was less than seen during a single active year in modern times. The few spots that were seen were mostly single spots, and at low latitudes.

The evidence for the Maunder minimum has been carefully reviewed and extended by Eddy (1976) and appears incontrovertible; further, based on ^{14}C data from tree ring samples, it appears that such minima have not been uncommon in the past. It may be crudely estimated from Eddy's results that the Sun spends perhaps 10^{-1} of its time in a much less active state. Eddy's data also indicate that the Sun may spend a similar fraction of its time in a more active state than the present one, such as during the time of the "Grand Maximum (around 1500). One might suspect then, that a similar fraction of the stars in Wilson's (1978) variability survey, or in Vaughan and Preston's (1980) neighborhood survey, could be in an unusually inactive or active state. As the rotation-activity connection becomes more firmly established over the coming years, such anomalous behavior may stand out among stars with otherwise similar spectral types, ages, and rotation rates.[1]

The observations of long-term variations in the envelope of solar activity, as epitomized by the Maunder minimum, as well as the new data on activity cycles for other main sequence stars with different spectral types, ages, and rotation rates, give important new observational constraints to dynamo theories. We now very briefly review some aspects of current dynamo theory in order to place these new observations into that context. For a fuller treatment, see e.g. Parker (1979).

Basically, it is thought (although not universally accepted) that dynamo action in the Sun is produced by the interaction of rotation, differential rotation, and convection on magnetic fields in the hydrogen convection zone. Differential rotation is produced by convection operating in the presence of rotation. The differential rotation in turn causes an initially poloidal field, passing through the convection zone, to be stretched azimuthally, thus creating a stronger, and primarily toroidal, field in the interior. Upon reaching a critical field strength, the magnetic flux tubes become buoyant and are carried upward by convective motions. The Coriolis force due to solar rotation causes the rising convective motions to be cyclonic, and induces a further twisting of the lines of force to give them again a poloidal component, but in the opposite direction from the initial poloidal field. Upon penetrating the surface, the emerging flux creates all the phenomena of solar activity. In addition, because of its reversed polarity it tends to oppose, and ultimately overcome and reverse, the initial poloidal field component. The result is a "reversing dynamo," whose properties (period, amplitude, structure of the butterfly diagram, etc.) are governed in a complicated way by the detailed run of convection and angular velocity with depth.

An important parameter governing the efficacy of the dynamo is the Rossby number $R_\theta = u_c/L\Omega$ where u_c is a characteristic convective velocity, L a characteristic convective length, and Ω the angular velocity in the convection zone. The Rossby number is simply the ratio of the rotation period to the convective turnover time; if it is small enough ($R_\theta \lesssim 1$), Coriolis forces can play the required role described above in recreating poloidal flux from toroidal flux. Although we cannot directly measure the Rossby number in the solar convection zone, much less other stars, we can use observed or inferred properties of the convection zone (i.e. mixing length and calculated convective velocities) and observed surface rotation rates of stars to estimate how it might vary from star to star. Thus, for example, if the rotation-age connection is valid, one might expect the Rossby number to increase systematically as a star ages and its rotation decreases, i.e. as it evolves vertically downward in the activity-color plane defined by Figure 2, so that at some point the dynamo becomes less effective. While it would be stretching speculation to associate this point with the gap in Figure 2, it is clearly of the greatest importance to measure rotational velocities for stars on either side of the gap.

Another important parameter in dynamo theory is the differential rotation with depth, $d\Omega/dr$. Very little is known of depth dependence of angular velocity in the Sun, although the first results of "solar seismology" (Deubner, Ulrich, and Rhodes 1979; see also Fossat 1981) indicate $d\Omega/dr < 0$ at least in the outer 30,000 km of the convection zone; this <u>increase</u> of angular velocity with depth is in accord with the requirements of the dynamo theory sketched above (Parker 1979). At present we have no solid evidence concerning differential rotation with depth or latitude in other stars. It is to be hoped that the promising techniques of solar seismology can eventually be applied to other solar-type stars.

A third important parameter in dynamo theory is the depth of the convection zone. It is well-known that this varies strikingly along the zero-age main sequence, being zero for stars earlier than about F5, about 20-30% of the stellar radius for stars like the Sun, and penetrating through the entire star for stars less massive than about $0.4 M_\odot$ (i.e. later than about M2). In addition the depth of the hydrogen convection zone varies for an individual star as it evolves. For example, evolutionary calculations indicate that in the Sun the fractional depth of the hydrogen convection zone has decreased by about 10% over its 4.5×10^9 year lifetime (Demarque 1980). As stars evolve, the changing depth of the hydrogen convection zone not only can change the region over which dynamo action can operate, but also, by penetrating more or less deeply into the more rapidly rotating interior, can perhaps change the average differential rotation gradient within that region.

Obviously our knowledge of rotation, and its depth gradient, in other stars is presently far too primitive to permit detailed predictions of how dynamo action should vary with mass and age of stars in the H-R diagram. Rather we may expect, as is usually the case, that observations will lead the way. Specifically, perhaps recent observations such as that of the "gap" in Figure 2, or the incidence, amplitude, and periodicity of activity cycles in solar-type stars of different spectral types and ages, should provide useful new insights and constraints on solar dynamo theory.

4. SOLAR AND STELLAR CORONAE

The message from Skylab's ATM solar observations, and from the OSO-8 satellite, is clear and unambiguous: the structure and heating of the solar corona is primarily due to the magnetic field. Acoustic waves, long thought to be the primary energy source for the corona, are now understood to play a negligible part in coronal heating, even if they may still be important for heating the low chromosphere (Athay and White 1978,1979). Rather the photospheric magnetic field, as it interacts with motions in the convection zone, stores and transmits energy to the corona,

and thus determines the structure, heating, and time development
of coronal features.

Images of the X-ray corona reveal that there are two
fundamental classes of structure (see e.g. Vaiana and Rosner,
1978):

(a) Closed magnetic structures, in which field-aligned
emission features loop up from low elevations to a maximum
height and return; the footpoints invariably lie in
photospheric field regions of opposite polarity. These
features range from the 3×10^3 km X-ray bright points
(which are in fact small unresolved loops) through active
region loops (L $\sim 3 \times 10^4$ km), to the large-scale closed
quiet Sun loops (L $\sim 3 \times 10^5$ km).

(b) Open magnetic structures, of which the only
clearly-identified examples are coronal holes -- regions in
which the magnetic field lines extend (or are pulled)
outward into the solar wind flow, and where the energy
entering the corona goes largely into accelerating the solar
wind plasma rather than heating a spatially confined plasma.

We discuss each of these features briefly below, with
particular reference to how they might be manifested on other
stars.

A. Closed Coronal Loop Structures and Dynamics

Active region loops are more or less discrete structures,
separated from their neighbors by regions of much lower
emissivity, and apparently much lower density. However the
corona above active regions must be nearly uniformly threaded by
magnetic flux. (Since the magnetic pressure greatly exceeds the
gas pressure above an active region, the magnetic pressure $B^2/8\pi$
must vary rather smoothly throughout the corona above an active
region.)   Thus it appears that some flux tubes are filled with a
higher-density plasma than are their neighbors. A likely
possibility is that the heating rate at the base of active
regions is highly non-uniform; and that larger amounts of energy
are deposited in that coronal flux tubes whose footpoints lie in
regions of enhanced heating. The excess energy over what can be
radiated away by the plasma within the flux tube is carried back
downward by thermal conduction, heating the base of the flux tube
and causing dense plasma from near the footpoints to expand
upward into the tube. In this fashion those flux tubes overlying
regions of enhanced heating fill up with material at an electron
density n_e far above ambient, and ultimately reach a point where
the radiation from this material (which varies as n_e^2) is
sufficient to carry off all the energy deposited. At that point

the energy conducted downward is just used up in radiation, and
none penetrates all the way to the chromosphere to "evaporate"
more material up into the loop. The result is an equilibrium
situation where the geometry of the loop is controlled by the
pre-existing field structure, and the density by the total heat
input. The thermal structure is determined by the requirement of
energy equilibrium, such that the divergence of the conductive
flux plus the radiative losses just balance the heat input at
every point. Because thermal conduction is so efficient in the
corona, the equilibrium thermal configuration is rather
insensitive to the detailed distribution of energy input with
position along the loop: conduction smooths out the temperature
profile, giving rise to a near-isothermal region with maximum
temperature T_{max} centered at the top of the loop, and a steep
transition zone to chromospheric temperatures at each footpoint.
Analytic or numerical calculations of the structure of loops in
static equilibrium, in which heat input is balanced by conduction
and radiation, have been carried out by a number of authors
(Landini and Monsignori-Fossi 1975, Rosner, Tucker, and Vaiana
1978, and others; see Withbroe (1981) for a review). These
calculations yield relations between the geometry of a loop, i.e.
its length, and the pressure and temperature of the plasma
confined within it. Rosner, Tucker, and Vaiana, for example,
derived the relation in terms of the scaling law $T_{max} \approx 1400$
$(PL)^{1/3}$, where T_{max} is the temperature at the top of the loop and
P the plasma pressure within. The relation has no free
parameters; the constant of proportionality and the exponent
depend only on the atomic physics governing radiative losses and
thermal conduction. All three quantities, P, L, and T_{max}, may be
estimated independently from XUV or X-ray data on solar coronal
loops, and it is found that the scaling laws fit the solar data
very well.

    As gratifying as such a simple scaling law is, it does not
reveal the essential physics of the problem; namely the exact
nature of the heating mechanism. Various heating mechanisms are
proposed, such as current-driven instabilities (Tucker 1973,
Rosner et al. 1978), Alfven waves (Wentzel 1978, Ionson 1978), or
fast-mode mhd waves (Habbal, Leer, and Holzer 1979); see Hollweg
(1981) for a review. These mechanisms might have different
signatures if the heat deposition rate or the associated velocity
field could be observed with high spatial resolution along (and
across) the loop. However, because of the effect of smoothing by
thermal conduction, the overall structure of the loop as defined
for example by P, L, and T_{max}, is quite insensitive to the
details of the heating function.

    Walter et al. (1980) and Holt et al. (1979) have observed
X-ray emission from Capella and concluded that the data are
consistent with a magnetically confined corona, as in the Sun.

Using the scaling relation given above, Walter et al. (1980)
calculate the size and number of loops on various stars from the
observed emission measure and T_{max}; however they make the
assumption that all loops are of the same size and pressure, an
assumption clearly at odds with the solar data. Nevertheless,
one conclusion is quite interesting and subject to further test:
namely, some stars (e.g. Capella) are like the Sun in that active
regions cover only a relatively small fraction of the stellar
surface, with the X-ray emission dominated by a few loops. Other
stars, such as UX Ari, appear to be essentially totally covered
with active regions, with a large number of coronal loops. If
this is true, one might expect a much larger fractional
modulation of coronal (i.e. X-ray) or chromospheric (e.g. Ca II)
emission in Capella than UX Ari, due to time variations of
individual loops or to rotation.

The dynamics of solar coronal loop structures is governed by
the magnetic field, as well as is their structure. As we have
indicated, it appears that the density of plasma within loops
continually adjusts in response to changes of energy input,
presumably related in turn to changing stresses of photospheric
magnetic fields at their footpoint. Characteristic times for
loop evolution vary from minutes, for small loop displacements or
emission measure changes, to days, characteristic of the lifetime
of the general loop configuration over a given active region.
However, large rapid changes are also seen. For example, loops
have been observed suddenly to be evacuated (e.g. Levine and
Withbroe 1971), apparently in response to a sudden heating
decrease. In the integrated light of the Sun seen as a star,
however, loop fluctuations would produce only a rather modest X-
ray emission variation.

Much larger variations of the emission of the Sun as a star
result from the phenomenon of solar flares, in which significant
fractions of the entire non-potential energy stored in an active
region are liberated in times as short as a minute or less,
giving rise to localized enhancements of high-energy radiation by
many orders of magnitude. Solar and stellar flares are discussed
elsewhere in these proceedings (Hood and Priest 1981, Gibson
1981) and have been thoroughly reviewed elsewhere (e.g. Gershberg
1975); following Gershberg, we note here simply that stellar
flares in late-type dwarfs (UV Ceti, dMe, dKe) have many
properties in common with solar flares, namely: (a) transient
nature of the phenomenon, with fast rise times and slow decay
times; the characteristic times for dMe flares can actually be up
to an order of magnitude shorter than their solar counterparts,
(b) localized emission on the stellar disk (since low-excitation
species such as TiO do not disappear during a flare), correlated
with starspots in the sense that more flares tend to occur near
the minimum of the "photometric wave" attributed to starspots,

(c) similar emission spectra from optical to radio wavelengths, (d) similar temperatures and densities in the optically-emitting flare chromosphere, (e) frequency drifts in radio emission requiring propagation like solar type II bursts. The major difference between late-type stellar flares and solar flares is in their total energy release, which may exceed that of solar flares by 10^2 to 10^3. Whether this is related to stronger stellar magnetic fields, larger area of stellar spots, or some other reason is an important matter for which there is little hard information.

B. Open Magnetic Field Regions in Solar and Stellar Coronae.

Coronal holes are large regions of the solar corona with lower than average density and temperature; they appear as very prominent dark regions in soft X-ray images. These features, only recognized as such a few years ago have rapidly achieved great importance in solar coronal research because they are recognized to be the solar source of high-speed solar-wind streams, and perhaps of the entire solar wind. In the context of this Institute, they are therefore of particular interest because of the role that stellar winds may play in spindown and mass loss. Coronal holes and the solar wind are discussed in detail elsewhere in these proceedings (Zirker 1981), so we only briefly delineate their properties here.

Coronal holes appear to overlie regions of open magnetic field, as determined from potential field extrapolations from measured photospheric fields. The mechanical energy deposited in coronal holes, as opposed to that deposited in the "quiet" corona, is not reradiated significantly in coronal lines (as is seen by inspection of X-ray images) or conducted downward to the chromosphere (the transition zone temperature gradient appears to be greatly decreased in coronal holes). However, the outward expansion of the solar wind provides a sink of energy that could compensate for the decreased importance of radiative and conductive losses from holes. Thus, there is no evidence that the total heating rate is any different in coronal holes than in the quiet corona.

Coronal holes, easily detected from space observations or XUV or X-ray observations, are very difficult to detect from the ground, since they have few observable effects in the transition zone or below. It is possible to detect them from coronal emission that reaches the ground, for example from coronagraph data, but in this case they cannot be mapped easily because of projection effects. Radio observations can also crudely indicate the location of coronal holes. However, the best means for studying them from the ground appears to be the line He I 10830, whose excitation is strongly dependent on the coronal radiation field. Spectroheliograms in He I 10830 show a weakening of the absorption line in coronal hole regions where the coronal radiation field is also weakened. The effect is, however, rather subtle and it is doubtful that solar-like coronal holes could be detected from He I 10830 observations of other stars. If huge coronal holes were detectable in other stars, one would expect from the solar analogy that weakening of He I 10830 absorption would be anticorrelated with the presence of active regions (revealed for example through Ca II emission).

Coronal holes can sometimes reach very large sizes at the poles, so that the disk of a solar-like star viewed pole-on might appear well over half covered with open field regions. In that case He I 10830 modulation might be detectable as the coronal hole waxes and wanes.

Direct detection of solar-like winds emanating from stellar coronal holes would be extremely difficult, because of the very small mass flux of the solar wind. This situation is of course in marked contrast to the detection of high-mass flux winds in giants and supergiants, as discussed elsewhere in these Proceedings (Hartmann 1981b). Nevertheless, as discussed there, observations of stellar winds in stars with different values of surface gravity may greatly help to illuminate the physical mechanisms of the solar wind.

5. SUMMARY AND CONCLUSIONS.

Our Sun is a magnetic variable star. It is now amply clear that the Sun's magnetic field controls most, if not all, of the solar variability in the Sun seen as a star. Understanding that variability in either the Sun or other stars requires the spatially resolved data that only the Sun can provide, and already there is a tremendous wealth of information on the nature and origins of solar magnetic variability: sunspots, related luminosity effects, surface fields, chromospheric plages, coronal active regions, coronal holes, and the solar wind. This information is complementary to related stellar data which, while much scantier, gives important information on dependence on mass, rotation, age, and related stellar parameters.

A qualitative picture for the relation between rotation, age, and surface activity in late-type main-sequence stars was originally suggested some years ago, and with an important modification involving the role of magnetic fields in producing coronal heating, remains relevant today. In this picture the rotation and convection produce surface fields through dynamo action; convection and surface fields produce coronae and winds; the surface fields "stiffen" the corona, causing winds to

decouple from the star at large radii, with consequent large
angular momentum loss; and the resulting rotation decrease causes
a weakening of dynamo action, surface fields, coronae and winds.
Thus magnetism in late-type rotating stars is seen as a self-
extinguishing process.

Many parts of this picture will no doubt be painted over in
years to come, as detailed processes are clarified by better
observations of both solar and stellar phenomena, and more
sophisticated theoretical studies based on these observations.
Other parts of the picture, presently blank, will be filled in.
Nevertheless, it is likely that in its various repaintings the
basic composition of the picture will remain the same.

Even now we can point to parts of the picture which may well
be filled in during coming years, by further solar observations,
related stellar observations, and theory. Some of these are:

A. Solar Observations.

A) Further studies of solar magnetism as a clue to the
dynamo. Examples are understanding how concentrated flux knots
form at the surface and what this implies for their subsurface
geometry, studies of ephemeral active regions and how they evolve
over the cycle, studies of both eruption and decay of surface
flux at all spatial scales, and studies of how sunspot and plage
magnetic fields affect solar luminosity, as a clue to their
interaction with convective energy flow.

b) Studies of large-scale surface velocity fields on the
Sun. These include large-scale convective-rotation flow (giant
cells, meridional circulation, etc.), surface rotation and
differential rotation and how it varies with the activity cycle.

c) Detailed study of closed coronal structures. A main
goal is to understand the precise roles of both convective motion
and magnetic fields in coronal heating, including identification
of specific mechanisms of energy transport and dissipation (e.g.
currents, mhd waves, etc.) Very high-resolution data, such as
can be provided by the Spacelab Optical Telescope (SOT), may be
of considerable help here.

d) Similar studies of open coronal structures. The goal is
to identify the precise energy source that accelerates the solar
wind and ultimately to determine its dependence on convective and
magnetic parameters.

e) More complete studies of the present solar wind, and in
particular, its braking torque. These studies would involve in-
ecliptic and out-of-ecliptic interplanetary velocity and mass

flux data, in relation to surface structure; coronal hole variability over the activity cycle makes it important to determine these parameters averaged over the cycle. If possible, the inferred braking torque should help us to infer whether the Sun was initially a fast or slow rotator; this should be compared with more complete stellar data on whether the Sun is really anomalous in its rotation.

f) Determination of the angular velocity in the solar interior. The depth variations throughout the solar convective zone, and perhaps deeper, should be pursued, using the new technique of solar "seismology" described in these Proceedings (Fossat 1981). Such data should be obtained as a function of latitude, using spatially-resolved observations, and also extended over an activity cycle to determine time variability. A possible source of information on the deep interior may be from precision tracking measurements of spacecraft passing close to the Sun and subject only to gravitational forces (i.e. NASA's proposed "Starprobe" mission). Such measurements can yield the quadrupole moment of the solar gravitational field to an uncertainty $\lesssim 10^{-8}$ (Reasenberg and Shapiro 1978), sufficient to determine whether, as suggested by Roxburgh (1976,1978), the bulk of the Sun below 0.6 $R_\odot$ is rotating at about twice the surface speed. Such a rapidly rotating interior could quite possibly transport angular momentum to the surface layers, thereby significantly affecting surface rotation and the dynamo.

g) Studies, using "proxy" data from geophysical or lunar or meteoritic sources, of the young Sun. Terrestrial ^{14}C studies should continue to yield insights into overall activity levels, and Maunder Minimum or Grand Maximum-like periods of low or high activity. Study of fossil data on trapping of gases on lunar grains may indicate whether the solar wind velocity or flux was much higher in the remote past, whether solar flare incidence was higher in the past, and possibly even whether the early interplanetary magnetic field was higher. (See Newkirk 1980 for an up-to-date review.)

h) Studies of the luminosity variability of the Sun. The sensitivity of "solar constant" monitoring from spacecraft is now sufficient to detect luminosity variations due to the effects of solar activity (e.g. sunspots and plages). This provides a new technique to study how magnetic fields interact with convection and convective energy flow, so that indirectly at least, we can hope to gain important information on the structure and dynamics of subsurface field structures (e.g. depth of concentration, rate of buoyant rise; see Foukal and Vernazza 1979).

B. Stellar Observations.

a) Stellar rotation measurements should be carried out for many more late-type main-sequence stars, including those in the Vaughan and Preston (1980) neighborhood survey; such data will answer the specific question whether the gap in Figure 2 is related to a rapid decrease in rotation at a particular epoch, and more generally will delineate the rotation versus age relation in much-needed detail for stars of different spectral types.

b) The study of activity cycles so well begun by Wilson (1978) should be continued for these and a larger sample of stars (including important southern hemisphere stars, like α Cen). In conjunction with rotation studies, we will be able to determine what, if any, connection exists between rotation rate and activity cycle period or amplitude.

c) Stellar magnetic fields should be detected and studied in late-type stars; the technique proposed by Robinson (1980) appears promising. It is important to learn whether surface fields and Ca II emission or other chromosphere-corona indicators are as well-correlated on other stars as they are on the Sun, and also important to infer the spatial coverage of such fields on other stars; Robinson's technique already provides a step in this direction. Detailed comparisons between stellar fields and photometrically-inferred star spots will determine whether these features are indeed magnetic in nature.

d) The dependence of rotation and activity on age should be tightened by further study of stars of common age (in galactic clusters, or in binary systems) or individual stars whose age is well-determined. One goal is to clarify the statistical relation between rotation, activity, and age to the point where it can clearly be decided whether or not the Sun has anomalous rotation and activity for its age.

e) Evidence for Maunder-type minima in other stars should be sought; a likely way to proceed is to search for anomalous activity-rotation relations among stars of the same age (e.g. formed in the same galactic cluster) and spectral type.

f) The technique of solar "seismology" (see Fossat 1981) may well be applicable to other stars in coming years; if so, this will give further information on the structure (e.g. depth) of stellar convection zones.

g) Stellar coronae are now being mapped throughout the H-R diagram in X-rays; the extraordinary strength of X-ray emission from dM stars must be due to magnetic heating (Vaiana et al. 1980) and gives support to the idea that magnetic heating is important in all late-type dwarfs. Such observational studies should be extended, and applied to more detailed models of closed coronal structures than are possible with the present limited X-ray data.

h) Further studies of cool stellar winds in supergiants and giant stars, while not a direct analogy to the hot thermally-driven solar wind, will help show how winds are produced under varying conditions of surface gravity and effective temperature; to the extent that a common mechanism may underlie wind production of all types, this may further illuminate the acceleration of the solar wind (see Hartmann and MacGregor 1981b).

C. Theory

The central theme of this review involves the relation between magnetism and rotation in convecting stars, i.e. solar and stellar dynamos. Important useful directions for dynamo theory are:

a) More realistic theories of generation of the observed solar magnetic field are needed, consistent with new observational results (including coronal holes, bright points, photospheric flux knots, photospheric circulation patterns, rotation, differential rotation with latitude and depth, the Maunder minimum) as well as the already well-known data included in previous attempts to build dynamo theories.

b) An evolutionary description of the dynamo is required, consistent with data on stars of different ages (including for example the apparent sudden jump from high to low levels of chromospheric activity, and the appearance of clear activity cycles only after the jump has been made).

c) Scaling laws should be obtained to indicate how basic parameters of the dynamo, such as its period or amplitude, depend on rotation rate, differential rotation, depth of the convective zone, strength of convection, or other stellar parameters. The goal is, of course, to explain the observed variation of period or amplitude of the cycle along the main sequence, and with age.

d) Sunspots are not yet well understood theoretically, and of course the theory of spots on other stars is in a much more rudimentary state. As new information on the size, magnetic field, or other properties of spots on other stars becomes available, it will be important to develop theoretical concepts that will reveal the similarities and differences between solar and stellar spots.

We have also been concerned in this review with

chromospheric and coronal manifestations of surface activity.
Given the existing data plus new data to be expected on
chromospheric and coronal activity, theoretical questions arise
such as:

e) What is the specific mechanism(s) for heating of
confined coronal structures, and how does it depend on strength
and geometry of surface fields, as well as motions in the
convection zone? Scaling relations should be developed to
predict how heating varies among late-type stars on (and off) the
main sequence. In addition, a more definitive theory should be
sought about how the temperature, as well as emission measure, of
coronal structures varies among late-type stars.

f) How are stellar winds generated in late-type main
sequence stars? The goal is to predict theoretically how their
mass and angular momentum flux depends on their age and position
in the HR diagram.

The above abbreviated lists show that the interdisciplinary
field of solar-stellar research can make major contributions in
coming years to our understanding of the physics of the Sun and
the stars. At the same time, as is amply indicated by the
remaining articles in these Proceedings, a firm foundation has
already been laid on which to build our new understanding of the
Sun as a star.

NOTES

[1]Hartmann, Londono, and Phillips (1979) point out from Harvard
plate archives that the mean spot star HD224085 remained
essentially constant in mean light for about 40 years before
beginning photometric fluctuations that could be attributed to
starspots; they suggest that spot stars may in general pass
through long periods of inactivity.

[2]It may be shown (Durney and Latour 1978) that the Rossby number
is related to the usual dynamo number D by $D \sim R_0^{-2}$. Thus the
requirement for dynamo action may be stated as $D \gg 1$.

[3]Some evidence has been reported (Vogt 1975) for period changes
in the spot star BY Dra which could be attributed to spots
erupting at different latitudes on a differentially rotating
star; however the data at present are too scanty to permit solid
conclusions; see Hartmann and Rosner (1979). Numerical
experiments carried out on strings of Wolf sunspot numbers
demonstrates that the solar differential rotation would be
extremely difficult to detect from the disk-integrated sunspot
number (Selinger 1980).

REFERENCES

Athay, R.G., and White, O.R.: 1978, "Chromospheric and Coronal Heating by Sound Waves," Astrophys. J., **226**, p. 1135.

Athay, R.G, and White, O.R.: 1979, "Chromospheric Oscillations Observed with OSO-8. IV. Power and Phase Spectra for C IV," Astrophys. J., **229**, 1147.

Babcock, H.W., and Babcock, H.D.: 1955, "The Sun's Magnetic Field 1952-1954," Astrophys. J., **121**, p. 349.

Blanco, C., Catalano, S., Marelli, E., and Rodono, M.: 1974, "Absolute Fluxes of K Chromospheric Emission in MS Stars," Astron. Astrophys., **33**, p. 257.

Bumba, V., Ranzinger, P., and Suda, J.: 1973, Bull. Astr. Inst. Czech., **24**, p. 22.

Bumba, V., and Howard, R.: 1965, "The Development of Active Regions on the Sun," Astrophys. J., **141**, p. 1492.

Demarque, P.: 1980, private communications.

Deubner, F.-L., Ulrich, R.K., and Rhodes, E.J., Jr.: 1979, "Solar P-Mode Oscillations as a Tracer of Radial Differential Rotation," Astron. Astrophys., **72**, p. 177.

Durney, B.R., and Latour, J.: 1978, "On the Angular Momentum Loss of Late-Type Stars," Geophys. Astrophys. Fluid Dyn., **9**, p. 241.

Eddy, J.A.: 1976, "The Maunder Minimum," Science, **192**, p. 1189.

Eddy, J.A., Gilman, P.A., and Trotter, D.E.: 1976, "Solar Rotation During the Maunder Minimum," Solar Phys., **346**, p. 3.

Fossat, E.: 1981, "Solar and Stellar Oscillations," these Proceedings.

Fossat, E.: 1981, these Proceedings.

Foukal, P., and Vernazza, J.: 1979, "The Effect of Magnetic Fields on Solar Luminosity," Astrophys. J., **234**, p. 707.

Frazier, E.N.: 1970, "Multichannel Magnetograph Observations," Solar Phys., **14**, p. 89.

Gershberg, R.E.: 1975, "Flares of Red Dwarf Stars and Solar Activity," IAU Symposium No. 67, p. 47.

Gibson, D.M.: 1981, "Stellar Analogs of Solar Microwave Phenomena," these proceedings.

Golub, L., Davis, J.M., and Krieger, A.S.: 1979, "Anticorrelation of X-ray Bright Points with Sunspot Number, 1970-1978," Astrophys. J., $\underline{229}$, p. L145.

Habbal, S., Leer, E., and Holzer, T.: 1979, "Heating of Coronal Loops by Fast Mode MHD Waves," Solar Phys., $\underline{64}$, p. 287.

Hall, D.S.: 1981, "The RS CVn Binaries," these Proceedings.

Hartmann, L., Londono, C., and Phillips, M.J., 229, p. 183.: 1979, "On the Long-term Variability of the K2e Star HD 224085," Astrophys. J.

Hartmann, L., and Rosner, R.: 1979, "Stellar Luminosity Stability: Luminosity Variations and Light-curve Period Changes in BY Dra Stars," Astrophys. J., $\underline{230}$, p. 802.

Hartmann, LH.: 1981a, "Stellar Spots -- Physical Implications", these Proceedings.

Hartmann, L.H.: 1981b, "Observations and Theory of Mass Loss in Late-type Stars," these Proceedings.

Hollweg, J.V.: 1981, "Solar Active Regions: Mechanisms of Energy Supply" in **Active Regions,** Orrall, F.A., ed., Univ. of Colorado Press, in press.

Holt, S.S., White, N.E., Becker, R.H., Boldt, E.A., Mushotzsky, R.F., Serlemitsos, P.J., and Smith, B.W.: 1979, "X-Ray Line Emission from Capella," Astrophys. J. $\underline{234}$, p. L65.

Hood, A. and Priest, E., "Solar Flares: MHD Instabilities," these Proceedings.

Ionson, J.: 1978, "Alfvenic Surface Waves and the Heating of Coronal Loops," Astrophys. J., $\underline{226}$, p. 650.

Kraft, R.P.: 1967, "Dependence of Rotation on Age Among Main Sequence Stars," Astrophys. J., $\underline{150}$, p. 551.

Landini, M., and Monsignori-Fossi, B.C.: 1975, "A Loop Model of Active Coronal Regions," Astron. Astrophys., $\underline{42}$, p. 213.

Leighton, R.B.: 1959, "Observations of Solar Magnetic Fields in Plage Regions," **Astrophys. J.,** $\underline{130}$, p. 366.

Levine, R.H., and Withbroe, G.L., "Physics of an Active Region

Loop System," Solar Phys., $\underline{51}$, p. 83.

Mochnacki, S.W.: 1980, "Magnetic Braking of Stars. I. Did the Sun Ever Rotate Rapidly?" Preprint.

Mullan, D.J.: 1973, "Sunspots, Supergranules, and the Depth of the Solar Convection Zone," Astrophys. J., $\underline{186}$, p. 1059.

Newkirk, G., Jr.: 1980, "Solar Variability in Timescales of 10^5 Years to 4×10^9 Years," Proc. Conf. on the Ancient Sun, Pergamon Press, in press.

Parker, E.N.: 1979, **Cosmic Magnetic Fields: Their Origin and Their Activity**, Clarendon Press, Oxford, England.

Parker, E.N.: 1981, "Spontaneous Concentration of Magnetic Fields," these Proceedings.

Perrin, M.-N., Hejlesen, P.M., Cayrel de Strobel, G., and Cayrel, R.: 1977, "Fine Structure of the HR Diagram in the Solar Neighborhood," Astron. Astrophys., $\underline{54}$, p. 779.

Ramsey, L.W., and Nations, H.L.: 1980, "HR 1099 and the Starspot Hypothesis," Astrophys. J., $\underline{239}$, L121.

Reasenberg, R.D., and Shapiro, I.I.: 1978, "Possible Measurements of J_{20} with a Solar Probe," in **A Closeup of the Sun**, JPL Pub. No. 19.

Robinson, R.D., Jr.: 1980, "Magnetic Field Measurements on Stellar Sources: A New Method," Astrophys. J., $\underline{239}$, p. 961.

Robinson, R.D., Worden, S.P., and Harvey, J.W.: 1980, "Observations of Magnetic Fields on Two Late-Type Stars," Astrophys. J., $\underline{236}$, p. L155.

Rosner, R., Golub, L., Coppi, B., and Vaiana, G.S.: 1978, "Heating of Coronal Plasma by Anomalous Current Dissipation," Astrophys. J., $\underline{222}$, p. 317.

Rosner, R., Tucker, W.H., and Vaiana, G.S.: 1978, "Dynamics of the Quiescent Solar Corona," Astrophys. J., $\underline{220}$, p. 643.

Roxburgh, I.W.: 1976, "The Internal Structure of the Sun and Solar-type Stars," in IAU Symposium No. 71, **Basic Mechanisms of Solar Activity**, Reidel, Dordrecht, p. 453.

Roxburgh, I.W.: 1978, "The Importance of Determining the Solar Quadrupole Moment," in A Closeup of the Sun, JPL Pub. No. 11.

Rucinski, S.M.: 1979, "Sizes of Spots in Spotted Stars," Acta Astron., **29**, p. 203.

Selinger, J.: 1980, private communication.

Severny, A., Wilcox, J.M., Sherer, P.H., and Colburn, D.S.: 1970, "Comparison of the Mean Photospheric Magnetic Field and the Interplanetary Field," Solar Phys., **15**, p. 3.

Sheeley, N.R.: 1967, "The Average Profile of the Solar K Line During the Sunspot Cycle," Astrophys. J., **147**, p. 1106.

Skumanich, A.: 1972, "Timescales for Ca II Emission Decay, Rotational Braking, and Li Depletion," Astrophys. J., **171**, p.565.

Skumanich, A., Smythe, C., and Frazier, E.N.: 1975, "Multichannel Observations of the Ca II Emission Network," Astrophys. J., **20**, p. 747.

Smith, M.A.: 1979, "Rotational Studies of Lower MS Stars," P.A.S.P., **91**, p. 737.

Stenflo, J.O.: 1976, "Small-scale Solar Magnetic Fields," in IAU Symposium No. 71, **Basic Mechanisms of Solar Activity**, Reidel, Dordrecht, p. 69.

Tucker, W.H.: 1973, "Heating of Solar Active Regions by Magnetic Energy Dissipation," Astrophys. J., **186**, p. 285.

Vaiana, G.S., and Rosner, R.: 1978, "Recent Advances in Coronal Physics," Ann. Rev. Astr. Astrophys., **16**, p. 393.

Vaiana, G.S., et al.: 1980, "Results from an Extensive Einstein Stellar Survey, "Astrophys. J., in press.

Vaughan, A.: 1980, "Comparison of Activity Cycles in Old and Young Main Sequence Stars," P.A.S.P., **92**, p. 392.

Vaughan, A., and Preston, G.W.: 1980, "A Survey of Chromospheric Ca II H and K Emission in Field Stars of the Solar Neighborhood," P.A.S.P., **92**, p. 385.

Vogt, S.S.: 1975, "Light and Color Variations of the Flare Star BY Draconis," Astrophys. J., **199**, p. 418.

Walter, F.M., Cash, W., Charles, P.A., and Bowyer, C.S.: 1980, "X-Rays from RS CVn Systems: A HEAO I Survey and a Coronal Model," Astrophys. J., **236**, p. 212.

Weber, E.J., and Davis, L., Jr.: 1967, "The Angular Momentum of the Solar Wind," Astrophys. J., 148, p. 217.

Wentzel, D.G.: 1978, "Heating of the Solar Corona: A New Outlook," Rev. Geophys. Spa. Phys., 16, p. 757.

White, O.R., and Livingston, W.: 1980, Astrophys. J., in press.

Wilson, O.C.: 1966, "Stellar Chromospheres," Science, 151, p. 1487.

Wilson, O.C.: 1968, "Flux Measures of the Center of Stellar H and K Lines," Astrophys. J., 153, p. 221.

Wilson, O.C.: 1978, "Chromospheric Variations in Main Sequence Stars," Astrophys. J., 226, p. 379.

Withbroe, G.L.: 1981, "Physics of Static Loop Structures and Scaling Laws," in Active Regions, Orrall, F.Q., ed., University College Press, in press.

Zirker, J.B.: 1981, "Coronal Holes and Solar Mass Loss," these Proceedings.

THE SPONTANEOUS CONCENTRATION OF MAGNETIC FIELD
IN THE PHOTOSPHERE OF THE SUN

E. N. Parker*

Department of Physics
University of Chicago, Chicago, Illinois 60637   USA

Abstract. Magnetic fields at the surface of the sun exhibit the
remarkable property of fragmentation and concentration into separate
tubes of 1-2 kilogauss. In newly emerging active regions the
separate tubes cluster together to produce the sunspot phenomenon,
with the magnetic field compressed further, to 2-4 kilogauss.
Magnetic fields are concentrated by compressive forces which must
equal the magnetic pressure $B^2/8\pi$ in the flux tubes. In the
surface of the sun $B^2/8\pi$ exceeds the dynamical pressure of the
observed fluid motions by a factor of ten or more, and exceeds any
other pressure differences known to observation or theory except
those produced by cooling of the gas over several vertical scale
heights. There is, however, no evidence of cooling in the individual flux tubes.

   The first part of the review explores the basic physics of
flux tubes so as to define the problem before going on to list
the presently known effects that contribute to concentration of
fields. The second part of the review takes up the curious clustering of flux tubes to form sunspots, exploring the anti-cluster-
ing forces (from the tension in the magnetic lines of force) and
the clustering effect of the aerodynamic forces exerted on the
separate tubes before and after clustering.

---

* This work was supported in part by the National Aeronautics and
Space Administration under grant NGL 14-001-001.

R. M. Bonnet and A. K. Dupree (eds.), Solar Phenomena in Stars and Stellar Systems, 33–58.
Copyright © 1981 by D. Reidel Publishing Company.

I.   INTRODUCTION

    The average magnetic field extending through the surface of
the sun is of the order of ten gauss. In active regions, the
average may be as high as a hundred gauss. One of the most remarkable astronomical discoveries of the last two decades is the
break up and compression of the general field into separate flux
tubes of 1-2 kilogauss. The individual tubes are 200-400 km in
diameter and are separated by distances ten times larger. The
diameter of the individual tube is below the resolution of the
magnetograph, so the discovery of the separation and the extraordinary compression to 1.5 kilogauss was the result of some
intricate detective work over many years (Sheeley 1967; Beckers
and Schröter 1968; Livingston and Harvey 1969, 1971; Frazier
1970, 1972; Sawyer 1971; Frazier and Stenflo 1972; Howard and
Stenflo 1972; Stenflo 1973, 1978; Chapman 1974; Harvey 1977;
Ramsey, Schoolman, and Title 1977; Tarbell and Title 1977; Wiehr
1978; Tarbell, Title, and Schoolman 1979). We may be sure that
there is a lot that yet remains unknown concerning the fine structure of the field in the solar photosphere. The individual flux
tubes are located in downdrafts at the corners of supergranules
and in the dark intergranular lanes (Leighton, Noyes, and Simon
1962; Leighton 1963; Noyes and Leighton 1963; Simon and Leighton
1964; Beckers and Schröter 1968; Frazier 1970. See also the discussion in Ramsey, Schoolman, and Title 1977; Tarbell, Title, and
Schoolman 1979).

    The magnetic flux tubes rising through the surface of the sun
in a developing active region have the additional property of
clustering together to form pores and sunspots, further compression
bringing the field up to 3000 gauss or more at the visible surface
(Beckers and Schröter 1968; Vrabec 1979; Zwaan 1978).

    None of these concentration effects were anticipated on
theoretical grounds and indeed a complete explanation is still
lacking in spite of the progress in theoretical exploration since
their discovery. The break up of the field into separate bundles
of flux at the surface of the sun is evidently to be understood as
a direct consequence of the granules and supergranules. The field
is carried with the conducting fluid and concentrated in the converging flows at the surface where down drafts are formed (Parker
1963, 1973, 1979b; Weiss 1964; Clark 1966, 1968). It is the compression of the separate tubes to 1-2 kilogauss, to magnetic pressures far in excess of the dynamical pressure $\frac{1}{2}\rho v^2$ of the fluid
motions, that is so puzzling.

    Piddington (1975, 1976a,b) argues that the flux tubes are
concentrated by twisting, and that they are wrapped around each
other deep in the convective zone of the sun to form a large rope.

The individual strands of the rope are imagined to have separated
in the upper expanded regions of the rope where the rope first
comes through the surface. Hence, at the surface we see first
the separated flux tubes, followed by their coalescence to form
a single large tube (sunspot) as this "magnetic tree" structure
rises through the surface. Zwaan (1978) has developed a detailed
scenario based on the compression of the field at depth by the
external turbulent pressure. He too starts with the idea that
the tubes are clustered together to form a tree-like structure
far below the surface, so that the formation of an active region
is a consequence of the emergence of the tree through the surface.
Meyer, Schmidt, Weiss, and Wilson (1979) suggest that a deep convective upwelling plays a major role in clustering the individual
tubes to form a sunspot. Parker (1979a) argues that a subsurface
downdraft may be responsible. Altogether, it is clear that there
are many different effects that may contribute to the behavior of
the magnetic flux tubes. The effects must first be recognized and
then evaluated in the context of conditions in the convective
region beneath the surface of the sun.

It is the purpose of this review to summarize the basic
physics of flux tubes, to see how far we can get with hard
theoretical fact, and to see how much is still conjecture. We
begin, then, with the question of the concentration of individual
small flux tubes to 1.5 kilogauss. This is the general state of
the magnetic fields at the surface of the sun, in both quiet and
active regions. It may be supposed that the magnetic fields in
many other stars exhibit the same extraordinary tendencies. We
cannot resolve them, of course, so a necessary condition for
understanding their observed behavior is the development of a
solid theory for what we can see in the sun. We conjecture that
the concentration of fields in the sun is a consequence of the
strong convection immediately beneath the surface. But we cannot extrapolate this idea to other stars until we know just how
the concentration is accomplished by the convection.

To keep the exposition simple and to make the physics clear,
the discussion is limited to the equilibrium of slender flux
tubes, in which the radius $R(s)$ of the tube varies only slowly
with distance s along the tube, $\partial R / \partial s \ll 1$. The individual flux tubes in the sun are estimated to have radii of the order
of $1 - 2 \times 10^2$ km at the visible surface where the pressure scale
height is 180 km. The characteristic scale height for the tube
radius is four times the pressure scale height, or some 500 km.
Hence we expect that $\partial R / \partial s$ is generally less than one, even if
it is not extremely small. The slender flux tube representation
$(\partial R / \partial s \ll 1)$ omits the effect of tension along the field,
thereby making the best (and simplest) case for the equilibrium
compression. The interested reader may readily include the effect
of tension in the results if desired. It will be found that the

tension changes the numerical results for equilibrium only slightly
and the principles not at all. The principal contribution of the
tension is toward destabilization of the equilibria that we shall
discuss. The tension and $\partial R/\partial s$ together contribute the hydromagnetic exchange instabilities (Parker 1955, 1975b, 1970a; Meyer,
Schmidt, and Weiss 1977). But there is not time to go into the
question properly here. We will be content to treat the equilibrium of compressed slender tubes of magnetic flux.

II. EQUILIBRIUM CONDITIONS AND THE ROLE OF TURBULENCE

        Magnetic fields must be forcibly confined if they are not to
disperse. Consider the static equilibrium of an isolated slender
tube of magnetic field $B(s)$ across a tube of radius $R(s)$. The
magnetic pressure $B^2(s)/8\pi$ within the tube must be balanced by the
excess external gas pressure (Parker 1955). Thus if $p_e(s)$ is
the external gas pressure and $p_i(s)$ the internal gas pressure,
it follows that

$$B^2(s)/8\pi = p_e(s) - p_i(s).$$ (1)

The magnetic pressure is transferred to the gas via the electric
currents flowing around the surface of the tube, of course, but
the currents are themselves only a passive agent, transmitting
forces between fluid and field but supporting no stresses themselves.

        It is possible to twist a flux tube so that it possesses an
aximuthal field $B_\phi(\varpi,s)$ in addition to the longitudinal field
$B(\varpi,s)$, where ϖ represents distance from the axis of the tube.
The boundary condition at the surface of the tube is essentially
eqn. (1) again. The magnetic pressure must be confined by the
excess of p_e over p_i ,

$$B_\phi^2(R,s) + B^2(R,s) = 8\pi \left[p_e(s) - p_i(s) \right].$$ (2)

For a fixed total magnetic flux along the tube, the effect of
twisting the tube is to (a) increase the radius of the tube, (b)
decrease the outer longitudinal field, (c) increase the mean square
total field, (d) increase the longitudinal field near the axis of
the tube, and (e) decrease the mean longitudinal field across the
tube (cf. Parker 1979b, Chap. IX). The twisting leaves the mean
square longitudinal field invariant. It can be shown that the
twisting tends to propagate along the tube away from the concentrated regions into the expanded portions (Parker 1975c) causing
the expanded portions to expand further, and diminishing the mean
longitudinal field there. Only a little twisting is required to
initiate the kink instability, causing the tube to go into a spiral
and eventually tight loops (Parker 1975c).

Inasmuch as the observations deal with something akin to the mean longitudinal fields, twisting is counterproductive, in spite of its popularity, in concentrating the field. Of course, we do not expect the flux tubes in the turbulent sun to be free of twisting. Indeed, the twisting concentrates in the expanded portions of the tubes that we observe above the surface of the sun. It is not surprising, then, to see filamentary structure in the photosphere with spiralling of as much as $20°$ or $30°$ on occasion. More remarkable, perhaps, is the larger number of active regions showing so little spiralling. In any case, the compression of the field in individual tubes must be carried out in spite of the fact that the tubes are sometimes twisted. The static equilibrium of the single twisted tube has been worked out formally in the literature (Parker 1979b, Chap. IX) and need not be repeated here.

The flux rope, made up of two or more twisted tubes wound around each other is a much more difficult question, for several reasons, not the least of which is the general absence of any equilibrium. The rope is subject to internal neutral point annihilation until either the winding of the tubes about each other has been destroyed or until the tubes have lost their individual twisting, whichever happens first (Parker 1972, 1976c, 1979b, Chap. XIV). So the properties of the composite rope are dynamical rather than static in nature, and are relevant to activity rather than the simple equilibrium pursued here.

Now if we go back to eqn. (1), it is evident that the magnetic pressure $B^2/8\pi$ can be enhanced only by increasing the ambient external pressure P_e and/or by diminishing the internal pressure P_i . The pressures P_e and P_i may be in part a consequence of turbulent motions as well as the thermal motions of the individual atoms, of course. The granule velocities at the surface of the sun are of the order of 3 km/sec, compared to the speed of sound of about 7 km/sec at $5600°$ K. The dynamical pressure of the turbulence is $\frac{1}{3}\rho v^2 \cong 10^4$ dynes/cm^2, equivalent to only about one sixth of the pressure $B^2/8\pi$ of the observed fields, in the neighborhood of 1.5 kilogauss. Zwaan (1978) has suggested that the flux tubes deep in the sun may be compressed by the external turbulence, where ρv^2 is much larger (as a consequence of larger ρ) than at the surface. The idea is that (a) the thermal pressures inside and outside the tube are the same and (b) there is no turbulent pressure inside. In that case the net pressure difference is just the dynamical pressure of the external turbulence, compressing the field to some value near equipartition,

$$B^2/8\pi \cong \frac{1}{2}\rho v^2$$

The question is implicit in all of the dynamical mechanisms reviewed in this lecture. What is the effect of strong turbulence on a weak flux tube? Does the turbulence confine and compress the flux tube until $B^2/8\pi$ becomes comparable to $\frac{1}{2}\rho v^2$, or does

the turbulence disperse the tube, or does the turbulence not compress the field? We are inclined to the view that true turbulence,
fluctuating with time and position in a chaotic manner, disperses
the flux tube (cf. Parker 1979b, Chap. XVII). On the other hand,
steady circulation through closed convective cells may have quite
the opposite effect (cf. Parker 1979b, Chap. XVI), depending upon
how long the topological coherence of the fluid motion is maintained.

To take a first look at the problem, consider the circumstance
that the flux tube is buffetted symmetrically on both sides by the
turbulence so that the field is compressed rather than dispersed.
To provide a clear picture that can be formulated in precise
mathematical terms imagine an infinite expanse of homogeneous,
incompressible fluid. Throughout some finite central region (see
Figure 1), the fluid is in turbulent motion on a scale $l \equiv 1/k$
with a mean kinetic energy density $\frac{1}{2}\rho v^2$. The turbulent
region is surrounded by static fluid with pressure P_0, equal to
the total pressure in the region of turbulence. Thus if the mean
fluid pressure in the region of turbulence is P_1, and the mean
turbulent pressure is $\frac{1}{3}\rho v^2$, it follows that $P_0 = P_1 + \frac{1}{3}\rho v^2$
in order that the static gas confine the region of turbulence.

Consider, then, the effect on a flux tube extending across
the region of turbulence. Denote by p_0 the fluid pressure within
the flux tube. The pressure p_0 extends uniformly along the tube
through both the turbulent and nonturbulent regions. The external

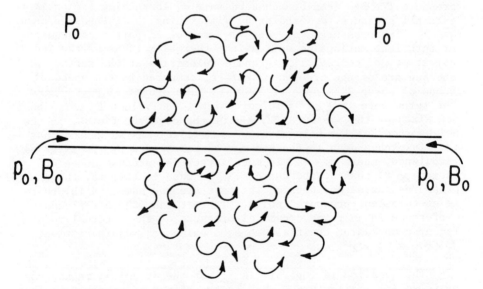

Fig. 1: A sketch of a long thin flux tube extending
through a localized region of turbulence.

pressure exerted on the tube is $P_1 + \frac{1}{3}\rho v^2$ in the region of turbulence and P_e in the static region. Since these two are equal, it follows from (1) that the field is compressed to the same pressure in the region of turbulence or outside. The turbulence has no effect on $B^2/8\pi$.

In fact the problem is a little more complicated than this because of the longitudinal motion $u(s,t)$ of the fluid within the tube, caused by the buffetting of the external turbulence and the associated Bernoulli effect.

Supposing that the tube is buffetted equally on both sides, the dynamics of the tube in the turbulent region is subject to exact analysis for any specified turbulent fluctuation (Parker 1974b, 1979b, §10.4). Denote by B_o the field strength in the static region, so that

$$B_o^2/8\pi = P_e - P_o.$$
(3)

The problem is to provide an exact calculation of the field $B(s,t)$ in the region of turbulence in terms of the fluctuating pressure $P(s,t)$ applied by the turbulence to the surface of the tube. A complete solution of the dynamical equations in the idealization of a slender tube provides the fluctuating fluid velocity $u(s,t)$ along the tube. The Bernoulli effect of $u(s,t)$ provides a significant reduction of fluid pressure and a corresponding enhancement of the mean square field within the tube. It is readily shown (Parker 1979b, eqn. (10.13)) that the mean square (time averaged) field in the tube is

$$\langle B^2 \rangle/8\pi = B_o^2/8\pi + \frac{1}{2}\rho\langle u^2 \rangle$$
(4)

everywhere along the tube. The time averaged field pressure is enhanced by the Bernoulli effect by $\frac{1}{2}\rho\langle u^2 \rangle$, as we would expect.

However, we are interested in the mean field rather than the rms field. It is necessary to average along the tube in addition to the average over time, denoting the results by a double angular bracket, $\langle\!\langle B \rangle\!\rangle$. The calculation is more complicated and comes out in terms of integrals of functionals of $P(s,t)$. To provide a concrete example, note that the result is independent of the shape of the cross section of the tube, so long as the tube is slender, so we choose a slab of thickness

$$Y(s,t) = Y_o(1 + n\cos\omega t\cos ks)$$
(5)

to keep the calculation as simple as possible. The parameter n lies between zero and one, and $kY_o \ll 1$. This form of $Y(s,t)$ is assumed to apply to the central part of the turbulence, converging then to Y_o as the edge of the turbulence is approached, with

$\Upsilon = \Upsilon_o$ outside the region.

Recalling that the mean external pressure exerted on the tube is P_o everywhere along the tube, the general dynamical equations (see eqn. (10.35) Parker 1979b) reduce to the expression

$$\frac{\langle\!\langle B \rangle\!\rangle^2}{8\pi} = \frac{B_o^2}{8\pi}\left\{1 + \left[\frac{2}{\pi}K(n)\right]^2 - \frac{2}{\pi}\frac{d}{dn}nK(n)\right\} + \frac{1}{2}\rho\langle\!\langle u^2 \rangle\!\rangle \tag{6}$$

for the square of the mean field in the region of turbulence, where $K(n)$ is the complete elliptic integral of the first kind and

$$\langle\!\langle u^2 \rangle\!\rangle = \frac{2\omega^2}{\pi k^2}\int_0^{\pi/2}d\theta\,\tan^2\theta\left[\frac{1}{(1-n^2\cos^2\theta)^{\frac{1}{2}}} - 1\right]. \tag{7}$$

If n is significantly less than one, then

$$\frac{\langle\!\langle B \rangle\!\rangle^2}{8\pi} \simeq \frac{B_o^2}{8\pi} + \frac{1}{2}\rho\frac{n^2}{4}\left[\frac{\omega^2}{k^2}\left(1 + \frac{3}{16}n^2\right) - V_A^2\left(1 + \frac{23}{16}n^2\right)\right], \tag{8}$$

upon neglecting terms sixth order in n, where V_A is the Alfven speed in B_o. Putting the phase velocity ω/k equal to the rms turbulent velocity v, and supposing that B_o is a weak field $(V_A^2 \ll v^2)$, the result is

$$\frac{\langle\!\langle B \rangle\!\rangle^2}{8\pi} = \frac{B_o^2}{8\pi} + \frac{1}{2}\rho v^2\frac{n^2}{4}, \tag{9}$$

so that there is only a small enhancement of the mean field above the value B_o that would prevail in the absence of the external turbulence. Evidently, the much larger enhancement of the mean square field, by $\frac{1}{2}\rho\langle\!\langle u^2 \rangle\!\rangle$, arises from the <u>fluctuations</u> of the field as much as any increase of the mean.

If $B_o^2/8\pi$ is considerably less than $\frac{1}{2}\rho v^2$, then n may not be small, of course. For n near one,

$$\frac{\langle\!\langle B \rangle\!\rangle^2}{8\pi} \simeq \frac{1}{2}\rho\left[v^2\left(1 - 2/\pi\right) - \frac{2n^2 V_A^2}{1 - n^2}\right].$$

If B_o is so small that $V_A^2/(1-n^2)$ can be neglected, then

$$\langle\!\langle B \rangle\!\rangle^2/8\pi \simeq (1 - 2/\pi)\frac{1}{2}\rho v^2 \simeq 0.36\,\frac{1}{2}\rho v^2, \tag{10}$$

and the Bernoulli effect within the tube provides compression of a weak field, B_o, up to a fraction 0.6 of the equipartition value.

Finally, note that if B_o is strong, so that $V_A = \omega/k$, then n is smaller than one and (8) reduces to

$$\frac{\langle\langle B \rangle\rangle^2}{8\pi} \cong \frac{B_o^2}{8\pi}\left(1 - \frac{5}{16}n^4 + \cdots\right). \qquad (11)$$

The mean field is slightly <u>reduced</u> by the effect of the external turbulence, although the rms field is increased according to (1). In summary, it appears that there is no direct compression of a flux tube by the dynamical pressure of the external turbulence and only a modest compression at best as a consequence of the internal Bernoulli effect caused by the turbulence.

So far as the concentrated flux tubes in the sun are concerned, it is obvious that the turbulence near the surface of the sun has a kinetic energy density $\frac{1}{2}\rho v^2$ much too small to accomplish the compression through the Bernoulli effect. The equipartition field is only about 600 gauss, and the Bernoulli effect can accomplish only a fraction of that. Stronger turbulence at some depth does not solve the problem of the surface fields because (1) is a local condition that must be satisfied at each point along the tube. The problem is to evacuate the gas from the interior of the flux tube.

III. EVACUATION OF FLUX TUBES

Consider what effects may pump the gas out of a flux tube to reduce p_i below the ambient external pressure p_e, thereby compressing the magnetic field B. The magnitude of the problem, as posed by the observations of the sun, may be seen from the fact that the ambient gas pressure at the surface of the sun is approximately $p_e \cong 1.4 \times 10^5$ dynes/cm^2 (3×10^{-7} gm/cm^3 at 5600° K), while the magnetic pressure of 1.5 kilogauss is 0.9×10^5 dynes/cm^2. It follows from (1) that the gas pressure inside the tube is of the order of 0.5×10^5 dynes/cm^2 or $p_i \cong 0.4\, p_e$. The interior must be strongly evacuated.

Several possibilities spring to mind. For instance, suppose that the magnetic fields in the sun are closed locally on themselves, so that the gas within them has no access along the lines of force to the ambient gas. Imagine that the fields are all in the form of closed loops, with length L and cross section A. The net tension in each tube is $\mathfrak{I} = A B^2/4\pi$, exerting a transverse force $\mathfrak{I}/\mathcal{R}$ per unit length where the tube has a local radius of curvature $\mathcal{R}$. The aerodynamic drag per unit length as a consequence of transverse motion $v_\perp$ of the external gas is $\frac{1}{2}\rho v_\perp^2 (A/\pi)^{1/2} C_D$, where C_D is the drag coefficient, (Goldstein, 1938). It follows that the aerodynamic drag overpowers the tension in the tube whenever

$$\tfrac{1}{2}\rho v_\perp^2 \left(A/\pi\right)^{\frac{1}{2}} C_D \gtrsim A \, B^2/4\pi\mathcal{R}.$$

Hence the length of the tube may be increased by the fluid provided
only that

$$\frac{B^2}{8\pi} \lesssim \tfrac{1}{2}\rho v_\perp^2 \, \frac{C_D}{(4\pi)^{\frac{1}{2}}} \left(\frac{\mathcal{R}^2}{A}\right)^{\frac{1}{2}}. \tag{12}$$

For slender flux tubes, $\mathcal{R}^2/A$ may be a very large number, so that
$B^2/8\pi$ is not limited to the equipartition value. If, for instance, the stretching takes place at constant ρ , then conservation of volume within the tube of length L yields $A \propto L^{-1}$.
Conservation of flux AB yields $B \propto L$, so that the field increases with the length of the tube, up to the limit set by (12)
(Parker 1976a). The gas pressure inside follows from (1).

There are two obvious difficulties with the application of
this scheme to the sun. The first is the question of whether the
flux tubes are closed on themselves. There is no reason to think
that they are, and if they are not, then there is nothing that
prevents more fluid from streaming in along the field to fill the
partially evacuated interior, so that LA is not conserved. The
second is the magnetic buoyancy that brings the tube so rapidly
to the surface, at nearly the Alfven speed, long before the random
stretching of the tube has been able to increase the field. It
must be remembered that the fields of 1-2 kilogauss are maintained
for days or weeks at the visible surface of the sun.

Suppose, then, that the magnetic fields do not form closed
systems. Assume instead that the local convective cells represent
closed circulation systems. It is well known that the steady closed
circulation of conducting fluid excludes fields from the interior
of the motion, compressing the fields into the periphery of the
pattern of circulation (Parker 1963, 1979b, Chap. XVI; Weiss 1966;
Parker 1966). One would expect the field to be compressed to some
value near equipartition with the fluid motions, $B^2/8\pi = O(\tfrac{1}{2}\rho v^2)$.
As a matter of fact the recent work of Weiss and collaborators
(Galloway, Proctor, and Weiss 1977, 1978; Peckover and Weiss 1978;
Galloway and Moore 1979; Weiss 1980) has disclosed the remarkable
fact that the magnetic field may be compressed to values for above
equipartition with the kinetic energy density $\tfrac{1}{2}\rho v^2$. This achieve-
ment comes about through the effect of the viscosity and may be
understood as follows. Consider the fluid streaming with velocity
U over the layer of compressed field. Denote by $h(x)$ the thickness of the layer of field at a distance x downstream from where
the fluid first contacts the layer of field. The fluid at x has
been in contact with the field for a time t of the order of x/U.
The rate of diffusion of the field across the stream of fluid is
of the order of $2\eta/h(x)$, where η is the resistive diffusion
coefficient. Hence for an observer moving with the fluid the rate

of increase of h with time is

$$dh/dt \cong 2\eta/h.$$

In the fixed frame of reference this is

$$U\, dh/dx \cong 2\eta/h,$$

from which it follows that

$$h(x) \cong \left(4\eta\, x/U\right)^{\frac{1}{2}}$$

This is nothing more than the standard boundary layer thickness $(4\eta t)^{\frac{1}{2}}$ for a diffusion coefficient η. We suppose that η is sufficiently small $(\eta \ll \nu)$ that the magnetic boundary layer is very thin compared to the viscous boundary layer.

For large magnetic Reynolds numbers, the field is strong and the Lorentz force impedes the flow of the fluid at the boundary, so that a viscous boundary layer is set up, with thickness $H(x) = (4\nu t)^{\frac{1}{2}} = (4\nu x/U)^{\frac{1}{2}}$ large compared to the thickness $h(x)$ of the magnetic field. In effect, the inertia and kinetic energy of the fluid in the viscous boundary layer are coupled by the viscosity to the fluid in the thin magnetic boundary layer. Hence some fraction of energy $H(x)\frac{1}{2}\rho v^2$ is brought to bear on the field so that the field may be compressed until

$$h(x)\, \frac{B^2(x)}{8\pi} \cong H(x)\, \frac{1}{2}\rho v^2.$$

Thus, in order of magnitude, B may become as large as

$$B^2/8\pi \cong \frac{1}{2}\rho v^2 \left(\nu/\eta\right)^{\frac{1}{2}} \tag{13}$$

when $\nu \gg \eta$.

Unfortunately, some questions arise when we apply this powerful effect to the flux tubes in the solar photosphere. The molecular kinematic viscosity in the photosphere is of the order of $1 \text{ cm}^2/\text{sec}$ while the molecular resistive diffusion coefficient is of the general order of $10^7 \text{ cm}^2/\text{sec}$. Thus, ν is small rather than large compared to η. Obviously, then, the effect depends upon the eddy viscosity and diffusivity. But the eddy diffusion is strongly suppressed when $B^2/8\pi$ becomes as large or larger than $\frac{1}{2}\rho v^2$. Hence, it is not clear how the effect can be put to work.

A similar difficulty arises with an effect called turbulent pumping (Parker 1974a, 1979b, §10.5). Imagine a vertical flux tube extending up through the convective zone to the surface of

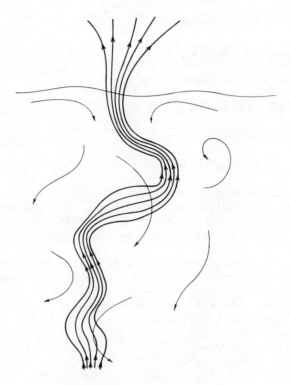

Fig. 2:   A sketch of a flux tube in a turbulent
          downdraft.

the sun, as illustrated in Figure 2, surrounded by a convective
downdraft. This is the general circumstance in which the flux
tubes in the photosphere are observed. Suppose   that the field
is relatively weak, with $B^2/8\pi \ll \frac{1}{2}\rho v^2$, so that the tube is
strongly massaged and distorted by the turbulence in the surrounding downdraft, as sketched in Figure 2. The turbulent eddies are
carried along in the downdraft outside the tube and, coupling
strongly with the fluid inside, drag the fluid down along the
tube. The upper end of the tube opens into the tenuous chromosphere so that only a small inflow of gas can take place at the
upper end. The result is that the upper portion of the tube is
evacuated until the increased pressure gradient becomes strong
enough to oppose the turbulent coupling to the external fluid.
In order of magnitude, the Reynolds stress across the surface of
the magnetic field is equal to $\rho U w$ where U is the speed of the
downdraft immediately outside the surface and w is the transverse
component of the small scale turbulent velocity at the surface.
Then if λ is the eddy size, the downward force exerted on the
fluid within the field is of the order of $\rho U w/\lambda$ per unit
volume. Equilibrium of the fluid within the field is achieved,

then, when the vertical pressure gradient increases to

$$\frac{dp_i}{dz} = -\rho_i g - \rho_i U_w/\lambda .$$
(14)

For the simple case of an isothermal atmosphere, with characteristic thermal velocity $a^2 = kT/M$, we have $p_i = \rho_i a^2$ and

$$\frac{1}{\rho_i}\frac{d\rho_i}{dz} = -\frac{g}{a^2} - \frac{U_w}{a^2\lambda}$$

so that the pressure at some height z above the base level $z = 0$
is

$$p_i(z) = p_i(0)\exp\left(-\frac{gz}{a^2}\right)\exp\left(-\int_0^z dz'\frac{U_w}{a^2\lambda}\right)$$

$$\cong p_e(z)\exp\left(-\int_0^z dz'\frac{U_w}{a^2\lambda}\right),$$
(15)

assuming that $p_i(0) \cong p_e(0)$. Reduction of $p_i(z)$ below the
static value of $p_e(z)$ depends, then, upon the integral of
$U_w/a^2\lambda$ over height.

Models of the solar convective zone (cf. Spruit 1974) show
that the Mach number of the convective motions is small, 0.1 or
less, and increases with height, reaching 0.2 - 0.4 only in the
top few hundred km. Since U_w/a^2 is essentially the square of
the Mach number, it is evident that the integration over height,
with λ comparable to the scale height, accumulates very little
except perhaps in the upper most level of the convective zone
immediately below the photosphere. It was pointed out earlier
that $p_i(z)$ must be reduced to less than half $p_e(z)$ in order that
1.5 kilogauss be realized at the photosphere. But this would make
$B^2/8\pi$ some 5-10 times larger than the turbulent energy density,
in which case the turbulent coupling to the interior of the tube
would be rendered ineffective (Parker 1979a). Hence it is not
obvious that turbulent pumping can achieve the necessary concentration of field with the standard values attributed to the turbulent
velocity beneath the photosphere.

It appears that we must turn from turbulent effects to other
devices. There are some interesting dynamical effects that permit
the slow evolution of a flux tube toward ever more concentrated
fields in the presence of an equipartition flow $\frac{1}{2}\rho v^2 = B^2/8\pi$
along the tube (Parker 1976b, 1979b, §§10.6-10.8). But there is
no evidence for such flow, which would be 4-5 km/sec in 1.5 kilogauss at the visible surface of the sun. So we will not go into
the effect here. The only remaining possibility appears to be a
reduced temperature of the gas inside the flux tube.

IV. CONCENTRATION OF FIELD THROUGH COOLING

Consider the possibility of cooling the gas within the magnetic flux tube below the temperature of the surrounding gas. In that way the scale height $\Lambda = a^2/g$ is reduced inside (where a is the characteristic thermal velocity) and gravity evacuates the upper end of the tube (Parker 1955, 1979b, Chap. X). If the ambient scale height is $\Lambda(z) = kT(z)/Mg$, then the pressure in the external gas declines with height according to

$$p_e(z) = p_e(0) \, exp\left(- \int_0^z \frac{dz'}{\Lambda(z')}\right) \tag{16}$$

A temperature reduction $\Delta T(z)$ within the flux tube yields

$$p_i(z) = p_i(0) \, exp\left(- \int_0^z \frac{dz'}{\Lambda(z') - \Delta\Lambda(z')}\right), \tag{17}$$

where $\Delta\Lambda(z) = k \, \Delta T(z)/Mg$. Assuming that $\Delta\Lambda \ll \Lambda$ so that the denominator in the integrand can be expanded in $\Delta\Lambda/\Lambda$, and assuming that the field is so weak at the base level $z = 0$ that $p_i(0) = p_e(0)$, it follows that

$$p_i(z) \cong p_e(z) \, exp\left(- \int_0^z \frac{dz' \Delta\Lambda(z')}{\Lambda^2(z')}\right). \tag{18}$$

To produce 1.5 kilogauss at the photosphere requires the reduction of p_i to about $0.4 \, p_e(z)$, so the integral should be approximately equal to one if cooling is to effect the compression of the field.

The magnetic flux tubes are found exclusively in downdrafts, at the corners of supergranules and in the dark intergranular lanes, as noted in the Introduction. It is a fact that in the sun, and other relatively cool stars, the temperature gradient immediately below the visible photosphere is significantly steeper than the adiabatic gradient. Hence any downflowing material tends to be much cooler than its surroundings, accounting for the darkness of integranular lanes, etc. Observations suggest that the gas within the flux tubes is also flowing down, with velocities perhaps as high as 1-2 km/sec (Deubner 1976). Hence the gas within the tube is strongly cooled. The important point is that the gas flowing downward within the tube is insulated from the convective heat exchange that moderates the adiabatic cooling of downdrafts outside. Hence the gas within the tube follows closely along the adiabat and is soon significantly cooler than its surroundings. Figure 3 is a plot of the temperature reduction below the ambient value at the indicated depth for a volume of gas moving downward from the visible surface under adiabatic conditions based on Spruit's (1974) model of the convective core.

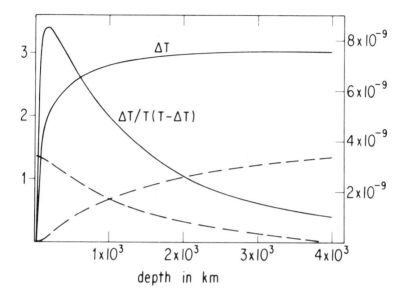

Fig. 3: The curve labeled ΔT is a plot of the temperature reduction (with increasing depth below the surface
of the sun) of a downward adiabatic flow of gas. The
curve $\Delta T/T(T-\Delta T)$ is plotted in terms of the correspond-
$\Delta\Lambda/\Lambda(\Lambda-\Delta\Lambda)$ cm^{-1} on the scale on the right-hand
side. The dashed curve sloping down to the right is the
integral of $\Delta\Lambda/\Lambda(\Lambda-\Delta\Lambda)$ upward from a depth of 4000
km, while the dashed curve sloping up to the right is
the integral downward from the surface.

The curve labeled $\Delta T/T(T-\Delta T)$ represents the integrand,
$\Delta\Lambda/\Lambda(\Lambda-\Delta\Lambda)$ cm^{-1}, measured on the scale on the right-hand side.
The dashed curve sloping down to the right represents the integral
of $\Delta\Lambda/\Lambda(\Lambda-\Delta\Lambda)$ upward from a depth of 4000 km. The important
point is that the value of the integral is approximately one at
the surface $z=0$, providing the reduction of gas pressure p_t to
$0.4 \, p_e$. The expected field strength is then about 1.5 kilogauss.
We suggest that this superadiabatic effect is the cause of the
concentration of fields in the photosphere of the sun.

     It is obvious that updrafts should have the opposite effect
of downdrafts, dispersing the magnetic field as a consequence of
the increased temperature within the updraft. Once the field is
dispersed, the usual turbulent mixing moderates ΔT so that nothing
much happens. Stability considerations (Unno and Ando 1979; Spruit
and Zweibel 1979) suggest that downdrafts are dynamically favored
over updrafts. We should add here that, once established, a downdraft becomes very strong because of the thermal insulation provided by the magnetic field that the downdraft pulls in around it.

We should think of the whole effect as a self-concentrated magnetic-downdraft needle piercing the surface of the sun. The magnetic field is the cocoon that insulates the downdraft from the surrounding heat and the downdraft is the driving force that reduces the internal temperature and concentrates the field. The sun is stuck full of these needles, which occur whenever (a) there is present a general magnetic field of several gauss and (b) the ambient temperature gradient is sufficiently superadiabatic.

There are a number of questions that should be investigated further, of course. For instance, the very difficult observations of the downdraft inside the flux tube have not yet been carried to the point where a direct measure of the velocity is available. The magnetic inhibition of heat transfer in the flux tube has been estimated (Parker 1978) and appears to be more than adequate to preserve the adiabatic state of the downdraft, but more work needs to be done. There should be a theoretical lower limit on the diameter of a flux tube that can be concentrated by the adiabatic cooling of a downflow.

The discussion has addressed itself to the equilibrium compression of the field thus far. Obviously, the dynamical stability of the equilibrium is the next problem to be explored. The evidence for the association of spicules with the individual flux tubes (Beckers 1968; Gibson 1973; Parker 1969, 1979a; Wentzel and Solinger 1967) suggests that the equilibrium may be a dynamic one rather than the simple static scenario presented here. Observations may be of help in this question. Then there is the old question of how the gas is able to flow across the lines of force into the upper end of the field above the photosphere in order to feed the downflow that is indicated by the observations. The hydromagnetic exchange instability suggested by Giovanelli (1977) may be adequate for the purpose, but the problem is a difficult one and has not yet been treated adequately (Parker 1978, 1979b, §10.11.4). Finally, we should say that any alternatives to the concentration of field by the superadiabatic effect should be explored, because a significant part of the argument in favor of the superadiabatic effect is the absence of alternatives. It is gratifying to know that other authors at this meeting are pursuing a variety of possibilities.

If the suggestion is correct that the separate flux tubes are concentrated by the adiabatic cooling in a downflow, then it would appear that the magnetic fields of all stars with significant convective zones immediately below the photosphere should show the same effect. In particular, the cooler stars, such as the M-dwarfs, should show a very strong concentration of the individual flux tubes. The effect may contribute to the vigorous activity of the cooler stars.

V.   CLUSTERING OF FLUX TUBES TO FORM SUNSPOTS

This last section deals with the peculiar tendency of the separate flux tubes to cluster together, while they are emerging through the surface of the sun, to form pores and sunspots. As noted in the Introduction, some authors have suggested that the fields are concentrated to many kilogauss and formed into a tree-like structure far below the surface of the sun. The formation of a sunspot group is then nothing more than the rising of the magnetic tree through the surface of the sun. It is the lower trunk of the tree, below the point where the branches connect together, that forms the final mature sunspot. Our own view is that the fields may be concentrated far below the surface of the sun if there is some effect there to concentrate them. For instance, one might derive a very large value for the turbulent kinetic energy density $\frac{1}{2}\rho v^2$ from some mixing length theory, by assuming that the turbulent heat transport is relatively ineffective. Then an appeal to the Bernoulli effect described in the preceding section might be made to give fields of suitable strength. But the results would not be particularly convincing. As indicated in the section above, the only effects, of which we are aware, for compressing the field to a kilogauss or more, operate near the surface. The preassembled tree structure may exist far below the surface, if there is some effect to hold the branches to the trunk. Our task, it would seem, is to explore the physics of fluids and flux tubes to see what are the possibilities.

Consider some of the basic effects operating in and around a sunspot. Biermann (1941) pointed out many years ago that the strong vertical magnetic field of a sunspot inhibits the convective transport of heat, evidently accounting for the reduced heat flux observed in the sunspot umbra. The blocking of heat transport causes a pileup of heat underneath, of course (Parker 1974c; Spruit 1977; Eschrich and Krause 1977; Clark 1979) which has the effect of dispersing the magnetic field rather than concentrating it (Parker 1974c, 1979b, §10.10.2) and which produces a bright ring around the outside of the boundary of the sunspot. Detailed calculations by Spruit (1977), Eschrich and Krause (1977), and Clark (1979) show that the extremely effective convective heat transport at depths of 10^9 km or more would spread the heat so broadly around the spot as to make the bright ring undetectable. The real problem lies in the dispersal of the magnetic field. We require a reduction of temperature below the ambient value to accomplish the observed strong concentration of field (Parker 1955).

Some years ago Parker (1974c) revived Danielson's (1965; Savage 1969; Moore 1973) suggestion that convective overstability may convert a large fraction of the heat flux underneath a sunspot into Alfven waves. The Alfven waves then propagate away along the

field and dissipate elsewhere. Danielson originally made the
suggestion to account for the missing heat flux through the sunspot. Parker was more concerned with the local pile up of heat
beneath the umbra, which disperses the field. (The missing heat
flux is no longer the problem, with the calculations that show
how broadly it is dispersed.) The convective instability would
have to operate as a heat engine of high efficiency if it were
to remove completely the pile up of heat (Parker 1974c, 1975b,
1979h; Roberts 1976; Boruta 1977; Thomas 1978). Unfortunately,
contrary to earlier expectations, it appears that most of the
waves are emitted downward (Parker 1979h) so that there is little
chance of an observational determination of their strength (see
Beckers 1976 and Beckers and Schneeberger 1977). It is our opinion
that the pile up of heat must be largely avoided in any successful
theory of the sunspot. We will come back to this problem a little
later.

In another direction, there is the obvious problem of over-
coming the tension in the individual flux tubes as they are pushed
together to form a sunspot at the photosphere, (or to form a tree
at some depth below the surface). The tension along the lines of
force spreading out from a newly forming sunspot tends to pull the
tubes away from the sunspot. Evidently some force overpowers the
tension and pushes the tubes together to produce or maintain the
sunspot. Meyer, Schmidt, and Weiss (1977) have pointed out that
the magnetic buoyancy of the tubes tends to stabilize the field
holding the tubes together where they flare out across the penumbra
at the surface of the sun. But it remains to explain what stabilizes the vertical column of field beneath the umbra, and it remains to explain what pushes the separate tubes together in the
first place, or what keeps the branches from stripping off the
tree if one prefers that picture. In this direction, Meyer,
Schmidt, Weiss, and Wilson (1979) suggest that there is a powerful
converging flow surrounding the field of the sunspot at a depth
of several thousand kilometers. They suggest that the dynamical
pressure of the converging flow is responsible for maintaining the
integrity of the sunspot. They suggest that the converging flow
gives rise to an upwelling around the outside of the field and
produces the outflow observed at the surface around a mature sunspot. There is no evident outflow during the early formative
stages.

Alternatively, Parker (1979a) suggested that a subsurface
converging flow and downdraft might work better than the updraft
for a number of reasons. We suggest, too, that in place of the
conventional idea of a single broad flux tube beneath a single
sunspot (see Figure 4) the separate flux tubes that are pushed
together to form the spot in the first place may maintain their
separation, if not their initial identity, below the surface of
the umbra. In that case, the converging flow and downdraft

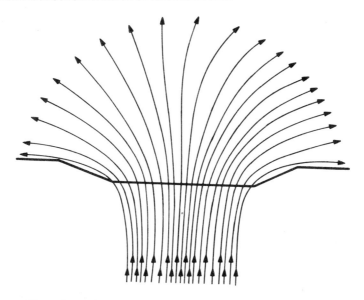

Fig. 4:   The conventional sketch of the magnetic
configuration of a sunspot.

penetrate through the entire region of field, as illustrated in
Figure 5. We would argue, then, that the initial "spontaneous"
clustering of the flux tubes observed at the surface is in response
to the unseen converging flow and downdraft a thousand or more km
below the surface. It would be the same converging flow and downdraft continuing to hold the tubes in the tightly packed cluster
for the life of the sunspot. The downdraft has the attractive
feature that in flowing down between the separated tubes of flux
(see Figure 5) it reduces the upward flow of heat which otherwise
would raise the gas temperature above normal (in view of the reduced flow of heat through the umbra). We suggest that a strong
downdraft flowing between the separate tubes may carry away the
heat that would otherwise pileup somewhere beneath the sunspot
(Parker 1979a,g). Thus the downdraft between the separated flux
tubes below the umbra prevents the field from being dispersed by
the heat. Indeed it seems possible that the convective removal
of heat by the downdraft, perhaps in combination with the generation of Alfven waves (Parker 1979h), may reduce the temperature
sufficiently below the ambient value to compress the field to the
observed 3 kilogauss of the mature sunspot. It must be remembered
that the tentative explanation for compression to 1.5 kilogauss in
the separate tubes is based on the downdraft observed inside each
tube. There is no downdraft inside the tubes once they are pushed
together to form a sunspot. Some other more powerful effect must
take over the cooling to account for the increase to 3 kilogauss.
It may be the general downdraft beneath the sunspot.

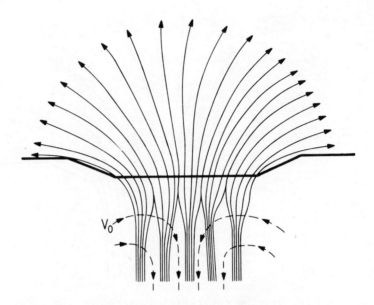

Fig. 5: A suggested magnetic configuration of a sunspot, in which the field separates into many filaments
below the visible surface, in the presence of a converging flow and downdraft (dashed lines).

The bright umbral dots (Beckers and Schröter, 1968; Loughhead,
Bray, and Tappere, 1979; Bumba and Suda, 1980) are a curious and
fascinating aspect of a sunspot. We have suggested (Parker 1979a,
i) that they are a direct manifestation of the field-free gas
beneath the umbra (Figure 5). A simple linear treatment of the
gas motions in the space between the separate tubes (Parker 1979h,
i) shows convective overstability in the form of strong vertical
oscillations. On this basis, the umbral dots may be understood
as an upward intrusion of the field-free gas to the visible surface.
The dots may be a glimpse of the field-free gas that is concealed
a few hundred km below the visible surface of the umbra.

We have explored the dynamical properties of separate flux
tubes to develop some understanding of how they might be expected
to behave in various circumstances of gas flow. The calculations
show that the tubes are not stable against splitting into two or
more separate tubes unless twisted in some small degree (Parker
1979b, §8.7, 9.8; Tsinganos 1980), so there are some interesting
unanswered questions on the shape of the cross section of the
individual flux tubes, perhaps related to the filigree structure.

The aerodynamic drag per unit length on a tube of circular or
elliptical cross section is $\frac{1}{2}\rho v_{\perp}^{2}(A/\pi)^{\frac{1}{2}}C_{D}$, as already noted,
where $v_{\perp}$ is the transverse component of the fluid velocity, A is

the cross sectional area, and C_D is the drag coefficient (Goldstein 1938). The drag depends upon the kinetic energy density. Now the convective envelope of the sun is stratified vertically, the density increasing by a factor of ten in the first 10^3 km beneath the visible surface, a factor of 10^2 in the first 3×10^3 km, 10^3 in the first 7×10^3 km, 10^4 in the first 2×10^9 km, and more than a factor of 10^5 to the bottom of the convective zone. If we think of a closed convective pattern, $\nabla \cdot (\rho \underline{v}) = 0$, extending over a depth of some thousands of km, it is evident that the flow in one direction at higher levels (where ρ is small) contains far more kinetic energy than the return flow in the opposite direction at a lower level where ρ is large. The only way to avoid this conclusion is to suppose that the return is strongly concentrated into a thin layer at the lower level. To demonstrate this point, note that a given mass flux m_1 in a layer of thickness h_1 , where the density is ρ_1 and the velocity is v_1 , can be written $m_1 = \rho_1 v_1 h_1$. The return flow at some lower level is $m_2 = \rho_2 v_2 h_2$, say. Conservation of mass requires $m_1 = m_2$ if the two flows have the same horizontal width. The total drag F on a tube extending vertically across a layer is proportional to $\rho v^2 h$, so with $m_1 = m_2$ it follows that the ratio of the drag in the two layers is

$$ F_1 / F_2 = \rho_2 h_2 / \rho_1 h_1 . $$

With the subscript 2 designating the lower level, it follows that ρ_2 may be very much larger than ρ_1 , so that the aerodynamic drag is larger in the upper flow. Only if h_2 is so small that $h_2 < h_1 \rho_1 / \rho_2$ could it be otherwise. We are not aware of any reasons for believing that h_2 is so small in the convective cells in the sun as to compensate for ρ_2 . Hence it appears that the upper levels should dominate (Parker 1979c). It was on this basis that we first suggested that the flux tubes are herded together into a sunspot by a converging flow associated with a downdraft rather than an updraft. The downdraft places the diverging flow at lower levels where it is less effective. And, of course, the downdraft has the added attraction of suppressing the upward flow of heat. The downdraft eliminates this fundamental problem, providing both a converging force to cluster the separate tubes as well as a reduced temperature to concentrate the field once the sunspot is formed. The idea that the sunspot is held together by a subsurface downdraft is consistent with the observed fact that all the small magnetic features on the sun are confined to downdrafts. We suggest that the observed outflow around a sunspot once it has formed may be a local cell of counterflow above the main converging flow and downdraft beneath the surface.

It must be remembered, of course, that there is no direct observational evidence for a subsurface converging flow of suffi-

cient strength to accomplish the compression of field into a sunspot, whether accompanied by an updraft or a downdraft. Nor is
there any <u>direct</u> evidence for the separation of flux into a forest
of distinct tubes beneath the surface of the sunspot. So we should
not cease to seek alternative explanations for the formation of
sunspots.

To move on, we have explored the aerodynamic lift on tubes
with circular or elliptical cross section in a nonuniform flow of
ideal fluid (Parker 1978, 1979b, §8.9, 1979d,e; Parker and Tsinganos
1979) demonstrating the mutual attraction of flux tubes in a large-
scale flow. The attraction is strong enough to account for much
of the observed tendency for emerging flux tubes to cluster together
in a growing active region. It is also obvious that if the tubes
were pressed together, so that there is no longer fluid flowing
between them, the attraction ceases. With the hypothesis that
the field is divided into separate tubes in a converging flow
beneath the sunspot (Fig. 5), the mutual hydrodynamic attraction
of neighboring tubes, and the aerodynamic drag on the individual
tubes, both press the cluster of tubes together to maintain the
sunspot.

Tsinganos (1979) has shown that an isolated flux tube may
be pushed by the aerodynamic lift (in the absence of drag) into
any one of a number of points of symmetry in a convection cell.

The phenomenon of convective propulsion (Parker 1979f) is
an important agent in manipulating flux tubes beneath the photosphere. Convective propulsion generally has the effect of amplify-
ing the motion of each tube relative to the background fluid. One
may think of it as a negative aerodynamic drag. Convective propulsion enhances the magnetic buoyancy, suggesting that the rate of
rise of a flux tube through the convective zone of the sun, or
other star, may be larger than we have estimated previously
(Parker 1975a, 1979b, §8.7). The rate of rise was already calculated to be rather large, comparable to the Alfven speed in the
tube, leading to difficulty in understanding how the magnetic
field remains long enough in the lower convective zone to be
amplified by the solar dynamo. We do not know how to resolve
that difficulty.

Altogether, it appears to this reviewer that the formation
of a sunspot is a complicated dynamical problem involving the
mutual interaction of hydrodynamic convection, hydromagnetic
flux tubes, and the massive heat transfer through the upper layers
of the convective zone. The growing awareness in the astronomical
community of the magnitude of the problem provides hope that it
may one day reach a satisfactory conclusion. At the present time
it may be stated that we have ideas, but we cannot be certain
about the compression of field to kilogauss strengths in the
individual flux tubes. We have ideas, but we certainly cannot be

certain as to why the flux tubes cluster together to form sunspots. I am confident that hard work in observation and theory can shake out the loose ends to give a self-consistent picture before too many more decades slip by.

When the time comes that we understand the concentration of fields on the sun with more confidence, we will be in a position to make statements about what is happening on other stars.

REFERENCES

Beckers, J.M. 1968, Solar Phys. 3, 367.

Beckers, J.M. 1976, Astrophys. J. 203, 739.

Beckers, J.M. and Schneeberger, T.J, 1977, Astrophys. J. 215, 356.

Beckers, J.M. and Schröter, E.H. 1968, Solar Phys, 4, 142, 165, 303.

Biermann, L. 1941, Vlerteljahrsschr, Astr. Gesselsch. 76, 194.

Boruta, N, 1977, Astrophys. J. 215, 369.

Bumba, V, and Suda, J. 1980, Bull. Astron. Inst. Czechosl. 31, 101.

Chapman, G. 1974, Astrophys, J. 191, 255.

Clark, A, 1966, Phys. Fluids 9, 485.

Clark, A. 1968, Solar Phys. 4, 386.

Clark, A, 1979, Solar Phys, 62, 305.

Danielson, R.E. 1965, Stellar and Solar Magnetic Fields, IAU
    Symposium No. 22 (ed. R, Lust) (Amsterdam, N.Holland) p, 315.

Deubner, F. 1976, Astron. Astrophys, 47, 475.

Eschrich, K.O, and Krause, F, 1977, Astron, Nachr, 298, 1,

Frazier, E.N, 1970, Solar Phys. 14, 89.

Frazier, E.N. 1972, Solar Phys. 26, 130.

Frazier, E.N. and Stenflo, J.O. 1972, Solar Phys, 27, 330.

Galloway, D.J. and Moore, D,R. 1979, Geophys. Astrophys, Fluid 12, 73.

Galloway, D.J., Procter, M.R.E., and Weiss, N.O. 1977, Nature,
    266, 686.

Galloway, D.J., Procter, M.R.E., and Weiss, N.O. 1978, J. Fluid
    Mech.87, 243.

Gibson, E.G. 1973, The Quiet Sun (NASA Sp-303); Washington, DC:
    U.S. Government Printing Office) p. 236.

Giovanelli, R.G. 1977, Solar Phys. 52, 315.

Goldstein, S. 1938, Modern Developments in Fluid Dynamics (Oxford,
    Clarendon Press) Vol. 1, pp. 13-19; Vol. II, p. 417.

Harvey, J.W. 1977, Highlights of Astronomy (ed. E.A. Muller) 4,
    Part II, 223.

Howard, R. and Stenflo, J.O. 1972, Solar Phys. 22, 402.

Leighton, R.B. 1963, Ann. Rev. Astron. Astrophys. 1, 19.

Leighton, R.B., Noyes, R.W. and Simon, G.W. 1962, Astrophys. J.
    135, 474.

Livingston, W. and Harvey, J. 1969, Solar Phys. 10, 294.

Livingston, W. and Harvey, J. 1971, Solar Magnetic Fields, IAU
    Symposium, No. 43 (ed. R. Howard) p. 51, Reidel, Dordrecht.

Loughhead, R.E., Bray, R.J., and Tappere, E.J. 1979, Astron.
    Astrophys. 79, 128.

Meyer, F., Schmidt, H.U., and Weiss, N.O. 1977, Mon. Not. Roy.
    Astron. Soc. 179, 741.

Meyer, F., Schmidt, H.U., Weiss, N.Y., and Wilson, P.R. 1979,
    Mon. Not. Roy. Astron. Soc. 169, 35.

Moore, R.L. 1973, Solar Phys. 30, 403.

Noyes, R.W. and Leighton, R.B. 1963, Astrophys. J. 138, 631.

Parker, E.N. 1955, Astrophys. J. 121, 491.

Parker, E.N. 1963, Astrophys. J. 138, 552.

Parker, E.N. 1964, Astrophys. J. 140, 1170.

Parker, E.N. 1972, Astrophys. J. 174, 499.

Parker, E.N. 1973, Astrophys. J. 186, 693, 665.

Parker, E.N. 1974a, Astrophys. J. 189, 563.

Parker, E.N. 1974b, Astrophys. J. 190, 429.

Parker, E.N. 1974c, Solar Phys. 36, 249; 37, 127.

Parker, E.N. 1975a, Astrophys. J. 198, 205.

Parker, E.N. 1975b, Solar Phys. 40, 275, 291.

Parker, E.N. 1975c, Astrophys. J. 201, 499.

Parker, E.N. 1976a, Astrophys. J. 204, 259.

Parker, E.N. 1976b, Astrophys. J. 210, 810, 816.

Parker, E.N. 1976c, Astrophys. Space Sci. 24, 279.

Parker, E.N. 1978, Astrophys. J. 221, 368.

Parker, E.N. 1979a, Astrophys. J. 230, 905.

Parker, E.N. 1979b, Cosmical Magnetic Fields (Oxford: Clarendon Press).

Parker, E.N. 1979c, Astrophys. J. 230, 914.

Parker, E.N. 1979d, Astrophys. J. 231, 250.

Parker, E.N. 1979e, Astrophys. J. 231, 270.

Parker, E.N. 1979f, Astrophys. J. 232, 282.

Parker, E.N. 1979g, Astrophys. J. 232, 291.

Parker, E.N. 1979h, Astrophys. J. 233, 1005.

Parker, E.N. 1979i, Astrophys. J. 234, 333.

Parker, E.N. and Tsinganos, K.C. 1979, Phys. Fluids 22, 1847.

Parker, R.L. 1966, Proc. Roy. Soc. A 291, 60.

Peckover, R.S. and Weiss, N.O. 1978, Mon. Not. Roy. Astron. Soc. 182, 189.

Piddington, J.H. 1975, Astrophys. Space Sci. 34, 347.

Piddington, J.H. 1976a, Astrophys. Space Sci. 40, 73; 45, 47.

Piddington, J.H. 1976b, Basic Mechanisms of Solar Activity, IAU Symposium No. 71 (ed. V. Bumba and J. Kleczek) p. 389.

Ramsey, H.E., Schoolman, S.A., and Title, A.M. 1977, Astrophys. J. Letters 215, L41.

Roberts, B. 1976, Astrophys. J. 204, 268.

Savage, B.D. 1969, Astrophys. J. 156, 707.

Sawyer, C. 1971, Solar Magnetic Fields, IAU Symposium No. 43 (ed. R. Howard) p. 316, Reidel, Dordrecht.

Shelley, N.R. 1967, Solar Phys. 1, 171.

Simon, G.W. and Leighton, R.B. 1964, Astrophys. J. 140, 1120.

Spruit, H.C. 1974, Solar Phys. 34, 277.

Spruit, H.C. 1977, Solar Phys. 55, 3.

Spruit, H.C. 1979, Solar Phys. 61, 363.

Spruit, H.C. and Zweibel, E.G. 1979, Solar Phys. 62, 15.

Stenflo, J.O. 1973, Solar Phys. 32, 41.

Stenflo, J.O. 1978, Rep. Prog. Phys. 41, 865.

Tarbell, T.D. and Title, A.M. 1977, Solar Phys. 52, 13.

Tarbell, T.D., Title, A.M. and Schoolman, S.A. 1979, Astrophys. J. 229, 387.

Thomas, J.H. 1978, Astrophys. J. 225, 275.

Tsinganos, K.C. 1979, Astrophys. J. 231, 260.

Tsinganos, K.C. 1980, Astrophys. J. 239, No. 2.

Unno, W. and Ando, H. 1979, Geophys. Astrophys. Fluid Dyn. 12, 107.

Vrabec, D. 1974, Chromospheric Fine Structure, IAU Symposium No. 56 (ed. R.G. Athay), p. 201, Reidel, Dordrecht.

Weiss, N.O. 1964, Mon. Nat. Roy. Astro. Soc. 128, 225.

Weiss, N.O. 1966, Proc. Roy. Soc. A. 293, 310.

Weiss, N.O. 1980, (in publication).

Wentzel, D.G. and Solinger, A.B. 1967, Astrophys. J. 148, 877.

Wiehr, E. 1978, Astron. Astrophys. 69, 279.

Zwaan, C. 1978, Solar Phys. 60, 213.

INTERNAL STRUCTURE OF THE SUN AND STARS

Ian W. Roxburgh

Dept. of Applied Mathematics
Queen Mary College, University of London

Abstract. The current theory of solar and stellar structure is reviewed with particular emphasis on the basic physics, simplifying assumptions and approximations made in deriving models of the sun and stars.

I. INTRODUCTION

The observational astronomer measures the properties of the radiation reaching his detectors, for some stars very detailed information is available, for most the measurements are of the flux of radiation in certain wavebands usually in the blue (4400 Å) and the visual, or yellow (5480 Å). These observations can be used to calculate the total rate of energy output, the luminosity L, and the effective temperature of the stellar surface, T_{eff}. If stars were black body emitters this calculation would be easy, in reality the calculation is complicated requiring a detailed knowledge of the structure of stellar atmospheres. One of the principal observational results is that the majority of stars lie on the main sequence, that is they satisfy a relation which is approximately :

$$\frac{L}{L_\odot} = (\frac{T_{eff}}{T_\odot})^6$$

where L_0 = 3.86 10^{33} ergs/sec is the solar luminosity and T_0 = 5800 is the effective temperature of the sun.

Of course there are stars that do not satisfy this relation

59

R. M. Bonnet and A. K. Dupree (eds.), Solar Phenomena in Stars and Stellar Systems, 59–74.
Copyright © 1981 by D. Reidel Publishing Company.

- high luminosity low effective temperature red giants, and low
luminosity high effective temperature white dwarfs, as well as
a wide variety of stars that do not fall into any of these groups.
However, in this article I shall concentrate on the main - sequence since the sun and stars like it are main sequence stars.

The other important observation is of stellar masses. These
are determined from the properties of double systems using
Newton's laws. The principal observational result being that
main - sequence stars satisfy the mass-luminosity and mass radius
relations which are approximately :

$$\frac{L}{L_\odot} = (\frac{M}{M_\odot})^4 \quad , \quad \frac{R}{R_\odot} = (\frac{M}{M_\odot})^{0.7}$$

Finally we need to know what stars are made of. There are,
of course, variations in composition but most stars seem to be
about 75 % hydrogen 25 % helium, the abundance of everything else
(principally carbon nitrogen and oxygen) varying between 0.1
and 4 %.

II. PHYSICS OF STELLAR INTERIORS

Stars are in equilibrium because the internal pressure is
sufficiently large to support the star against its own self gravity, the energy is produced in the high temperature central regions by the fusion of light elements to form heavier elements,
and that energy is transported from centre to surface by radiation and in some cases also by convection. In the outer layers
the optical depth decreases and the radiation escapes into space.
The nuclear fusion reactions, hydrogen forming helium in main
sequence stars, produces a change in chemical abundance and therefore a change in the internal structure, the total energy produced, and the radius of the star. These basic ideas are able to
explain the majority of observations on the properties of stars.

Before quantifying the models it is worth emphasising the
assumptions that go into the construction of such models. Firstly
we normally assume that stars are initially homogeneous in composition, secondly in most models we assume spherical symmetry
ignoring any effects of magnetic fields or rotation. Then in the
usual models mass loss is ignored, and finally it is assumed that
the stellar material is unmixed except in those regions which
are unstable to convection, in which case we assume efficient
mixing. All these simplifying assumptions can be, and have been
relaxed in more sophisticated models.

The basic equations governing such simple stellar models

can be expressed as :

Equation of State :

$$P = \frac{\mathcal{R} \rho T}{\mu} + \frac{aT^4}{3} \quad (+ \text{ degeneracy}) \qquad (2.1)$$

Hydrostatic Support :

$$\frac{dP}{dr} = - \frac{GMr}{r}\rho \ , \ \frac{dMr}{dr} = 4\pi r^2 \rho \qquad (2.2)$$

Radiative Transport :

$$L_r = - \frac{16\pi acr^2}{3\kappa\rho} \ T^3 \ \frac{dT}{dr} \qquad (2.3)$$

Energy Production :

$$\frac{dL_r}{dr} = 4\pi \ r^2 \epsilon \ \rho \qquad (2.4)$$

Molecular Weight :

$$\mu = \frac{4}{8X + 3Y + 4Z} \approx \frac{4}{(3+5X)} \qquad (2.5)$$

Energy Generation :

$$\epsilon = \epsilon(X, \ Y, \ Z_i, \ \rho, \ T) \qquad (2.6)$$

Opacity :

$$\kappa = \kappa(X, \ Y, \ Z_i, \ \rho, \ T) \qquad (2.7)$$

Convective Transport :

$$L_c = f4\pi r^2 T\rho(\frac{GMr\rho}{r^2})^{1/2} (\frac{dS}{dr})^{3/2} (\frac{P}{dP/dr})^2 \qquad (2.8)$$

Here P is the pressure, ρ the density, T the temperature, r the distance from the centre of the star, M_r the mass in a sphere of radius r, L_r the energy carried by radiation crossing a sphere of radius r per unit time, X is the fraction by mass of hydrogen, Y the fraction by mass of ^{4}He, Z the fraction by mass of other elements, and Z_i the fraction by mass of each of these other elements. ϵ is the energy produced by nuclear fusion per unit mass per unit time and κ the opacity per unit mass. G, a, c, $\mathcal{R}$, are the constant of gravity, Stefan's radiation constant, the velocity of light and the gas constant respectively. For completeness I also give the expression for the energy transported by convection, L_c, that is used in building stellar models. A derivation of this expression using the so called mixing length theory is given below. S is the entropy per unit mass and f a 'fudge factor' undetermined by the theory and adjusted to obtain agreement between the observed and predicated values of the solar radius.

The pressure P has three significant contributions whose importance depends on the density and temperature ; the first term in equation 2.1 is just the ordinary gas pressure, the second term is the pressure due to radiation, and under conditions of high densities the matter becomes degenerate. Degeneracy pressure is important in white dwarfs, neutron stars and the cores

of evolved stars but is not of importance in models of main sequence stars. The expression for the molecular weight μ, the
average mass of a particle in terms of the mass of the hydrogen
atom, is obtained by assuming complete ionisation and that, on
average, the atomic weight of heavy elements, Z_i is twice their
atomic number.

The opacity, κ, of stellar material is very difficult to
calculate since there are contributions from scattering by electrons, free-free and bound-free transitions and bound-bound
transitions in atoms that are not completely ionised. Most stellar opacities are calculated by a group at the Los Alamos Laboratories (Cox (3), Huebner et al (6), Huebner (7)) who will
provide tables of calculated values for different compositions,
temperatures and densities. These tables are then used in numerical solutions of the equations (2.1) to (2.8). However, in
order to understand the properties of stellar models, it is convenient to make simple analytic approximations to these numerical
values of the opacity, such fits give :

$$\kappa = 0.2 \ (1 + X) \ \text{(high temperature, low density)} \tag{2.9}$$

$$\kappa = 10^{3.1} \ \rho^{0.4} \ T^{-2.8} \text{(solar condition, X=0.74, Z=0.02)} \tag{2.10}$$

$$\kappa = 7.10^{-18} \ \rho^{0.75} \ T^{5.75} \text{(cool surface conditions)} \tag{2.11}$$

The first is just the contribution from electron scattering,
the second is dominated by bound-free transitions and the third
is dominated by the contributions from the negative hydrogen ion
H^- that forms in cool stellar surface layers, and is an approximation to numerical values for the solar surface.

The nuclear reactions are also difficult and complicated to
evaluate since there are many ways of converting hydrogen to helium, and many cross sections have to be determined experimental-
ly and theoretically. However, the basic process in hydrogen
burning is :

$$4 \ ^1H \rightarrow \ ^4He + 2e^+ + 2\nu \ (26 \ \text{MeV}), \tag{2.12}$$

and there are two principal mechanisms, the proton-proton chain
and the carbon nitrogen cycle, the former being the 'simple'
fusion of light elements to form heavier elements, the latter
using carbon as a catalyst. Again for pedagogical purposes we
can make a simple power law fit to the detailed calculations
(Schwartzschild, 11) and these are :

$$\epsilon_{pp} = 2.6 \ 10^{-33} \ X^2 \ \rho \ T^{4.5} \tag{2.13}$$

$$\epsilon_{cn} = 7.9 \ 10^{-78} \rho \ XZT^{16}, \tag{2.14}$$

where again units are in cms, gm, sec, and $^\circ$K. The proton-proton chain is dominant in stars like the sun with central temperature of the order of 15.10^6 $^\circ$K. In more massive stars (M > 1.5 M$_\odot$) the central temperature is higher and the C-N cycle dominates.

In certain periods of a star's life gravitational energy release is also important. If the star is in thermal equilibrium with the energy radiated balancing the energy produced, the star only changes due to the gradual conversion of hydrogen to helium. However, if these conditions are not satisfied the star will contract thereby releasing gravitational energy. The gravitational energy of a homogeneous sphere of mass M and radius R is

$$V = - \int \frac{GM_r \, dM_r}{r} = - \frac{3}{5} \frac{GM^2}{R}, \qquad (2.15)$$

and so a decrease in radius leads to a decrease (negative increase) in V. If the star has a luminosity L (energy per unit time) then a typical time scale on which gravitational energy can sustain the energy loss of the star is :

$$t_{kh} = 3 \frac{GM}{RL} \sim 3,10^7 \text{ ys.} \qquad (2.16)$$

This is called the Kelvin-Helmholtz time scale and is the time scale for thermal readjustment of a star.

In fact not all the gravitational energy is available to sustain the luminosity. On integrating equation (2.2) throughout the star we find :

$$\int 3 \frac{R}{\mu} T \, 4\pi \, r^2 \, \rho dr = - \int \frac{GM_r}{r} \, dM_r = V, \qquad (2.17)$$

and therefore a decrease in V also results in an increase in mean temperature. For a fully ionised star, one half of the gravitational energy release goes into increasing the temperature, the other half is radiated away. We are now in a position to understand the main features of stellar evolution.

Initially a star is large, diffuse and the internal temperature is low. The star, therefore, contracts, gravitational energy release both compensates for the energy radiated and heats up the star. When the central temperature is high enough (10^7 $^\circ$K) hydrogen burning begins and the contraction ceases ; this takes a thermal relaxation time of the order of 10^7 years. The star then burns hydrogen until this fuel is exhausted, the central regions then contract and heat up until the temperature is high enough to burn helium to carbon (10^8 $^\circ$K). This process continues until some other effect enters (e.g. degeneracy pressure) that causes the evolution to stop, or until the star explodes.

III. SIMPLE STELLAR MODELS

The first stellar model was constructed by Homer Lane (8) with the assumption that the sun was fully convective and adiabatic. Thus the radiative transfer equation is replaced by the adiabatic condition :

$$P = K\rho^{5/3},$$ (3.1)

where K is a constant. Such models were generalised to other power law relations between pressure and density. There are polytropes, the properties of which are examined in considerable detail in Chandrasekhar's book 'Stellar Structure' (1).

Eddington (5) then studied completely radiative models and achieved the first real success in explaining observations. Cowling (4) then developed the point source model which remains a reasonable good approximation to present models of star somewhat more massive than the sun.

The properties of these models can be understood from simple order of magnitude approximations. From the equation of hydrostatic support (2.1) it follows that for a mass M and radius R the typical magnitude of the density and pressure and temperature are:

$$\rho \sim \frac{M}{R^3}, \quad P \sim \frac{GM}{R^4}, \quad T \sim \frac{\mu}{R} \frac{GM}{R} .$$ (3.2)

If we follow the same procedure with equations (2.3) and (2.4) and use the power law approximations (2.10) for the capacity and (2.14) for the energy generation we find :

$$L \propto M^{5.4} \cdot P^{-1.7} \quad , \quad L \propto \frac{M^{18}}{R^{19}}$$ (3.3)

and hence

$$M \propto R^{0.73} \quad , \quad L \propto M^{4.13} .$$ (3.4)

These are fairly good fits to the observed radius-mass and luminosity-mass relations. If we now define the effective temperature T_{eff} by Stefan's law

$$L = 4 \pi R^2 \sigma T_{eff}^4 ,$$ (3.5)

Then on eliminating M from the two equations (3.4) we find

$$L \propto T_{eff}^{6.2} ,$$ (3.6)

which is again a reasonable approximation to the observed main sequence relation.

If the same order of magnitude estimates are used for smaller mass stars with energy coming from the proton-proton chain (2.13) we find

$$L \propto \tilde{M}^5 \quad , \quad R \propto M^{0.21} \tag{3.7}$$

which is not such a good fit particularly to the radius-mass relation. The reason is that we have not included the effect of convection.

IV. CONVECTION

IV.1   The Onset of Convection

In regions of a star where the temperature gradient is large the fluid may be unstable to convection. In principle this should be determined using a full stability analysis, but since the viscosity of stellar material is so small a simple adiabatic bubble analysis is adequate.

Let r be a vertical coordinate and g be the local acceleration due to gravity, acting downwards. If a bubble of fluid originally at r with density $\rho(r)$ undergoes a small adiabatic displacement to r + dr where it is in pressure equilibrium with its surrounding it will take up a new density ρ^*. If this is greater than the density of the surrounding matter it will sink back, if not it will be accelerated upwards and the fluid will be unstable. There will be unstability if

$$\rho^* = \rho(r) \left[\frac{P\ (r+\ dr)}{P(r)} \right]^{1/\gamma} < \rho\ (r + dr) \tag{4.1}$$

which on using the gas law and Taylors theorem give the Schwartzschild condition for the onset of convection

$$\frac{T}{P} \cdot \frac{dP}{dr} \cdot \frac{dr}{dT} \ < \ \frac{\gamma}{\gamma-1} \tag{4.2}$$

where γ is the ratio of specific heats ($\gamma = 5/3$ for a monatomic gas).

The pressure gradient is determined from the hydrostatic condition (2.2). Convection will set in when

$$\left| \frac{dT}{dr} \right| > \frac{\gamma-1}{\gamma} \cdot \frac{\mu g}{}.$$

Thus in regions where the temperature gradient is large the fluid will be convectively unstable. Moreover although in most regions of a star $\gamma = 5/3$, in those sub-surface layers where hydrogen is ionised.γ is only slightly greater than unity, so again we should expect convective instability. In main sequence stars there are then two important convective zones. For stars on the C-N cycle a lot of energy is produced in a small region requiring a steep temperature gradient. If it were to be carried by radiation, such stars have convective cores. For stars like the sun the opacity in the surface regions comes principally from the negative hydrogen ion, this becomes very large as T increases and as is clear from equation (2.3) if κ is large the temperature gradient will be large and therefore convectively unstable. Thus we expect surface convective zones in cool solar type stars.

Having deduced that convection will occur we need a theory to model the heat transport and here is one of the basic difficulties of stellar structure. Any estimate of the ensuing convection shows that the Reynolds number is very large ($\sim 10^{14}$) since stars are so large, and in these conditions the flow will be turbulent. We therefore need a theory of turbulent convection and of course we do not have one. Turbulence is still an unsolved problem even in much simpler conditions than the interior of stars. The astronomer therefore, uses the mixing length theory originally developed by Taylor and Prandtl.

IV.2   The Mixing Length Theory

If we assume that the turbulence is statistically steady and average the energy equation we obtain :

$$\text{div} \left[\overline{C_\rho \rho u T} + \overline{\rho u^2 u} + \overline{F_r} \right] = \overline{\varepsilon \rho} \qquad (4.3)$$

where u is the turbulent velocity, F_r the radiative energy flux and all variables are fluctuating quantities. The mixing length theory then neglects the flux of kinetic energy and estimates the enthalpy flux writing

$$F_c = \overline{C_p \rho u T} \qquad \overline{C_p T \rho_1 u} \qquad (4.4)$$

The correlation $\rho_1 u$, where ρ is the fluctuating part of the density, is then estimated by assuming that blobs of fluid accelerate from rest, conserving entropy and mixing with their surroundings after travelling a distance ℓ, the mixing length. This gives

$$\sim \frac{1}{C_p} \frac{\ell dS}{dr} = \ell \left[\frac{1}{T} \cdot \frac{dT}{dr} - \frac{\gamma-1}{\gamma} \frac{1}{P} \cdot \frac{dP}{dr} \right] \quad , \quad u^2 = g \frac{\rho_1}{\rho_0} \ell \cdot \qquad (4.5)$$

The mixing length ℓ is taken as a multiple of the pressure scale height $P/(dP/dr)$ which on using the hydrostatic equation gives

$$F_{conv} = f C_p T V_s \left[\frac{\gamma}{\gamma-1} \frac{1}{(n+1)} - 1 \right]^{3/2} , \qquad (4.6)$$

$$u^2 = f V_s^2 \left[\frac{\gamma}{(\gamma-1)} \frac{1}{(n+1)} - 1 \right] \quad , \quad n+1 = \frac{T}{P} \frac{dP}{dT} , \qquad (4.7)$$

where V_s is the local sound speed. This expression for F_{conv} is the same as that in equation (2.8). The factor f is unknown and is taken as an adjustable parameter and is chosen so that models of the present sun have the observed radius (f $\simeq$ 0.25).

Having described the 'theory' used by the astronomers I should emphasise that there is no reason I know of for believing that it gives an adequate description of reality, and in my opinion any results that depend on the details of the mixing length theory should be treated with caution, nay scepticism ! Let me just list a few problems.

1) The theory only allows for convection (i.e. fluid motion) in convectively unstable regions. Observations of convection in the earth's atmosphere and in the laboratory show motion penetrating into the stable regions. This overshooting may be very important in determining the extent of chemical mixing and thus the details of stellar evolution.

2) The mixing length theory neglects the flux of kinetic energy, yet the theory itself shows that this is of the same order as the enthalpy flux, indeed if the downwards motion is in thin columns the kinetic energy flux can dominate.

3) The theory is a local theory but it could be that the whole convective region is closely coupled and that local parameters are not so important.

4) The assumption of a statistically steady state may be incorrect, the short term averages changing in an irregular way and the convection being intermittent.

It is unfortunately easier to criticise that it is to construct better theories. Turbulence remains an unsolved problem and at the moment all one can do is to make different assumptions and see the effects on the models (c.f. Roxburgh (9) (10)). What is needed is some break through in theoretical analysis or some more detailed multiscale numerical calculations. My own estimates of convective overshooting suggested that it could increase the main sequence lifetime of stars by some 50 %.

Fortunately some results may be almost theory independent For example, in all but the cool surface layers of stars the total flux is very much less than the maximum flux $C_p \rho T V_s$ and hence to a very good approximation we can deduce from equation (4.6) that convective zones are very nearly adiabatic :

$$\frac{T}{P} \frac{dP}{dT} = \frac{\gamma}{\gamma - 1} \tag{4.8}$$

This adiabatic region may, however, include regions that are convectively stable since they will be mixed by convective overshooting.

IV.3   Outer Convective Zones

The opacity in conditions similar to those in the surface layers of the sun have been calculated by the Los Alamos Group (Heubner (6)). For pedagogical purposes I made a power law approximation to those giving :

$$\kappa = \kappa_o P^{1/2} T^8. \tag{4.9}$$

Since the surface layers are thin, we can take a plane parallel approximation and write the equation of hydrostatic support as

$$\frac{dP}{d\tau} = \frac{g}{\kappa} \quad , \quad \tau = \int_z^\infty \kappa_\rho dz, \quad T^4 = \frac{3}{4} T_{eff}^4 (\tau + 2/3) \quad , \tag{4.10}$$

where τ is the optical depth, T_{eff} the effective temperature and the last result gives the variation of temperature with optical depth obtained from the simple Eddington approximation for grey atmospheres. These equations can be integrated to give

$$\rho = (\frac{4g}{\kappa_o T_{eff}^8})^{2/3} (\frac{\tau}{\tau + 2/3})^{2/3} \quad , \quad \frac{TdP}{PdT} = \frac{16}{9\tau} . \tag{4.11}$$

For $\tau \to 0$ the atmosphere is stable but as τ increases the Schwarzschild criterion is satisfied, and the surface is unstable to convection at an optical depth $\tau = 32/45$. Radiation from this optical depth can easily escape from the stars so we should be able to see convection on the sun. Of course we do see the solar granulation.

The properties of the outer convective zone depend on whether convection is immediately efficient, that is whether

$$\sigma \, T_{eff}^{\,4} < f \, C_p \rho \, TV_s = f \, C_p^{1/2} \, PT^{1/2} \tag{4.12}$$

or not. If this condition is satisfied the convective zone is immediately adiabatic and the value of the entropy can be obtained his in turn is enough to determine the base of a convective envelope. On the other hand if this condition is not satisfied, convection is inefficient and most of the energy is still carried by radiation even though the layer is unstable ; at some greater depth inside the star the condition (4.12) will be satisfied (since P and T increase inwards) and the rest of the convective one will be adiabatic. This highlights the difficulties of using the mixing length theory since for solar type stars the radiative flux and the maximum convective flux are nearly equal and the unknown factor f is critical in determining where convection becomes efficient, and hence the value of the entropy in the adiabatic bulk of the convective zone. It is this value of the entropy that determines the depth of the convective envelope and the radius of the star.

If we take the expressions (4.10) and (4.11) for T and P we can determine the optical depth   at which convection is first efficient, this is given by

$$\sigma \, T_{eff}^{\,4} = f \, C_p^{1/2} \, \left(\frac{3}{4}\right)^{1/2} \left(\frac{4}{\kappa_0}\right)^{2/3} \left(\frac{g^{2/3}}{T_{eff}^{\,35/6}}\right) \left(\frac{\tau_a}{\tau_a + 2/3}\right) (\tau_a + 2/3)^{1/8},$$

$$\tag{4.13}$$

and hence we can determine

$$K_a = \frac{P_a}{T_a^{\,5/2}} = \left(\frac{4}{3}\right)^{5/8} \left(\frac{4}{\kappa_0}\right)^{2/3} \left(\frac{g}{T_{eff}^{\,47/6}}\right)^{2/3} \left(\frac{\tau_a}{\tau_a + 2/3}\right)^{2/3} \frac{1}{(\tau_a + 2/3)^{5/8}} \cdot \tag{4.14}$$

If $\tau_a < 32/45$, the value at which convection sets in, then convection is immediately efficient, and equation (4.14) gives the value of K when τ_a is replaced by 32/45. We can see that we expect different results depending on whether τ_a is less than or greater than 32/45.

Having determined the value of K_a at the top of the adiabatic region we can determine the value after ionisation is complete by noting that the entropy per unit mass is

$$S = \frac{X}{m_H} \frac{K}{} \left[\frac{5}{2} (1 + x + \delta) + \frac{X}{KT} + \frac{3\ell n}{2} \left(\frac{2\pi m_H}{h^2} \right) \right. \tag{4.15}$$

$$\left. + \delta \ell n \left(\frac{8\pi m_H}{h^2} \right) + x \ln \left(\frac{2\pi M}{h^2} \right) + (1 + x + \delta) \ln \left(\frac{(kT)^{3/2}(1+x+\delta)}{P} \right) \right],$$

where X is the ionisation energy of hydrogen, x is the fraction
of hydrogen ionised δ = Y/4X, h is Planck's constant and k is
Boltzman's constant. Since S is constant in the adiabatic region
on equating the values for x = 0 and x = 1, we find the value of
$P/T^{5/2}$ after ionisation is complete, for X = 0.74, Y = 0.24,
δ = 0.08 and

$$K = \frac{P}{T^{5/2}} = 2.01 \quad \frac{2 \quad m_e}{h^2} (ke)^{5/2} \quad 0.48 \quad K_a^{0.52} . \tag{4.16}$$

Hence on using the result (4.14) for K_a, the value of K after
ionisation is complete can be determined in terms of T_{eff} and g.

Once the value of K is known the structure and depth of the
convective zone are easy to calculate. The equation of hydrostatic support becomes

$$\frac{dT}{dr} = - \frac{2}{5} \frac{g\mu}{\mathcal{R}} \quad , \quad T \simeq \frac{2}{5} \frac{\mu}{\mathcal{R}} g h , \tag{4.17}$$

where h is the depth into the convective zone. The convective
zone ends when the total flux σT_{eff}^4 can be carried by radiation
in the adiabatic zone (neglecting overshooting)

$$\sigma T_{eff}^4 = - \frac{4ac}{3} \frac{T^3}{\kappa \rho} \frac{dT}{dr} \tag{4.18}$$

If we again take a power law approximation to the opacity κ
= $\kappa_o \, \rho^{1/2} \, T^{-2}$, then after some elementary algebra we can find
the depth T of the convective zone, the density at the base and
the mean square velocity in the deep parts of the zone.

The detailed results depend critcially on the value of the
'fudge factor' f in the mixing length theory but for a given
value of f we can take the limits for large T_{eff} when $\tau_a \gg$ 32/45
and for small T_{eff}, τ_a = 32/45, and obtain the dependence of the
properties of the convective zone on T_{eff} and g.

$$\text{High } T_{eff}$$ $$\text{Low } T_{eff}$$

$$K \propto g^2 T_{eff}^{-26} \qquad\qquad K \propto g^{1/3} T_{eff}^{-4}$$

$$\rho_b \propto g^3 T_{eff}^{-45} \qquad\qquad \rho_b \propto g^{2/3} T_{eff}^{-56/11}$$

$$h \propto g^{-1/3} T_{eff}^{-33} \qquad\qquad h \propto g^{-13/11} T_{eff}^{-8/11}$$

$$u^2 \propto g^{-2} T_{eff}^{33} \qquad\qquad u^2 \propto g^{-0.04} T_{eff}^{6}$$

The results show that for high T_{eff} the convective zone is
very thin. As T_{eff} decreases the zone becomes very much deeper
but for low T_{eff} the rate of increase in depth is much slower.
For low enough values of T_{eff} the convective zone reaches all
the way to the centre, $h \simeq R$, this occurs for stars with masses
< 0.4 $M_\odot$. For the sun the convective zone covers about the outer
30 % by radius.

V. A SURVEY OF STELLAR STRUCTURE AND EVOLUTION

V.1 Pre-main sequence evolution

The details of star formation are not yet well understood,
stars are believed to form from interstellar clouds and the collapse phase ceases when the material ionises and the internal
pressure halts the collapse. The properties of such stars is again
not well understood, they are likely to be fully convective have
extensive chromospheres coronae and winds and be surrounded by
debris left over from the collapse phase. At this early stage
the stars have large radii and low internal temperatures so their
energy comes from gravitational contraction. We have seen earlier
that such contraction leads to an increasing temperature, so
except for the low mass stars the convection dies away in the
central regions, the surface temperature rises, and eventually
the centre becomes hot enough to burn hydrogen to helium and the
stars stop contracting. This is the main sequence phase.

V.2 Main Sequence Stars

For stars with masses less than about 1.5 $M_\odot$ the central
temperature is moderately low $\simeq 10^7$ °K and they derive their energy from hydrogen burning on the proton-proton chain. Larger mass
stars derive their energy from the Carbon-Nitrogen cycle. This
C-N cycle is very temperature sensitive, so most of the energy is
produced in a small region near the centre. This in turn requires

a relatively steep temperature gradient to transport the energy
causing these regions to be convectively unstable. The proton-
proton chain on the other hand is less temperature sensitive, a
large fraction of the mass contributes to the energy generation
thus the temperature gradient is not so steep and the centre is
stable to convection. Larger mass stars (> 1.5 $M_\odot$) have convective cores, smaller mass stars do not.

In the surface regions the situation is reversed. From the
mass-luminosity and mass-radius relations we find that along the
main sequence

$$T_{eff} \propto M^{0.65} ,$$

so that larger mass stars have higher effective temperatures. If
these are large enough the H^- ion does not form, the opacity
does not increase with increasing temperature, and the surface
layers are stable to convection. When the temperature is low
enough for the analysis of section IV to be appropriate we have
seen that the depth of the convective zone increases as the effective temperature (and therefore the mass) decreases. The mass (or
effective temperature) at which surface convective zones first
appear is not very easy to determine but is probably about 2 $M_\odot$,
although larger mass stars may have subsurface convective regions
in the ionisation zones where the effective ratio of specific
heats γ is decreased to near unity and convective instability
is that much easier.

V.3 Post Main Sequence Evolution

Only minor changes in a star's structure occur while it is
converting hydrogen to helium, the main feature being an increase
in the luminosity due to an increase in mean molecular weight as
helium is produced from hydrogen. The major changes occur when
the hydrogen in the centre is exhausted. The central regions then
have to contract drawing on their gravitational energy. During
this contraction the density and temperature both increase. For
smaller mass stars (< 2.5 $M_\odot$) the density becomes so large that
degeneracy pressure halts the collapse, the core then cools and
hydrogen is burnt in a shell surrounding an isothermal degenerate
core. The rest of the star has to expand in order to maintain
hydrostatic equilibrium at the shell, and the star becomes a red
giant.

For larger mass stars the main sequence central temperatures
are higher, and the densities lower ($T_c \propto M^{0.3}$, $\rho c \propto M^{-1}$) so that
when their cores contract they reach helium burning temperatures
before degeneracy becomes important. Such stars also expand and
have cooler surface temperatures.

The details of subsequent phases of stellar evolution depend both on the mass and on the degree of mass loss when the
star has cooler surface regions. This mass loss can be quite
large, large enough for a significant fraction of the total mass
to be lost. It is generally believed that the smaller mass stars
eventually ignite helium, evolve back to be red giants and then
eject their envelopes to become planetary nebulae, the remnant
core cooling down to become a white dwarf.

The larger mass stars are believed to go through further
stages of chemical evolution until the core collapses resulting
in a supernova explosion which either completely shatters the
star or leaves behind a remnant neutron star.

VI. CONCLUSIONS

The basic structure of solar type stars is well understood.
Internal pressure supports the star against its own self gravity,
energy is produced in the central regions by the fusion of hydrogen to helium on the proton-proton chain, and the energy is transported from the central regions by radiative energy transport.
In the surface layers the temperature is sufficiently low for H⁻
ions to be formed, the opacity therefore increases with increasing temperature and the surface is unstable to convection. In
the sun this convective zone occupies about 30 % of the radius.
As hydrogen is (and has been) converted to helium the luminosity
and central temperature have increased (by about 30 %).

The problems are in the details. Just what is turbulent
convection like, does it penetrate into stable regions, if so
does it make any significant change in the models ? Is the sun
stable or does it go through periods of instability, how important are magnetic fields and rotation, and why is the neutrino
flux so low? These and other problems will be discussed by myself
and other authors elsewhere in this volume.

REFERENCES

1. Chandrasekhar, S., 1939, Stellar Structure, Univ. Of Chicago
   Press
2. Clayton, D., 1968, Principles of Stellar Evolution and Nucleo-
   synthesis, Mc Graw Hill, New York
3. Cox, A.N., 1965, in Stellar Structure, ed. Aller and McLaughlin,
   Univ. Of Chicago Press
4. Cowling, T.G., 1936, Monthly Notices Roy. Astr. Soc., 96, 42
5. Eddington, A.S., 1921, Ziet. fur Astrophys., 7, 351
6. Huebner, W.F., Merts, A.L., Magee, N.H. and Argo, M.F., 1977
   Astrophysical Opacity Library, Los Alamos Scientific

Laboratory Manual LA-6570-M
7. Huebner, W.F., 1968, Proceedings of an Informal Conference on
   the Status and Future of Solar Neutrino Research, ed. G.
   Friedlander, Brookhaven National Laboratories BNL 50879.
8. Lane, J. Homer, 1969, American J. Science, 50, 57
9. Roxburgh, I.W., 1976, in Basic Mechanisms of Solar Activity,
   ed. Bumba and Kleczeck, Reidel, Dordrecht.
10. Roxburgh, I.W., 1978, Astronomy and Astrophysics, 65, 281
11. Schwartzschild, M., 1958, Structure and Evolution of the
    Stars, Princeton University Press.

SOLAR AND STELLAR OSCILLATIONS

Eric Fossat

Département d'Astrophysique
Faculté des Sciences de Nice
06034 NICE CEDEX, France.

Abstract. We try to explain in simple words what a stellar oscillation is, what kind of restoring forces and excitation mechanisms can
be responsible for its occurence, what kind of questions the theoretician asks to the observer and what kind of tools the latter
is using to look for the answers. A selected review of the most
striking results obtained in the last few years in solar seismology and the present status of their consequences on solar models
is presented. A brief discussion on the expected extension towards
stellar seismology will end the paper. A selected bibliography on
theory as well as observations and recent papers is also included.

INTRODUCTION

This paper does not intend to be an extensive review on the
theory or the observations of solar and stellar oscillations. It
is just to be regarded as an introductive paper on solar and
stellar seismology, trying to present simple answers to such
simple questions as what is a stellar oscillation, what causes
an oscillation, what is to be expected by the theorists from the
observers. It will also summarize the main techniques of solar
observations being used in this field and try to present the
present status of the obtained results.

Seismology is interested mostly in the internal structure
of the star, or the sun. Therefore, and for the purpose of this
paper, we will forget the existence of all photospheric, chromospheric and coronal complicated features, and a star will just
be regarded as a sphere of gas. The observers have not many possibilities of getting direct information from the interior of this

R. M. Bonnet and A. K. Dupree (eds.), Solar Phenomena in Stars and Stellar Systems, 75–98.
Copyright © 1981 by D. Reidel Publishing Company.

sphere of gas, simply because this interior cannot be seen, at
any wavelength of the electromagnetic spectrum.

The measurement of the solar neutrino flux was in fact the
first observational result providing direct information about the
solar interior. The incompatibility between the measured flux and
the one predicted from the standard solar models has demonstrated
how uncertain this standard model was, and the importance of such
an uncertainty has very quickly become quite obvious, due to the
consequences that a possibly needed change in the stellar interior
theory could have over all astronomy.

In 1975, the almost simultaneous publication of the calculation and the observation of the same solar oscillatory pressure
modes by Ando and Osaki (1975) and Deubner (1975) was the real
start of what has been called since solar seismology. It consists
in observing global pulsations of this big spherical resonator
that the sun is, and to use the results as probes of the internal
structure through modeling and calculation of eigen values.

I. WHAT IS A STELLAR (OR SOLAR) OSCILLATION ?

A star is like a musical instrument, which can resonate with
various modes and tones. Let us compare this resonator with a
guitar string. The latter has a fundamental sound, in which case
it resonates with one wave, and all possible overtones, each one
being completely defined by one parameter, for example the number
n of nodes along the string. Consider now the star, for comparison. It is a volume (three dimensions) instead of a straight
line (one dimension) and then a given overtone will be defined by
the eigen functions of a simple system of three ordinary differential equations with boundary conditions at the center and surface of the star, which will determine the complex frequency as
eigen value. As a consequence, there are three parameters which
define one given mode. Two of them define the surface behaviour
(which is of interest for observers). The distribution of surface
amplitudes is given by the spherical harmonic function Y_ℓ^m , where
ℓ (called the degree) is the number of lines of modes around the
spherical surface (ℓ = 0, 1, 2, 3 ...) and m defines how those
lines are distributed over this surface (distribution between lines parallel to the equator and meridional lines - $\ell \leq m \leq +\ell$).
As an example, the figure 1 shows the three possible surface patterns of surface amplitude in the case ℓ = 2. If ℓ = 0, the amplitude is purely radial. ℓ = 1 is called the dipole, ℓ = 2 the
quadrupole, ℓ = 3 the octopole oscillation, ... There is a third
parameter n, called the order, the absolute value of which is
roughly the number of nodes along a radius, between the center and
the surface. Following a classification proposed by Cowling(1941),
the mode with n=0 is called the f mode, the modes with n < 0

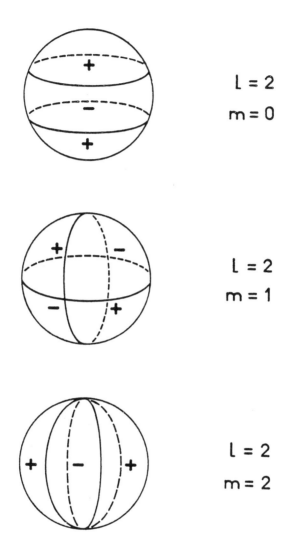

Figure 1.   In the case ℓ = 2 (quatrudpole oscillation), three
            different surface patterns are possible, with exactly
            the same period in absence of rotation.

are denoted g modes and the modes with n > 0 p modes. We will
see further the physical meaning of this notation.

   In absence of rotation, the temporal eigen frequency ν
(often referred by $\omega = 2\pi\nu$) is a function of ℓ and n only with a
degeneracy in m. In case of a rotation, there is a "Zeeman
splitting" of a given frequency into $(2\ell + 1)$ components. Such a
splitting, if it can be measured, is a powerful tool for investigating the rotation as a function of depth.

II. RESTORING FORCES

   The spherical harmonics functions Y_ℓ^m and the number of
radial nodes n define the three-dimensional geometry of oscillatory modes. But the existence of our oscillatory mode requires
the presence of a restoring force. The gas of a star is subject
to many forces, two of them, pressure and gravity, being of
prevailing interest for our purpose. If this gas is disturbed
from an initial stable state, these forces will act to return
it towards this initial state. If the dissipation is sufficiently
small, it will overshoot this initial state and will oscillate
around its initial position.

   The simplest restoring force is compressibility, because the
gas pressure is isotropic. Another restoring force is gravity,
which gives rise to modes driven by buoyancy. A convective instability is a simple example of an unstable initial state from
which the gravity will not act as a restoring force. If compressibility prevails as a restoring force for a given mode, it is
called a pressure mode. If gravity provides the prevailing restoring force, the mode will be called a gravity mode. A few
simple examples show easily that one or the other of these two
restoring forces will prevail. In a convective zone, pressure
modes are possible, but not gravity modes. Another example is
the case of a pure radial motion. Since the change in the gravitational force is inward in the compressed phase and outward in
the expanded phase, gravity cannot be a restoring force for the
radial pulsation, while compressibility obviously is. On the
other hand, if the motion of a gas parcel is sufficiently slow,
the gas parcel will remain in pressure equilibrium with its
surroundings because waves propagating at the sound speed in the
parcel are able to reestablish pressure equilibrium before it
has moved appreciably. This means that very slow oscillations
will not be driven by compressibility. For any star there is a
frequency cut-off for pressure modes, which is the frequency of
the fundamental radial f-mode $(\ell = m = n = 0)$. All possible oscillatory modes with longer periods must be gravity modes.

In simple stellar models like polytropes of low index, the
p-modes in the classification of Cowling (n > 0) are pressure
modes and the g-modes (n < 0) are internal gravity modes. This
physical significance can be lost in more complicated models,
but the Cowling's classification is in general still used for
mathematical convenience.

III. EXCITATION MECHANISMS

If the gas is disturbed from an initial stable state, the
restoring forces will make it oscillate around this initial
state. But what was the cause of the disturbance, or what will
balance the dissipation to prevent the amplitude to be very
quickly damped, in other words, what is the physical process
which can excite or drive the oscillation. This has been the matter of considerable debate among theoreticians. I will briefly
summarize this debate by dividing, following Gough (1979), the
driving mechanisms into two classes : those which are really an
excitation, and require in some way an interaction with the horizontal fluctuations produced by motions others than that associated with the oscillation itself (for example a convective turbulence, or another oscillating mode) ; and those which balance the
damping to make the oscillation overstable, in which case energy
is tranferred from the mean environment to the oscillations. The
latter is a continuous interaction between the oscillation and
its environment, and in this class, the Kappa-mechanism (amplification of the compressibility in the presence of a gradient of
opacity) which has been known for a long time to play an important role in the radial pulsations of Cepheids and RR Lyrae, has
been proposed by Ando and Osaki (1975) as the possible dominant
driving process for the solar five-minute oscillation. The former
class of excitation mechanism could be much more stochastic and
although both forms of energy transfer probably take place, this
only difference can help the observer to distinguish which one
prevails.

IV. WHAT IS NEEDED FROM OBSERVERS ?

The problem is to put as much constraint as possible on the
construction of a theoretical model of the sun or a star to improve the fit of this model to the real sun or star.

The high degree and high order normal modes can as well
be regarded as waves locally trapped in the upper layers of the
star (Gough, 1980, Leibacher and Stein, 1980) and therefore they
are sensitive only to these upper layers. An access to information
related to deep layers will be obtained, in seismology, only
through the observation of very low degree oscillating modes. The

ideal information that the theorist expects from his colleague
observer is the observation and the identification (in ℓ and n)
of a large number of different modes, p-modes and g-modes, with
for each one the accurate measurement of the frequency, the time
behaviour of amplitude and phase and if possible the rotational
m-splitting.

Let us see briefly now what are the tools used for this
research by the observers in the solar case. In view of checking
the current standard solar models, I will then briefly review
the major results obtained so far.

V. METHODS OF OBSERVATION

A given oscillatory mode will materialize at the solar surface as gas displacements, pressure fluctuations, and then temperature fluctuations as well. Therefore, three different approaches have been used so far by the observers.

V.1. Doppler shift measurements

Using all possible kinds of spectroscopic techniques available, these measurements are sensitive to vertical velocities
at the disk center and mostly to horizontal velocities near the
solar limb. High degree oscillating modes will be seen by high
spatial resolution Doppler mapping of the solar surface. In
principle, such a mapping obtained as a function of time should
give access to any mode, of high or low degree. But in practice,
the senstivity of the measurement is limited by different kinds
of noise sources (photon noise, telluric noise and for a large
part solar noise due to a background of large scale convectice
velocities) and it does not give access to low degree modes.
These can be observed by integrating the Doppler shift measured
over the entire solar disk (observation of the sun "as a star").
Such an integration makes a very strong spatial filtering and
therefore, only the very low degree modes may be looked for
($\ell < 4$). It has the advantage of reducing to the minimum possible
all noise from solar origin, but on the other hand it is more
sensitive to telluric atmospheric noise, due to the inhomogeneities
of transparency fluctuations in front of the rotating sun (Grec
and Fossat, 1977, Fossat and Grec, 1978). An alternative
which has been developed in Crimea (Severny, Kotov and Tsap,
1976) consists of using the outer part of the solar disk as a
reference for the Doppler measurement of a inner circular part
between 60 and 70 %, in diameter (close to 50 % in brightness).
Such a differential measurement eliminates most of the instrumental drifts and the large trend due to the earth rotation, but it
does not reduce the sensitivity to telluric transparency effects.

Another alternative which was suggested by Grec and Fossat would be to observe the integrated sunligh reflected by the moon. The much lower velocity of the lunar rotation would reduce the sensitivity to atmospheric transparency flucturations by a large amount.

V.2. Intensity fluctuations measurements

Most of the variable stars are first, and often only observed through their variation of luminosity with time. So, brightness fluctuations are also a possible approach in the search for solar oscillations. The sun is not known to be a variable star because there is no significant change of its total brightness with time, but high resolution mapping with time can give access to high degree modes, and high accuracy measurements of the total brightness on the sun itself, or of its light reflected by planets (Deubner, 1977) can be used to look for low degree modes. The same differential method of observation (central part with respect to outer annulus) can also be used with the same kinds of advantages and limitations.

V.3. Astrometry

If the solar surface goes up and down due to the presence of a global ascillation, such an oscillation should be observable through changes with time of a given diameter. Such an observational approach has been developed in Arizona by Hill and co-workers (see bibliography) using an equipment which was originally designed to measure the solar oblateness. The problems of this method are its high sensitivity to atmospheric noise through differential refraction fluctuations (Fossat et al., 1977, 1980) and the difficulty of identifying a diameter oscillation with a given (ℓ, n) global mode (Gough, 1980, Christensen-Dalsgaard, 1980). On the other hand, this method can be sensitive to temperature changes which affect the shape of the limb darkening function but are not associated with measurable velocities (Hill, 1980).

VI. WHAT IS EXPECTED FROM PRESENT THEORY ?

The region between 0.7 ($\pm$ 0.1) and 0.9999 solar radius is convectively unstable, Because in this case, the buoyancy changes from a restoring to a destabilizing force, oscillatory g-modes are evanescent in this convective zone. Thus there are two classes of gravity modes : internal gravity modes, trapped below the convective zone, and surface waves, trapped in the atmosphere. Internal g-modes cannot have periods shorter than about 40 minutes (Christensen - Dalsgaard et al, 1980). They can possibly be observed at the solar surface after an exponential decay of the amplitude through the convective zone. Atmospheric modes may have periods as short as 4 minutes, with short horizontal wavelength. But obviously, being trapped in the solar atmosphere, they will not help to probe models of the solar interior.

The p-modes of low order exist with substantial amplitude throughout the entire radius of the sun. The modes of high degree are more confined in the surface layers. From a seismological point of view, the modes with low ℓ are the most interesting. The figure 2 displays the frequencies of a few low degree modes as calculated by Gough (1980) using a standard solar model including an atmosphere.

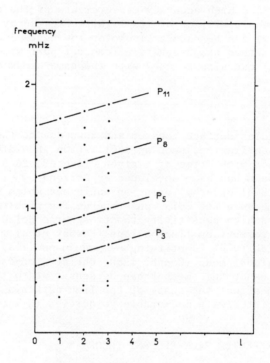

Figure 2.   From Gough (1980, private communication), the fre-
            quencies of a few low-order, low-degree modes are
            already indicative of the discrete lines which will be
            observed at high ℓ-values (with resolution in ℓ not
            sharp enough to resolve each individual degree) and of
            the discrete spectrum which will be observed in full-
            disk observation (with spatial filtering which limits
            the investigation to $\ell < 4$).

The five-minute solar oscillation first discovered by
Leighton in 1960 has been shown later to have typical horizontal
wavelengths of about 10 to 30 Mm (1 Mm = 10^6 m). This means that
many wavelengths exist along a solar circumference and, if regarded as global modes, these oscillations correspond to very high
ℓ values. In this case, ℓ being the number of lines of nodes on
the solar surface, the relation $\lambda\ell \overset{\sim}{\cdot} 2\ \pi r$ gives $140 < \ell < 420$
for the above values of λ. A high resolution Doppler mapping being
necessarily limited to a fraction of the solar surface, the horizontal wavelength resolution obtained by space-time Fourier
analysis will not be sharp enough to distinguish every 1-value,
but as we will see below, the discrete lines are observable.

Very low degree modes can be looked for in full disk observations. Such an integration represents an amplitude filtering
which depends on the given mode and on the direction of its axis
with respect to the line of sight. Mean values of this filtering
effects have been calculated by Christensen-Dalsgaard and Gough
(1980). For ℓ = 0, 1, ... they are respectively (approximately)
0.7, 0.9, 0.7, 0.2, 0.04, ... The very fast decrease of the
filter transmission for $\ell > 3$ indicates that the modes that the
observer can more likely be able to detect are those for which
$\ell \leqslant 3$. For these modes of low degree ℓ and in the case of high
order n, the frequencies are approximately given by (Tassoul,
1980) :

$$\nu_{n,\ell} \simeq (n + \frac{\ell}{2} +\varepsilon)\ \nu_o - (\ \ell(\ell+1) + \delta)\ A\nu_o^2\ \nu_{n,\ell}^{-1}$$

where ν_o, ε, δ and A are constants of the equilibrium model. The
assumption is $\ell << $ n and for such modes, the approximation is
quite good and the second term is only a small correction to the
first. Just neglecting this second term leads to two simple approximations :

$$\nu_{n+1,\ell} - \nu_{n,\ell} \overset{\sim}{\cdot} \nu_o$$

$$\nu_{n,\ell+2} \overset{\sim}{\cdot} \nu_{n+1,\ell}$$

These relations can be seen on the figure 2 where it is
clear that apart from very low n-values, successive harmonics
of the same degree ℓ display an equidistance in frequency. The
second relation indicates that the two sequences of odd or even
ℓ have almost coincident frequencies, and the frequencies of
modes with odd ℓ lie approximately midway between modes with
even ℓ. It is then expected that a full disk Doppler observation
should supply a signal which has a discrete power spectrum, and

this can be regarded as the hope of a powerful use of solar, and possibly later stellar seismology.

The theoretical predictions for the excitation of all these p-modes have been the subject of a very controversial debate. If the prevailing driving process is a stochastic excitation by the turbulence, one might expect the mode energy to depend on frequency alone (Christensen-Dalsgaard and Gough, 1980). There has been some observational evidence that it is so, but more direct and quantitative information is still needed from observation.

VII. MAIN PRESENT RESULTS

I will just present a selection of the most striking results of observations obtained so far and which have been regarded as informative in solar seismology.

VII.1 The five-minute ridges resolved

Fifteen years after the discovery of the five-minute oscillation by Leighton (1960), the first incontrovertible identification of this oscillation as a mixture of high degree p-modes was obtained by Deubner in 1975 when for the first time, the famous k-ω diagram (two-dimensional power spectrum, in space and time) was resolved in discrete ridges from high resolution Doppler observations made at the German Observatory of Capri (Italy). This was the real birthdate of solar seismology and almost immediately, the comparison of this exciting result with the calculation of the same modes by Ando and Osaki (1975) was used as a probe of models in the upper part of the convective zone. The figure 3 shows two recent k-ω power spectra obtained by Duvall and Harvey (private comm.) and by Deubner, together with the theoretical modes calculated by Ando and Osaki. Many careful investigations since 1975 have shown that a better agreement could be obtained using a standard model of the solar envelope corrected by increasing the depth of the convective zone (see bibliography).

For these high ℓ, low n modes, the equatorial solar rotation in one period is not very small compared with the horizontal wavelength and then, there is a substantial rotational splitting of the discrete lines. This can be seen as the asymmetry between the negative and positive frequency parts of the power spectrum of the figure 3. Taking advantage of the fact that the different modes penetrate more or less deeply inside the convective zone, the accurate measurement of this splitting by Deubner et al (1979) has been used to investigate the speed of rotation as a function of depth. So far, this investigation has been limited to a few tens of megameters (between 10 and 20 % of the depth of

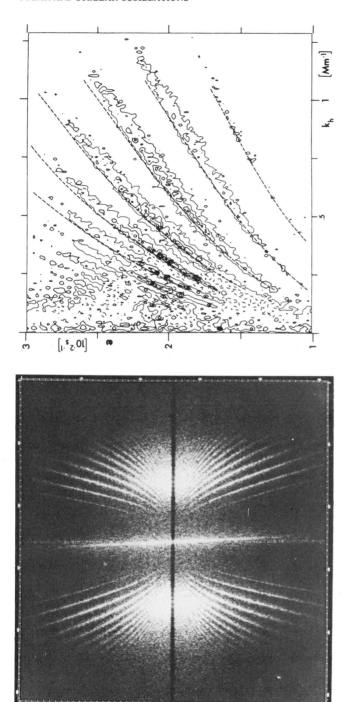

Figure 3.   Two different examples of the K–ω power spectrum which resolve the discrete ridges corresponding to high–ℓ, low–n oscillating normal modes of the sun. The left one, from Duvall and Harvey at Kitt Peak (1980, private communication), shows the asymmetry between positive and negative frequency parts, due to rotational splitting. The right one, from Deubner et al. (1979), shows the comparison with the same modes as theoretically calculated by Ando and Osaki (1975).

the convective zone), but it has shown a significant increase of
rotational speed with depth, and it has overall proved to be
very promising.

VII.2. Solar Diameter Oscillations

Observing oscillations with periods substantially longer
than five minutes have proved to be very ackward because such
oscillations have apparently much smaller amplitudes and also
because with any observational technique, all sources of noise
(from optics, electronics, earth atmosphere or the sun itself)
have an increasing power with decreasing frequency. The signal
used for Fourier analysis being of finite duration, the power
spectrum will always have a spiked structure due to the random
fluctuations around a mean broad curve, as well as due to a real
presence of spectral peaks in the solar signal. The very strong
controversy around the interpretation of the solar diameter
oscillations observed in Arizona by Hill and co-workers is a
good illustration of how difficult for the observer is the statistical analysis of such a small signal buried in noise. Using
a new definition of the solar edge by means of the Fourier transform of the limb darkening function, Hill, Stebbins and Brown
(see bibliography) have measured a given solar diameter as a
function of time and the figure 4, from Brown et al (1978) shows
a power spectrum of this signal. In the range 0-5mHz, the
authors attribute about 50 % of the power to noise and the remaining 50 % to spectral peaks indicating the observation of
solar global oscillations. On the basis of statistical consideration and comparison with Doppler results, Grec and Fossat
(1977) and Dittmer (1978) suspected that more likely, 100 % of
this power could be attributed to noise. Fossat et al (1978, 1980)
have measured the order of magnitude of noise produced by earth
atmosphere in these diameter measurements, and the figure 5
shows their conclusion. New arguments based upon long term phase
coherency at a given frequency have been used to prove the solar
origin of spectral peaks (Hill and Caudell, 1978, Caudell et al,
1980), but they are also controversial (Grec and Fossat, 1979,
Christensen-Dalsgaard and Gough,1980). The discrepancy between
interpretations has so far not been settled. In any case, the
interpretation ot these observations in terms of solar seismology
would present some difficulties because no direct determination
of the ℓ-values has been possible.

VII.2. Radial and Low-ℓ Modes in the Five-Minute Range

The difficulties faced by the observer looking for long
period solar pulsations can be illustrated (because he is
"listening" to acoustic vibrations of the sun) by imagining that
the is trying to listen to a quatuor of Beethoven played by a very
bad quality radio receiver, the amplifier being set at a very low

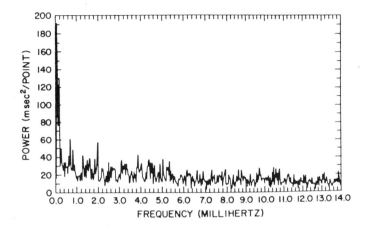

Figure 4.   Average spectrum of eleven daily power spectra of
            solar diameter fluctuations measured in Arizona by
            Brown et al. (1978). The flat mean power present at
            frequencies higher than 6 mHz is regarded as noise,
            while all the power above this level between 0 and
            6 mHz is interpreted as a spiked spectrum due to global
            solar oscillations.

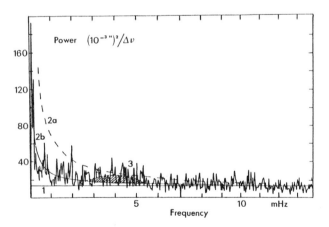

Figure 5.   The interpretation of the power spectrum of figure 4
            by Fossat et al (1980) (1) would be white noise spec-
            trum, (2a) the typical contribution of the earch at-
            mosphere through differential refraction, (2b) the
            minimum of this atmospheric contribution and (3) the
            contribution of the five-minute oscillation, which is
            more concentrated around three-minues at the limb's
            altitude in the solar atmosphere.

power and located in the neighbour room behind a concrete wall
in which somebody is drilling a hole. To simulate the day-night
cycle, imagine moreover that the amplifier power is turned on and
off periodically. It is not trivial in these conditions to identify Beethoven rather than Mozart, Eric Satie or the Pink Floyds.
The way to do it is to improve the quality of the receiver (very
stable and sensitive spectroscopic techniques), of the amplifier
(low spatial frequency filtering by full-disk integration), to
reduce the noise in the concrete wall (observations in high
altitude sites, where good atmospheric transparency prevails).

     Using two potassium cell optical resonance spectrometers
both at Teneriffe (Canaries islands) and Pic du Midi (French
Pyrenees), Claverie et al (1979), while looking for long-period
oscillations, by means of full disk Doppler measurements, were
the first to discover the discrete power spectrum that we predicted a few pages earlier, from the figure 2, in the five-minute
range. The figure 6 shows this discrete power spectrum which was
indeed the first result of observation which could be interpreted
as low degree global pulsation of the sun. According to this
result, theoretical speculations started very quickly (too quickly) to conclude that the standard solar model had to be dramati-
ally modified towards a much lower helium abundance (Gough, 1979,
Isaak, 1979). More careful investigations have shown later that
this was premature. I will come back to this discussion a little
bit further.

     The last observational problem, i.e. the day-night cycle,
found an original solution with the use of the South Pole site.
Located at a barometric altitude of 3400 m in a very desertic
climate, this site, which suppresses the day-night alternation
and the Doppler shift trend due to the earth rotation, and offers
others unique advantages, was for the first time used for astronomical research in December 1979 - January 1980 by Grec, Fossat
and Pomerantz (1980).More than 200 hours of observations were
obtained in extremely good conditions, including an almost continuous run of 120 hours (five days). At the time of this A.S.I.
only this 120 hours run has been analyzed and the figure 7 shows
its power spectrum, which is, in the five-minute range, in very
good agreement with the figure 6. However,the 120 hour record
gives in this case a sufficient frequency resolution to show the
individual components in the groups corresponding to odd or even
ℓ-values. Dividing the 2.4-4.6 mHz frequency range into segments
of 136 μHz (mean separation measured between successive n-
harmonics) and adding these segments together gives the average
fine structure of the spectral peaks in this frequency range. The
figure 8 clearly shows peaks corresponding to $\ell = 0$, 1, 2 and 3
as well as a faint indication of a peak which might correspond to
$\ell = 4$. The width of these peaks would indicate that the corresponding oscillating modes are damped with a Q-value of a few

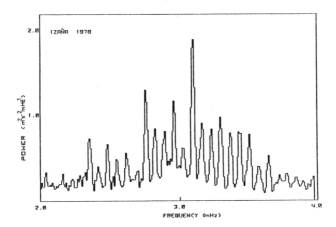

Figure 6.   In the five-minute range, the discrete power spectrum
            of full disk Doppler measurements as resolved for the
            first time by Claverie et al. (1979).

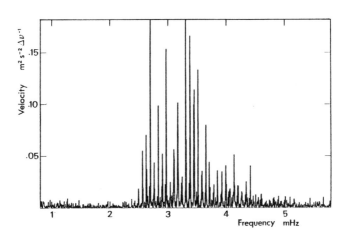

Figure 7.   Same as figure 6, from a 120 hours almost continuous
            data sample collected at the geographic South Pole
            by Grec, Fossat and Momerantz (1980).

hundred. This is confirmed by making a separate Fourier analysis
of shorter segments of data, for example 12 hours long. In this
case, the height of a given peak clearly shows a damping with a
e-folding time of typically 2 days (Q ≃ 600), and a sudden excitation which, typically, happens 2 or 3 times during the five-day-
run. This two-day damping contradicts a recent result of Claverie
et al who found a consistent phase coherency over more than ten
days. I personnaly think that the almost direct observation of
the exponential decrease of a peak amplitude with time looks
very convincing. This behaviour of damping and occasional excitation strongly favours the stochastic excitation by turbulence.

On the other hand, the figure 8 also shows that the peak
corresponding to ℓ = 0 is significantly narrower than the three
others. Because only the radial modes are not subject to rotational splitting, it is attempting to interpret this difference
as the presence of a rotational splitting of a few μHz in the
modes ℓ = 1, 2, 3, which is not resolved. It is hoped that the
analysis of the total amount of the South Pole data, which covers
more than 10 days, will eventually resolve this splitting. Douglas Gough pointed out that the departure from strict frequency
equidistance of peaks over the full spectral range used in the
figure 8 could be responsible of a broadening which could depend
on ℓ. A very careful examination of the power spectrum, including this departure from equidistance, has in fact shown that
the width difference between ℓ = 0 and others is indeed significant. It has also shown that this non-equidistance can be measured
quite accurately (the separation varies from about 134 μHz to
138 μHz in the above spectral range, and it is not the same
for all four modes), and this considerably reduces the degree of
freedom in the construction of solar envelope models.

All these last results are very preliminary but it already
appears that a pretty good fit, at least for the mean characteristics of the results over the whole spectral range, is obtained
by using a standard solar model including the standard H.S.R.A.
atmosphere. The first calculations made after the publication of
the discrete spectrum by Claverie et al and which concluded to
the need of low helium abundance solar models were based upon
standard interior solar models without atmosphere. It happens that
the absence of an atmosphere in the model is not important for
the calculation of modes of very low degree, but this is not true
any longer in the five-minute range, where a significant part of
the mode energy is concentrated just below the solar surface. The
present status of this question is that the best fit with observations is obtained with standard abundances solar models, with
however a convective zone somewhat deeper than expected before
(about 0.3 solar radius), which is also in good agreement with
the implication of results of observations in the high ℓ range.

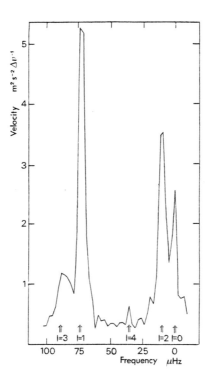

Figure 8.   (From Grec et al.,ˉ 1980). A superposed frequency
             analysis of the frequency range between 2.4 and 4.8 mHz
             reveals the average shape of spectral lines displayed
             in the power spectrum of figure 7. The horizontal axis,
             labeled in μHz, indicates where the asymptotic theory
             predicts the positions of ℓ = 1, 2, 3 and 4 modes if
             one assumes that the one at the extreme right is ℓ = 0.
             The extremely good agreement leaves no room for doubt.
             Note that the natural width of each separate mode
             indicates a Q value of the order of 600 (author's
             caption).

VII.4. Long Period Solar Oscillations

     The occurence of longer periods oscillations (from ten
minutes to a few hours) has been reported, in the last ten years,
by many observers in brightness, Doppler shifts or diameter measurements and also in radio wavelengths (see bibliography). But,
as I pointed out earlier, these long-period measurements are
very difficult and most results can be interpreted just as spurious effects. In the power spectrum of the full-disk Doppler

data from the South Pole (figure 7), there is some preliminary
evidence of significant peaks in the 0.7 - 2.2 mHz frequency
range, with amplitudes of 2 to 4 cm/s, several orders of magnitude below the amplitudes expected from a naive interpretation of
solar diameter oscillations observations. It has not been possible, yet, to identify the lowest order solar modes, radial and
non-radial, with a good degree of confidence.

At even longer periods, a special comment must be made on
the 160 minute oscillation. This periodicity was discovered in
1975 by Valeri Kotov in the differential (central part of the
disk, minus outer annulus) Doppler measurements made at the
Crimean solar tower (Severny, Kotov, and Tsap, 1976). At that
time, many observers regarded this result as quite suspicious
because the period was one ninth of a day and therefore any kind
of diurnal drift could easily give rise to the presence of this
harmonic of 24 hours in the Doppler signal. Starting in 1976,
almost the same technique of observation was used at the Stanford
Observatory, and after some hesitation (the observed amplitude is
far below 1 ms^{-1} and at that long period, it takes a long time of
integration to get it significantly out of noise), it was concluded that a reasonably good agreement was obtained in amplitude
and phase at this period. Even this agreement was not strongly
convincing because the two observatories are separated by almost
exactly 4 x 160 minutes in longitude and a diurnal atmospheric
drift could still explain the results (Fossat and Grec, 1978).
However, a drift in phase of about half an hour per year was observed more and more significantly and consistently at both
observatories. This indicated that the real period was 160.01
minutes and was used to rule out the atmospheric interpretation.

In this low frequency range, the power spectrum of the South
Pole data (figure 9) does not show any significant peak at this
peculiar period. But doing a superposed epoch analysis at this
period exactly as was done with the Crimean and Californian data
gives the result of the figure 10, which shows an impressively
good agreement in amplitude and in phase.

At the present level of this investigation, the solar origin
of this 160 minute oscillation seems to be the most reasonable
explanation of the agreement between the three different results
of observations. The question is now what is the nature of this
oscillation ? The period is far too long for a pressure mode. If
it is a gravity mode, why only one isolated period ? And then,
why such a good agreement between observations made by different
techniques (full-disk and differential Doppler measurements)
which have different spatial filter transmissions ? And why such
a long term phase coherence (Q > 2 10^4) ?

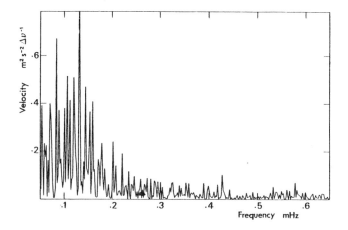

Figure 9. (From Grec et al., 1980). The low frequency part
           of the power spectrum shown on Figure 7. Its spiked
           structure is consistent with statistical departure
           around a broad, continuous spectrum. A spiked spectrum
           of oscillatory g-modes, a broad-band solar convective
           spectrum as well as a broad-band telluric spectrum
           could separately be responsible for this result
           (author's caption).

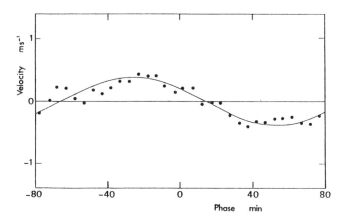

Figure 10. (From Grec et al., 1980). The superposed epoch analysis
           of the same data sample used for obtaining the power
           spectrum of figure 9. The points represent the South
           Pole data, and the solid line is the average based
           upon the observations obtained at the Crimean Observ-
           atory and Stanford (author's caption).

The South Pole power spectrum shows at low frequencies other
peaks which have no a priori reason to be regarded as less significant than the one at 160 minutes. A detailed comparison with
the results of Crimea and Stanford is to be made at all periods
in this low frequency frange, and this may open a new field of
interest in helio seismology. However, it will not be of considerable benefit for theory until an identification in ℓ and n
appears to be possible.

VII.5. Stellar Seismology

Many variable stars are known to be oscillating with radial
or low degree non-radial modes. Therefore, for those kinds of
stars, like βCephei, early-type stars, δ Scuti (a review can be
found in Unno et al, 1979), the seismologic study of internal
structure has been existing for a long time. However, the identification of a given spherical harmonic with a stellar oscillation is in most cases very problematic and a big uncertainty
prevails in general. In view of recent results of observations
of the sun as a star, the possibility arises of using stellar
seismology for investigating the solar-type stars, which are not
known as variable stars. Ando and Osaki (1977) and Unno et al
(1979) have shown that many stars should be subject to non-radial
oscillations of small amplitude in a well defined temporal frequency range, similar to the solar five-minute oscillations. The
periods, obviously, would depend strongly on the size of the star
but in any case, the same kind of discrete power spectrum as
observed in integrated sunlight can be expected.

The problem is far from trivial for the observers. The light
flux coming from the brightest stars is about 10^{11} smaller than
the solar flux and measuring Doppler shift fluctuations below
1 ms^{-1} in a stellar spectrum has not been possible yet. Moreover,
the brightest stars are in most cases giant stars, for which the
expected periods are much longer than five minutes and make the
observational problem even more difficult. However, this challenging problem looks to be worth being attacked by observers in
the next decade in view of the great expected theoretical benefit.

For example, preliminary investigations made on Arcturus
with a modified sodium optical resonance spectrophotometer have
shown that the solar five-minute power spectrum could be resolved
with a few night of integration if the sun was removed far
enough to have a magnitude zero, provided a telescope of at
least five-meter diameter can be used at full time during about
a week. It may be hoped that the technological difficulties not
yet solved in this preliminary observation will find a solution
in the next few years. On the other hand, Myron Smith (1980,
private communication), working with a standard spectroscopic
equipment at Mc Donald Observatory in Texas, has obtained on the

same Arcturus a r.m.s. noise of about 6 ms^{-1} during an hour of good seing. Other attempts have been made, for example by Traub et al (1978) using a spectrophotometer made of three Fabry-Perot in series, and once again the noise level was beaten down to several meters per second.

It seems presently quite reasonable to think that the limit of 1 ms^{-1} in the sensitivity of stellar Doppler shift fluctuations measurements will be attained in the next few years, and therefore a very rich amount of information is expected from seismology of non-variable stars.

SELECTED BIBLIOGRAPHY

Reviews of Solar Oscillations (theory)

1. Christensen-Dalsgaard, J., 1980, "Solar Oscillation, Theory"
   Fifth European Meeting of the IAU, Liège (Belgium)
2. Cowling, T.G., 1941, MNRAS, 101, 367
3. Eckart, C., 1960, "Hydrodynamics of Ocean and Atmospheres"
   (Pergamon Press)
4. Gough, D.O., 1977, Proc. IAU, Colloq. 36, ed. R.M. Bonnet
   and Ph. Delache, G. De Bussac, Clermont-Ferrand
5. Gough, D.O., 1978, Pleins feux sur la physique solaire,
   ed. J. Rösch, CNRS, Paris
6. Gough, D.O., 1980, Lecture notes in Physics, 125, 279
   (ed. H. Hill and W. Dziembowski, Springer, Heidelberg)
7. Leibacher, J.W., and Stein, R., 1980, NASA/CNES monograph,
   "The Sun", chapter III.A
8. Lighthill, M.J., 1978, "Waves in Fluids" (Cambridge University
   Press)
9. Hill, H.A., 1978, The New Solar Physics (ed. J. Eddy,
   Westview, Colorado)
10. Stein, R.F. and Leibacher, J.W., 1974, Ann. Rev. Astron.
    Astrophys. 12, 407.

Reviews of Solar Oscillations (Observations)

1. Beckers, J.M., and Canfield, R.C., "Motions in the Solar
   Atmosphere", Sacramento Peak Observatory, Nov. 18, 1975
2. Deubner, F., 1980, "Solar Oscillations, Observations".Fifth
   European Meeting of the IAU, Liège, Belgium
3. Hill, H.A., 1978, The New Solar Physics, Chapter 5 (J. Eddy,
   ed. Westview, Colorado)
4. Stenflo, J.O., 1980, "Solar Oscillations, Observations",
   Fifth European Meeting of the I.A.U., Liège, Belgium

Reviews of Stellar Oscillations

1. Cox, J.P., 1976, Annual Rev. Astron. Astrophys., 14, 247
2. Ledoux, P., 1974, Stellar Instabilities and Evolution, I.A.U.
    Symposium n° 59, 135 (Reidel, Dordrecht, Holland)
3. Ledoux, P., and Walraven, T., 1958, Handbuch der Physik, 51,
    Chapter IV (Springer, Berlin)
4. Traub, Mariska and Carleton, 1978, Ap. J., 223, 583
5. Unno, W., Osaki, Y., Ando, H., and Shibahashi, H., 1979,
    Non Radial Oscillations of Stars (University of Tokyo Press)

Observations of Solar Oscillations - Selection of recent papers

1. Beckers, J.M. and Ayres, T.R., 1977, Ap. J. Letters, 217, L69
2. Brookes, J.R., Isaak, G.R. and Van der Raay, H.B., 1976
   Nature, 259. 92
3. Brown, T.M., 1980, Ap. J. Letters, submitted
4. Brown, T.M., Stebbins, R.T., and Hill, H.A., 1978, Ap. J.
   223, 324
5. Caudell, T.P., Knapp, J., Hill, H.A. and Logan, J.D., 1980
   "Lecture Notes in Physics", 125, 206 (Springer, Heidelberg)
6. Claverie, A., Isaak, G.R., Mc Leod, C.P., Van der Raay, H.B.,
   and Roca-Cortes, T., 1979, Nature, 282, 591
7. Claverie, A., Isaak, G.R., Mc Leod, C.P., Van der Raay, H.B.,
   and Roca-Cortes, T., 1980, Astron. Astrophys., in press
8. Deubner, F.L., 1972, Solar Phys. 22, 263
9. Deubner, F.L., 1975, Astron. Astrophys. 44, 371
10. Deubner, F.L., 1977, Astron. Astrophys., 57, 317
11. Deubner, F.L., Ulrich, R.K., and Rhodes, E.J., 1979, Astron.
    Astrophys. 72, 177
12. Dittmer, P.H., 1978, Ap. J. 224, 265
13. Fossat, E., and Ricort, G., 1975, Astron. Astrophys. 43, 243
14. Fossat, E., Grec, G., and Slaughter, C., 1979, Astron.
    Astrophys. 77, 151
15. Fossat, E., Grec, G., Kotov, V.A., Severny, A.B., and
    Tsap, T.T., 1971, MNRAS, in press
16. Frazier, E.N., 1968, Zeit. Astrophys., 68, 345
17. Grec, G., and Fossat, E., 1977, Astron. Astrophys. 55, 411
18. Grec, G., Fossat, E., and Pomerantz, M., 1980, Nature, 288
    5791, 541
19. Hill, H.A., and Caudell, 1979, MNRAS, 186, 327
20. Hill, H.A. and Stebbins, R.T., 1975, Ap. J., 200, 471
21. Hill, H.A., Stebbins, R.T., and Oleson, J.R., 1975, Ap. J.
    200, 484
22. Hill, H.A., Rosenwald, R.D., and Caudell, T.P., 1978,
    Ap. J., 225, 304
23. Isaak, G.R., 1980, Nature, 283, 644
24. Kotov, V.A., Severny, A.B., and Tsap, T.T., 1978, MNRAS
    183, 61

25. Leighton, R., 1960 (IAU Symposium n° 12 (Nuovo Cimento Suppl. 22, 1961)
26. Livingston, W., Milkey, R., and Slaughter, C., 1977, Ap. J. 211, 281
27. Lubow, J.H., Rhodes, E.J., and Ulrich, R.K., 1980, Lecture Notes in Physics, 125, 300, Springer, Heidelberg
28. Musman, J., and Nye, A.H., 1977, Ap. J. 212, L95
29. Rhodes, E.J., Deubner, F.L., and Ulrich, R.K., 1979, Ap. J. 227, 629
30. Rhodes, E.J., Ulrich, R.K., and Simon, G.W., 1977, Ap. J. 218, 901
31. Rosenwald, R.O., and Hill, H.A., 1980, Lecture Notes in Physics, 125, 404
32. Scherrer, P.H., Wilcox, J.M., Kotov, V.A., Severny, A.B., and Tsap, T.T., 1979, Nature, 277, 635
33. Scherrer, P.H., Wilcox, J.M., Kotov, V.A., Saverny, A.B., and Tsap, T.T., 1980, Ap. J. Letters, 237, L 97
34. Severny, A.B., Kotov, V.A., and Tsap, T.T., 1976, Nature 259, 87
35. Severny, A.B., Kotov, V.A., and Tsap, T.T., 1978, "Pleins feux sur la physique solaire", CNRS edit.
36. Tanenbaum, A.S., Wilcox, J.M., Frazier, E.N., and Howard, R., 1969, Solar Phys. 9, 328
37. White, O.R., and Cha, M.Y., 1973, Solar Phys. 31, 23
38. Worden, S.P., and Simon, G.W., 1976, Ap. J. Letters, 210, L1

Theory of Solar Oscillations - Selection of recent papers

1. Ando, H., and Osaki, Y., 1975, PASJ, 27, 581
2. Ando, H., and Osaki, Y., 1977, PASJ, 29, 221
3. Berthomieu, G., Cooper, A.J., Gough, D.O., Osaki, Y., Provost, J., and Rocca,A., 1980, Lecture Notes in Physics 125, 307 (Springer, Heidelberg)
4. Boury, A., Gabriel, M., Noels, A., Scuflaire, R., and Ledoux, P., 1975, Astron. Astrophys. 41, 279
5. Christensen-Dalsgaard, J.,1980, MNRAS, 190, 765
6. Christensen-Dalsgaard, J., and Gough, D.O., 1976, Nature 259, 89
7. Christensen-Dalsgaard, J., Gough, D.O., and Morgan, J.G., 1979, Astron. Astrophys. 73, 121
8. Christensen-Dalsgaard, J., Dilke, F.W. and Gough, D.O., 1974, MNRAS, 169, 429
9. Christensen-Dalsgaard, J., and Gough, D.O., 1980, MNRAS in press
10. Christensen-Dalsgaard, J., and Gough, D.O., 1980, Nature, in press
11. Dilke, F.W., and Gough, D.O., 1972, Nature, 240, 262
12. Dziembowski, W., 1980, Lecture Notes in Physics, 125, 22

13. Dziembowski, W., and Pamjatnykh, A.A., "Pleins feux sur la
    Physique solaire," CNRS edit.
14. Gabriel, M., Scuflaire, R., Noels, A., and Boury, A., 1975,
    Astron. Astrophys. 40, 33
15. Goldreich, P., and Keeley, D.A., 1977, Ap. J., 211, 934
16. Goldreich, P., and Keeley, D.A., 1977, Ap. J., 212, 243
17. Gough, D.O., 1977, Ap. J. 214, 196
18. Gough, D.O., Pringle, J.E., and Spiegel, E.A., 1976, Nature
    246, 424
19. Iben, I., and Mahaffy, J., 1976, Ap. J. Letters, 209, L 39
20. Lamb, H., 1909, Proc. London Math. Soc. 7, 122
21. Leibacher, J.W., and Stein, R.F., 1971, Astrophys. Letters
    7, 191
22. Provost, J., 1976, Astron. Astrophys. 46, 159
23. Scuflaire, R., 1974, Astron. Astrophys. 36, 101
24. Scuflaire, R., Gabriel, M., Noels, A., and Boury, A., 1980
    Fifth European Meeting of the IAU, Liège
25. Shibahashi, H., and Osaki, Y., 1976, PASJ, 28, 199
26. Shibahashi, H., Osaki, Y., and Unno, W., 1975, PASJ, 27, 401
27. Tassoul, M., 1980, Ap. J., Suppl. in press
28. Ulrich, R.K., 1970, Ap. J., 162, 993
29. Ulrich, R.K. and Rhodes, E.J., 1977, Ap. J. 218, 521
30. Wolff, C.A., 1972, Ap. J. Letters, 177, L 87
31. Wolff, C.A., 1979, Ap. J., 227, 942

On Atmospheric Effects in Observations of Solar Oscillations

1. Clarke, D., 1978, "Pleins feux sur la physique solaire"
   CNRS ed.
2. Clarke,D., 1978, Nature, 274, 670
3. Fossat, E., Harvey, J., Hausman, M., and Slaughter, C., 1977
   Astron. Astrophys., 59, 279
4. Fossat, E., and Grec, G., "Pleins feux sur la physique solaire"
   CNRS ed.
5. Fossat, E., Grec, G., Harvey, J., 1981, Astron. Astrophys.
   in press
6. Grec, G., and Fossat, E., 1979, MNRAS, 188, 21 P
7. Grec, G., and Fossat, E., 1979, Astron. Astrophys. 77, 351
8. Grec, G., Fossat, E., Brandt, P., and Deubner, F., 1979,
   Astron. Astrophys. 77, 347
9. Kenknight, C., Gatewood, G.D., Kipp, S.L., and Black, D.,
   1977, Astron. Astrophys. 59, L 27.

EVIDENCE FOR CHROMOSPHERES AND CORONAE IN STARS: RECENT
OBSERVATIONS, SOME UNANSWERED THEORETICAL QUESTIONS, AND A
SPECULATIVE SCENARIO

Jeffrey L. Linsky*

Joint Institute for Laboratory Astrophysics, National
Bureau of Standards and University of Colorado,
Boulder, Colorado 80309 U.S.A.

1. INTRODUCTORY REMARKS

Recent ultraviolet spectra and X-ray flux measurements by
the International Ultraviolet Explorer (IUE) and Einstein X-ray
Observatory have had a major impact on our knowledge of which
regions of the HR diagram definitely contain stars with chromospheres and coronae and of the properties of these outer atmosphere layers. These observations are pointing to similarities
and differences of these outer atmosphere layers compared with
the solar atmosphere and the important roles played by magnetic
fields. These data are also reorienting our thinking about
physical processes occurring in the outer atmospheres of stars.

As a result of the successful IUE and Einstein experiments we
are living in a golden age of observational stellar astronomy. At
this time, however, we must not lose sight of our real goal -- an
understanding in depth of the physical processes that underlie
the extraordinarily interesting phenomena that we are beginning
to see. Hopefully, the present golden age of observations will
introduce a golden age of theoretical interpretation. I anticipate that a major accomplishment of this theoretical effort will

---

The U.S. Government has the right to retain a nonexclusive
royalty-free license in and to any copyright covering this
paper.

*Staff Member, Quantum Physics Division, National Bureau of
Standards.

R. M. Bonnet and A. K. Dupree (eds.), Solar Phenomena in Stars and Stellar Systems, 99–122.
Copyright © 1981 by D. Reidel Publishing Company.

be to understand the various phenomena which we now call chromospheres, coronae, and winds as aspects of one unified problem in
astrophysics. It is perhaps ironic that while the Sun now appears to provide a useful prototype for understanding phenomena
on a wider range of stars than previously thought, we are beginning to recognize that we must go back and study the Sun more
carefully to properly compare it with stars.

In this paper I will be using the terms chromosphere, transition region, and corona to characterize layers with different
physical properties in the outer atmospheres of stars. These
terms are more fully defined elsewhere (cf. Linsky 1980a,b), but
I summarize their meaning here so as to avoid subsequent confu-
sion. I use outer atmosphere as a generic term for these atmospheric layers in which significant input of energy occurs by
mechanisms other than absorption of the emergent photospheric
radiation field. These nonradiative heating terms include dissipation by mechanical waves of various types (cf. review by
Stein and Leibacher 1980), magnetic heating processes (cf. review by Wentzel 1980), thermal conduction, and even radiation
from hotter layers, for example X-rays from flares and coronae.

I use the term chromosphere as that region of the outer atmosphere where (1) the nonradiative heating term is sufficient to
produce a temperature inversion, and (2) the temperature gradient
is small compared to the local pressure scale height because optically thick continua and/or resonance lines are efficient cooling agents. For the Sun and stars cooler than spectral type A,
the Lyman continuum and resonance lines of H I, Ca II, and Mg II
play the roles of efficient cooling agents (cf. Avrett 1980),
acting as thermostats to produce a gradual temperature increase
with increasing height and decreasing density. In these stars
the top of the chromosphere occurs near 20,000 K when the last
available resonance line (Lyman α) becomes optically thin and the
temperature rises steeply with height (Thomas and Athay 1961,
Athay 1976). Chromospheres as I have defined the term could exist in O and B stars if there are other efficient cooling agents
available to produce small temperature gradients.

In analogy with the Sun, I define a transition region as
that portion of a stellar outer atmosphere in which the temperature scale height is much smaller than the pressure scale height.
The solar transition region covers the temperature range 3×10^4-
1×10^6 K, but the temperature range for other stars could be different as a result of qualitative differences in the local energy
balance. As a practical matter we can only study the portion below about 2×10^5 K as the hottest line generally seen in IUE
spectra is the N V 1240 Å blend and Einstein cannot distinguish
flux originating in $T < 10^6$ K plasmas from hotter coronal plasma
from the same source.

Also in analogy with the Sun, I define a _corona_ as that region of a stellar outer atmosphere in which the temperature gradients are small compared with the pressure scale height due
to such efficient cooling processes as X-ray and EUV radiation,
thermal conduction, or a thermally driven (Parker type) wind.
Each of these processes generally require hot plasmas with $T \gtrsim$
1×10^6 K. In the Sun these high temperatures are a consequence
of nonradiative heating processes that are largely or entirely
magnetic in character (cf. Tucker 1973, Withbroe and Noyes
1977). In this review I will consider how general this result
may be for other stars.

During the course of this review I will try to ask in
various ways some fundamental questions including:

(1) What are the nonradiative heating processes and how do
they depend on stellar parameters?

(2) What are the relations between chromospheres, transition regions, and coronae?

(3) What are the various roles played by magnetic fields?

The general topic of solar and stellar chromospheres has
been reviewed recently by Withbroe and Noyes (1977), Linsky
(1979, 1980a, 1980b), Ulmschneider (1979), and Jordan (1980).
For recent reviews of solar and stellar winds, the reader is
referred to Cassinelli (1979), Dupree and Hartmann (1980), Snow
and Linsky (1980), and Hartmann (1980), among others.

2. LOCATION OF CHROMOSPHERES IN THE H-R DIAGRAM

The identification of chromospheres in stars is primarily
done by detection of emission in the resonance lines of Ca II,
Mg II, and H I, although other prominent spectroscopic features
include emission or absorption in the H I Balmer series (particularly in M dwarfs) and He I 10830 Å, ultraviolet emission lines of
O I, Si II, and Fe II, and continua in the ultraviolet and infrared. The usefulness of these direct indicators as well as other
less direct diagnostics has been surveyed by Praderie (1976),
Linsky (1977, 1980b), and Ulmschneider (1979), among others.

To the best of my knowledge, one or more of the above chromospheric indicators has been found in essentially every star
between spectral types late F and at least early M that has been
observed with sufficiently deep exposures. The information contained in these data will be presented later, but at this time
I would like to discuss whether chromospheres cease to exist in
some or all stars at the hot and cool ends of the late-type star
portion of the H-R diagram.

2.1. The hot end of the late-type stars

The hottest stars showing emission in the Ca II H and K
lines are the F0 dwarf γ Vir N (Warner 1968) and the F0 supergiant, α Car (Warner 1966). Occasionally, Ca II emission
has been reported in the A 7 III proposed δ Scuti star γ Boo
(LeContel et al. 1970, Auvergne et al. 1979). Dravins, Lind,
and Särg (1977) show that transient emission occurs in the
δ Scuti stars when the photosphere has maximum outward acceleration due to shock waves. Careful studies of the Ca II lines
at high dispersion in the early A-type stars (e.g. Freire et al.
1978) and in A dwarfs in young clusters (Dravins 1980) show no
evidence of emission.

There is unfortunately a coincidence in spectral type between the disappearance of Ca II emission in the early F stars
and the rapid decrease in convective energy transport, increase
in stellar rotational velocity, and decrease in estimated acoustic wave generation (e.g. de Loore 1970) as one proceeds to hotter stars from the middle F-type stars. Since the dissipation
of acoustic waves generated by convective turbulence used to be
the only accepted theory of chromospheric heating, this coincidence was taken as conclusive evidence that chromospheres do not
exist in stars hotter than early F. There are two possible flaws
in this logic. First, chromospheres may be heated by mechanisms
other than acoustic waves (see below). Second, chromospheric K
line emission must be measured against a background (the photospheric K line wings) that increases extremely rapidly with increasing stellar effective temperature. Thus the K line emission feature contrast must disappear with increasing T_{eff} just
as stars disappear to the naked eye at dawn. Dravins (1976),
for example, shows that there is no decrease in K line surface
flux towards spectral type F0, and concludes that chromospheres
probably exist in hotter stars even though they become unob-
servable in the K line.

To pursue this question further, several investigators have
searched for emission lines in the ultraviolet where the photospheric background is fainter. For example, Evans et al. (1975)
reported observation of emission in the cores of the Mg II lines
in α Car (F0 Ib-II), and Blanco et al. (1980) found emission in
the Mg II and Lα lines of α Aql (A7 V). In their survey of A-
and F-type stars with IUE, Böhm-Vitense and Dettmann (1980) detected no Mg II emission in any A star and found that Mg II emission begins to the right of the cepheid instability strip for
supergiants and at about spectral type F2 for the dwarfs. They
argue that chromospheres cease to exist at early F-type at least
for luminous stars on the grounds that early F supergiants show
no Mg II emission whereas dwarfs of similar spectral type do. I
feel that this argument is inconclusive as the Mg II surface flux

may be smaller in the luminous F stars and thus not be visible
against the bright photospheric background. On the basis of
their IUE Mg II spectra, Linsky and Marstad (1980) argue that
chromospheres "disappear" but may well exist in the A-type
stars. In any case very different data are needed to answer
in any convincing way the question of whether chromospheres end
in the early F stars. One interesting diagnostic which should
be pursued is to look for 20 μm excess flux due to chromospheric
free-free emission as Morrison and Simon (1973) have reported
for Vega.

2.2. The cold end of the late-type stars

     Strong Ca II and Mg II emission has been detected in even
the latest dMe star in surveys by Giampapa et al. (1980) and
Linsky et al. (1980). The dM stars (M dwarfs without Balmer
line emission) generally show Ca II and Mg II surface fluxes
much weaker than the dMe stars. Although extremely difficult
to observe, very late dM stars observed by Liebert et al. (1979)
show no Hα emission and apparently no Ca II emission either.
Further observations are needed to determine whether these stars
indeed have no chromospheres.

     Chromospheric emission lines (e.g. Ca II, Mg II, and Fe II)
are typically seen in the M supergiants (cf. van der Hucht et al.
1979), but Dyke and Johnson (1969), Jennings and Dyke (1972), and
Jennings (1973) find an inverse correlation between K-line emission and polarization or infrared excess, two symptoms of grains
in a cool outer atmosphere. They find some M supergiants that
appear to show no Ca II emission at all, even though the photospheric background against which K-line emission must be measured is extremely faint in these stars. Some of these stars
show emission in neutral metal lines instead. These observations
suggest that the M supergiants with no Ca II emission probably
have chromospheres as the term is defined in §1, but that dust
is an efficient cooling agent such that the available nonradiative heating flux results in a chromospheric rise in temperature
too small to produce appreciable Ca II emission. Alternatively
the dust prevents any temperature rise (and thus no chromosphere),
and the neutral emission lines are produced in a cold extended
envelope by resonance fluorescence of photospheric radiation.

3. LOCATION OF TRANSITION REGIONS IN THE H-R DIAGRAM

     IUE has made it possible to observe emission lines formed
in transition regions of a large number of stars. The strongest
lines observed are typically C II 1335 Å, Si IV 1394,1403 Å,
He II 1640 Å, C IV 1549 Å and N V 1240 Å. Since most late-type
stars are too faint to be observed in high dispersion by IUE,

only integrated fluxes are available from the low dispersion
spectra. High dispersion line profiles have been obtained for a
few bright stars including Procyon (Brown and Jordan 1980) and
Capella (Ayres and Linsky 1980). It is important to recognize
that IUE cannot observe transition region plasmas hotter than
about 2×10^5 K. Spectroscopic experiments sensitive to EUV
(100 Å $\leqslant \lambda \leqslant$ 900 Å) are necessary. To the best of my knowledge,
most or all stars of spectral type late F and G (except for late
G supergiants) show transition region emission lines in sufficiently deep IUE spectra. Two important questions are whether
transition regions exist beyond the hot and cool end of this
region of the H-R diagram.

3.1. Stars hotter than middle F spectral type

In their survey of A and F stars with IUE, Böhm-Vitense and
Dettmann (1980) find a strong correlation between the presence
of Mg II emission and the presence of transition region and
chromospheric emission lines in the 1150-2000 Å spectral region. Thus the hottest stars in their survey with obvious
transition regions are F2 dwarfs and luminous stars just to
the right of the Cepheid instability strip.

In deep exposures of the 1150-2000 Å spectral range, Linsky
and Marstad (1980) find transition region and chromospheric
emission lines in α CMi (F5 IV-V) and β Cas (F2 IV), but none in
α Car (F0 Ib-II) and γ Boo (A7 III). They find that the surface
fluxes of typical transition region lines in the rapidly rotating
δ Scuti star β Cas are 30 times those of the quiet Sun. Lines
this bright would be barely detectable in α Car and undetectable
in γ Boo against the bright photospheric absorption line spectra
of these stars with IUE. Thus IUE cannot determine whether transition regions cease to exist in the early F stars, but the High
Resolution Spectrograph on Space Telescope, which will have better resolution, signal-to-noise, and low scattered light background, may be able to answer this question.

Freire (1979) has searched for C II λ1335 Å emission from
Vega (A0 V) in high resolution Copernicus spectra but finds no
evidence for emission. In a survey of supergiants of spectral
type B2 Ia-A2 Ia with IUE, Underhill (1980a) generally found
P Cygni C IV line profiles in the hotter stars and displaced
C IV absorption components in the cooler supergiants. Both
types of profiles are probably produced in the extended winds
of these stars. However, one star, HR 1040 (A0 Ia), perhaps
shows weak C IV emission features in one spectrum suggesting
the presence of hot transient plasma that could be a transition
region similar to that in the cooler stars. An infrared excess
also suggests the presence of 23,000 K plasma in the outer atmosphere of this star (Underhill 1980b).

Early-type stars more luminous than $M_{BOL} = -6$ typically show evidence of large mass loss by P Cygni-type line profiles or displaced absorption lines in the ultraviolet and excess radio emission. IUE and earlier Copernicus and rocket spectra also show superionization -- stages of ionization up to O VI in the wind far in excess of what would be expected in radiative equilibrium. These data, estimates of mass loss, and possible acceleration mechanisms have been reviewed by Cassinelli (1979), Lamers (1980), and Snow (1980), among others. These and X-ray data to be discussed shortly present clear evidence of nonradiative heating in the outer atmospheres of these stars, which is their physical connection to the transition regions and coronae in the late-type stars.

3.2. G supergiants, K, and M stars

G- and K-type dwarf stars typically show transition region emission lines with surface fluxes comparable to the quiet Sun as in α Cen A (G2 V) and α Cen B (K1 V) (Ayres and Linsky 1980), or comparable to bright solar active regions as in ξ Boo A (G8 V) and ε Eri (K2 V) (Hartmann et al. 1979). In their survey of M dwarfs with IUE, Linsky et al. (1980) find that dMe stars exhibit prominent transition region emission lines with surface fluxes up to 50 times the quiet Sun in C IV at least as late as spectral type dM6e. The dM stars (M dwarfs without Hα emission) show much weaker lines or none at all. For example, the star GL 380 (dM0) shows C IV, C II, and He II emission lines with surface fluxes comparable to the quiet Sun, but only upper limits on C IV and C II emission ≈0.3 times those of the quiet Sun are obtained for GL 411 (dM2). Further observations are needed to establish whether transition regions cease to exist or are just becoming very faint in the late dM stars.

In their initial survey of 22 late-type stars with IUE, Linsky and Haisch (1979) noted chromospheric emission lines in all of their 1150-2000 Å low dispersion spectra, but no obvious emission features at the wavelengths of prominent transition region lines in giants cooler than about spectral type K2 III and in late G-M supergiants. On the basis of this small sample of stars, they proposed a dividing line in the H-R diagram, nearly vertical through about spectral type K2 III, separating stars with and without transition region lines observed in their ultraviolet spectra. They noted two possible explanations for the apparent absence of spectral lines from hot plasma to the right of this "transition region dividing line" in the H-R diagram: either transition regions cease to exist, or the emission measure of hot plasma is too small to be seen in realistic exposure times against a background of scattered light and weak chromospheric lines like Fe II in the low dispersion IUE spectra.

Linsky and Haisch (1979) favored the first explanation on
the basis that the maximum temperature allowable for gravitational retention of a hot outer atmosphere is several $\times 10^5$ K for
K giants and $<10^5$ K for supergiants (Mullan 1976). Thus they
expected to see strong emission in lines formed near the temperature maximum (which should be an extended region) and no
lines from plasma at hotter temperatures. Also Reimers (1977)
and Stencel (1978) noted an abrupt shift in asymmetry of the
Ca II emission lines near the transition region dividing line,
which implies a rapid increase in mass loss to the right of this
dividing line. On theoretical grounds Mullan (1978) proposed
that winds should become large along a locus in the H-R diagram
(the so-called supersonic transition locus) near the transition
region dividing line. Given the near coincidence of the wind
and transition region dividing lines, Linsky and Haisch (1979)
argued, in analogy with solar coronal holes, that for stars to
the right of the dividing lines most of the available nonradiative energy goes into driving the wind and insufficient energy
remains to heat the outer atmospheres above chromospheric temperatures. Thus stars to the right of the transition region
dividing line do not have transition regions or coronae.

Whether this scenario is basically correct or not can and
should be examined carefully on both observational and theoretical grounds. Careful reexamination of the 1150-2000 Å spectra
of stars to the right of the proposed transition region dividing
line has resulted in upper limits for C IV and other lines for
α Ori (M2 Iab), ε Gem (G8 Ib), α Boo (K2 III), and α UMa (K0 II-
III). However, ε Sco (K2 III-IV) and α Ser (K2 III) appear to
show real but weak C IV emission, suggesting that the dividing
line (if real) may be slightly to the right (cooler) than originally proposed. Deep exposures of a number of stars near the
proposed dividing line are needed to explain its reality, its
precise location with the H-R diagram, and its intrinsic width.
Results of an IUE observing program by Stencel and Simon will be
reported at this meeting (Stencel 1980) and other programs are
under way.

Subsequent to the Linsky-Haisch paper, Stencel and Mullan
(1980) have shown that the Mg II lines like the Ca II lines show
a rather abrupt change in asymmetry along a dividing line in the
H-R diagram slightly to the left of the Ca II asymmetry line.
The Mg II data, unlike the Ca II data, must be corrected for
interstellar absorption components. When these corrections are
made, the Mg II asymmetry line is very close to the transition
region dividing line.

Ayres et al. (1980a) have presented stellar emission line
fluxes in a way that casts the transition region dividing line
in a different light. They divide emission line fluxes by the

bolometric luminosity of the star at the Earth to obtain a ratio, f_ℓ/ℓ_{bol}, which gives the fraction of the total stellar luminosity emitted in the particular line. They find that chromospheric emission line ratios are proportional to the Mg II line ratio, but that transition region line ratios are proportional to $(f_{Mg\ II}/\ell_{bol})^{1.5}$. In other words, those stars with bright Mg II lines, as measured by $f_{Mg\ II}/\ell_{bol}$, have extremely bright transition region lines, and the converse is also true. The question then arises as to whether transition region emission may seem to disappear along the transition region dividing line because of this effect. I doubt that this is the correct explanation because there is no large change in $f_{Mg\ II}/\ell_{bol}$ as one crosses the transition region dividing line and because the well observed star Arcturus shows transition region line upper limits far below predicted values.

Hartmann, Dupree, and Raymond (1980) show that the early G supergiants α Aqr (G2 Ib) and β Aqr (G0 Ib) exhibit both transition region emission lines and blue-shifted absorption features up to ~125 km s^{-1} in the Mg II lines, indicative of strong winds. The outer atmospheres of these stars and presumably others close to the transition region dividing line are thus "hybrid" in character. In addition, Dupree and Baliunas (1979) report that both the wind velocity and the emission line fluxes in α Aqr are variable. There are two reasonable explanations for the hybrid character of stars located near the transition region dividing line. Either the input of nonradiative energy goes into both driving the wind and heating a transition region; or the outer atmospheres of these stars have two very different components. One component, analogous to solar coronal holes, may have open magnetic fields and a cool wind; while the other component, analogous to solar active regions, has closed magnetic loop structures with hot plasma and a very weak wind. Stellar rotation may account for the variable amounts of each component. By contrast, β Dra (G2 II) exhibits bright transition region emission lines and no spectroscopic evidence for a wind, and ε Gem (G8 Ib) shows no apparent transition region lines but clear spectroscopic evidence for a large wind (Basri and Linsky 1980). I will return to this very interesting region of the H-R diagram later.

4. LOCATION OF CORONAE IN THE H-R DIAGRAM

Prior to Einstein, only a few of the very brightest stellar coronae were observed by HEAO-1, ANS, and rocket experiments. These sources included a number of RS CVn-type systems, Vega, several dMe flare stars during flares, and a few late-type stars. As previously noted, IUE is insensitive to plasmas hotter than 2×10^5 K and no evidence has yet emerged from IUE for extended regions at $T \leqslant 2 \times 10^5$ K in any stars that might be considered as having cool coronae rather than transition regions

as defined in §1. The major breakthrough in the discovery of
stellar coronae came with Einstein (cf. Giaconni et al. 1979 for
a description), primarily with the Imaging Proportional Counter
(IPC) experiment. These data were reviewed initially by Vaiana
(1980) and Linsky (1980c), and the first survey of Einstein
stellar coronal observations is now available (Vaiana et al.
1980). It is important to recognize that the IPC obtains only
very crude information on coronal temperatures, and the Solid
State Spectrometer (SSS), Crystal Spectrometer, and Objective
Grating experiments on Einstein can only observe a few of the
very bright stellar coronae. Thus X-ray spectroscopy of stellar
coronae must await the development of new space experiments.
In this review I will summarize the existing data, primarily in
Vaiana et al. (1980) and emphasize what I consider to be the
important points for theoretical interpretation.

4.1. O, B, and A-type stars

        Listed in Table 1 are ranges and mean values observed by
Einstein of the total X-ray flux, L_x (ergs cm^{-2} s^{-1}), the
L_x/L_{BOL} ratio, and the X-ray surface flux f_x (ergs cm^{-2} s^{-1})
for different spectral types and some interesting individual
stars. Also given are typical data for solar flares, active
regions, the mean quiet Sun (variable by about a factor of 3
over the solar cycle), and solar coronal holes (the darkest regions of the solar corona). I have estimated values of L_x/L_{BOL}
from the values of L_x/L_V cited by Vaiana et al. (1980) and estimates of the fraction of the stellar bolometric luminosity contained in the visual (V) band. The discovery that O and B stars
are bright X-ray sources (cf. Harnden et al. 1979, Long and
White 1980) with $\langle \log L_x \rangle \approx 32$ for dwarfs and 34 for supergiants
($\log L_{BOL} = 33.5$ for the Sun) was a real surprise to many astronomers who assumed that acoustic waves generated in convective zones are required for the heating of stellar coronae.
Both the crude IPC pulse light spectra (Snow et al. 1980) and
the Solid State Spectrometer data (Swank 1980) of a few bright
stars suggest that the X-ray spectra of OB supergiants consist
of a soft component (T $\approx$ 3 × 10^6 K), perhaps formed in the wind
itself, and a cut off hard component (T $\approx$ 10^7 K), presumably
formed in a corona at the base of the wind. This latter component is attenuated by the wind itself primarily at low energies.
Snow et al. (1980) also report that the cut off hard component
is highly variable with time.

        A-type stars exhibit a huge range in L_x/L_{BOL} from <-9 upper
limits to -4. Also reported observations of several bright A
and Ap stars by Einstein show that the X-ray emission from these
stars is extremely variable by factors of 100 or more (Cash
1980). Vega (A0 V) was originally detected by a rocket experiment (Topka et al. 1979), then not seen by the IPC, and finally

Table 1. SUMMARY OF EINSTEIN STELLAR OBSERVATIONS

Spectral Type or Specific Star	log L_x (range)	$\langle$log $L_x\rangle$	log L_x/L_{BOL} (range)	$\langle$log $L_x/L_{BOL}\rangle$	log f_x
OB Dwarfs[a]	31–33	32.5	−8 to −5	−6	6 to 7.6
A Dwarfs[b]	<26–31		<−9 to −4		3 to 7
F Dwarfs	28–30	29	−7 to −4	−5.5	5 to 7.5
GK Dwarfs	26–30	28	−7.5 to −4.5	−6	4.5 to 7
Solar Flare	31–33		−2 to −1		8.5
Solar Active Region		29.3		−4.3	6.5
Quiet Sun		27.8		−6	5
Solar Coronal Hole		26.8		−7	4
M Dwarfs[c]	26–30		−5 to −3	−4	4.5–7.5
Flares	28–31				
O Supergiants		34			
G Giants	28–30		−7.3 to −6		
α Aqr (G2 Ib) }				<−7	<4.7
β Aqr (G0 Ib) }					
M Supergiants[d]					
α Ori (M2 Iab) }	<29.3			<−8	<1
α Sco (M1 Ib) }					
RS CVn Systems	30.3–31.3		−5 to −2.5		

[a] Soft component (~3 × 10^6 K) and cut off hard component (~10^7 K).
[b] Highly variable (>100×).
[c] dMe stars >> dM stars.
[d] No detections.

observed as a very weak source by the High Resolution Imager (HRI)
in Einstein (Harnden 1980). Since the HRI is sensitive to 0.1 keV
photons whereas the IPC is not, Vega must be an extremely soft
source. One complication in studying A stars is that many appear
to have previously unknown K star companions which may contribute
all or most of the observed X-ray flux. Nevertheless, some A
stars are clearly real X-ray sources and stars with strong magnetic fields are not conspicuously bright or faint sources.

4.2. Late-type dwarfs

Two important results from the initial Einstein survey are
that the brightest X-ray emitting dwarfs have $\log L_x \approx 30$ independent of spectral type between about F0 and late M, and that the
range of luminosities is at least a factor of 10^2 at each spectral type. These data show clearly that T_{eff} is not a critical
parameter in determining the X-ray emission as deduced from simple
acoustic wave heating theory (Mewe 1979) and that some other
parameter must be critical. Correlations between L_x/L_{BOL} and
stellar rotation using the HEAO-1 data (Ayres and Linsky 1980),
Einstein observations of single G and K stars (Walter 1980), and
RS CVn-type synchronous binaries (Walter and Bowyer 1980), point
to stellar rotation as the critical parameter for X-ray emission.
Among the F and G dwarfs the maximum value of $L_x/L_{BOL} \approx 10^{-4}$,
similar to solar plages. This Einstein result, previously seen
in the HEAO-1 data of Walter et al. (1980a), suggests that 10^{-4}
is the maximum conversion efficiency of stellar luminosity into
coronal heating in stars similar to the Sun if it is proper to
think of coronal heating in these terms. The M dwarfs differ
from the warmer dwarfs as L_x/L_{BOL} can be as large as 10^{-3}. Also
among the M dwarfs the brightest X-ray sources tend to be close
binary systems or young dMe stars, which are presumably more
rapid rotators, while the stars with smallest L_x including stars
like Barnard's star (Vaiana et al. 1980) are the only unambiguously single stars and are presumably old. Using IPC observations of the bright nearby dM5e star Proxima Centauri, Haisch et
al. measure a temperature of $\sim 4 \times 10^6$ K for quiescent emission
and $\sim 17 \times 10^6$ K during the peak of a flare.

RS CVn binaries are detached systems generally consisting
of a late G dwarf primary and a K0 IV secondary with periods of
2-17 days (Hall 1976). Walter et al. (1980b) showed that for
these systems $\log L_x$ generally lies in the range 30.3-31.3, at
the top of the late-type star distribution of L_x and probably
not really a separate class of emitters. Since tidal forces
will induce synchronism of rotational and orbital periods on
evolutionarily short time scales (Zahn 1977), the active secondary stars in these systems have large (≥ 30 km s^{-1}) rotational
velocities. Dupree (1980) will discuss the outer atmospheres of
these and other close binary systems.

4.3. G-M giants and supergiants

Excluding spectroscopic binary systems, Vaiana et al. (1980)
report Einstein observations of several G giants with log L_x in
the range 28-30 and two early K giants, ε Sco (K0 III-IV) and
α Ser (K2 III), with log $L_x \approx 28$. These latter stars have log
$L_x/L_{BOL} \approx -7$, equivalent to a solar coronal hole. Cooler bright
K giants like α Boo (K2 III) and α Tau (K5 III) and G-M supergiants have not been detected. Upper limits on log L_x/L_{BOL} for
the early G supergiants α Aqr (G2 Ib) and β Aqr (G0 Ib) are ≈ -7,
and for α Ori (M2 Iab) and α Sco (M1 Iab) are ≈ -8 (Vaiana et
al. 1980, Ayres et al. 1980b). Since the M supergiants are much
cooler than the Sun, upper limits on the X-ray surface fluxes
are $\sim 10^3$ times lower than the darkest regions in the solar
corona (coronal holes). These and other recent data have led
Ayres et al. (1980b) to propose a dividing line in the H-R
diagram separating stars with and presumably without coronae
located near the previously discussed transition region and
Ca II and Mg II asymmetry dividing lines. It is important to
recognize the difficulties inherent in confirming this proposed
coronal dividing line. First, it is impossible to prove the
nonexistence of a corona by the absence of detected X-ray emission. One can always argue that a corona exists with an emission measure too small to observe. Second, the IPC on Einstein
is insensitive to photons less energetic than about 0.25 keV and
thus has difficulty detecting X-rays from cool (T < 1-2 $\times 10^6$ K)
coronae. The HRI is sensitive down to 0.1 keV, but is also insensitive to X-rays from coronae at T < 1 $\times 10^6$ K. Thus the
Einstein data cannot exclude the possibility of cool corona to
the right of the coronal dividing line.

5. WHAT ARE THE IMPORTANT STELLAR PARAMETERS DETERMINING THE PROPERTIES OF STELLAR CHROMOSPHERES, TRANSITION REGIONS, AND CORONAE?

Throughout this talk I have hinted at what the important
parameters determining properties of outer atmosphere layers
might be. It is clear that stellar effective temperature,
gravity, and chemical composition play little if any role.
Stellar age presumably does play a major role, but only indirectly through correlations with other parameters. Here I
would like to present the arguments for several important
parameters in a more systematic way.

5.1. Varying magnetic fields

Within the last decade, solar astronomers have realized
that the solar magnetic field has a totally unexpected, highly
inhomogeneous structure consisting of ≈ 1500 gauss randomly

oriented flux tubes with subarcsecond diameters in the photosphere and small $(10^{-3}-10^{-4})$ filling factors, except in active
regions. Why the Sun has such a "curious" magnetic field structure is an important question for theoretical investigation, but
the existence of this field determines the structure and energy
balance of the solar outer atmosphere in fundamental ways. For
example, the spatial correspondence of strong photospheric magnetic fields (clumps of flux tubes) with bright overlying chromospheric emission (say in the Ca II lines) and downflows in the
supergranulation network (e.g. Skumanich et al. 1975) imply that
the magnetic field plays important roles in supplying or channeling nonradiative heating, defining the plasma geometry, and
controlling velocity fields in the chromosphere. Similarly,
many investigators have noted a strong correlation of bright X-
ray emission with the coronal magnetic field extrapolated from
the measured photospheric field. Many investigators have
pointed out that the solar corona contains two basic features:
open field regions called coronal holes, from which the high
speed wind streams emerge, and where the corona is coolest; and
closed flux tubes confining hot plasma. Consequently, the magnetic field determines both the geometry and energy balance of
coronal structures.

Until recently there was indirect evidence, but little
direct evidence for magnetic fields in late-type stars. Kelch
et al. (1979) did show that chromospheric models for G-M dwarfs
with unusually bright Ca II K line surface fluxes for their
spectral types exhibit properties similar to solar plages and
are likely covered with structures comparable in chromospheric
emission and perhaps also in field strength to solar plages.
Also the presence of starspots in BY Draconis stars (cf. Kunkel
1975) and in RS CVn-type binary systems (cf. Eaton and Hall
1979) has been deduced from periodic photometric variability.
If starspots are analogous to sunspots, they indicate the
presence of large scale concentrations of strong magnetic
fields.

Anderson et al. (1976) have presented direct evidence for a
40 kG longitudinal field in a starspot on BY Dra based on Zeeman
spectrograms taken when a large spot was presumably present on
the stellar limb. Recently Robinson (1980) has proposed measuring solar-like randomly oriented fields in late-type stars by
deconvolving the Zeeman triplet pattern in unpolarized light
from lines of different Landé g factors. Robinson et al. (1980)
have used this technique to measure fields of 2.6 ± 0.4 kG covering roughly 30% of the surface of ξ Boo A (G8 V) and fields of
1.9 ± 0.4 kG covering perhaps 10% of the surface of 70 Oph A
(K0 V). Walter et al. (1978) measured an X-ray luminosity of
$\log L_x = 30.3$ for ξ Boo A, some 30 times that of the quiet Sun,
but comparable to the Sun entirely covered by plages (Vaiana

and Rosner 1978). Very recently, Marcy (1980) has remeasured
the magnetic field of ξ Boo A using the Robinson technique with
a null result, suggesting either a variable field or an error in
the original measurement.

Thus the correlation of magnetic fields with coronal X-ray
emission is on firm ground for the Sun and reasonably certain
for late-type stars. However magnetic A stars are not bright X-
ray sources. Since these stars appear to have simply structured
constant magnetic fields which may inhibit convection and other
turbulent motions, time variations, presumably due to turbulent
motions, are needed to heat the outer atmospheres of stars.

5.2. Rotation

During the 1960's a number of authors (cf. Wilson 1966 and
Linsky 1977 for reviews) found a statistical correlation of decreasing Ca II K line strength with increasing age and decreasing rotational velocity for solar type stars. F and cooler main
sequence stars apparently spin down with age due to the loss of
angular momentum by coronal winds. Skumanich (1972) proposed
that both the stellar rotational velocity and Ca II emission
strength decrease with age proportional to $t^{-1/2}$, suggesting a
causal relation between rotational velocity and chromospheric
heating.

It is now feasible to obtain more precise rotation-stellar
activity relations. For example, Zahn (1977) has shown that binary systems with less than 20 day periods should become synchronous rotators on short time periods. On this basis, Carrasco et
al. (1980) argue that as a class the dMe stars, which have larger
chromospheric radiative loss rates and therefore larger heating
rates, are more rapidly rotating than the dM stars. Walter and
Bowyer (1980) find that L_x/L_{BOL} is proportional to Ω (the stellar angular velocity) for RS CVn systems over a factor of 100 in
Ω. Bopp (1980) and Ayres and Linsky (1980) argue that rapid rotation not tidal forces per se is the cause of the photometric
symptoms of spots in EQ Vir (dK7e) and Capella Ab (F9 III), since
EQ Vir is a single star and Capella Ab is a nonsynchronous rotator in a widely separated system.

Fourier deconvolution of high signal-to-noise line profiles
has led to precise recent measurements of stellar rotational velocities (cf. review by Smith 1979). With such data, Ayres and
Linsky (1980) find a good correlation of L_x/L_{BOL} with measured
rotational velocities, and Walter (1980) finds that $L_x/L_{BOL} \sim \Omega$
in a sample of 15 G-K dwarfs observed by Einstein. These data
clearly point to Ω as a key parameter in determining stellar X-
ray emission. The likely mechanism for this connection in F and
cooler stars is almost certainly dynamo action (Parker 1958),

which amplifies magnetic fields deep in the stellar convection
zone. Buoyancy brings these fields to the surface where interaction with turbulent motions can lead to heating by many processes (cf. Chiuderi 1980).

5.3. Convection zone parameters

Rotation alone cannot be the sole parameter determining the
rate at which dynamo regenerated fields reach the surface of a
dwarf star as rapidly rotating dMe stars with $v_R \approx 10$ km s^{-1}
have values of L_x/L_{BOL} an order of magnitude larger than rapidly
rotating F dwarfs with larger rotational velocities. In general, dynamo processes operate because of the interaction of
rotation or differential rotation with convection. Thus some
parameter of the convection zone should be significant. Gilman
(1980) argues that the efficacy of dynamo processes should depend on the ratio of the convective turnover time ($t_c \sim d_c$,
where d_c is the convective zone depth) to the rotational time
scale ($t_R \sim \Omega^{-1}$). In this case the creation of magnetic fields
should be proportional to the inverse Rossby number, and this
proportional to Ωd_c. Skumanich (1980) argues that this formula
oversimplifies the essential physics of dynamo processes.
Future theoretical studies will settle this point, but it is
clear that L_x/L_{BOL} is larger for stars with deeper convective
zones but similar Ω. Thus, one might expect L_x/L_{BOL} to be
proportional to $\Omega^\alpha d_c^\beta$ or $\Omega^\alpha (d_c/R_x)^\beta$, with $\alpha \sim 1$ and $\beta > 0$.

6. A SPECULATIVE SCENARIO FOR STELLAR CHROMOSPHERES/TRANSITION REGIONS/CORONAE

I would now like to describe an admittedly speculative scenario which brings together the diverse data on chromospheres,
transition regions and coronae into a single conceptual framework. The unifying theme of this scenario is that magnetic
fields play a major role in the nonradiative heating of outer
atmospheres, but it would be a mistake to require magnetic
fields to be the sole cause of heating. This scenario is a
modification of that described previously by Linsky (1980c)
and Rosner and Vaiana (1980) to take into account recent observational and theoretical developments, but stellar atmospheres are almost certainly more complex than described below.

6.1. O, B, and A-type stars

Both the IPC and the SSS data strongly suggest that X-ray
emission from OB stars with strong winds is characterized by
two components. The soft component ($T \approx 3 \times 10^6$ K) is likely
formed in the wind itself as a consequence of an instability in
radiatively-driven winds. Lucy and White (1980) and Castor

(1980) have proposed that this component is heated by the dissipation of bow shocks at the front of cooler blobs of gas that
are preferentially accelerated by photospheric radiation and
plow through the slower-flowing gas. Since the hard component
is cut off, that is attenuated at lower energies, it is likely
formed at the base of the wind in a corona, and the attenuation
is likely by the material in the wind itself. The great time
variability of this hard component as well as the observed
variability in P Cygni line profiles on time scales of hours
suggests a stochastic rather than a constant heating source. I
would like to propose that this heating process may involve the
annihilation of magnetic fields (either remnant from the proto-
stellar nebula or dynamo regenerated during a convective premain
sequence phase) by the highly turbulent velocity fields. Between
the photosphere and corona there should be a transition region,
but emission lines formed there may not be visible against the
extremely bright photospheric background in the ultraviolet.
Since there is essentially no neutral hydrogen present in these
stars, there may be no optically thick efficient cooling source
available to permit a chromosphere as defined in §1.

The existence of extreme variability and weakness of the X-
ray flux and softness of the X-ray spectra in those A-type stars
like Vega for which late-type companions can be ruled out, suggests a heating source in the A stars that barely works. By
this I mean that the heating source only supplies enough energy
per unit surface area to heat the outer atmosphere to barely
10^6 K, and the heating source is thus effective only some of the
time and/or over limited portions of the star. At first sight
this may appear strange as many A stars have measured kilogauss
magnetic fields. But these magnetic fields must be highly
ordered (perhaps simple dipoles) and stable to be measured by
Zeeman techniques. Also these fields may suppress convection
and thus make the photosphere very stable. In the absence of
turbulence and time variability, there is no way of converting
fields to heat. However, small amounts of turbulence over
limited regions of these stars could produce MHD waves which
will damp eventually in closed magnetic loops and heat some of
these loops to 10^6 K some of the time. Rotational modulation
and variability in the turbulence source would then lead to the
observed X-ray emission properties.

6.2. Late-type dwarfs

In dwarf stars of spectral types F-M we are likely seeing
phenomena produced by the same heating processes as in the Sun.
For the upper chromosphere, transition region, and corona the
magnetic field must be playing a dominant role either through
magnetoacoustic waves (cf. Stein 1980, Ulmschneider 1980) or
perhaps annihilation of magnetic fields directly (Withbroe and

Noyes 1977). In the lower chromosphere purely acoustic waves
may be important in quiet regions on the Sun and stars, but magnetoacoustic waves likely heat solar plages and active stellar
chromospheres. There is near universal agreement that the magnetic fields in F-M dwarfs result from dynamo process, but no
agreement yet as to the nature of the dynamo itself.

We have previously noted observational work which shows
that the L_x/L_{BOL} ratio of late-type dwarfs and binaries is proportional to Ω and increases towards later spectral type (deeper
convective zones) for stars of the same Ω. This suggests that
the rate of emergence of magnetic flux per unit area, dB/dt,
might depend on the product $dB/dt = \Omega^{\alpha}(d_c/R*)^{\beta}$, where $\alpha \approx 1$ and
$\beta > 0$. Figure 1 is a schematic description of how dB/dt might
behave as a function of spectral type and age. What is plotted
is Ω for young stars on the zero age main sequence (ZAMS) and
those about to leave the ZAMS (old stars), and the $d_c/R*$ ratio
as a function of spectral type. The schematic behavior of dB/dt
is given for young and old stars as a function of spectral type
and age. The figure shows the rapid turn-on of dynamos in the
early F stars, its decrease into the G stars (due to the decrease
in Ω at all ages), and its increase in the M stars (due to the
increase of $d_c/R*$ towards later spectral type while values of Ω
are generally no less in the M stars than in the late G stars).

MAIN SEQUENCE STARS

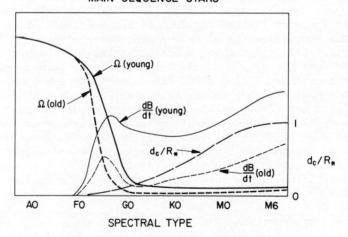

Figure 1. A schematic representation of dB/dt (the rate of emergence of magnetic flux per unit area) as a function of Ω (the
stellar angular velocity), $d_c/R*$ (the ratio of the convective
zone depth to stellar radius), spectral type, and age. Young
stars are those on the zero age main sequence and old stars are
those about to leave the main sequence.

Also dB/dt is far larger in young than old stars at each spectral type. Since the M dwarfs have very deep convective zones or are fully convective, dB/dt depends only on Ω and thus on age. Along the main sequence the X-ray surface flux f_x should be a function of dB/dt. Thus L_x/L_{BOL} should be proportional to a function of dB/dt times a term that takes into account that T^4_{eff} decreases rapidly towards cooler stars.

How chromospheric and transition region heating depends on dB/dt theoretically is unclear (but see Stein 1980 for recent work). Ayres et al. (1980) have shown empirically that heating rates in transition regions are proportional to those in the chromosphere to the 1.5 power, and that coronal heating rates are proportional to those in the chromosphere to the power of 3. Linsky et al. (1980) also find that the heating rate of transition regions in active stars (normalized to σT^4_{eff}) increases a factor of 100 between solar plages and late dMe stars, whereas the corresponding chromospheric heating rate for the same sample of active stars does not change appreciably with spectral type. Thus a great deal of work needs to be done before we can go from even the schematic dB/dt picture in Figure 1 to an understanding of heating rates.

As previously noted, rapid rotation can result from youth (rotational braking has not proceeded for a long time) or tidal synchronism in short period RS CVn, BY Dra, and W UMa binary systems. Thus synchronous binary systems act as if they are young stars even though many are evolved to subgiants or giants.

6.3. G-M giants and supergiants

I have previously presented evidence that transition region emission lines do not usually appear in giants cooler than about K2 III and in supergiants cooler than about G5 Ib. Also Einstein has not detected X-rays for giants cooler than about K2 III or in any G-M supergiants, and that the X-ray surface flux upper limits in some cases are far smaller than even solar coronal holes. Chromospheres clearly exist, however, into the M stars of all luminosity classes and perhaps even to the coolest luminous M stars.

I suspect that the apparent absence of transition regions and coronae to the right of dividing "lines" in the H-R diagram near spectral type K2 III are real, and not an artifact of insufficiently deep exposures, for the following reasons.

(1) In the absence of closed magnetic fields, a star cannot retain an outer atmosphere when the kinetic energy of a particle exceeds the gravitational potential energy ($kT_{crit} \geqslant GM_*m_p/R_*$). When this condition is violated explosive mass loss should occur (Parker 1980). In fact, slow mass loss occurs in the Sun even

though $T_{cor} \approx 1/5 \ T_{crit}$. For α Boo (K2 III) $T_{crit} \approx 3 \times 10^5$ K and for α Ori (M2 Iab) $T_{crit} \approx 1 \times 10^5$ K. Even if $T_{cor} \approx T_{crit}$, these stars should not emit appreciable X-rays. There is no evidence that the chromospheric heating rate is any different statistically for stars to the left and right of the transition region and coronal dividing "lines." This suggests, but in no way proves, that mechanical energy, which heats transition regions and coronae to the left of the dividing line, instead goes into another mode to the right. Asymmetry dividing "lines" in the same region of the H-R diagram suggest that this mode is rapid mass loss. Under such conditions Mullan (1978) has argued that the wind energy will exceed likely magnetic field energy densities ($\frac{1}{2} \rho v^2 \geqslant B^2/8\pi$), in which case the wind will drag out the field and prevent formation of closed loop structures. The argument is thus self-consistent, but it needs to be worked out in detail.

(2) One condition for thermal stability of an atmosphere is that it respond to a small increase in T by an increase in radiative losses (i.e. $d\Lambda/dt > 0$, cf. Athay 1976). Typical computed cooling curves for solar abundance plasmas (i.e. McWhirter et al. 1977) show $d\Lambda/dt < 0$ for 10^7 K $\gtrsim$ T $\gtrsim$ 1-2 $\times 10^5$ K and $d\Lambda/dt \approx 0$ for 1-2 $\times 10^5$ K $\gtrsim$ T $\gtrsim$ 1-2 $\times 10^4$ K. It is likely that cooling by conduction and thermally-driven winds stabilizes the atmosphere for T $\gtrsim 5\times10^5$ K, but the whole temperature regime $2\times10^4 - 5\times10^5$ K may be thermally unstable or only marginally stable. Thus one would expect coronae hotter than ~5 $\times 10^5$ K (if there is sufficient heating) or a cool outer atmosphere (T $\lesssim 2 \times 10^4$ K), but few if any stars in the unstable intermediate regime. For this region I suggest that the "hybrid" stars like α Aqr and β Aqr, which lie along the dividing "lines," are hybrid in the sense of having two components; an open field cool component in which the mechanical energy input goes into wind flow, and a closed field hot component in which the mechanical energy input goes into heating a transition region and a corona with T_{cor} perhaps in the range $5\times10^5-1\times10^6$ K. These two components coexist on the same star as coronal holes and active regions exist side-by-side on the Sun. Rotation may modulate the importance of these two components with time. Clearly this argument is schematic and should be worked out in detail to test its validity.

7. SOME UNANSWERED THEORETICAL QUESTIONS

I conclude this talk by restating what I consider to be some important questions for future theoretical investigations.

(1) What are the nonradiative heating mechanisms operative in the outer atmospheres of stars? How do they operate in detail and how do they depend functionally on different stellar parameters?

(2) How does the rotation-activity connection work in detail?

(3) Why do the transition region, corona, and asymmetry dividing "lines" occur in the H-R diagram and what should their widths be?

(4) How do magnetic fields control all or almost all phenomena in the outer atmospheres of stars?

(5) Why does the relative input of nonradiative energy into chromospheres, transition regions, and coronae change drastically with spectral type?

I wish to thank Drs. T. R. Ayres, G. S. Basri, W. Cash, T. Simon, R. Stencel, and G. S. Vaiana for discussions and communication of unpublished Einstein and IUE data. I thank the NATO Advanced Study Institute and NASA through grants to the University of Colorado for their generous support of this work.

REFERENCES

Anderson, C.M., Hartmann, L.W., and Bopp, B.W.: 1976, Astrophys. J. 204, p. L51.
Athay, R.G.: 1976, "The Solar Chromosphere and Corona: Quiet Sun" (Dordrecht: Reidel).
Auvergne, M., Le Contel, J.-M., and Baglin, A.: 1979, Astron. Astrophys. 76, p. 15.
Avrett, E.H.: 1980, this volume.
Ayres, T.R.: 1979, Astrophys. J. 228, p. 509.
Ayres, T.R. and Linsky, J.L.: 1980, Astrophys. J., in press.
Ayres, T.R., Marstad, N.C., and Linsky, J.L.: 1980a, submitted to Astrophys. J.
Ayres, T.R. et al.: 1980b, preprint.
Baliunas, S.L., Avrett, E.H., Hartmann, L., and Dupree, A.K.: 1979, Astrophys. J. (Letters) 233, p. L129.
Basri, G.S. and Linsky, J.L.: 1980, submitted to Astrophys. J.
Blanco, C., Catalano, S., and Morilli, E.: 1980, in "Proceedings of the Second Year of IUE," Tubingen, in press.
Böhm-Vitense, E. and Dettmann, T.: 1980, Astrophys. J. 236, p. 560.
Bopp, B.W.: 1980, in "Highlights in Astronomy" 5, p. 847.
Brown, A. and Jordan, C.: 1980, in "Proceedings of the Second Year of IUE," Tubingen, in press.
Carrasco, L., Franco, J., and Roth, M.: 1980, Astron. Astrophys., in press.
Cash, W.: 1980, private communication.
Castor, J.I.: 1980, private communication.
Cassinelli, J.P.: 1979, Ann. Rev. Astron. Astrophys. 17, p. 275.
Chiuderi, C.: 1980, this volume.

de Loore, C.: 1970, Astrophys. Space Sci. 6, p. 60.
Dravins, D.: 1976, in "Basic Mechanisms of Solar Activity," eds.
    V. Bumba and J. Kleczek (Dordrecht: Reidel), p. 469.
Dravins, D.: 1980, Astron. Astrophys. in press.
Dravins, D., Lind, J., and Särg, K.: 1977, Astron. Astrophys.
    54, p. 381.
Dupree, A.K.: 1980, this volume.
Dupree, A.K. and Baliunas, S.: 1979, IAU Circ. No. 3435.
Dupree, A.K. and Hartmann, L.: 1980, in "Stellar Turbulence,"
    eds. D.F. Gray and J.L. Linsky (New York: Springer-Verlag),
    p. 279.
Dyke, H.M.and Johnson, H.R.: 1969, Astrophys. J. 156, p. 389.
Eaton, J.A. and Hall, D.S.: 1979, Astrophys. J. 227, p. 907.
Evans, R.G., Jordan, C., and Wilson, R.: 1975, M.N.R.A.S. 172,
    p. 585.
Freire, R.: 1979, Astron. Astrophys. 78, p. 148.
Freire, R., Czarny, J., Felenbok, P., and Praderie, F.: 1978,
    Astron. Astrophys. 68, p. 89.
Giaconni, R. et al.: 1979, Astrophys. J. 230, p. 540.
Giampapa, M.S., Worden, S.P., Schneeberger, T.J., and Cram,
    L.E.: 1980, Astrophys. J., in press.
Gilman, P.A.: 1980, in "Stellar Turbulence," eds. D.F. Gray and
    J.L. Linsky (New York: Springer-Verlag), p. 19.
Hall, D.S.: 1976, in "Multiple Periodic Variable Stars," ed.
    W.S. Fitch (Dordrecht: Reidel), p. 287.
Harnden, F.R., Jr.: 1980, private communication.
Harnden, F.R., Jr. et al.: 1979, Astrophys. J. (Letters) 234, p.
    L51.
Hartmann, L.: 1980, this volume.
Hartmann, L., Davis, R., Dupree, A.K., Raymond, J., Schmidtke,
    P.C., and Wing, R.F.: 1979, Astrophys. J. (Letters) 233, p.
    L69.
Hartmann, L., Dupree, A.K., and Raymond, J.C.: 1980, Astrophys.
    J. (Letters) 236, p. L143.
Jennings, M.C.: 1973, Astrophys. J. 185, p. 197.
Jennings, M.C. and Dyke, H.M.: 1972, Astrophys. J. 177, p. 427.
Jordan, C.: 1980, in "Highlights of Astronomy," Vol. 5
    (Dordrecht: Reidel), p. 533.
Kelch, W.L., Linsky, J.L., and Worden, S.P.: 1979, Astrophys. J.
    229, p. 700.
Kunkel, W.E.: 1975, in "Variable Stars and Stellar Evolution,"
    eds. V.E. Sherwood and L. Plaut (Dordrecht: D. Reidel), p.
    15.
Lamers, H.J.G.L.M.: 1980, in "The Universe at Ultraviolet
    Wavelengths: The First Two Years of IUE," eds. R.D. Chapman
    and A. Boggess (Greenbelt: Goddard), in press.
LeContel, J.M., Praderie, F., Bijaoui, A., Dantel, M., and
    Sareyan, J.P.: 1970, Astron. Astrophys. 8, p. 159.
Liebert, J., Dahn, C.C., Gresham, M., and Strittmatter, P.A.:
    1979, Astrophys. J. 233, p. 226.

Linsky, J.L.: 1977, in "The Solar Output and Its Variation," ed.
   O.R. White (Boulder: Colorado Assoc. Univ. Press), p. 477.
Linsky, J.L.: 1979, in Report of Commission 36 in Trans. IAU
   XVII A, Part 2, p. 197.
Linsky, J.L.: 1980a, in "Stellar Turbulence," eds. D.F. Gray and
   J.L. Linsky (New York: Springer-Verlag), p. 248.
Linsky, J.L.: 1980b, Ann. Rev. Astron. Astrophys. 18, p. 439.
Linsky, J.L.: 1980c, invited presentation, HEAD/AAS Meeting on
   X-Ray Astronomy, Cambridge, MA, Jan. 1980 (to be
   published).
Linsky, J.L., Bornmann, P.L., Carpenter, K.G., Wing, R.F.,
   Giampapa, M.S., and Worden, S.P.: 1980, submitted to
   Astrophys. J.
Linsky, J.L. and Haisch, B.M.: 1979, Astrophys. J. (Letters)
   229, p. L27.
Linsky, J.L. and Marstad, N.C.: 1980, in "The Universe at
   Ultraviolet Wavelengths: The First Two Years of IUE," eds.
   R.D. Chapman and A. Boggess (Greenbelt: Goddard), in press.
Long, K.S. and White, R.L.: 1980, Astrophys. J. (Letters) 239,
   p. L65.
Lucy, L.B. and White, R.L.: 1980, Astrophys. J., in press.
Marcy, G.W.: 1980, preprint.
McWhirter, R.W.P., Thonemann, P.C., and Wilson, R.: 1975,
   Astron. Astrophys. 40, p. 63.
Mewe, R.: 1979, Space Science Rev. 24, p. 101.
Morrison, D. and Simon, T.: 1973, Astrophys. J. 186, p. 193.
Mullan, D.J.: 1976, Astrophys. J. 209, p. 171.
Mullan, D.J.: 1978, Astrophys. J. 226, p. 151.
Parker, E.N.: 1958, Astrophys. J. 128, p. 664.
Parker, E.N.: 1980, this volume.
Praderie, F.: 1976, Mem. Soc. Astron. Italiana 47, p. 553.
Reimers, D.: 1977, Astron Astrophys 57, p. 395.
Robinson, R.D.: 1980, Astrophys. J., in press.
Robinson, R.D., Worden, S.P., and Harvey, J.W.: 1980, Astrophys.
   J., in press.
Rosner, R. and Vaiana, G.S.: 1980, in "Proceedings of the Inter-
   national School of Astrophysics at Erice," eds. G. Setti
   and R. Giaconni, p. 129.
Skumanich, A.: 1972, Astrophys. J. 171, p. 565.
Skumanich, A.: 1980, this volume.
Skumanich, A., Smythe, C., and Frazier, E.N.: 1975, Astrophys.
   J. 200, p. 747.
Smith, M.: 1979, P.A.S.P. 91, p. 737.
Snow, T.P., Jr.: 1980, in "Proceedings of the Second Year of
   IUE," Tubingen, in press.
Snow, T., Cash, W., and Grady, C.: 1980, Astrophys. J.
   (Letters), in press.
Snow, T.P. and Linsky, J.L.: 1980, in "Proceedings of the Con-
   ference on Astronomy and Astrophysics from Spacelab," eds.
   P.L. Bernacca and R. Ruffini (Dordrecht: Reidel), p. 291.

Stein, R.F.: 1980, preprint.
Stein, R.F. and Leibacher, J.W.: 1980, in "Stellar Turbulence,"
    eds. D.F. Gray and J.L. Linsky (New York: Springer-Verlag),
    p. 225.
Stencel, R.E.: 1978, Astrophys. J. (Letters) 223, p. L37.
Stencel, R.E.: 1980, this volume.
Stencel, R.E. and Mullan, D.J.: 1980, Astrophys. J., 238, p.
    221.
Swank, J.: 1980, preprint.
Thomas, R.N. and Athay, R.G.: 1961, "Physics of the Solar
    Chromosphere" (New York: Interscience).
Topka, K., Fabricant, D., Harnden, F.R., Jr., Gorenstein, P.,
    and Rosner, R.: 1979, Astrophys. J. 229, p. 661.
Tucker, W.H.: 1973, Astrophys. J. 186, p. 285.
Ulmschneider, P.: 1979, Space Sci. Rev. 24, p. 71.
Ulmschneider, P.: 1980, this volume.
Underhill, A.B.: 1980a, Astrophys. J. (Letters) 235, p. L149.
Underhill, A.B.: 1980b, in "The Universe at Ultraviolet
    Wavelengths: The First Two Years of IUE," eds. R.D. Chapman
    and A. Boggess (Greenbelt: Goddard), in press.
Vaiana, G.S.: 1980, invited presentation, HEAD/AAS Meeting on X-
    Ray Astronomy, Cambridge, MA, Jan. 1980 (to be published).
Vaiana, G.S. and Rosner, R.: 1978, Ann. Rev. Astron. Astrophys.
    16, p. 393.
Vaiana, G.S. et al.: 1980, submitted to Astrophys. J.
van der Hucht, K.A., Stencel, R.E., Haisch, B.M., and Kondo, Y.:
    1979, Astron. Astrophys. Suppl. 36, p. 377.
Walter, F.M.: 1980, submitted to Astrophys. J.
Walter, F.M. and Bowyer, S.: 1980, submitted to Astrophys. J.
Walter, F.M., Cash, W., Charles, P., and Bowyer, S.: 1980b,
    Astrophys. J. 236, p. 212.
Walter, F., Charles, P., and Bowyer, S.: 1978, Astron. J. 83, p.
    1539.
Walter, F.M., Linsky, J.L., Bowyer, S., and Garmire, G.: 1980a,
    Astrophys. J. (Letters) 236, p. L137.
Warner, B.: 1966, Observatory 86, p. 82.
Warner, B.: 1968, Observatory 88, p. 217.
Wentzel, D.G.: 1980, in "The Sun as a Star," NASA-CNRS Monogr.
    Ser. Vol. VI A, preprint.
Wilson, O.C.: 1966, Science, 151, p. 1487.
Withbroe, G.L. and Noyes, R.W.: 1977, Ann. Rev. Astron.
    Astrophys. 15, p. 363.
Zahn, J.P.: 1977, Astron. Astrophys. 57, p. 383.

CONVECTION ZONES IN SUN AND STARS

Jean-Paul Zahn

Observatoire de Nice

INTRODUCTION

The convection zones in the Sun and the stars have been in the
center of the discussions of two recent meetings: that of Nice
in 1976 (IAU Colloquium 38), devoted specifically to the problems
of stellar convection, and that of London (Ontario) in 1979
(IAU Colloquium 51), which dealt more broadly with stellar turbulence. In both meetings, there was an active exchange between
people respectively working in stellar and solar physics, much
in the spirit of this summer course. Most of the topics I will
mention have been treated in more depth during these meetings,
and I therefore encourage those who have not yet done so to browse
through their proceedings.

Here we shall ask ourselves why do exist convection zones, how
are they generally described by the mixing length procedure,
why is there a need for a better treatment and what are the main
results of the more recent hydrodynamical approaches. We shall
conclude with a brief review of the relevant observations.

WHY CONVECTION ZONES EXIST.

In the absence of motion, a non-magnetic, non-rotating star
has a spherical symmetry and its radial structure is determined
by the laws of hydrostatic equilibrium

$$dP/dr = - \rho g = - \rho \ GM/r^2 \qquad (1)$$

and of radiative equilibrium

R. M. Bonnet and A. K. Dupree (eds.), Solar Phenomena in Stars and Stellar Systems, 123–136.
Copyright © 1981 by D. Reidel Publishing Company.

$$K \ dT/dr = - F = - L/4\pi r^2 \ .$$ (2)

The classical notations have been used for the radial coordinate r, measuring the distance from the center of the star, and
for the physical variables pressure, density and temperature
(P, ρ, T). g is the value of the local gravity and G the gravitational constant, K the radiative conductivity, F the energy
flux, M and L the mass and the luminosity. The variation with r
of the latter two is given by

$$dM/dr = 4\pi r^2 \rho$$ (3)

$$dL/dr = 4\pi r^2 \rho \ \epsilon \ ,$$ (4)

with ϵ being the energy generation rate, which, together with K
and ρ may be expressed in terms of the state variables P and T.
It is an easy task to integrate the fourth order system (1) - (4),
once the boundary conditions have been specified. This yields a
model of a star in radiative equilibrium, that is uniquely determined by (say) the total mass, for a given chemical composition
(which in the general case is a function of r). Many such models
have been constructed as soon as the gross properties of the stellar material had been established. They were so successful in explaining most of the available observations that it was taken for
granted that all stars should be in radiative equilibrium
(Eddington 1926).

That they were not necessarily so became apparent in the
thirties, when more was learned about the opacity of the stellar
plasma and about the nuclear reactions (and thus about K and ϵ).
It has been known for a while already that a radiative temperature stratification would be stable for small perturbations only
as long as

$$(\partial \ln T/\partial \ln P)_{rad} < (\partial \ln T/\partial \ln P)_{ad}$$ (5)

This stability criterion (Schwarzschild 1906), a derivation of
which is given in all textbooks, is established by imagining that
a parcel of stellar matter undergoes a vertical displacement,
during which it remains in pressure equilibrium with its surroundings and does not exchange heat with them. If this criterion is
not satisfied, the unstable layer becomes the source of motions
which, when vigorous enough, are of turbulent character, as can
be observed in the laboratory and in our atmosphere. The quanti-
ty $\nabla_{ad} = (\partial \ln T/\partial \ln P)_{ad}$, which is called the adiabatic temperature gradient, is governed by the thermodynamical properties
of the matter; for a fully ionized gas, $\nabla_{ad} = 0.4$. As for the
radiative gradient, it can be evaluated from Eq. (1) and (2) as

$$\nabla_{rad} = (\partial \ln T / \partial \ln P)_{rad} = \frac{1}{4\Pi \ G \ K} \ \frac{L}{M} \frac{P}{\rho T} \ . \tag{6}$$

This expression helps to identify the two circumstances under which the radiative gradient becomes greater than the adiabatic one, thereby creating an unstable situation. Firstly, the ratio L/M, which is nearly constant in the outer layers of a star, increases towards its center because the energy generation rate is an increasing function of the density and, even more so, of the temperature. If ε is sensitive enough to the temperature, the stellar core is bound to become convective. Secondly, a sharp increase in the opacity (and thus a decreases in the conductivity K) can also push the radiative gradient ∇_{rad} above its critical value. This situation tends to occur in the outer layers of a star, as we will see next.

ENVELOPE CONVECTION.

The precise value of ∇_{rad} at each depth of a radiative model must be determined by performing the integration of the differential system stated above. Near the surface of a star, this system reduces fortunately to the second order, since both M and L are very nearly constant there. Let us simplify it even further by neglecting the r^{-2} variation of the gravity g and of the flux F, and let us introduce the optical depth τ, which is related to r by

$$d\tau = -\chi \rho \ dr \tag{7}$$

(χ being the Rosseland mean opacity). The equation of hydrostatic equilibrium can then be formally integrated as

$$P = \int_{o}^{\tau} g/\chi \ d\tau \ = g\tau \ <\chi^{-1}> \tag{8}$$

where $<\chi^{-1}>$ is the mean of χ^{-1} over the interval $(0,\tau)$. The equation of radiative transfer, of which Eq. (2) is the asymptotic form valid in the deep interior, is known to have the solution

$$T^4 = 3/4 \ T_e^{\ 4}(\tau + 1/\sqrt{3}) \quad \text{for} \quad \tau \gg 1, \tag{9}$$

T_e being the effective temperature (see for instance Chandrasekhar 1960 or Unsöld 1955). Therefore, taking the logarithmic derivatives of Eq. (8) and (9),

$$\nabla_{rad} = \frac{1}{4} \frac{\tau}{\tau + 1/\sqrt{3}} \frac{<\chi^{-1}>}{\chi^{-1}} \tag{10}$$

If the opacity were constant, ∇_{rad} would increase monotonically from zero to its asymptotic limit 0.25, and it would thus never reach the critical value $\nabla_{ad} = 0.4$ (solid line in Fig. 1). But when χ increases strongly enough at some depth of the envelope, the local χ^{-1} is much smaller than the mean $<\chi^{-1}>$, and the radiative gradient becomes supercritical (dashed line in Fig. 1).

In many stars, the opacity actually increases with depth in this way, because the rising temperature excites more and more energy levels of the atoms present, thereby increasing the probability of photon absorption until the atoms are completely ionized. Since hydrogen is the major constituant of normal stars, the opacity goes through a sharp maximum when about half of the hydrogen is ionized. At the same time, the adiabatic gradient ∇_{ad} is substantially lowered because most of the energy, in an adiabatic compression, serves for the ionization and contributes much less to the temperature increase. Due to both effects, all late type stars, which are "cold" enough to contain partially ionized hydrogen, possess a convective envelope.

THE MIXING LENGTH TREATMENT.

When convection sets in, according to the Schwarzschild criterion, the temperature stratification can no longer be inferred from the radiative equilibrium, and the model builder must replace Eq. (2) by another one, which will predict this gradient in a convective region. Up to now, only approximate methods are available to estimate this convective temperature gradient, the most commonly used being, by far, the mixing length treatment.

The energy flux carried by the convective motions consists of two parts [1]: the enthalpy flux and the kinetic energy flux. The first is given by

$$F_c = \overline{\rho C_p T' w} \quad , \tag{11}$$

assuming, for simplicity, that one deals with a perfect gas. (T' is the temperature fluctuation, w the vertical velocity and C_p the heat capacity at constant pressure; the bar means that a horizontal average has been taken). As for the kinetic energy flux

$$F_K = 1/2 \overline{\rho V^2 w} \quad , \tag{12}$$

(V being the total velocity), it is usually neglected in the mixing length treatment.

In this treatment, the temperature fluctuations are estimated through

$$T' \sim (|dT/dr|_{ad} - |dT/dr|) \, \ell \quad , \tag{13}$$

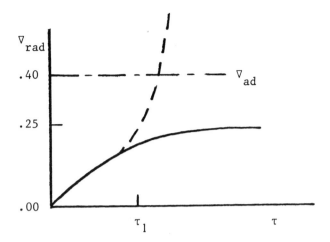

Figure 1. The radiative temperature gradient ∇_{rad} versus the optical depth τ in two cases: a) the opacity is constant (solid line), b) the opacity increases rapidly around $\tau = \tau_1$ (dashed line).

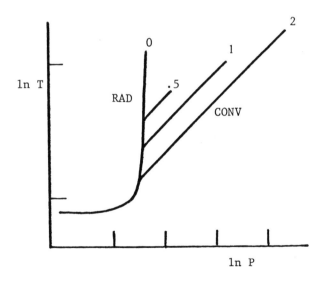

Figure 2. Envelope models in the (ln P, ln T) plane. The numbers refer to the mixing length parameter α. Notice that the extent of the convection zone increases with α.

ℓ being the mixing length, which can be visualized as the mean free path of the convective elements (or eddies). Introducing u, a typical value of the vertical velocities, one gets the follo- wing expression for the enthalpy flux – which is assumed to stand for the total convective flux –

$$F_c = (\rho \, C_p \, T/H) \; u.\ell \; (\nabla_{ad} - \nabla) \quad ; \tag{14}$$

H is the pressure scale height $|dr/d \ell n \, P|$ and ∇ the actual tem- perature gradient d ℓn T/d ℓn P.

The velocity u is estimated in the same way as the tempera- ture fluctuations:

$$u^2 \sim g \, \ell \rho'/\rho \tag{15}$$

with

$$\rho'/\rho \sim (\nabla_{ad} - \nabla) \; \ell/H \quad , \tag{16}$$

and therefore only the mixing length remains to be specified. We refer to Bierman (1977) and Gough (1977) for the reasons which led to the choice which is most commonly made nowadays, namely

$$\ell = \alpha \, H \tag{17}$$

with α being a constant parameter of order unity. The convective flux is thus approximated by

$$F_c = \rho \, C_p \, T \, (gH)^{1/2} \, \alpha^2 \, (\nabla_{ad} - \nabla)^{3/2} \quad , \tag{18}$$

an expression which depends only on the local values of the tem- perature gradient ∇ and on the two thermodynamic variables P and T for example. The total energy flux can now be written as the sum of the convective flux above and of the radiative flux that is still present, namely KT ∇/H, and it is easy to solve that equation for the temperature gradient ∇.

This is how the mixing length procedure as been introduced in astrophysics by Biermann (1933, 1938) and Öpik (1938). It is essentially in this form that it is still in use, were it not for an improvement brought by Vitense (1953) to take into account the heat diffusion.

When applied to the Sun, the method predicts that the tempe- rature gradient is nearly adiabatic in the unstable region, ex- cept in a tiny layer whose thickness depends on the parameter α and over which ∇ switches from ∇_{rad} to ∇_{ad}. Due to the steep radiative gradient (itself a consequence of the sharp increase in opacity, as we have seen), the adiabat which is finally attained critically depends on the choice of α, as does the extent of the convection zone (Fig. 2). Since the properties of a solar model (luminosity, radius) are very sensitive to the size of the convective region, it is possible to calibrate the parameter α by comparing a given model with the observations. The outcome is that for the generally admitted chemical compositions, α is of order unity [2]. Many astrophysicists have taken this result as a proof of the relevance of the mixing length treatment and have extrapolated this value of α to other stars without encountering much problems when comparing their models with the observations. Thus, one may ask, why bother to improve the theory of stellar convection?

THE NEED FOR A BETTER TREATMENT OF CONVECTION.

To answer that question, let us examine what are the effects to be expected from the convective motions. There are essentially three of them.

Firstly, one of the main properties of the convective instability is to transform thermal energy into kinetic energy, of which most is dissipated again in heat in the convection zone itself. However a fraction of the convective energy (and momentum) can escape from the unstable domain in the form of waves and can therefore be dumped in other parts of the star. This mechanism has been invoked for the heating of chromospheres and coronae, and some astrophysicists hold it responsible for the driving of stellar winds.

Secondly, the convective motions transport not only matter and energy, but also magnetic field and angular momentum; they are thus the probable cause of the magnetism of a star and of its so-called activity.

Thirdly, as a consequence of the energy transport, convection remodels the temperature stratification.

Only this latter effect appears to be adequately described by the mixing length treatment, as we have seen above, once the parameter α has been calibrated against the observations. The properties of the convective velocity field are too badly known through this procedure to allow for even the crudest predictions concerning the generation of waves. On the other hand, this procedure has been originally designed to handle turbulent transport, and for that reason it is worth spending some time examining how this problem is treated.

As is well known, the transport of a given physical agent, whose concentration is s, is governed by an equation of the type

$$\partial s / \partial t + \underline{V} . \nabla s = \{...\} \tag{19}$$

where the right hand side contains all source, sink and diffusion terms. For the sake of simplicity, the quantity s is assumed to be a scalar here, such as the specific entropy for example. To simplify even further, let us consider the extreme case where those terms on the right hand side are negligible, implying that the quantity s is conserved. If the velocity $\underline{V}$ were known at each point in space and time, it would be a relatively simple matter to follow the evolution of the quantity s. But, when the flow is more or less turbulent, all that is available are some estimates (measures, at best) concerning the statistical properties of the fluid motions, and it is no longer possible to solve the "deterministic" equation (19). When the motions are sufficiently disorganized - let us say turbulent, without opening here the everlasting debate among specialists on what turbulence really is - then one often observes that the quantity s just diffuses away. It is therefore tempting to follow the mixing length procedure, and to replace the advection operator V in Eq. (19) by a diffusion operator, which leads to the Laplace-type equation :

$$\partial s / \partial t - \nabla . K_t \nabla s = 0 \quad . \tag{20}$$

The diffusion coefficient K_t is given by $u\ell$, where u is a typical velocity of the turbulent eddies and ℓ their mean free path, or mixing length. The expression given above for the enthalpy flux (Eq. 14) can be derived precisely in this way from the exact equation of conservation of heat (or entropy).

The mixing length approach has scored some successes, but its limitations are well known to fluid dynamicists, especially when the advected agent is a vector field and/or when the motions have some degree of organization. (There are cases, for instance, where s is found to diffuse against its gradient). Such an organization is certainly present in thermal convection, where the driving force acts only in one direction, the vertical. But it should be mentioned that even in the ideal case of homogeneous isotropic motions, the mixing length treatment does not escape criticism (see Zahn 1980).

That is why more refined procedures have been designed to deal with turbulent flows and much of the specialized literature is devoted to them (for a classical and comprehensive treatise, see Tennekes and Lumley 1971). The most commonly used, which involve high order closures of the equations, have not yet been applied, as such, to stellar convection mainly for two reasons. Firstly, the astrophysical case is much more complicated than the laboratory case: the motions spread over many scale heights, the stellar matter undergoes ionization, the diffusivity of heat is much larger than that of momentum, etc., ... It would be a formi-

dable task to include all these ingredients in a numerical code
of the type currently used in fluid dynamics. Secondly, these
closure theories involve a whole set of dimensionless parameters
which are calibrated in the laboratory, but whose universal character is not established enough to confidently allow their extrapolation to the stellar conditions.

In the meanwhile, two lines of action have developed among
the astrophysicists. One is to endeavour to salvage the mixing
length procedure by fixing its most feeble points (see Gough 1977).
The other is to stay as close and as long as possible to the fluid
dynamics equations and to clearly identify the approximations that
one is obliged to make in order to render the problem tractable.
It is this the latter approach that will be described next.

HYDRODYNAMICAL APPROACHES.

It is not necessary to rewrite here the classical equations
which govern the fluid motions in a star; they may be found in
various papers and textbooks (for instance Ledoux and Walraven
1958, Gough 1969, Latour et al. 1976, Zahn 1980). To solve them,
the straightforward method is, of course, to set up a numerical
code working with the finest grid possible in three dimensional
space (Graham 1977). An alternate method is to treat the horizontal dimensions in Fourier space, on the assumption that the
motions are more or less homogeneous on horizontal surfaces; it
is the so-called modal approach (Gough et al. 1975, Latour et al.
1976). In any case, only a small set of scales can be simulated
with the presently available computers, due to still insufficient
speed and memory. However, and this is very important, one can
choose these scales to be those which transport most of the convective flux. Therefore the results concerning the large-scale
properties of the convective region are based on much firmer
grounds than with the mixing length procedure. This remains true
even though one cannot yet avoid the use of this procedure, or of
a similar one, when it comes to estimate the effects of the smaller scales of motions. The way these sub-grid scale effects are
treated generally is to introduce eddy transport coefficients, as
explained above, but with a diffusivity (and thus a mixing length)
characterizing the small scales. In fact, when the large scales
are well chosen, such that they transport most of the energy, the
eddy transport of heat can be neglected and just remains the eddy
transport of momentum, for which the fluid dynamicists have derived various treatments.

Many calculations have been carried out in the Boussinesq
approximation, in which the compressibility of the fluid is neglected except in the buoyancy force (see Spiegel 1971). They
served to test the validity of the methods used and especially of

the truncations that are made in keeping only a few scales. But
their results cannot be easily transposed to the stellar case, in
which the convective regions often span several density scale
heights. This is why several attempts have been made in the last
few years to simulate compressible convection; their main characteristics are summarized in Table 1 below.

TABLE 1.- Main investigations in compressible convection.

Problem treated	Method used	Authors
Polytropic atmosphere between horizontal plates	Grid (x, z) Grid (x, y, z) 1 and 2 Mode (x, y), Grid (z) 1 Mode (x), Grid (z) Analytical (x, z)	Graham 1975 Graham 1977 Massaguer and Zahn 1980 Van der Borght 1980 Depassier and Spiegel 1980
Sun	1 Mode (x), Grid (z) 1 Mode (x), Grid (z)	Van der Borght 1975, 1979 Nordlung 1980 Nelson and Musman 1977, Nelson 1978
A type star	1 Mode (x, y), Grid (z)	Toomre et al. 1976
F type star	1 Mode (x), Grid (z)	Nelson 1980

Only nonlinear calculations have been reported.
z is the vertical coordinate, x and y are the
horizontal ones. For instance, the designation
1 Mode (x), Grid (z) refers to a two-dimensional
calculation with one mode in x and a grid in z.

These calculations are probably far from having elucidated
all the problems and they all use some assumptions that remain to
be justified. But even at this stage, they shake some accepted
ideas on stellar convection. Let us state the most salient results;
for a detailed account, we refer to the original papers.

1.   The convective cells extend over several scale heights and

do not break up in the vertical as is assumed in the mixing
length treatment (Graham 1975, 1977; Massaguer and Zahn 1980).
2.   The pressure fluctuations are of the same order as the density and temperature fluctuations and this leads to an inversion
of the buoyancy force in the upper part of the unstable domain
(Massaguer and Zahn 1980; Glatzmaier and Gilman 1980; Latour et
al. 1981).
3.   Convective overshooting into the stable adjacent layers may
be important; it can, for instance, link two neighbouring unstable zones (Toomre et al. 1976; Nelson 1978, 1980; Nordlung 1980;
Latour et al. 1981).

     Preliminary as they are, these results question the mixing
length procedure which is broadly used in stellar structure theory and they should encourage more astrophysicists to participate
in this search for a better description of stellar convection.

OBSERVATIONS BEARING ON STELLAR CONVECTION.

     What are the observations with which one can compare the
theoretical predictions? During the Nice Colloquium, Böhm (1977)
and Böhm-Vitense (1977) reviewed them thoroughly; they came to
the conclusion that only a few observational tests were precise
enough to permit the choice between this or that theory. During
the last few years, the most important progress has been scored
in the field of the solar oscillations. The pioneering investigation by Deubner (1975, 1979) concerning the small scale oscillations has now been extended to the global scale (Claverie et al.
1980; Grec et al. 1980). The latter observations, which were
over 120 hours without interruption (see Fossat's lecture), have
reached such a precision that little doubt is left that the solar convection zone has indeed a thickness of about $2 \ 10^5$ km
(Christensen-Dalsgaard and Gough 1980). One is on the verge of
being able to measure the degree of adiabacy of the convection
zone.

     The velocity field associated with the supergranulation is
also better known thanks to observations which are carried out
both in space and from the ground (November et al. 1979). Some
evidence has been gathered for the existence of a mesogranulation,
whose scale is intermediate between granulation and supergranulation (Toomre 1979). There is no doubt that these observations will
provide crucial tests for the hydrodynamical models.

     For the stars, what is available reduces to more or less
indirect proofs of the existence of outer convection zones. The
presence of a chromosphere or a corona may be such a proof, if
one takes for granted that a convection zone is the only possible
source for the heating of those outer regions (see the lecture

by Linsky). The slow rotation of the late type dwarfs (when they
are single) has also been ascribed to present or past magnetic
activity and therefore to the presence of a substantial convection
zone (see Skumanich). There is also some hope in using the global
high-period oscillations, as for the Sun; they should become
detectable soon in the brightest stars.

Another observable parameter, which also yields some information on the existence of convection zones, has received little
attention so far: it is the eccentricity of the orbits of close
(but detached) binary stars. The tidal interaction between the
components of a binary system tends to circularize the orbit and
this phenomenon is characterized by a time scale which depends on
the dissipative processes that are at work. The turbulent viscosity in convective envelopes appears to be by far the most efficient
mechanism and therefore a close binary which contains one component (or two) that has such an envelope is much more likely to
have a circular orbit than a system with two stars possessing
radiative envelopes (Zahn 1966, 1977). The data on eclipsing binaries confirms this; they show that all systems with secondaries
of less than about 1.6 $M_{\odot}$ have circular orbits, and one can therefore conclude that up to that mass the considered stars possess
a convective envelope. The only exception is α CrB, but it confirms the rule: its period is extremely long for an eclipsing
binary (17.36 days) and the time scale for circularization
($\sim 10^{12}$ years) is much larger than the age of the Universe. This
test has been made originally on a relatively small sample of
close binaries; it would be worthwhile to repeat it with the ri-
cher and better data that are now available and to extend it to
the giants and subgiants.

Notes.

(1)   The so-called viscous flux is negligible here.

(2)   We oversimplify on purpose here: the actual comparison involves also the age of the model; furthermore, other dimensionless parameters come into play, but those are of lesser importance. Also, α is of order unity in the Biermann-Vitense formulation but it is not in that of Öpik (see Gough and Weiss 1976).

BIBLIOGRAPHY.

Problems of Stellar Convection (IAU Coll. 38), ed. Spiegel, E.A.
        and Zahn, J.P., Lecture Notes in Physics 71 (Springer,
        Heidelberg).
Stellar Turbulence (IAU Coll. 51), ed. Gray, D.F. and Linsky,
        J.F., Lecture Notes in Physics 114 (Springer, Heidelberg).

Biermann, L. 1933, Z. Astrophys. 5, 117
Biermann, L. 1938, Astron. Nachr. 264, 395
Biermann, L. 1977, Problems of Stellar Convection, 4
Böhm, K.H. 1977, Problems of Stellar Convection, 103
Böhm-Vitense, E. 1977, Problems of Stellar Convection, 63
Chandrasekhar, S. 1960, Radiative Transfer (Dover, New York)
Christensen-Dalsgaard, J., Gough, D.O. 1980 (preprint)
Claverie, A., Isaak, G.R., McLeod, C.P., Van der Raay, H.B.,
     Rocca-Cortes, T. 1980, Nonradial and Nonlinear Stellar Pulsa-
     tion (ed. Hill, H.A. and Dziembowski, W.A.), 181 (Springer,
     Heidelberg)
Depassier, M.C., Spiegel, E.A. 1980 (to appear in Astron. J.)
Deubner, F.L. 1975, Astron. Astrophys. 44, 371
Deubner, F.L., Ulrich, R.K., Rhodes, E.J. Jr 1979, Astron.
     Astrophys. 72, 177
Eddington, A.S. 1926, Internal Constitution of Stars, 98 (Dover)
Glatzmaier, G.A., Gilman, P.A. 1980 (preprint, submitted to
     Astrophys. J.)
Gough, D.O. 1969, J. Atmos. Sci. 26, 448
Gough, D.O. 1977, Problems of Stellar Convection, 15
Gough, D.O., Spiegel, E.A., Toomre, J. 1975, J. Fluid Mech. 68,
     695
Gough, D.O., Weiss, N.O. 1976, Mon. Not. R. Astron. Soc. 176, 589
Graham, E. 1975, J. Fluid Mech. 70, 689
Graham, E. 1977, Problems of Stellar Convection, 151
Grec, G., Fossat, E., Pommerantz, M. 1980 (to appear in Nature)
Latour, J., Spiegel, E.A., Toomre, J., Zahn, J.P. 1976, Astro-
     phys. J. 207, 233
Latour, J., Toomre, J., Zahn, J.P. 1981 (preprint, submitted to
     Astrophys. J.)
Ledoux, P., Walraven, Th. 1958, Handbuch der Physik, vol. 51
Massaguer, J., Zahn, J.P. 1980, Astron. Astrophys. 87, 315
Nelson, G.D. 1978, Solar Phys. 60, 5
Nelson, G.D. 1980, Astrophys. J. 238, 659
Nelson, G.D., Musman, S. 1977, Astrophys. J. 214, 912
Nordlung, A. 1980, Stellar Turbulence, 17
November, L., Toomre, J., Gebbie, K.B., Simon, G.W. 1979, Astro-
     phys. J. 227, 600
Öpik, E.J. 1938, Publ. Obs. Astron. Univ. Tartu 30, n°3
Schwarzschild, K. 1906, Nachr. Königl. Ges. Wiss., Göttingen, n°1
Spiegel, E.A. 1971, Ann. Rev. Astron. Astrophys. 9, 323
Tennekes, H., Lumley, J.L. 1972, A First Course in Turbulence
     (M.I.T. Press)
Toomre, J. 1980, Highlights of Astronomy 5, ed. Wayman, P.A.
     (Reidel) 571.
Toomre, J., Zahn, J.P., Latour, J., Spiegel, E.A. 1976, Astro-
     phys. J. 207, 545
Unsöld, A. 1955, Physik der Sternatmosphären (Springer, Berlin)
Van der Borght, R. 1975, Mon. Not. R. Astron. Soc. 173, 85
Van der Borght, R. 1979, Mon. Not. R. Astron. Soc. 188, 615

Van der Borght, R., 1980 (preprint)
Vitense, E. 1953, Z. Astrophys. $\underline{32}$, 135
Zahn, J.P. 1966, Ann. Astrophys. $\underline{29}$, 565
Zahn, J.P. 1977, Astron. Astrophys. $\underline{57}$, 383
Zahn, J.P. 1980, Stellar Turbulence, 1

RESULTS FROM THE CENTRAL HYADES SURVEY

R.A. Stern,[*] J.H. Underwood,[*] M.C. Zolcinski,[**] and
S.K. Antiochos[**]

[*] Jet Propulsion Laboratory, Pasadena, California
[**] Stanford University, Stanford, Alifornia.

Abstract. As participants in the HEAO-2 (Einstein Observatory) guest
investigator program, we have conducted a soft x-ray survey of
the central $\sim5°$ of the Hyades cluster. Using the Imaging Proportional Counter (IPC) on Einstein, over $\lesssim40$ discrete sources were
detected, most of which have bright ($V\sim10$) stars as optical counterparts. X-ray luminosities (L_x) for these stars are in the
range $10^{28.5}-10^{30}$ erg s^{-1}. From the fraction of Hyades detected
($\sim50\%$), and the observed range of L_x, we may infer that stellar
coronae are common phenomena in the cluster. The level of x-ray
emission among the late F and early G stars in the Hyades is
~30 times that of the sun viewed as a star. The relative youth
and thus higher rotational velocities of the Hyades dwarfs compared to the sun is probably the most significant factor in producing such enhanced coronal activity. Comparison of our survey
results with data in other wavelength regions and with other
stellar surveys on Einstein is beginning to prove fruitful in
understanding physical processes in coronae.

I. INTRODUCTION

    The recent rapid expansion in our knowledge of the frequency
and properties of stellar coronae has come almost exclusively
from x-ray observations. In particular, the CfA survey of stellar
coronae with the Einstein Observatory (1), and other coronal surveys (2,3) have demonstrated the pervasive nature of x-ray emission in virtually all stellar spectral types and a wide range of
luminosity classes. While most of the observations referred to
above have concentrated on stars of a given spectral class, nearby stars, or stars brighter than a certain limiting magnitude,

R. M. Bonnet and A. K. Dupree (eds.), Solar Phenomena in Stars and Stellar Systems, 137–144.
Copyright © 1981 by D. Reidel Publishing Company.

an alternative approach is to survey a star cluster, with the
concomitant advantages of known stellar distances, equal ages,
and homogeneous composition, which optical astronomers have long
used to better understand the characteristics of stellar populations. The Hyades star cluster has been extensively studied in
visible light; this same cluster is an excellent choice, too,
for an x-ray survey, because of its proximity (d$\sim$45pc), and lack
of visible absorption (which suggests little x-ray absorbing gas)
in the direction of the cluster (4). Since the cluster contains
A2 and later type main sequence stars, four K0 giants, and a number of white dwarfs, it also provides an excellent sample for
studying coronae of stars significantly younger ($\tau \sim 4-8 \times 10^8$ yr)
than the sun.

II. OBSERVATIONS

Observations of twenty-seven fields in the central region
of the Hyades were carried out with the IPC on Einstein from
March 1979 to March 1980. Exposure times ranged from 1000-4000
seconds for each observing field of $\sim 1° \times 1°$ extent. The IPC is
sensitive to x-rays of energies from 0.15-4.5 keV, and yields
information on source intensity, variability, and to some extent,
spectrum; in this contribution we will discuss only data on the
first of these properties. Details of the design and operation
of the Einstein Observatory may be found elsewhere (5).

In Figure 1 we show the location of the observing fields
superimposed on a reproduction of Becvar's (6) stellar atlas. A
composite image of the actual survey fields is shown in Figure
2; most of the individual stellar sources are clearly discernible
in this photograph.

Figure 3 is an H-R diagram for the cluster stars in our observing fields, indicating those detected in soft x-rays. Although the fraction of Hyads (V$\sim$16) detected for the entire survey is $\sim$50%, virtually all of the late F and early G main sequence stars have been detected above our sensitivity threshold
of $\sim 10^{-13}$ erg cm^{-2} s^{-1}. This roughly corresponds to an x-ray
luminosity, L_x, of $10^{28.5}$ erg s^{-1} at the Hyades distance of 45pc.

After correcting for individual stellar distances using published proper motion data (7), we have plotted in Figure 4 L_x vs.
the color index, B-V. For purposes of comparison, we have also
included the range of observed L_x for the sun viewed as a star.
The inescapable conclusion is that the typical solar type star in
the Hyades is emitting soft x-rays at a level of $\sim$30 times that
of the active sun. The factor of 10 or so range of L_x at solar
type is probably consistent with our observing the Hyades star
at various stages in their activity cycle: we will attempt to

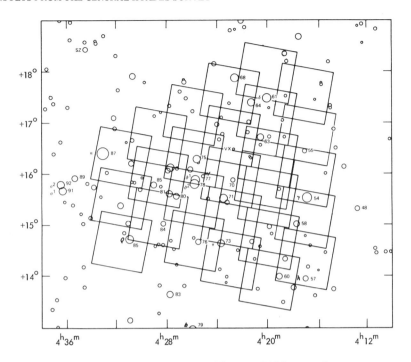

Figure 1. Location of survey fields in 1950 coordinates. Size of
circles indicates visual magnitudes.

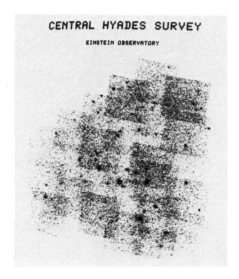

Figure 2. Composite x-ray image of central Hyades region.
Darker areas on image are brighter in x-rays.

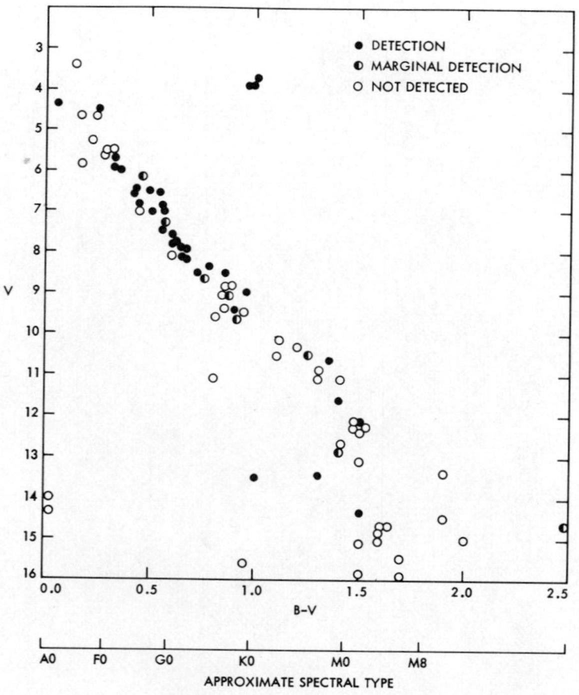

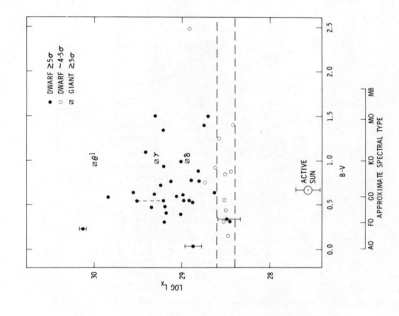

Figure 3. H-R diagram for Hyads in survey fields, with x-ray detections indicated "Marginal" is ~4– 5σ above adjacent background (see Figure 4)

Figure 4. X-ray luminosity (0.15–4.5 keV) of Hyads detected. Typical ±1σ errors are indicat- ed. Dashed line region is approximate sensiti- vity limit.

confirm this hypothesis with series of follow-up observations of the cluster on Einstein scheduled in the coming year. Some evidence for coronal variability already exists in the case of BD + 14°690, a G0 star for which we have observations 6 months apart. The star, however, is a spectroscopic binary (P = 4d.), making the interpretion of variability somewhat complicated. Another noteworthy star is the fast (v sini $\sim$200 km s^{-1}) rotator 71 Tau, which, although of spectral type A8-F0, is the most luminous x-ray source in the cluster. In this case, at least, it appears as though a well developed convection zone is not a necessary condition for the formation of a corona.

In Figure 5 we have computed the ratio log L_x/L_{bol} (where L_{bol} is the stellar luminosity) for the main sequence Hyads detected, and have plotted histograms of this parameter as a function of spectral type. The solar value of this parameter is indicated for comparison. The distinction between late A- early F stars, which are expected to possess little or no convective envelope, and the solar type stars (F8-G5) with convective outer atmospheres, is seen in the figure, but in a rather gradual way. The majority of the K0 and later type stars lie below our detection threshold, and thus we cannot estimate a typical log L_x/L_{bol}; however, the maximum value of this parameter is evidently an increasing function of spectral type, suggestive of the influence of the convection zone properties of these stars.

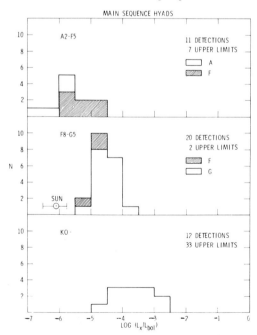

Figure 5. Histogram of x-ray to bolometric luminosity ratio.

Of the four known Hyades giants, three were observed during the survey, and all three were detected at varying levels of L_x, as can be seen in Figure 4. IUE observations of these same three giants appear to show a correlation in the strength of Si IV and N V high temperature transition region lines with the strength of x-ray emission (8).

III. DISCUSSION

Several points of comparison are worth making between the Hyades survey and studies of other stellar coronae. First, the factor of 30 in L_x and hence L_x/L_{bol} between the solar type Hyads and the sun is undoubtedly an indication of the strong influence of rotation on coronal activity; such a relation for chromospheric activity was originally pointed out by Wilson, Kraft, Skumanich and others (9,10,11). The Hyades G stars are similar to the sun in their other properties (12), but the equatorial rotational velocity of the sun is ~ 2.1 km s^{-1}, while for solar type stars in the Hyades it is ~ 10 km s^{-1} (13,14). If one naively looks for a scaling relation, the level of coronal activity, as measured by L_x/L_{bol}, appears to depend upon the square of the stellar angular velocity, Ω. This form of the dependence has been suggested earlier for highly active coronae of single stars observed on HEAO-1 (15), and is consistent with other recent x-ray data (16). On the other hand, a linear dependence of L_x/L_{bol} with Ω has been proposed for coronae of the RS CVn stars, as well as for a sample of other single stars (3). It is perhaps too early to suggest that a simple scaling relation exists; however, the correlation of coronal activity and stellar rotation is firmly established by the Einstein x-ray data. The dependence of L_x and L_x/L_{bol} on Ω will be an important input to theories of coronal heating, since in standard dynamo models, the rate of generation of magnetic flux is proportional to Ω.

A second area of comparison is with the results of the Einstein/CfA stellar survey (1). Although one must be somewhat cautious, since the CfA stellar sample of late-type stars with known luminosity class is not large at this time, a number of points can be made. In particular, both surveys establish the prevalence of stellar coronae throughout the main sequence and giant regions of the HR diagram, and, to a lesser extent, the lack of strong coronae among the white dwarfs (One important point to note is that although only two Hyades white dwarfs were observed - and not detected - with our survey, many more white dwarfs were in our observing fields. Upper limits to x-ray fluxes from these objects will be reported elsewhere). In the case of late-type main sequence stars, the F dwarfs in each sample have about the same mean log L_x, i.e. ~ 29, but the Hyades G stars also emit at roughly this level, as contrasted with the CfA survey G stars which have a mean log L_x of only $\sim$28. It should be pointed out that the number of stars in the CfA G dwarf category is only $\sim$7, with a spread of about 3 orders of magnitude; the spread is only about 1 order of magnitude for the Hyades G dwarfs, of which there are 17. Also, the fraction of undetected G dwarfs in the CfA survey is not tabulated, although this would tend to exaggerate the discrepancy. In the case of the K and later type main sequence stars, both surveys are sensitivity limited, but it is clear that the maximum L_x for these late type dwarfs is $\sim$10^{30} erg s^{-1} with indications of a wide spread in L_x, perhaps several orders of magnitude.

Part of the difference between the Hyades and CfA surveys must be the fact that the CfA survey is sampling stars of differing ages, in light of the age-rotation-activity correlation mentioned earlier.

If the difference between F and G stars in the field and the Hyades is confirmed, it could be the result of the lack of strong rotational braking in the early F stars hence a lack of strong dependence of coronal activity on age, unlike in the late F and early G stars, where the age-rotation effect is strong. The inclusion of a larger sample of stars in the CfA survey will do much to clear up the true difference between the field star and cluster populations.

Other cluster surveys are planned or are being carried out with Einstein. Of special interest is the survey of very young stars in the Orion nebula, in which $\sim$ 120 stellar sources have been identified. Though the fraction of F and later type stars in the nebula detected as x-ray sources is only about 25-50%, the maximum L_x for the late-type Orion stars is $\gtrsim$10$^{31.5}$, substantially higher than seen in either the CfA or Hyades surveys (2). Again, the relative youth of the Orion stars, and hence their rapid rotation, is most likely the key factor in the enhanced coronal activity.

More extensive cluster surveys should help us in decoupling the influences of rotation, convection, and other factors in coronal heating processes. The significant contributions made by such soft x-ray observations to our understanding of the solar corona and of stellar coronae have, we believe, been clearly demonstrated by the results of the Central Hyades and other stellar surveys.

We acknowledge useful discussions with A. Skumanich. This work was performed at the Jet Propulsion Laboratory under NASA contract NAS 7-100, and with much help from F. Seward, R. Harnden, P. Henry, L. Van Speybroeck and others associated with the Einstein Observatory.

REFERENCES

1. Vaiana, G.S. et al., submitted to the Astrophys. J., March
    1980.
2. Chanan, G.A., 1980, private communication.
3. Walter, F.M., 1980, private communication.
4. Allen, C.W., 1973, Astrophysical Quantities, Athlone Press,
    London, p. 278.
5. Giacconi, R., et al., 1979, Astrophys. J., 230, 540.
6. Becvar, A., 1974, Atlas Eclipticalis, Sky Publishing,
    Cambridge.
7. Hanson, R.B., 1975, Astron. J., 80, 379.
8. Baliunas, S.L., Hartmann, L., and Dupree, A.K., 1980, in
    "The Universe at Ultraviolet Wavelengths - The First Two
    Years of IUE", NASA, in press.
9. Wilson, O.C., 1966, Astrophys. J., 144, 695.
10. Kraft, R.P., 1967, Astrophys. J., 150, 551.
11. Skumanich, A., 1972, Astrophys. J., 171, 565.
12. Cayrel de Strobel, G., 1980, in "Star Clusters", J.E. Hesser,
    ed., IAU Symposium: No. 85, D. Reidel, Holland, p. 91.
13. Kraft, R.P., 1965, Astrophys. J., 142, 681.
14. Soderblom, D., 1980, Ph. D. Thesis, University of California,
    Santa Cruz.
15. Skumanich, A., 1979, B.A.A.A.S., 11, 624.
16. Ayres, T.R., and Linsky, J.L., 1980, submitted to Astrophys.
    J.

COMPRESSIBLE CONVECTION IN A ROTATING SPHERICAL SHELL

Gary A. Glatzmaier and Peter A. Gilman

High Altitude Observatory
National Center for Atmospheric Research
Boulder, Colorado

Abstract. Giant cell stellar convection is modeled by solving the
fluid equations for a compressible, rotating, spherical, fluid
shell. A large part of the motivation is to understand the maintenance of the two major ingredients in solar dynamo theory, that
is helicity and differential rotation. An anelastic approximation
filters out sound waves but permits the investigation of the
effects of large density stratifications in slightly superadiabatic stellar envelopes. Various rotation rates, convection zone
depths, density stratifications, boundary conditions, viscosities,
and conductivities are considered. The results of first order
numerical calculations for the onset of convection are discussed
with emphasis on the structure of the most unstable modes. Left
(right) handed helical motion dominates in the northern (southern)
hemisphere. Also, as the stratification increases, the horizontal
dimension of the most unstable modes decreases, the prograde phase velocity increases, and the buoyancy force does more negative
work in the upper part of the convection zone. Differential rotation is maintained by the transport of longitudinal momentum
and by the coriolis forces acting on the meridional circulation.
Second order numerical calculations provide profiles of the differential rotation and meridional circulation induced by the
first order perturbations. Results of these calculations for the
most unstable modes show that either equatorial acceleration, as
observed on the sun, or equatorial deceleration can be maintained
depending on the rotation rate, density stratification, viscosity,
and conductivity. Small viscous diffusion relative to thermal
diffusion is required for equatorial acceleration in rapidly
rotating, highly stratified convection zones.

R. M. Bonnet and A. K. Dupree (eds.), Solar Phenomena in Stars and Stellar Systems, 145–172.
Copyright © 1981 by D. Reidel Publishing Company.

1. INTRODUCTION

Differential rotation on the sun is generally agreed to be induced by the influence of rotation on convection. The magnitude of this influence can be estimated by the ratio of the coriolis frequency to the turnover frequency or growth rate of convective cells (1), (2). This ratio is $\lesssim 10^{-3}$ for granules, $\sim 10^{-1}$ for supergranules, and $\gtrsim 1$ for convective cells that reach from the bottom to the top of the solar convection zone, i.e. giant cells (3). In other words, because of their small length and time scales, granules are not significantly affected by rotation, and supergranules by perhaps only a small amount, whereas the coriolis forces should significantly influence the size and shape of giant cells. Therefore giant cell convection would appear to be the best candidate for maintaining differential rotation on the sun.

In the past most hydrodynamical models of giant cell, stellar convection, especially those dealing with rotating, spherical shells or spheres, have employed the Boussinesq approximation in which there is no basic density stratification (4)-(14). Today we would like to describe how large density stratifications can influence giant cell convection in stars. However this work represents only the first step for compressible, rotating, spherical shell models because it is limited to linear, single mode calculations. Since these results are being published in more detail in the Astrophysical Journal, only a summary of some of the major effects will be given here. This talk will focus mainly on those effects that have a significant influence on helicity and differential rotation, the two major ingredients in solar dynamo theories. After a brief description of the model, I will discuss the first order solutions emphasizing the structure of the giant cell convection. Then I will describe the second order solutions in terms of the axisymmetric differential rotation and meridional circulation that is induced by the first order, nonaxysymmetric convection.

2. MODEL DESCRIPTION

A number of simplifying assumptions are made which are consistent with the objective of modeling giant cell solar convection. The top of the model convection zone corresponds to a surface approximately 3% of the solar radius below the photosphere. Based on mixing length theories (15), this model convection zone is below the hydrogen and helium ionization zones and is everywhere only slightly superadiabatic. As a result the fluid is taken to be a perfect gas of constant composition, i.e. there is no partial ionization. Also the smaller scale granule and supergranule convection in the thin, highly superadiabatic

layer just below the photosphere is not being modeled here in line
with the assumption that large scale differential rotation is
maintained by the interaction of giant cell convection and rotation. In addition, the gravitational field is assumed inversely
proportional to the radius squared. That is, the self-gravitation
of the convection zone is neglected since its mass is a small
percentage of the solar mass. Centrifugal acceleration is also
neglected since on the sun it is approximately five orders of
magnitude smaller than the gravitational acceleration. A Newto-
nian fluid is assumed, therefore the viscous stress tensor is
proportional to the rate-of-strain tensor. The diffusion of momentum and heat is parameterized as turbulent, linear diffusion
since small scale (unresolvable) convective turbulence dominates
over radiative and conductive diffusive processes in the solar
convection zone. The kinematic viscosity and the thermometric
diffusivity are assumed at most functions of radius. Also, the
diffusive energy flux is assumed proportional to the superadiabatic temperature gradient consistent with the assumption of
turbulent diffusion of heat.

We employ the anelastic approximation which enables us to
model convection in large density stratifications without dealing
with sound waves. As a result, this is not a fully compressible
model. However, it is compressible on a time scale much larger
than the acoustic time scale of the fluid. The basic assumption
in our formulation is that the convective velocities are much
smaller than the sound speed or, equivalently, the basic reference state is only slightly superadiabatic. The advantage is
not having to model several time scales that differ by many
orders of magnitude. Since the characteristic period of global
acoustic oscillations is of the order of one hour on the sun
compared to a rotation period of the order of one month, it is
assumed that these fast moving sound waves, which are filtered
out via the anelastic approximation, have little influence on
the maintenance of the large scale solar differential rotation.
In addition, based on mixing length theories, the temperature
gradient in the solar convection zone, except near the very top,
departs no more than about one part in 10^4 from the adiabatic
value.

With the above assumptions, a formal scale analysis is
performed on the nonlinear fluid equations. All dependent variables are expanded in a power series in a small parameter
representing the magnitude of the departure from an adiabatic
state or, equivalently, representing the Mach number of the
convective velocity. Only terms up to first order in the expansion are retained. As a result, sound waves are filtered out (16).

The anelastic perturbation equations involve coefficients
that depend on the chosen reference state which is assumed time

independent and a function of radius only. The reference state
is determined by specifying the depth of the convection zone,
the number of density e-folds across the zone (N_ρ), and radial
dependence of the thermometric diffusivity $\bar{x}$ (r) which determines
the entropy gradient as a function of radius. To leading order
in the expansion, the reference state temperature, density, and
pressure are adiabatic. Five different density stratifications,
corresponding to $N_\rho = 10^{-2}$, 1, 3, 5 and 7, have been studied for
a convection zone depth of 40 % of the stellar radius. In addition, some calculations have been done for a 20 % depth. We have
chosen the radial dependence of x to represent two extreme cases:
constant thermometric diffusivity and constant dynamic conductivity. In the former case x is constant and $(\nabla - \nabla_{AD})$ is inversely
proportional to the density ; in the latter case   is inversely
proportional to the density while $(\nabla - \nabla_{AD})$ is a constant. The
nondimensional reference state density $\bar{\rho}$ and the temperature θ
are plotted vs. radius in Figure 1 for the five density stratifications for the 40 % depth. The corresponding profiles for
$(\nabla-\nabla_{AD})$ are also plotted in Figure 1 for the constant $\bar{x}$ case.

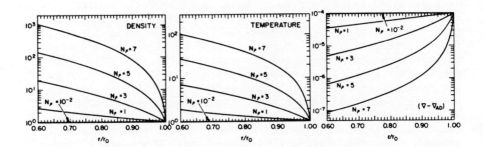

Fig. 1.   Reference density, temperature, and $(\nabla - \nabla_{AD})$
plotted vs. radius for a 40 % depth with density e-folds
$N\rho = 10^{-2}$, 1, 3, 5, and 7. The density and temperature are scaled
by their values at the top of the zone. The $(\nabla- \nabla_{AD})$ profiles
are for the constant $\bar{x}$ case and are assumed to be 10^{-4} at the
top.

The anelastic perturbation equations also depend on three
nondimensional parameters. The Rayleigh number R is a measure of
how effective buoyancy, which drives the convection, is relative
to viscous and thermal diffusion which hinder convection. The
Prandtl number P is a measure of viscous diffusion relative to
thermal diffusion. The coriolis effect relative to the viscous
diffusion effect is measured by the Taylor number T. As a
result, TP/R is a measure of the coriolis effect relative to the

buoyancy effect which, as mentioned in the introduction, should be $\gtrsim 1$ for giant cell convection on the sun.

Finally, boundary conditions must be specified. Impermeable, stress free velocity boundary conditions were always assumed. However, several combinations of temperature boundary conditions have been tested including constant temperature and constant diffusive heat flux.

3. FIRST ORDER SOLUTIONS

3.1. Solution technique

The effects of compressibility at the onset of stationary convection in a stratified, rotating, spherical shell have been studied via the linear, anelastic fluid equations. Stationary convection here means in the global sense ; that is, the total kinetic energy of the convection zone remains constant. The linear, anelastic perturbation equations are described in detail by Glatzmaier and Gilman (17). Briefly, the divergence of the mass flux vanishes and the relative pressure perturbation equals the sum of the relative density and temperature perturbations. The net force density on the fluid is the sum of the pressure, buoyancy, coriolis, and viscous force densities. The convergence of the diffusive heat flux determines the rate of change of the thermal energy density.

The spatial and temporal dependences of the perturbations are separated employing a Fourier expansion, $e^{-i\omega t}$, of the time dependence and a spherical harmonic expansion, $Y_\ell{}^m$, of the spatial dependence. Consequently the onset of stationary convection is characterized by $Im\omega = 0$. For $m \neq 0$, the $Re\ \omega$ in general does not vanish, so the phase of the convection pattern propagates in longitude with respect to the rotating coordinate system. The result of these expansions is a set of coupled, complex, differential equations for the radially dependent coefficients. The coriolis force produces ℓ coupling, however there is no m coupling for these linear equations. The reference state, the boundary conditions, and the parameters P, T, and m must be specified. Then the system is solved as a double eigenvalue problem (ω and R) with a generalized Newton-Raphson technique after the series of spherical harmonics is truncated. The number of spherical harmonics used is determined by a strict convergence criterion. Small wave numbers (m), large rotation rates (T), and large density stratifications (N_ρ) require more spherical harmonics ; the maximum number used was 22.

3.2. Mode structure

The most unstable mode for a given density stratification, rotation rate, viscosity, conductivity, convection zone depth, and set of boundary conditions is the one that has the smallest Rayleigh number at the onset of convection. Since modes with longitudinal wave numbers close to the critical wave number of the most unstable mode grow the fastest at supercritical Rayleigh numbers and will probably dominate in nonlinear, multimode calculations, I will focus on the properties of these modes.

Several effects have been discussed for Boussinesq convection also exist for compressible convection (17), (18). The most unstable modes are single cells extending from the bottom to the top of the convection zone since this is a more efficient method of releasing gravitational potential energy than via multiple cells in radius. Also, these modes peak in the equatorial region because the stabilizing influence of the coriolis force is most easily overcome there. The longitudinal dimension of these cells tends to be comparable to the radial dimension in order to minimize viscous resistance. However, their longitudinal dimension decreases as a) viscous diffusion decreases relative to thermal diffusion,   b) the rotation rate increases, c) constant dynamic diffusivities are replaced by constant kinematic diffusivities, d) constant heat flux boundaries are replaced by constant temperature boundaries, or e) the convection zone depth decreases.

Consider, in Figure 2, the mass flux vectors and thermodynamic perturbation contours plotted in the equatorial plane for the most unstable modes at $T = 10^5$, constant temperature boundaries, constant and equal diffusivities, and a 40 % depth. The $N_\rho = 10^{-2}$ case is on the top row ; the $N_\rho = 5$ case on the bottom row. Solid (broken) contours represent positive (negative) perturbations. Notice how, in general, light, hot fluid rises and heavy, cold fluid sinks. Also, at this high rotation rate, the fluid tends to flow along pressure contours so the pressure gradient can offset the coriolis force.

At the density stratification increases the longitudinal dimension of the most unstable modes decreases. The combination of the coriolis, pressure gradient, and buoyancy forces tends to tilt the flow patterns in the constant latitude surfaces which results in less efficient conversion of gravitational potential energy. Smaller longitudinal dimensions reduce this tilt. Consequently the critical wave number increases with density stratification.

In addition, as the density stratification increases, the velocity becomes enhanced in the upper part of the zone for the case of constant kinematic viscosity $\bar{\nu}$ and in the lower part for

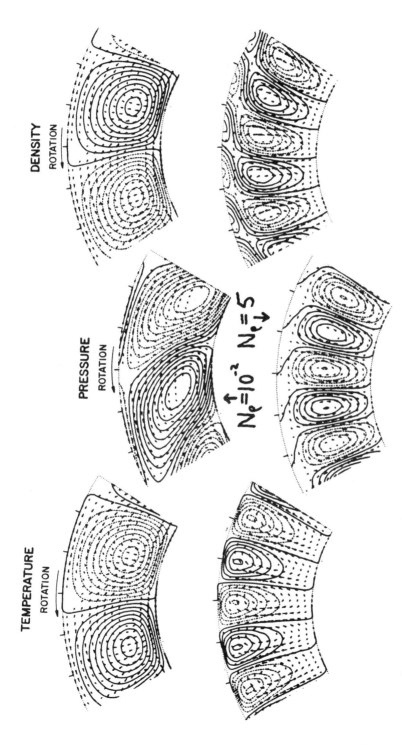

Fig. 2. Thermodynamic perturbations plotted in the equatorial plane for the most unstable modes for $T = 10^5$, $P = 1$, constant temperature boundaries, constant $\bar{s}$ and $\bar{x}$, 40 % depth. The top row is the $N\rho = 10^{-2}$ case; the bottom row the $N\rho = 5$ case. Arrows represent mass flux; solid (broken) contours represent positive (negative) thermodynamic perturbations.

the case of constant dymanic viscosity $\overline{\rho\nu}$(18). The mass flux is
more uniformly distributed when the velocity is enhanced in the
upper part of the zone. However, when the kinematic viscosity is
assumed inversely proportional to the density the viscosity suppresses the velocity in the upper part of the zone. Graham and
Moore (19) found this effect to be small in their nonrotating,
plane parallel model. But we have found this effect to be dramatic at high rotation rates as is illustrated in Figure 3. There
the velocity vectors are plotted in the equatorial plane for the

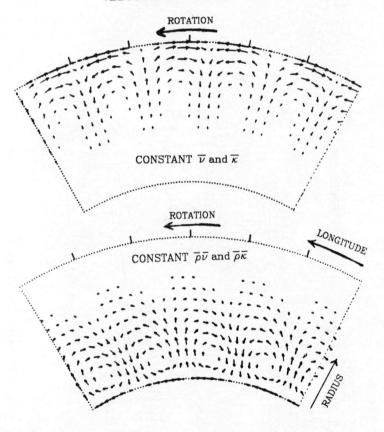

Fig. 3. Velocity vectors plotted in the equatorial plane for
the most unstable modes for T = 10^5, P = 1, N_ρ = 5, constant temperature boundaries, 40° depth. The constant $\overline{\nu}$ and $\overline{\kappa}$ case is on
the top ; the constant $\overline{\rho\nu}$ and $\rho\kappa$ case on the bottom.

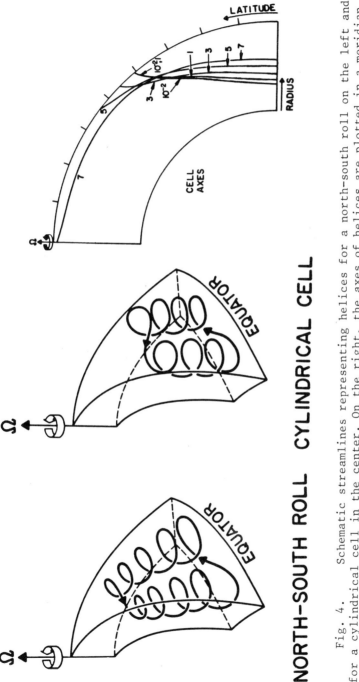

Fig. 4. Schematic streamlines representing helices for a north-south roll on the left and for a cylindrical cell in the center. On the right, the axes of helices are plotted in a meridian plane for the most unstable modes for $T = 10^5$, $P = 1$, constant temperature boundaries, constant $\bar{v}$ and $\bar{\varkappa}$, 40 % depth, and the five different stratifications $N_\rho = 10^{-2}$, 1, 3, 5 and 7.

The density stratification and rotation rate also influence the three dimensional structure of the cells. At low rotation rates the north-south rolls are helices as illustrated on the left in Figure 4. Left-handed helices develop in the northern hemisphere because the coriolis forces tend to expand the cross sections of counterclockwise north-south rolls ; as a result latitudinal pressure gradients force the fluid in the counterclockwise cells toward the equator, in the direction of increasing forces that tend to contract the cellular cross section and so there is flow toward the pole. In addition, due to conservation of mass, fluid at low latitudes spills over from the equatorward helices to the poleward helices and vice versa at high latitudes producing, between these alternating north-south helices, radial left-handed helices of rising (sinking) clockwise (counterclockwise) vortices. Likewise right-handed helices develop in the southern hemisphere. Consequently, helicity $(V.\nabla \times V)$ tends to be negative (positive) in the northern (southern) hemisphere.

For high rotation rates $(T \gtrsim 10^4)$ and small density stratifications $(N_\rho \sim 0)$ the coriolis forces, which are perpendicular to the rotation axis, overwhelm the components of buoyancy forces parallel to the rotation axis producing cylindrical cells with axes parallel to the rotation axis. This effect has been discussed by Roberts (5) and Busse (7) for a Boussinesq fluid in terms of the Taylor-Proudman theorem. The curl of the linear, anelastic equation of motion is useful for understanding under what conditions the Taylor-Proudman theorem applies.

$$\frac{\partial}{\partial t}(\nabla \times \bar{\rho}v) = 2\Omega\bar{\rho}\frac{\partial v}{\partial z} + 2\Omega v\frac{\partial \bar{\rho}}{\partial z} + \begin{bmatrix} \text{buoyancy} \\ \text{torque} \end{bmatrix} + \begin{bmatrix} \text{viscous} \\ \text{torque} \end{bmatrix} \quad (1)$$

Here the rotational frequency Ω is in the z direction. The buoyancy and viscous torques tend to cancel. If the time derivative is small and the density relatively constant, the Taylor-Proudman theorem applies, i.e. the velocity does not vary significantly in the z direction, and as a result cylindrical cells develop parallel to the rotation axis as illustrated in the center of Figure 4. The axes of these cells are also illustrated in the meridian plane on the right in Figure 4 for the constant $\bar{v}$ and $\bar{x}$ case. However, for large stratifications the relative density variation with z is large above mid-latitude near the top of the zone. Consequently, by equation (1), the relative velocity variation with z is also large in this region which accounts for the bend in the cell axis as illustrated on the right of Figure 4 most unstable modes at $N_\rho = 5$, $T = 10^5$, and $P = 1$. The constant r case is on the top and the constant ρ case is on the bottom. However, this velocity enhancement decreases as viscous diffusion decreases relative to thermal diffusion, i.e. as P decreases.

Helicity is of considerable interest because it is the basis of "cyclonic turbulence" which is a key factor in mean field solar dynamo theories (20), (21). Physically, helical fluid motion twists toriodal magnetic field lines producing poloidal magnetic fields. Our model can not resolve small scale turbulence but does provide profiles of helicity due to large scale convective motions. Although several interesting helicity profiles are possible, the most unstable modes, as mentioned above, tend to have negative (positive) helicity in the northern (southern) hemisphere. Longitude averaged helicity profiles in the meridian plane are plotted in Figure 5 for the most unstable modes at $T = 10^5$. The plot on the left is for $N_\rho = 10^{-2}$; the one in the center is for $N_\rho = 5$ with constant $\bar{\rho}\bar{v}$ and $\bar{\rho}\bar{x}$; the one on the right is also for $N_\rho = 5$ but with constant $\bar{v}$ and $\bar{x}$. Notice how the helicity peaks in the region where the convective velocity is enhanced.

3.3. Phase propagation

The first order, nonaxisymmetric solutions are locally time dependent because the real part of the frequency does not vanish. The phase velocity in longitude of the convective pattern is $\text{Re}\omega/m$. As in the Boussinesq case, the wave number corresponding to the peak real frequency is close to the critical wave number. In addition, the real frequency increases with rotation rate and decreases with viscosity. But as the density stratification increases the magnitude of the real frequency increases significantly and, unlike the Boussinesq case, the phase propagation is always prograde, i.e. in the direction of rotation.

Using the linear, anelastic vorticity equation, Glatzmaier and Gilman (17) describe how the time rate of change of the vorticity is determined by the sum of the pressure, buoyancy, viscous, stretching, and compressibility torques. The stretching torque determines the magnitude and direction of the phase propagation for small stratifications. However, for $N_\rho \gtrsim 3$, the compressibility torque dominates.

Glatzmaier and Gilman (22) analytically solve a simple analog to this problem in which a form of potential vorticity is conserved. Linear perturbations in this system take the form of prograde propagating vorticity waves induced by a compressibility torque. Their frequencies depend on the longitudinal wave number and the density stratification in the same way as the convective modes of the numerical model. This is illustrated in Figure 6 where the prograde frequencies are plotted vs. longitudinal wave number for several density stratifications ; both the analytical and the numerical results are displayed. If such convective modes

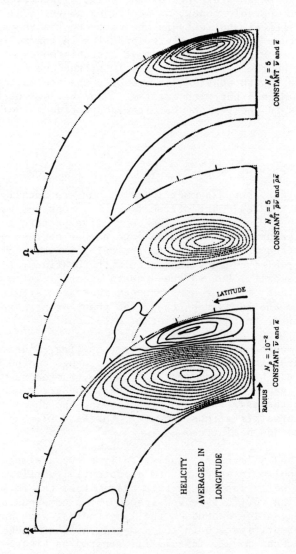

Fig. 5. Helicity in the northern hemisphere averaged in longitude and plotted in a meridian plane for the most unstable modes for T = 10^5, P = 1, constant temperature boundaries, 40 % depth. Three cases are illustrated. Broken contours represent negative (left-handed) helicity ; solid contours represent positive (right-handed) helicity.

exist on the sun they would propagate on the order of 10 % faster
than the basic rotation rate.

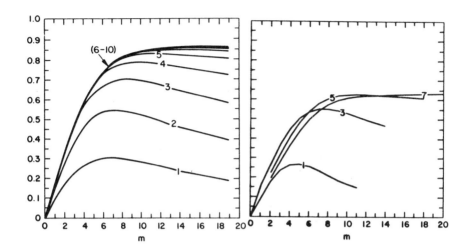

Fig. 6.   Prograde frequency scaled by the rotational
frequency and plotted vs. the longitudinal wave number for
several density stratifications N_ρ . The analytic results are on
the left ; the numerical results with T = 10^4, P = 1, constant
temperature boundaries, constant $\bar{\mathcal{J}}$ and $\bar{\mathcal{X}}$, and 40 % depth are
on the right.

     The effect of the compressibility torque can easily be
understood by considering the velocity vectors in the equatorial
plane plotted in Figure 7 for N_ρ = 3, T = 10^3. The vorticity
(∇ x V) in the outward (inward) direction is indicated by solid
(broken) contours in the top plot. The generation rate of this
vorticity is illustrated in the lower plot. In these plots the
rotational frequency Ω is outward (normal to the equatorial plane).
Consider the giant cell to be composed of fluid columns, parallel
to the rotation axis, all revolving about the axis of the giant·
cell. As a particular column rises toward the upper surface it
expands in the plane normal to the rotation axis. The resulting
tangential coriolis forces torque the expanding fluid column in
the opposite direction to Ω, generating negative vorticity relative to the rotating frame as illustrated in the bottom plot of
Figure 7. Likewise positive relative vorticity is generated in
sinking columns. Therefore, since rising fluid columns are on
the prograde side of negative vorticity cells and on the retrograde side of positive vorticity cells, the cellular pattern

propagates with a prograde phase velocity. The larger the rotation
rate and the density stratification the greater the compressibility torque and so the faster the mode propagates relative to
the rotating frame in the direction of increasing longitude. The
dependence of the frequency on the wave number is discussed in
detail in Glatzmaier and Gilman (22) and basically has to do
with the magnitude of the radial shear of the longitudinal velocity relative to the longitudinal shear of the radial velocity.

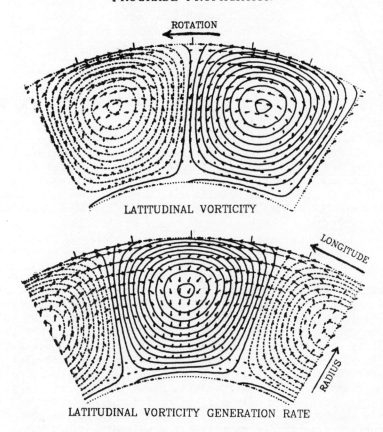

Fig. 7.   Latitudinal component of vorticity (top)
and vorticity generation rate (bottom) plotted in the equatorial
plane for the most unstable mode for T = 10³, P = 1, N_ρ = 3,
constant temperature boundaries, constant $\bar{\jmath}$ and $\bar{x}$, 40 % depth.

3.4. Kinetic energy balance

    Large scale kinetic energy density is locally time depen-
dent due to the phase propagation ; but the average in longitude
is time independent for these linear stability solutions. Consequently the sum of the average pressure, buoyancy, and viscous
work densities per time vanishes. An example of the work densities averaged in both longitude and latitude is illustrated in
Figure 8 for N_ρ = 5, T = 10^3, m = 6, constant $\bar{\nu}$ and $\bar{\varkappa}$, P = 1,
and constant heat flux top and bottom boundaries. The velocity
and thermodynamic perturbation contours are also plotted in the
equatorial plane for this solution. Notice how the average pressure work is positive near the bottom and the top of the zone,
because the fluid flows down the pressure gradient, and negative
in the middle of the zone because there the fluid flows against
the pressure gradient. In addition, the pressure perturbation
does work on expanding and contracting fluid. The viscous force
in general converts large scale kinetic energy into small scale
turbulent energy ; so the viscous work is negative. The buoyancy
force, for the most part, converts gravitational potential energy
into large scale kinetic energy ; so the buoyancy work is mainly
positive. That is, in general, light (heavy) fluid rises (sinks).
But for N_ρ > 0 there is a reversal of the density perturbation in
the upper part of the zone (see Figure 2 for N_ρ = 5) which results in negative buoyancy work there. This never occurs in
Boussinesq convection.

    Massaguer and Zahn (23) have also found this negative buo-
yancy work near the top surface in their nonrotating, plane parallel, anelastic models. They argue that, in the linear problem,
it is due to constant temperature boundary conditions which,
because of the equation of state, forces the density perturbation
to have the same sign as the pressure perturbation near the top
boundary. However we find this effect also present for constant
heat flux boundaries as illustrated in Figure 8. Notice how the
radial gradient of the temperature perturbation vanishes at the
boundaries while the temperature perturbation peaks there with
the same sign as the pressure perturbation. Nevertheless, a contribution of the same sign by the density perturbation is still
required to satisfy the equation of state. In addition, we find
that the buoyancy force does negative work in a small region
between multiple cells in radius where no boundary condition is
applied. Therefore the temperature boundary condition cannot be
the dominant reason for the negative buoyancy work.

    In order to not have this reversal of the density pertur-
bation in the upper part of the zone the magnitude of the temperature perturbation would have to be larger there than that of
the pressure perturbation and the pressure perturbation would
have to peak at the top surface instead of below it in order to

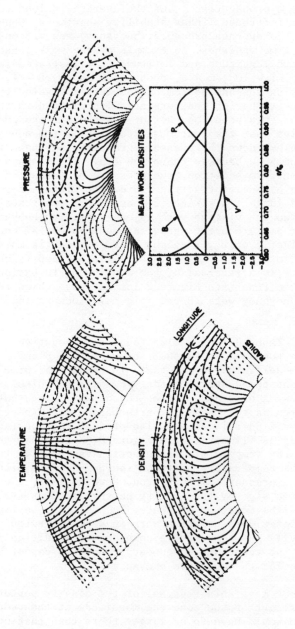

Fig. 8. Temperature, density, and pressure perturbations contours and velocity vectors plotted in the equatorial plane for the most unstable mode for $T = 10^5$, $P = 1$, $N_\rho = 3$, constant temperature boundaries, constant $\bar{\sigma}$ and $\bar{\kappa}$, 40 % depth. Also, the average buoyancy (B), pressure (P), and viscous (V) work densities per time are plotted vs. radius for this case.

oppose the radial flow which would then be driven by the buoyancy
force near the top. But, although the equation of state could be
satisfied, apparently the equation of motion and the thermodynamic equation could not simultaneously be satisfied for a stratified zone due to the adjusted thermodynamic perturbations. No
simple argument has been found that explains this effect.

3.5. Momentum transport

The flux of longitudinal momentum in both radius and latitude helps to maintain an axisymmetric differential rotation.
First consider the radial flux of longitudinal momentum ($\bar{\rho}uw$).
Here $\bar{\rho}$ is the density, u is the longitudinal velocity, and w is
the radial velocity. For small density stratifications at low
rotation rates the flow from the counterclockwise north-south
rolls spills over to the adjacent clockwise rolls at low latitudes. As a result the radial momentum flux directed inward, i.e.
$\bar{\rho}uw < 0$, at low latitudes is much stronger than that directed
outward, i.e. $\bar{\rho}uw > 0$. At high latitudes the flow spills over
from the clockwise to the counterclockwise rolls; so the radial
momentum flux tends to be directed outward but has a much smaller
magnitude because the velocities are small compared to their magnitude at low latitudes. At high rotation rates, for small stratifications, the prograde tilt associated with the cylindrical
cells produces a stronger outward radial momentum flux (see the
mass flux vector plots of Figure 2 for $N_\rho = 10^{-2}$). Gilman (12)
found this same effect for the Boussinesq case. A similar effect
exists for large density stratifications with constant $\bar{\rho}\bar{v}$ and
$\bar{\rho}\bar{x}$, although the flux peaks in the lower part of the zone (see
Figure 3). This similarity is reasonable because the scale height
in the lower part of the stratified zone is relatively large and
constant. However, for large stratifications with constant $\bar{v}$ and
$\bar{x}$, the momentum flux peaks in the upper part of the zone where
the scale height is small and decreases rapidly with radius.
Consequently, due to the slight retrograde tilt (see Figure 2 for
$N_\rho = 5$), the radial momentum flux is mainly directed inward at
low latitudes for all T. The profiles of the longitudinally averaged radial flux of longitudinal momentum for the most unstable
modes at $T = 10^5$ are displayed in the lower part of Figure 9 for
$N_\rho = 10^{-2}$, $N_\rho = 5$ with constant $\bar{\rho}\bar{v}$ and $\bar{\rho}\bar{x}$, and $N_\rho = 5$ with constant $\bar{v}$ and $\bar{x}$. When the mean radial momentum flux is outward it
increases with radius in the lower part of the zone and decreases
with radius in the lower part of the zone and decreases with
radius in the upper part ; that is the mean radial flux of longitudinal momentum diverges in the lower part and converges in the
upper part of the zone. Therefore, based on just this outward
radial flux profile, the axisymmetric angular velocity should
increase with radius. Likewise, a mean radial flux directed inward, supports an angular velocity profile that increases with
depth.

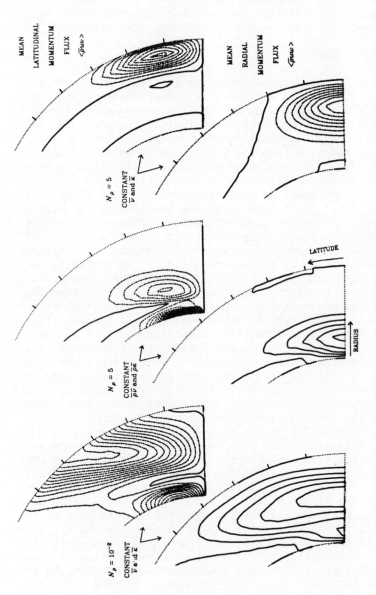

Fig. 9   Latitudinal momentum flux averaged in longitude (top row) and radial momentum flux averaged in longitude (bottom row) plotted in a meridian plane for the most unstable modes at $T = 10^5$, $P = 1$, constant temperature boundaries, 40 % depth. Broken contours represent latitudinal flux toward the equator (top row) and inward radial flux (bottom row). Three cases are illustrated. For a given case, the values of the contour levels are the same for the latitudinal and radial fluxes.

However the angular velocity profile also depends on the latitudinal flux of longitudinal momentum ($\overline{\rho uv}$). Here v is the latitudinal velocity. At low rotation rates stronger latitudinal momentum flux near the top (bottom) of the zone is directed toward the equator (pole) because the fluid in the counterclockwise (clockwise) rolls spirals toward the equator (pole) (see the north-south roll of Figure 4). At high rotation rates for small stratifications the latitudinal momentum flux near the bottom of the zone is also directed toward the equator at low latitudes due to the prograde tilt and the forced latitudinal flow due to the surface curvature (see the cylindrical cell of Figure 4 and the upper left of Figure 9). Notice how the flux peaks along a line parallel to the axis of the cylindrical cells. Again, the same effect was found in Gilman's Boussinesq calculations (12). But, for large stratifications at high rotation rates, the mean latitudinal momentum flux profiles in the meridian plane (see upper right of Figure 9) resemble those at low rotation rates due to the north-south roll structure above mid-latitude. The velocity vectors in the constant radius plot of Figure 10 illustrate this equatorial transport of longitudinal momentum, i.e. $\overline{\rho}$ uv < 0, near the top surface. This agrees with Ward's (24) correlation of latitudinal and longitudinal sunspot motions. The velocity vectors of Figure 10 also illustrate the left-handed helical flow in the northern hemisphere ; fluid rises in the regions where the vectors diverge and sinks in the regions where they converge. In the upper part of the zone the equatorward mean momentum flux increases with latitude at low latitudes and decreases with latitude at high latitudes (see the top row of Figure 9) ; that is, the mean equatorward flux of longitudinal momentum converges at low latitudes and diverges at high latitudes. Based on just this latitudinal momentum flux profile, the axisymmetric angular velocity should decrease with latitude in the upper part of the zone.

The differential rotation that is induced depends on the sum of the convergences of the radial and latitudinal momentum fluxes as well as the coriolis force on the axisymmetric meridional circulation (13). Therefore second order calculations are required to determine the mean differential rotation at the onset of convection.

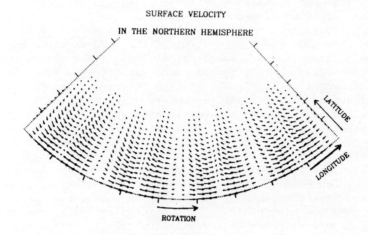

Fig. 10   Velocity vectors at the top boundary in the
northern hemisphere for the most unstable mode for $T = 10^5$,
$N\rho = 5$, $P = 1$, constant temperature boundaries, constant $\bar{\nu}$ and $\bar{x}$.

4. SECOND ORDER SOLUTIONS

4.1. Solution technique

      We want to solve for the second order axisymmetric flow
induced by the first order, nonaxisymmetric solutions. Therefore
we write the perturbations (relative the reference state) as sums
of their longitudinal mean <f> and their Fourier components f.
Then averaging in longitude the nonlinear, anelastic perturbation
equations results in a set of coupled, inhomogeneous, differential equations (25) for the mean, second order perturbations <f>
with the forcing terms <f_1 f_2> constructed from the first order
solutions.

      The <f> are expanded in Y_ℓ^0 since they are independent of
longitude. Consequently, the forcing terms must also be expanded
in Y_ℓ^0 :

$$<f_1 f_2> = \frac{1}{2}\mathrm{Real} \left| \left(\sum_{\ell_1 = m} C_1 Y_{\ell_1}^m \right) \cdot \left(\sum_{\ell_2 = m} C_2 Y_{\ell_2}^m \right) \right| = \frac{1}{2} \sum_{\ell=0} C_3 Y_\ell^0 \qquad (2)$$

Here C_1 and C_2 represent the radially dependent coefficients for
the first order solutions ; C_3 represents the radially dependent
coefficients for the forcing functions that are calculated by

invoking the orthogonality property of spherical harmonics and
Gaunt's integral formula.

The set of radially dependent, inhomogeneous equations is
then solved via a Newton-Raphson relaxation technique after trun-
cating the series of Legendre functions when additional terms
becomes negligible. Typically between 20 and 30 spherical harmonics (m = 0) are required.

4.2 Differential rotation and meridional circulation

Differential rotation is a key factor in mean field solar
dynamo theories ; it shears poloidal magnetic field lines producing toroidal magnetic fields. Differential rotation is also of
major interest because it can be used to calibrate giant cell
models. The observed equatorial acceleration at the solar surface
corresponds to peak longitudinal velocities that differ by as
much as 200 m/s from the average surface rotation rate whereas
the axisymmetric meridional circulation at the solar surface is
at least an order of magnitude smaller (24), (26), (27), (28).
Differential rotation is maintained by the transport in both radius and latitude of longitudinal momentum (as discussed above),
by the coriolis force on the meridional circulation, and by the
axisymmetric viscous force. Meridional circulation is maintained
by the transport in both radius and latitude of the radial and
latitudinal momenta, by the coriolis force on the differential
rotation, and by the axisymmetric buoyancy, pressure, and viscous
forces.

The case of constant and equal diffusivities, constant temperature boundaries, and a 40 % depth, will be considered first.
The mean differential rotation and meridional circulation induced
by the most unstable, first order solutions at $T = 10^5$ are plotted
in a meridian plane in Figure 11 for $N_\rho = 10^{-2}$, for $N_\rho = 5$ with
constant $\bar{\rho v}$ and $\bar{\rho \chi}$, and for $N_\rho = 5$ with constant $\bar{v}$ and $\bar{\chi}$. The
arrows represent the linear velocity in the meridian plane while
the angular velocity perturbation is represented by contours :
solid (broken) contours represent an angular velocity greater
(less) than that of the rotating frame.

Notice for this high rotation rate that the fluid in the
main meridional cell rises at low latitudes and sinks at high
latitudes. The meridional circulation profile for $N_\rho = 10^{-2}$ is
very similar to Gilman's nonlinear, multimode, Boussinesq results
for a 20 % depth (2) and for a 40 % depth (14). The differential
rotation profile for $N_\rho = 10^{-2}$ is an equatorial acceleration with
angular velocity increasing with radius. Notice how the angular
velocity at low latitudes is approximately constant on cylinders
coaxial with the rotation axis. Again this is typical of Gilman's
(2), (14) Boussinesq results when the Rayleigh number is not too

far above critical. But for large density stratifications (see
the N_ρ = 5 cases of Figure 11) the meridional circulation becomes enhanced near the equator and in the lower part of the zone
for constant $\overline{\rho v}$ and $\overline{\rho x}$ and in the upper part for constant $\overline{v}$ and
$\overline{x}$ as did the first order velocity profiles. In addition, the
differential rotation loses its "constant on cylinders" profile
and an equatorial deceleration is induced.

The ratio of the second order RMS longitudinal velocity to
the RMS meridional velocity (DR/MC) is a measure of the magnitude
of the differential rotation to the meridional circulation. This
ratio is indicated in Figure 11 for the respective solutions.
As discussed above, this ratio appears to be $\gtrsim 10$ for the sun unless the mean, solar surface velocities are not representative of
the mean flow in the entire convection zone. This ratio increases
with Taylor number ; hence the reason large Taylor number solutions have been emphasized here.

Consider Figure 11 again. The equatorial acceleration for
the N_ρ = 10^{-2} case is driven by the convergence of longitudinal
momentum flux in the equatorial region and near the upper surface
(see Figure 9) since the magnitude of the meridional circulation
is relatively small. The coriolis force on this differential
rotation is outward in the equatorial region and inward in at
mid-latitude. As a result, the meridional flow is outward in the
equatorial region and inward at mid-latitude. However, the DR/MC
ratio decreases as the density stratification increases for P = 1
(see Figure 11). As a result, the coriolis force on the meridional
circulation decelerates the longitudinal flow in the equatorial
region and accelerates it at mid-latitude. The DR/MC ratio is
larger for the constant $\overline{\rho v}$ and $\overline{\rho x}$ case in Figure 11 (N_ρ = 5)
because the convection is enhanced in the lower part of the zone
where the scale height is relatively large and constant ; hence
this case is more like the N_ρ = 10^{-2} case.

So far these second order results for large stratifications
look fairly discouraging due to the small DR/MC ratios and the
equatorial deceleration. The one parameter that has not yet been
varied in the second order solutions is the Prandtl number P,
i.e. the ratio of the viscous diffusion to the thermal diffusion.

First consider the N_ρ = 10^{-2} stratification for $(TP/R)^{1/2}$
~ 2.5, constant temperature top and bottom boundaries, and cons-
tant $\overline{v}$ and $\overline{x}$. The mean differential rotation and meridional circulation profiles are plotted in Figure 12 for P = 10^{-1}, 1 and 10.
Also indicated are the DR/MC ratios which illustrate how decreasing the viscous diffusion relative to the thermal diffusion while holding the ratio of the coriolis effect to the buoyancy effect
constant, increases the magnitude of the differential rotation
relative to the meridional circulation. Apparently viscous diffu-

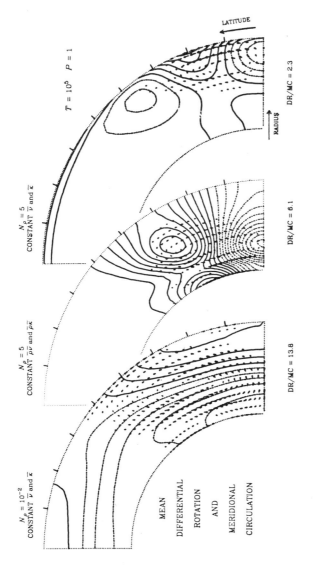

Fig. 11. Induced, mean angular velocity contours and mean meridional linear velocity vectors for the most unstable modes for T = 10⁵, P = 1, constant temperature boundaries, 40 % depth. Three cases are illustrated. Solid (broken) contours represent angular velocity greater (less) than that of the rotating frame. The magnitude of the differential rotation to that of the meridional circulation is indicated for each case.

sion inhibits the large shear of the mean longitudinal velocity
more than that of the mean meridional velocity. Notice, for P =
10^{-1}, how the differential rotation profile, resulting from the
convergence of longitudinal momentum near the equator and upper
surface, is "constant on cylinders" at low latitudes, and an equatorial acceleration. A polar vortex, which has not been observed
on the sun (29), exists for this P = 10^{-1}, $N_\rho = 10^{-2}$ case similar
to those found in Gilman's (14) nonlinear, multimode, Boussinesq
solutions. As the Prandtl number increases the meridional circulation increases relative to the differential rotation enhancing
the decelerating influence of the coriolis force acting on the
meridional circulation at low latitudes. By P = 10, an equatorial
deceleration profile exists that is similar to those for large
stratifications. Notice, for P = 10, how the coriolis force on
the meridional circulation is in the direction of decreasing
longitude in the equatorial and polar regions while in the direction of increasing longitude at mid-latitude where the peak angular velocity is produced.

    A similar effect occurs for large stratifications although
smaller Prandtl numbers are required to get large DR/MC ratios
and equatorial acceleration at high rotation rates. Consider the
N_ρ = 5 stratification for $(TP/R)^{1/2} \sim 1.8$, constant temperature
top and constant heat flux bottom, and constant $\bar{\rho}\bar{\upsilon}$ and $\bar{\rho}\varkappa$. The
Prandtl number effect on the differential rotation and meridional
circulation is illustrated in Figure 13 where the profiles for P
= 10^{-2}, 10^{-1}, and 1 are plotted. Again, decreasing the Prandtl
number decreases the coriolis force on the meridional circulation
by increasing the DR/MC ratio. Apparently, for P $\gtrsim$ 1, the large
radial gradient of the second order longitude velocity due to the
enhancement of the first order velocity near the bottom (or near
the top for the case of constant $\bar{\upsilon}$ and $\varkappa$) results in large viscous resistance of the mean differential rotation and hence small
DR/MC ratios. Consequently, smaller Prandtl numbers are required
for large stratifications to get more uniform first order velocity distributions and therefore smaller radial gradients of the
second order longitudinal velocity and larger DR/MC ratios. Also,
for large stratifications, high rotation rates, and constant $\bar{\rho}\bar{\upsilon}$
and $\bar{\rho}\varkappa$, the outward radial flux of longitudinal momentum increases as P decreases. However the resulting equatorial acceleration
profile for P = 10^{-2} is significantly different than the "constant on cylinders" profile at small stratifications (see Figures
12 and 13). Notice, for the P = 10^{-2} case, that the angular velocity increases with depth in the lower two thirds of the zone.
This ridge of peak angular velocity is mainly due to the large
convergence of the radial flux of longitudinal momentum which is
maximum at one third the depth of the zone below the top surface.
Also notice how there is no tendency for a polar vortex in this
very stratified case.

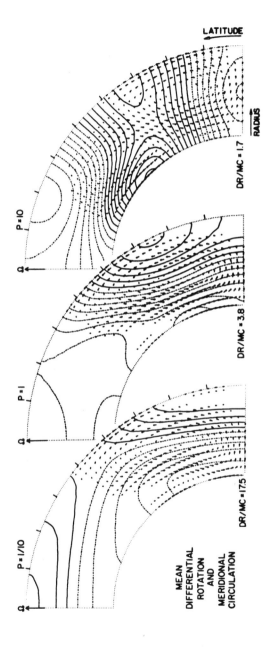

Fig. 12. Induced, mean angular velocity contours and mean meridional linear velocity vectors for $N_\rho = 10^{-2}$, $(TP/R)^{1/2} \sim 2.5$, constant temperature boundaries, constant $\bar{\nu}$ and $\bar{\mathcal{K}}$, 40 % depth. Three cases, corresponding to $P = 10^{-1}$, 1, and 10, are illustrated. The magnitude of the differential rotation to that of the meridional circulation is indicated for each case.

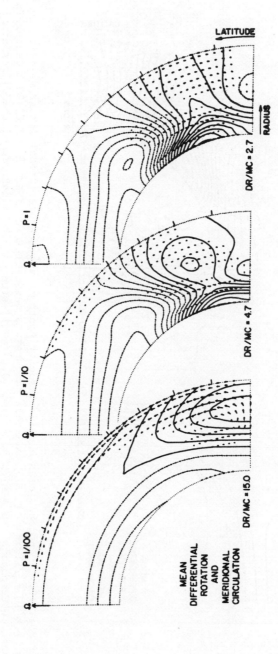

Fig. 13. Induced, mean angular velocity contours and mean meridional linear velocity vectors for $N_\Omega = 5$, $(TP/R)^{1/2} \sim 1.8$, constant temperature top and constant heat flux bottom boundaries, constant $\overline{\rho v}$ and $\overline{\rho}\propto$, 40 % depth. Three cases, corresponding to $P = 10^{-2}$, 10^{-1} and 1, are illustrated. The magnitude of the differential rotation to that of the meridional circulation is indicated for each case.

One might wonder if increasing the Taylor number would
have the same effect on the differential rotation, as decreasing
the Prandtl number. In Figure 13 the P = 10^{-2}, 10^{-1}, and 1 cases
have respectively Taylor numbers of 10^6, 10^5, and 10^4. Increasing
the Taylor number for the P = 10^{-1} case from 10^5 to 10^6 does not
significantly alter the differential rotation profile. Likewise,
decreasing the Taylor number for the P = 10^{-2} case from 10^6 to
10^5 results in only a slight change in the differential rotation
profile. Therefore, although increasing the Taylor number does
strengthen an equatorial acceleration for a small Prandtl number,
it is not evident from these linear calculations that it could
produce an equatorial acceleration for a large stratification
with P $\sim$ 1. Recall, for P $\sim$ 1, that increasing the Taylor number
increases the velocity enhancement, near the bottom of the zone
(for constant $\bar{\rho}\bar{v}$) and, as discussed above, this apparently inhibits equatorial acceleration.

One must remember that these results describe the initial
tendency of the mean flow induced by a single mode at the onset
of convection. Gilman (1977) found in his nonlinear, multimode,
Boussinesq calculations that an equatorial acceleration that
exists at modest (supercritical) Rayleigh number switches over
to an equatorial deceleration when the Rayleigh number increases.
Also, this switchover occurs closer to the stability boundary,
i.e. at smaller R, for smaller Taylor numbers. What happens in
the very supercritical, compressible case will have to be determined by nonlinear, multimode calculations.

4.3. Summary

Mean momentum transports induce a mean differential rotation which is influenced by the axisymmetric coriolis forces on
the mean meridional circulation. For small stratifications at
high rotation rates with P $\sim$ 1, the magnitude of differential
rotation is large compared to the meridional circulation; consequently the transport of longitudinal momentum, which is toward
the equator, produces an equatorial acceleration, and the coriolis
force on this differential rotation drives a meridional circulation with an outward mean flow at the equator. But, for large
stratifications at high rotation rates with P $\sim$ 1, the magnitude
of the meridional circulation is comparable to the differential
rotation ; consequently the coriolis force on the meridional
circulation drives a differential rotation which is an equatorial
deceleration near the upper surface. But with large Taylor numbers and small Prandtl numbers, i.e. large rotation rates and
small viscosities, the strong rotational influence results in
transport of longitudinal momentum toward the equator while the
relatively small meridional circulation limits the decelerating
effect of the coriolis force at low latitudes. Consequently,
under these conditions at the onset of convection for large stra-

tifications, longitudinal momentum transports by giant cell convection can dominate and induce an equatorial acceleration.

ACKNOWLEDGEMENTS

The National Center for Atmospheric Research is sponsored
by the National Science Foundation.

REFERENCES

(1) Davies-Jones, R.P. and Gilman, P.A., 1970, Solar Physics,
    12, pp. 3-22.
(2) Gilman, P.A., 1977, Geophys. Astrophys. Fluid Dyn., 8, pp. 93-135.
(3) Simon, G.W., and Weiss, N.O., 1968, Zs. f. Ap., 69, pp. 435-450.
(4) Roberts, P.H., 1965, Ap. J., 141, pp. 240-250.
(5) Roberts, P.H., 1968, Phil. Trans. Roy. Soc., A263, pp. 93-117.
(6) Busse, F.H., 1970a, Ap. J., 159, pp. 629-639.
(7) Busse, F.H., 1970b, J. Fluid Mech., 44, pp. 441-460.
(8) Busse, F.H., 1973, Astron. Astrophys., 28, pp. 27-37.
(9) Durney, B., 1968, J. Atmos. Sci., 25, pp. 771-778.
(10)Durney, B., 1970, Ap. J., 161, pp.1115-1127.
(11)Durney, B., 1971, Ap. J., 163, pp. 353-361.
(12)Gilman, P.A., 1975, J. Atmos. Sci., 32, pp. 1331-1352.
(13)Gilman, P.A., 1978, Geophys. Astrophys. Fluid Dyn., 11,
    pp. 157-179.
(14)Gilman, P.A., 1979, Ap. J., 231, pp. 284-292.
(15)Gough, D.O. and Weiss, N.O., 1976, Mon. Not. R. Astr. Soc.
    176, pp. 589-607.
(16)Gilman, P.A., and Glatzmaier, G.A., 1981, Ap. J. Suppl.,
    (in press) (Paper I).
(17)Glatzmaier, G.A. and Gilman, P.A., 1981a, Ap. J. Suppl.,
    (in press) (Paper II).
(18)Glatzmaier, G.A., and Gilman, P.A., 1981c, (in progress)
    (Paper IV).
(19)Graham, E., and Moore, D.R., 1978, Mon. Not. R. Astr. Soc.,
    183, pp. 617-632.
(20)Stix, M., 1976, IAU Symposium No. 71, pp. 367-388.
(21)Moffatt, H.K., 1978 "Magnetic Field Generation in Electrically
    Conducting Fluids", Cambridge Univ. Press.
(22)Glatzmaier, G.A. and Gilman, P.A., 1981b, Ap. J. Suppl.
    (in press) (Paper III).
(23)Massaguer, J.M. and Zahn, J.P., 1980, Astron. Astrophys.,
    (in press).
(24)Ward. F., 1965, Ap. J., 141, pp. 534-547.
(25) Glatzmaier, G.A. and Gilman, P.A., 1981d (in progress)(Paper V).
(26)Howard, R. and Harvey, J., 1970, Solar Physics, 12, pp. 23-51.
(27)Howard, R., 1971, Solar Physics, 16, pp. 21-36.
(28)Duvall, T.L., 1979, Solar Physics, 63, pp. 3-15.
(29)Beckers, J.M., 1978, Ap. J., 224, pp. L143-146.

ENERGY BALANCE IN SOLAR AND STELLAR CHROMOSPHERES

Eugene H. Avrett

Harvard-Smithsonian Center for Astrophysics

ABSTRACT

   The spectrum from a specific region of the sun or from a
star can be used to determine the temperature-density stratification and other physical properties of the emitting atmospheric
layers. We need such a description of the atmosphere in order
to understand the causes of solar activity and the nature of
various stellar phenomena. Computer programs are available for
solving the detailed radiative transfer and statistical equilibrium equations to calculate the spectrum emerging from an
optically thick gaseous medium of prescribed properties. It is
usually possible to adjust the properties of the atmospheric
model to obtain a spectrum which agrees with an observed one.
Current models of the solar atmosphere indicate that a meaningful
temperature-density stratification can be determined from the
spectrum: using relatively few atmospheric parameters to define
the model, good agreement can be obtained between the calculated
spectrum and the observed one. In this paper we consider the
chromospheric models corresponding to faint, average, and bright
components of the quiet sun, a solar plage and flare, and the
stars α Boo and λ And. In each case we calculate as a function
of depth the net radiative cooling rate per unit volume due to
the transitions of various atoms and ions (principally H, Mg II,
Ca II, and H$^-$). This radiative cooling must be balanced by the
same amount of mechanical heating at each depth. The chromospheric cooling rate, obtained from models based on observations,
provides a detailed constraint on theories which explain how the
chromosphere is heated.

R. M. Bonnet and A. K. Dupree (eds.), Solar Phenomena in Stars and Stellar Systems, 173–198.
Copyright © 1981 by D. Reidel Publishing Company.

I. INTRODUCTION

Considerable progress in the field of chromospheric energy balance has been made since the 1976 Nice symposium on the Energy Balance and Hydrodynamics of the Solar Chromosphere and Corona (Bonnet and Delache [5]). Improvements have been made in the calculation of atmospheric models, in theories of both mechanical energy dissipation and radiative losses, and in solar and stellar observations. These improvements have led to a better understanding of the outward increase of temperature in solar and stellar chromospheres.

Energy balance in a stellar chromosphere involves two main components: mechanical heating and radiative cooling. The topic of mechanical heating is reviewed in the paper by P. Ulmschneider that appears in this volume. The present paper deals mainly with the chromospheric energy losses by radiation, specifically the calculation of net radiative cooling rates from atmospheric models. The paper by C. Jordan following this one deals with energy balance in coronae and optically thin chromosphere-corona transition regions. References to earlier work are given in these two papers by Ulmschneider and by Jordan.

II. CHROMOSPHERIC MODEL CALCULATIONS

In this section we describe the general method used to obtain chromospheric models from observed spectra, and we indicate how the radiative losses are determined from such models. These calculations are carried out according to the following steps:

1. Assume a chromospheric temperature distribution as a function of geometrical height, column mass, optical depth, or some other equivalent parameter. Also assume a microvelocity distribution. We use the microvelocity both to determine the Doppler line widths and to define a contribution to the total pressure due to turbulence.

2. Solve the hydrostatic equilibrium equation and the coupled equations of statistical equilibrium and radiative transfer for the lines and continua of hydrogen and other atoms and ions to compute the ionization and excitation of these constituents and the internal radiation intensities.

3. Compute the emergent spectrum for the various lines and continua and compare with observations.

4. Readjust the temperature and microvelocity distributions of step 1 and repeat the calculations to get better agreement with the observations. In this way a model is established based on the available observations.

5. Finally, use this model atmosphere to compute the net radiative cooling rate (ergs cm^{-3} s^{-1}) at each depth. In a

steady state this net energy lost by radiation is equal to the
heating due to mechanical wave dissipation and any other nonradiative processes.

These model atmosphere calculations are based on a given
observed spectrum and lead to a stratification of the atmospheric parameters. When we consider particular solar features,
we ignore horizontal interaction between neighboring regions
emitting different spectra.

We now consider models corresponding to various components
of the solar atmosphere, and for this purpose summarize the
principal results obtained in a study of the quiet-sun EUV spectrum by Vernazza, Avrett, and Loeser (8), hereafter VAL-III. The
following seven figures are from that paper.

The observed spatially averaged disk-center continuum
intensity of the quiet sun in the wavelength range 40-140 nm is
shown in Figure 1. The prominent observed features here are the
wings of the Lα line, centered at 121.6 nm, and the carbon,
hydrogen, and helium continua starting at 110, 91, and 50.4 nm,
respectively. This region of the spectrum is filled with emission lines but we show only the underlying continuum. The upper
panel of Figure 1 shows the observed intensity, and the corresponding brightness temperatures are plotted in the lower panel.
The continuum at 140 nm is formed in the low chromosphere at
temperatures between 4000 and 5000 K. The general increase of
continuum brightness temperature with decreasing wavelength
corresponds to an increase in the chromospheric height at which
the spectrum is formed. However, there are significant differences between the temperature where the spectrum originates and
the observed brightness temperature because of departures from
local thermodynamic equilibrium.

Figure 2 shows the average quiet-sun temperature distribution obtained by the procedure described in steps 1-4 above.
The temperature is shown as a function of height h(km) and
column mass $m(g \ cm^{-2})$; h = 0 has been fixed at continuum
optical depth unity at the wavelength λ = 500 nm. The column
mass is equal to the total pressure divided by the surface
gravity g which has the value $2.74 \times 10^4 \ cm \ s^{-2}$ in the solar
case. We indicate in Figure 2 the depth regions where the
spectrum at different wavelengths is formed. The Lα wing at
1 Å and 5 Å from line center and the 109.8 nm carbon continuum
are formed in the extended chromospheric region at temperatures
between 5000 and 7000 K. The hydrogen Lyman continuum intensities at 90.7 nm and 70 nm are formed at the top of the
chromosphere where the temperature abruptly rises to values
exceeding 10^4 K. Figure 2 also shows the depths at which other
parts of the spectrum are formed; see VAL-III for details.

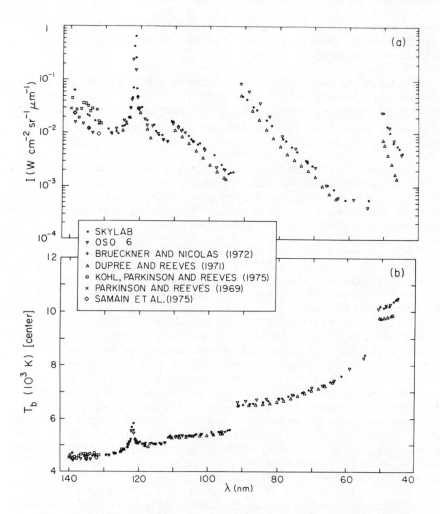

Fig. 1 - Observed values of the disk-center continuum intensity (upper) and the corresponding central brightness temperature (lower) in the wavelength ranges 40-140 nm. References in the figure are listed in the VAL-III paper.

The extent to which this model is consistent with the observed 40-140 nm continuum is shown in Figure 3 where we plot the calculated intensity as a function of wavelength along with the observed values. The distinction between the curves labeled "Model C" and "mean" will be explained later.

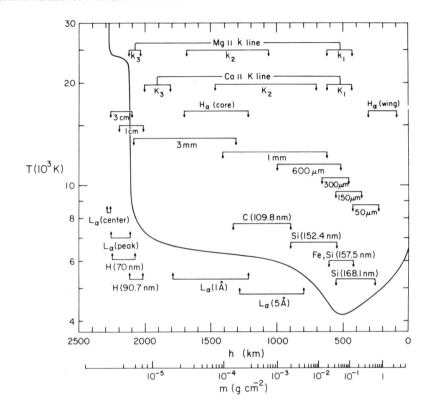

Fig. 2 - The average quiet-sun temperature distribution derived
from the EUV continuum, the Lα line, and other observations. The
approximate depths where the various continua and line components
originate are indicated.

This model is in reasonable agreement with spatially
averaged quiet-sun observations throughout most of the spectrum.
However, the quiet sun observed with high spatial resolution
exhibits a complex cell and network structure. The contrast
between the faintest and brightest areas is almost a factor of
10 at EUV wavelengths.

Figure 4 shows Skylab observations of six brightness components of the EUV continuum, identified with the following
regions:
    A. a dark point within a cell;
    B. the average cell center;
    C. the average quiet sun;
    D. the average network;

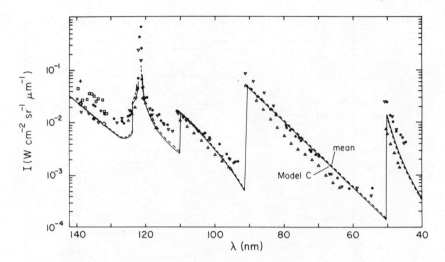

Fig. 3 - Comparison of calculated and observed EUV continuum intensities for the average quiet sun.

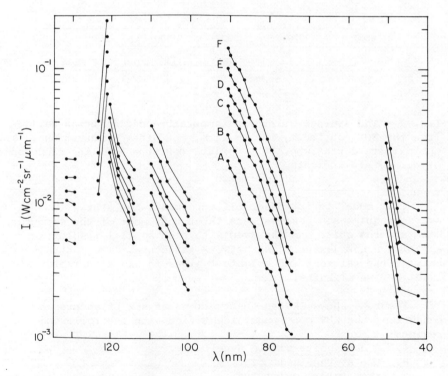

Fig. 4 - Observed EUV continuum intensities for quiet-sun
brightness components A-F.

    E. a bright network.element;
    F. a very bright network element.
These discrete component intensities have been chosen to represent
the observed intensity distribution which varies continuously at
each wavelength from one feature to another. The components
A through F have the area contributions 8%, 30%, 30%, 19%, 9%,
and 4%, respectively.

    We have constructed models from each separate intensity
distribution. Figure 5 shows the intensities in the range 70-135
nm (exclusive of Lα) computed from these six models, compared
with the observed values. See VAL-III for further comparisons.
The temperature distributions derived for components A through F
are shown in Figure 6. The assumed microvelocity distribution is
the same for all six models (Figure 11 of VAL-III).

    The six computed intensity distributions in Figure 5, each
weighted by its fractional contribution, gives the "mean" curve
plotted in Figure 3. Note that this weighted-mean intensity

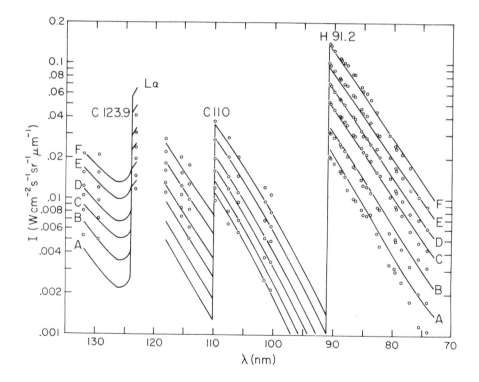

Fig. 5 - Comparison of computed and observed intensities in the
range 70-135 nm, exclusive of the Lα band which is shown in
Figure 13 of VAL-III.

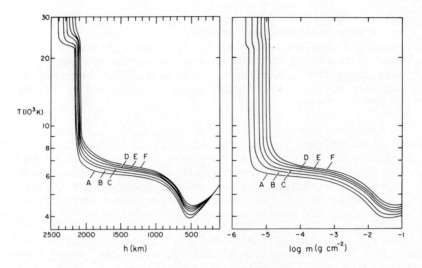

Fig. 6 - Temperatures as functions of height and of log m for Models A-F.

distribution is in close agreement with the Model C intensity distribution plotted in the same figure. The temperature curve in Figure 2 is the same as the one labeled "C" in Figure 6, and was derived from the component C intensities in Figure 4 rather than from the spatially averaged values in Figure 1. However, Figure 3 indicates that essentially the same results would be obtained either way.

The set of temperature distributions shown in Figure 6 correspond to a range of quiet-sun intensities, from the faintest 8% to the brightest 4% of the observed solar surface area. Note that the six temperature curves all resemble each other in showing a gradual increase in the 6000-7000 K range and a very rapid increase above 8000 K. Brighter components have higher temperatures everywhere above the photosphere and have transition regions (where T > 8000 K) located closer to the photosphere at higher pressures.

In the temperature range 2×10^4 to 3×10^4 K we have introduced plateaus to account for observations of the Lα and Lβ integrated intensities, the self reversals of these lines, and the short-wavelength intensities in the Lyman continuum. Gouttebroze et al. (6) demonstrate that the calculated Lβ emission line does not have a central absorption feature as observed unless a transition-region plateau is introduced. The models constructed by Basri et al. (3) to match Lα observations also have

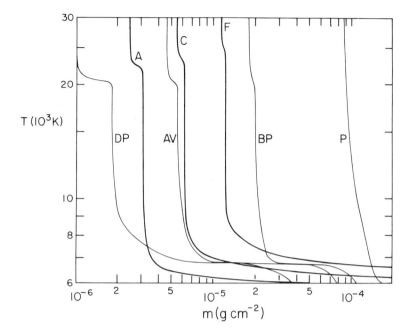

Fig. 7 - Transition-region temperatures as functions of mass
column densities. Our Models A, C, and F are compared with the
dark-point, average quiet sun, bright-point, and plage models of
Basri et al.

transition-region plateaus. Figure 7 shows the quiet-sun
temperature distributions found by Basri et al. for a typical
dark point, average region, and bright point, compared with our
Models A, C, and F. The figure also shows the temperature distribution for a typical plage, or active region, from the Basri
et al. analysis. Note that the plateau thickness diminishes from
the fainter to the brighter components. On the other hand, the
brighter-component plateaus occur at higher densities, causing
greater Lα radiative losses, as will be shown later.

Are these plateaus "real"? The answer is yes, in that the
line of sight passes through the indicated thickness of material
at the indicated temperatures. Otherwise, we cannot account for
the observed spectrum. However, it is not known whether
1) these plateaus are basic features of the temperature stratification of the transition region, or 2) they result from projections of chromospheric gas along magnetic field lines which are
intercepted by the line of sight.

The chromospheric cell and network pattern is determined by complex magnetic fields. The field lines tend to be oriented radially in bright network regions but to be inclined relative to the outward direction in cell centers. Lα observations by Bonnet et al. (4) show small-scale magnetic loops connecting bright points in the network such that the cell-center magnetic field is often parallel to the solar surface. The magnetic field inhibits thermal interaction of electrons and ions perpendicular to the field lines, and flux tubes containing cooler gas surrounded by hotter gas can extend upward from the network and across the fainter cell regions. The plateau thus could be a manifestation of the small scale magnetic field structure along the line of sight, without such a feature in the temperature variation along the magnetic field lines.

Another possibility is that the line of sight passes through spicules or prominence-like material, i.e., relatively cool gas extending along the field lines high into the corona, and that such overlying gas accounts for the required plateau material. In this case the plateau would be a manifestation of larger-scale structures along the line of sight, but again there might be no plateau in the temperature variation along the magnetic field.

These remarks are prompted by our attempts to understand the basic causes of the chromospheric temperature stratification. It may be difficult to identify a specific heating process occuring in the 2×10^4-3×10^4 temperature range that would account for the plateau in a plane-parallel atmosphere. Otherwise, the general characteristics of the chromospheric temperature distribution seem physically reasonable, as we will discuss later.

In this paper we calculate the net radiative cooling rates for quiet-sun Models A, C, and F, and for models of a plage and a solar flare. Figure 8 compares Skylab Lyman-continuum observations of a typical flare with those of the quiet sun. The number of counts is proportional to the intensity so that the flare spectrum in this wavelength range is over 100 times brighter than that of the quiet sun. Such observations are used to construct atmospheric models in the same way as before. Figure 9 from Machado et al. (7) shows the temperature variations with column mass for two flare models F1 and F2 compared with our quiet-sun model C (QS in the figure) and the plage model P of Basri et al. Models F1 and F2 represent faint and bright flares, respectively. See Machado et al. for details.

In the next section we calculate the radiative losses from quiet-sun models A,C and F, and from plage and flare models P and F1. Also, we show the radiative losses calculated from models representing the chromospheres of two stars: α Boo and λ And.

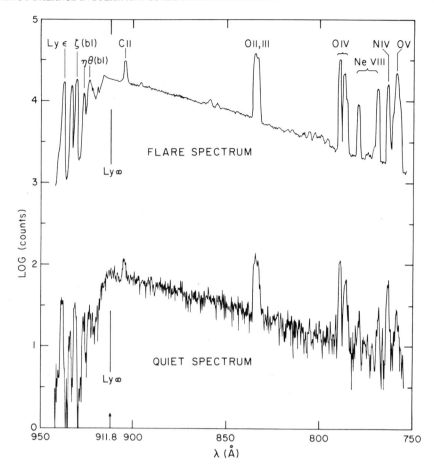

Fig. 8 - <u>Skylab</u> Lyman-continuum observations of the quiet sun and a flare; see Machado <u>et al</u>. (7).

III. RADIATION LOSS CALCULATIONS

     Given an atmospheric model for which we know as functions of depth the number densities of the various atomic energy states and the radiation intensity in the important lines and continua, we can calculate the net radiative cooling rates $\Phi_{u\ell}$ and Φ_m in ergs cm^{-3} s^{-1} as functions of depth for lines between levels u and ℓ, and the continua associated with levels m. The equations for $\Phi_{u\ell}$ and Φ_m are given in section IX of VAL-III.

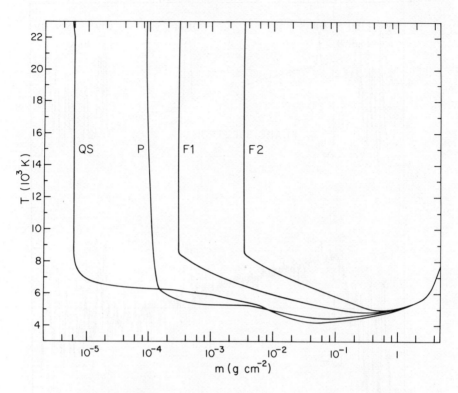

Fig. 9 - Temperature as a function of column mass for models of the average quiet sun, a plage region, and two solar flares.

Figure 10 shows the rates computed for the lines and continua of hydrogen in the case of Model C. Three general conclusions are apparent from these results: 1) the Lα line gives the most important hydrogen cooling rate in the temperature range $T > 10^4$ K, 2) the Lα cooling rate is substantial throughout the plateau region where $2x10^4 < T < 3x10^4$ K, and 3) the total contribution due to all hydrogen transitions, plotted as a heavy line in the figure, is small throughout the chromosphere, because net heating in the Balmer continuum approximately cancels the cooling due to Hα.

Part of the energy in the transition region that the Lα line could absorb is that carried by thermal conduction from the corona to the lower portions of the transition region. In Figure 10 we plot the negative of the conductive flux gradient F_c' , since this is a heating rate. The conductive heating in this case is clearly insufficient to match the Lα cooling except in the narrow layer where $T > 3x10^4$ K.

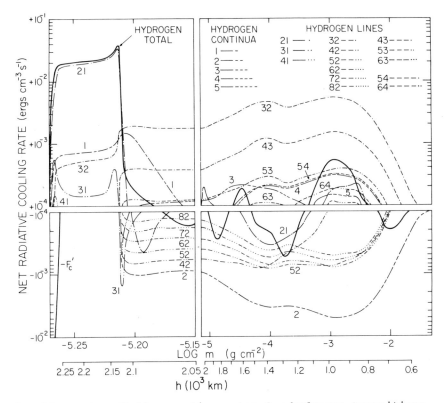

Fig. 10 - Net radiative cooling rates for hydrogen transitions, calculated from Model C. The negative of the conductive flux gradient is shown in the lower left panel.

In Figure 11 we plot the total hydrogen contribution along with the net radiative cooling rates due to the Ca II H, K, and infrared triplet lines, the Mg II h and k lines, the H⁻ ion, the Mg I b and Na I D lines, the lines Si II λ181.6 nm and He I λ58.4 nm, and the continua of Si I at 152.4 nm and 168.1 nm, Fe I at 157.5 nm and 176.8 nm, Mg I at 162.2 nm, and He I at 50.4 nm, all calculated from Model C.

We can simplify the results given in Figure 11 by combining the five Ca II lines, the two Mg II lines, and the bound-free and free-free H⁻ contributions. Figure 12 shows these three combined rates and the combined hydrogen rate plotted against height. The total curve in Figure 12 does not include the other contributions which appear in Figure 11. The remaining figures in this paper also show only the combined H, Ca II, Mg II, and H⁻ contributions.

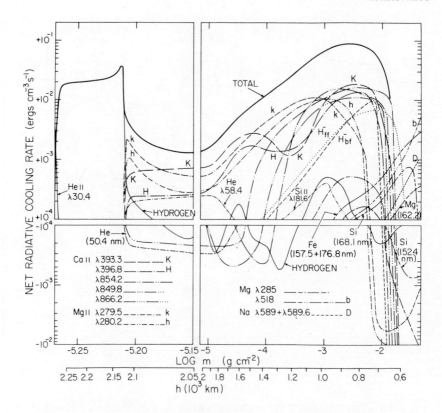

Fig. 11 - Net radiative cooling rates for Ca II, Mg II, H⁻, H, and other constituents, calculated from Model C.

The Model C temperatures at various heights are indicated at the top of Figure 12. The transition-region pressure at the temperature $T = 10^4$ K is given in the lower panel. The negative values of Φ in the temperature minimum region are presumably balanced by positive contributions from atoms and molecules which we have not included. Little further can be said about this negative Φ region without further study.

Above the temperature minimum the total rate reaches a maximum value, and then log Φ decreases almost linearly with height. When the temperature reaches the 7000-8000 K range where hydrogen starts to become ionized, Φ increases abruptly due to radiative losses in Lα. The rate maintains large values throughout the plateau region and then decreases abruptly as hydrogen becomes completely ionized.

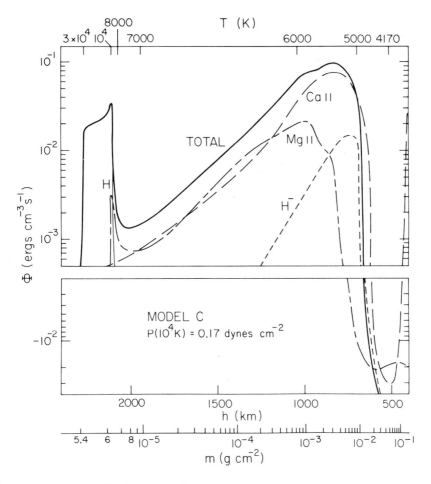

Fig. 12 - Net radiative cooling rates for Model C.

In a steady state, the calculated net radiative cooling
rate in the chromosphere between T $\sim$ 5000 K just above the
temperature minimum and T $\sim$ 7000 K near the base of the transition region must be equal to the distribution of local nonradiative heating. Conduction and other non-radiative energy
transport processes are negligible for such small values of the
temperature gradient. The chromospheric distribution of Φ,
obtained from models based on observations, provides a detailed
constraint on theories which explain how the chromosphere is
heated.

The large values of Φ in the Lα emitting region are more
difficult to interpret because of uncertainties in the nature

of the transition-region plateau. If the plateau is a basic
feature of the temperature stratification, then it seems necessary to have a large amount of mechanical heating in the
2×10^4 - 3×10^4 K temperature range since the shallow temperature
gradient there would inhibit thermal transport of energy from
the corona. If instead the plateau is caused by projections of
chromospheric gas along magnetic field lines which are intercepted by the line of sight, then the intrinsic variation of Φ
with height would have a much narrower Lα peak. Further theoretical work should help to resolve these questions.

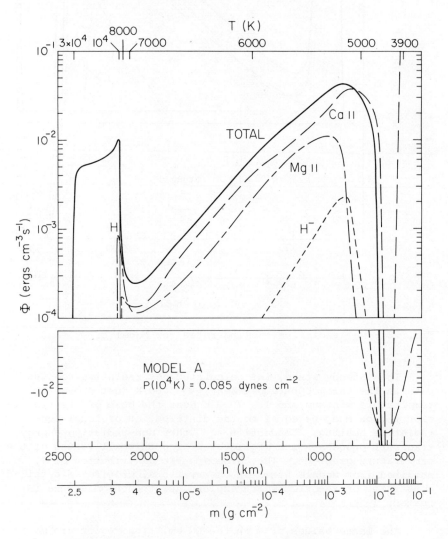

Fig. 13 - Net radiative cooling rates for Model A.

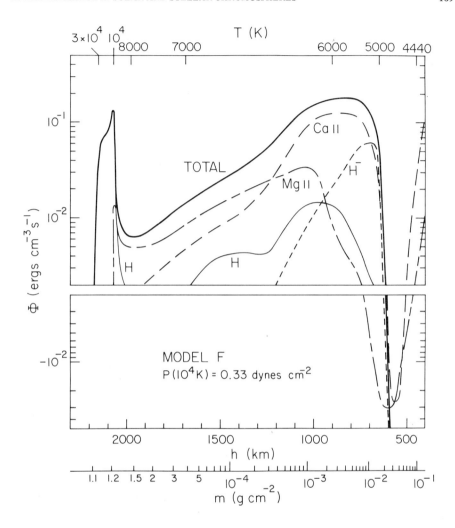

Fig. 14 - Net radiative cooling rates for Model F.

It is useful to compare the net radiative cooling rates
computed for the average quiet sun with those corresponding to
other solar regions. Figure 13 and 14 show $\Phi(h)$ from Models A
(dark cell center) and F (very bright network), respectively.
The values of $\Phi(h)$ from Model A are 2 to 3 times smaller than
those from Model C while the values from Model F are 2 to 3
times larger. The width and location of the outer $\Phi(h)$ maximum
feature corresponds to the width of the plateau and location of
the transition region in all three cases. The area under this
transition-region feature progressively increases in Models A,

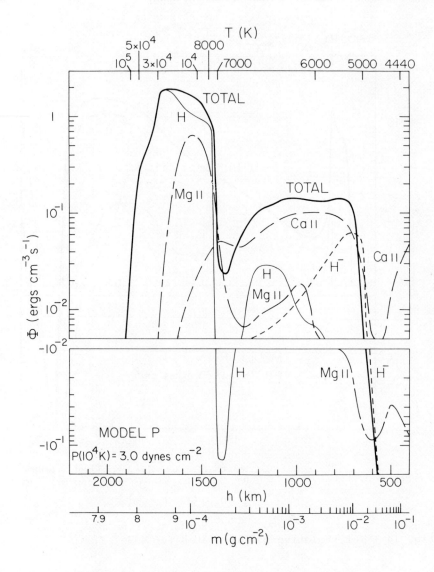

Fig. 15 - Net radiative cooling rates for Model P.

C, and F even though the plateau width decreases. Tabular
values of the contributions to $\int \Phi(h)\,dh$ will be given later in
this section.

We extend this sequence of calculations to include the
results from approximate plage and flare models. The temperature distribution we adopt for Model P is the plage $T(m)$ from

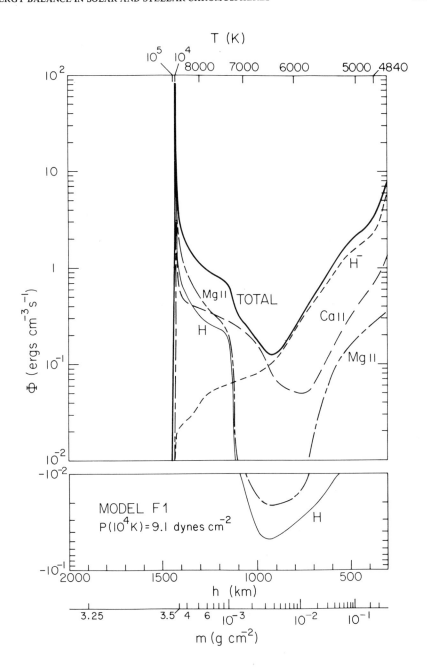

Fig. 16 - Net radiative cooling rates for Model F1.

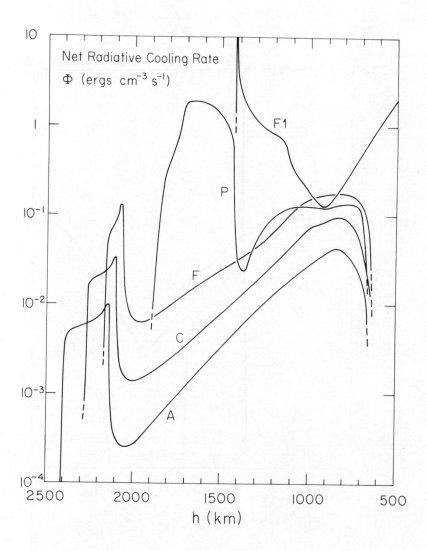

Fig. 17 — Comparison of the net radiative cooling rates for the five solar models.

Basri et al. in the transition region (Figure 7) and the T(m) from Model F in the chromosphere. We use Model F starting where the Basri et al. plage temperatures fall below those of Model F. As discussed in the VAL-III paper, the chromospheric temperatures derived by Basri et al. appear to be incorrect. The flare model we consider is the Machado et al. Model Fl shown in Figure 9.

The net radiative cooling rates calculated from Models P and Fl are shown in Figure 15 and 16. Given the hybrid nature of Model P and the fact that both Models P and Fl are based on a more limited set of observations than were used to determine the quiet-sun models, the detailed features of $\Phi(h)$ for Models P and Fl are probably less reliable than those calculated for Models A, C, and F. For example, the transition-region temperature distribution T(m) given by Basri et al. leads to a geometrical thickness between $T = 10^4$ K and $T = 3x10^4$ K which is greater than the corresponding thickness in Model F. The sequence of results from Models A, C, and F (and Fl) suggest that this region should have a smaller thickness. Note that there are net contributions to $\Phi(h)$ from hydrogen in the chromospheres of Models F, P, and Fl while in Models A and C hydrogen does not provide a net chromospheric contribution. The flare model differs qualitatively from the other ones in the temperature minimum region where $\Phi(h)$ has much larger values than those in the middle chromosphere.

Figure 17 shows a comparison of the functions $\Phi(h)$ computed from the five models. Models P and Fl exhibit some of the characteristics suggested by the progressive behavior of Models A, C, and F, but they also exhibit differences that need further study.

We feel that the $\Phi(h)$ distributions calculated from Models A, C, and F are reliable, at least relative to each other, since they were derived from a unified set of observed intensities, and these observations seem sufficient to establish the models with reasonable certainty. Thus we feel justified in presenting a table of the integrated components of Φ for these three quiet-sun models.

Table 1 gives $\int \Phi(h)\,dh$ in units of 10^5 ergs cm^{-2} s^{-1} for the Mg II h and k lines, the Ca II H, K, and infrared triplet lines, Lα, and the H$^-$ bound-free and free-free contributions. For Model F we also give the total hydrogen value which exceeds the contribution of Lα alone. Each integral is terminated at a specified depth, often where the rate changes from positive to negative. The last column in the table gives the temperature corresponding to the upper limit of integration. The integrated Model C line contributions are all about a factor of 2 larger than the corresponding Model A values, and the integrated Model F line contributions are about a factor of 2 larger than those for Model C. The H$^-$ integrated contribution increases by about a factor of 5 from Models A to C, and 5 from C to F. We do not tabulate the integrated rates for Models P and Fl because of the uncertainties in some of the detailed features of these models.

Table 1
Integrated Net Radiative Cooling Rates in Units of
10^5 ergs cm^{-2} s^{-1}

Model A	Mg II	h	1.8	5000
		k	2.5	"
	Ca II	H	2.3	4400
		K	3.3	"
		4/2	1.8	"
		5/2	2.7	"
		5/3	3.1	"
		Lα	1.5	8000
		H$_{bf}^-$	0.2	5000
		H$_{ff}^-$	0.6	4400
Model C	Mg II	h	4.3	5600
		k	5.2	"
	Ca II	H	4.9	4700
		K	6.4	"
		4/2	4.6	"
		5/2	5.5	"
		5/3	6.8	"
		Lα	3.4	8000
		H$_{bf}^-$	1.7	5000
		H$_{ff}^-$	2.2	4400
Model F	Mg II	h	7.7	5600
		k	11	"
	Ca II	H	9.9	4700
		K	14	"
		4/2	9.6	"
		5/2	9.7	"
		5/3	13	"
		Lα	5.9	8000
		H(total){	6.7	8000
			14	5000
		H$_{bf}^-$	8.7	5000
		H$_{ff}^-$	11	4440

To conclude this section we show the net radiative cooling rates computed from models of two stellar chromospheres having surface gravities much smaller than the solar value. Figure 18 shows Φ(h) computed from the α Boo model of Ayres and Linsky (1). In this model the chromospheric temperature rises gradually over an extended height range. The transition region pressure $P = 4.9 \times 10^{-4}$ dynes cm^{-2} occurs at a column mass 1×10^{-5} g cm^{-2}, which lies to the left of the values plotted in Figure 18.

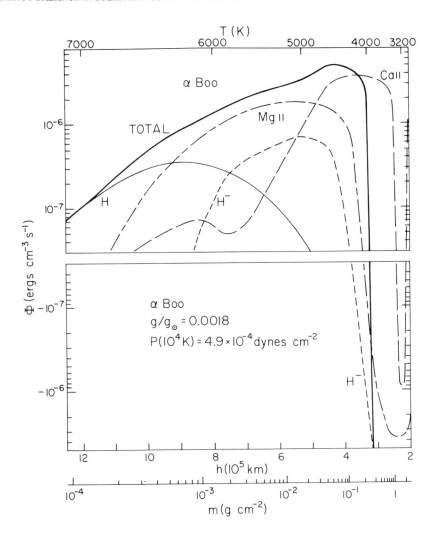

Fig. 18 - Net radiative cooling rates calculated for the chromosphere of α Boo.

Figure 19 shows $\Phi(h)$ from the chromospheric model D determined by Baliunas et al. (2) for λ And. In this case the transition region pressure is 1.1 dynes cm^{-2} which lies between the corresponding pressure values for solar models F and P. In addition, these results for λ And have a closer resemblance to those computed for the solar plage model rather than the quiet sun models.

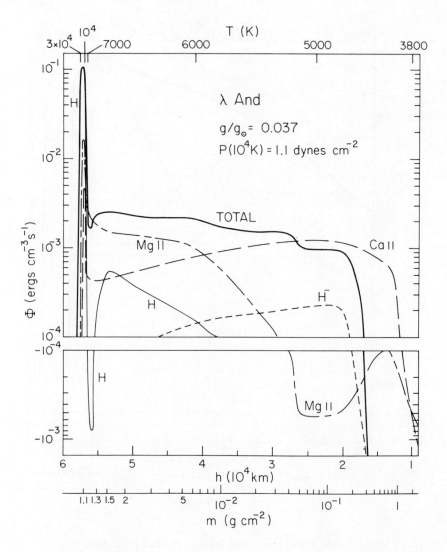

Fig. 19 - Net radiative cooling rates calculated for the chromosphere of λ And.

On the other hand, the α Boo model has much lower pressures throughout the chromosphere and has a computed Φ(h) distribution which generally resembles the results we obtain for quiet rather than active regions of the sun. Stellar chromospheres with low and high transition region pressures thus seem analogous in many ways to faint regions of the solar chromosphere and active regions, respectively.

CONCLUSIONS

We consider seven chromospheric models: those corresponding
to three different components of the quiet sun, a solar plage
and flare, and the stars α Boo and λ And. The temperature distribution for each model is adjusted so that the calculated
spectrum is in best agreement with the available observations.
The various number densities are determined by solving the
equations of hydrostatic equilibrium, statistical equilibrium,
and radiative transfer.

In all of the cases we consider, hydrogen is neutral in the
temperature minimum region, and the temperature increases with
height in the chromosphere rather gradually until hydrogen starts
to be ionized. Then the temperature increases abruptly to much
larger values, eventually causing hydrogen to be fully ionized.

The net radiative cooling rate $\Phi(h)$ throughout the chromospheric portion of each model is a direct measure of the non-radiative heating required to produce the chromospheric temperature
increase. The large values of Φ at the base of the chromosphere-
corona transition region are due principally to $L\alpha$ emission. We
interpret the abrupt temperature increase at the base of the
transition region to the high efficiency of radiative losses in
$L\alpha$ at intermediate hydrogen-ionization temperatures. Because $L\alpha$
emission causes Φ to be so large at these temperatures, the
thickness of the region must be limited so that $\int \Phi(h)\,dh$ does not
exceed the sum of direct non-radiative heating and the heating
due to the flow of energy from above, e.g., by conduction. A
steep temperature gradient not only limits the emission due to
$L\alpha$, but also enhances this flow of energy from above.

A temperature plateau in the transition region complicates
this interpretation since a shallow temperature gradient inhibits
conduction and presumably other ways of transporting energy from
high to low temperatures. Nevertheless, the plateau is "real"
in that it is a measure of the amount of hydrogen at intermediate-
ionization temperatures along the line of sight. Note that
$\int \Phi(h)\,dh$ for $L\alpha$ corresponds to the total emitted $L\alpha$ flux and
thus is observationally determined. As mentioned earlier,
two alternative explanations for the plateau are possible. The
first is that there is a localized source of heating in the lower
transition region that can balance the substantial radiative losses
in this region. The second is that the line of sight repeatedly
passes through transition layers associated with small-scale or
large-scale magnetic fields, that the observed $L\alpha$ emission is the
sum of the contributions from such layers, and that the basic
temperature stratification along the magnetic field lines does
not include a plateau.

The chromospheric portion of each model can be interpreted more easily, because the calculated $\Phi(h)$ should be equal to the mechanical energy input as a function of height. Thus the chromospheric distribution of $\Phi(h)$ provides a direct constraint on theories of chromospheric heating.

I am grateful to Rudolf Loeser and Sallie Baliunas for their help with the computational work reported here and to Lee Hartmann for many useful discussions. This research was supported by NASA Grant NSG-7054.

REFERENCES

(1)   Ayres, T.R. and Linsky, J.L.: 1975, Astrophys. J. 200, pp. 660-674.

(2)   Baliunas, S.L., Avrett, E.H., Hartmann, L., and Dupree, A.K.: 1979, Astrophys, J. (Letters) 233, pp. L129-L133.

(3)   Basri, G.S., Linsky, J.L., Bartoe, J.-D.F., Brueckner, G., and Van Hooser, M.E.: 1979, Astrophys. J. 230, pp. 924-949.

(4)   Bonnet, R.M., Bruner, E.C., Jr., Acton, L.W., Brown, W.A., and Decaudin, M.: 1980, Astrophys. J. (Letters) 237, pp. L47-L50.

(5)   Bonnet, R.M. and Delache, Ph.: 1977, The Energy Balance and Hydrodynamics of the Solar Chromosphere and Corona, Proceedings of IAU Colloquium No. 36, G. de Bussac Clermont-Ferrand.

(6)   Gouttebroze, P., Lemaire, P., Vial, J.C., and Artzner, G.: 1978, Astrophys. J. 225, pp. 655-664.

(7)   Machado, M.E., Avrett, E.H., Vernazza, J.E., and Noyes, R.W.: 1980, Astrophys. J. (in press).

(8)   Vernazza, J.E., Avrett, E.H., and Loeser, R.: 1981, Astrophys. J. Suppl. (in press).

ENERGY BALANCE IN SOLAR AND STELLAR CORONAE

C. Jordan and A. Brown

Department of Theoretical Physics, Oxford University

The methods by which observed EUV and X-ray fluxes may be used to derive models of stellar chromospheres and coronae are discussed. The importance of measuring the electron density through spectroscopic techniques is stressed.

Energy may be lost from each region of the atmosphere by radiation, conduction and mass motions. The radiation losses can be calculated from the emission measure distribution derived from line fluxes and the theoretical radiative power function, without knowledge of the temperature and density structure of the atmosphere.

The pressure is required in order to calculate the energy transferred by thermal conduction and mass motions. It is also needed if comparisons are to be made between the energy fluxes implied by observed non-thermal motions and those which could be carried by particular wave processes.

Some examples of the results obtained through applying the methods to particular stars are given. The broad conclusions which may be drawn so far are that main sequence stars have hot coronae with thermal conduction an important energy transfer process, whereas giants and supergiants may have either hot or cool coronae, depending mainly on the surface gravity and the presence of a strong stellar wind.

R. M. Bonnet and A. K. Dupree (eds.), Solar Phenomena in Stars and Stellar Systems, 199–225.
Copyright © 1981 by D. Reidel Publishing Company.

1.   INTRODUCTION

   The main aims of current work on stellar chromospheres and
coronae are to understand the processes by which these atmospheres are heated, and to find how these processes depend on
conditions in the sub-photospheric convective zones. In short -
why do some stars have hot coronae but not others? Observations
of the EUV and X-ray spectra of stars are now being carried out
with these goals in mind.

   The present discussion will be concerned with (a) the methods
by which the terms of the energy balance equation may be investigated and (b) results available so far for a range of late-
type stars. Many of the methods have been developed in the context of the solar atmosphere and are well-known to solar physicists,
but they are included to give a systematic account, beginning
with observation of emission line fluxes and profiles.

   The term 'corona' will be interpreted to include any region
where the emission lines are effectively thin and are excited
predominantly by electron collisions or recombination. Avrett
has discussed the more complex problems of stellar chromospheres
in the previous lecture.

   Section 2 gives the methods by which the temperature and
density structure can be found. Since it is essential to measure the electron density or pressure a range of possible diagnostic techniques is discussed.

   The processes by which energy is lost from the atmosphere,
or transferred within it are the subject of Section 3. Mass loss
is only briefly mentioned since it is discussed in other lectures
later in this volume.

   In Section 4 the methods are applied to a range of stars,
including the Sun, to show how the relative importance of
different processes changes with spectral type and luminosity
class.

2.   DERIVATION OF STRUCTURE FROM OBSERVED LINE FLUXES

(a)   The Emission Measure Distribution

   The methods used for finding the emission measure distribution, and relative abundances in the solar atmosphere were
pioneered by Pottasch(1).

   Each emission line flux observed at the earth, $F_\oplus$, can be
expressed as

$$F_{\oplus} = \tfrac{1}{2} hc \int_v N_e^2 A_{21} \, dV/4\pi \, d^2 \lambda. \qquad (1)$$

where hc/λ is the quantum energy, N_2 is the population density
of the excited level, A_{21} is the spontaneous transition probability, d is the distance to the star, and the factor $\tfrac{1}{2}$ is an
approximation to the fraction of photons emitted in the outwards
direction.

It is assumed that the line is optically thin. Provided
no other decay process competes with spontaneous decay in the
line, equation (1) is also valid for lines which have optical
depth, τ, greater than unity. The opacity will affect the line
profile but not the total flux.

In statistical equilibrium N_2A_{21} can be expressed in terms
of the excitation processes. In main-sequence stars the lines
from ions are predominantly excited through electron collisions,
but recombination (and radiative processes) may be important
for lines of neutral atoms. In giant and supergiant stars
recombination and other processes may be relatively more important.

Using collisional excitation as an illustration,

$$N_2 A_{21} = C_{12} N_e N_1 \qquad (2)$$

where C_{12} is the collisional excitation rate, N_e is the electron
density, N_1 is the lower level (usually the ground state) population. Statistical equilibrium should be valid, except perhaps
for a few metastable levels, although ionization equilibrium
may not be achieved (see below). Wherever the lower level
population itself depends on density some iteration using
approximate values of N_e may be necessary.

Expressing the collisional excitation rate in terms of an
average collisional strength, Ω_{12}, (2) gives

$$C_{12} = 8.6 \times 10^{-6} \Omega_{12} \exp(-W_{12}/kT_e)/T_e^{\tfrac{1}{2}} \, \omega_1 \qquad (3)$$

where ω_1 is the statistical weight of the lower level and W_{12}
is the excitation energy.

N_1 is expressed in terms of other quantities which can be calculated or measured,

$$N_1 = \frac{N_1}{N_{ion}} \cdot \frac{N_{ion}}{N_E} \cdot \frac{N_E}{N_H} \, N_H \tag{4}$$

where N_1/N_{ion} must be estimated, but for ions without metastable levels will be $\simeq 1.0$; N_{ion}/N_E is the fractional population of the ion, which can be calculated simply if ionization equilibrium is appropriate, N_E/N_H is the element abundance. If $T_e \gtrsim 2 \times 10^4$K, $N_H \simeq 0.8 \, N_e$; below this temperature N_e/N_H should be calculated as part of the iterative modelling procedure. For stars it cannot of course be automatically assumed that abundances are solar, but if these are adopted, abundance anomalies will be revealed in the resulting emission measure distribution.

Regarding ionization balance, Brown et al. (3) have pointed out that in the cool, low density chromospheres of giants and supergiants the times for ionization and recombination are long. In the presence of mass motions and temperature gradients ionization equilibrium may not be a valid approximation. Dupree et al. (4) have also discussed non-equilibrium in the context of the solar transition region.

Putting the temperature dependent parts of equations (3) and (4) as

$$g(T_e) = T_e^{-\frac{1}{2}} \exp(-W_{12}/kT_e) \, N_{ion}/N_E \tag{5}$$

and assuming, initially, that there is a spherically symmetric uniform atmosphere, gives for the line flux, at the star,

$$F_* = \frac{hc}{\lambda} \, 8.6 \times 10^{-6} \, \frac{\Omega_{12}}{\omega_1} \, \frac{N_E}{N_H} \, \int_{\Delta h} N_e^2 \, g(T) \, dh \tag{6}$$

where $F_* = F_\phi \, (2/\theta)^2$

and θ is the stellar angular diameter in radians. Δh is the region over which a line is predominantly formed. The effects of a non-uniform surface distribution are discussed below.

Since $\int_{\Delta h} N_e^2 \, g(T) \, dh$ may vary rapidly with T_e, particularly in the chromosphere, the common procedure of replacing g(T) by an average value and removing it from the integral should be examined. Several approaches are possible (eg. 1,5,6). Following

Jordan and Wilson (5), each $g(T)$ function is computed over a
wide range of temperature, say that over which $g(T)$ has decreased
by two orders of magnitude from its peak value; the percentage
of the area contained by $\log T_m \pm 0.15$ is calculated and then
the normalization factor required to approximate the region contained within $\log T_m \pm 0.15$ by a constant $g(T)$ is found. T_m is
the temperature of the peak of the $g(T)$ function. Then

$$\int_{\log T_1}^{\log T_2} g(T) \, d \log T = G \cdot g(T_m) \tag{7}$$

and G is the combined normalization constant. T_1 and T_2 are the
temperatures well outside $\log T_m \pm 0.15$. It is important to
adopt a constant temperature width rather than a constant fraction of $g(T_m)$ otherwise the different shapes of the $g(T)$ curves
will lead to errors in relative abundances. Once the initial
emission measure distribution is found, taking $G \cdot g(T_m)$ outside
the integral, an iteration should be performed to take into
account the variation of $\int_{\Delta h} N_e^2 \, dh$.

Alternatively $\int_{\Delta h} N_e^2 \, dh$ can be plotted as a function of T_e,
using the actual values of $g(T)$. These loci of the value of
$\int_{\Delta h} N_e^2 \, dh$ sufficient to produce the observed line emission place
strong constraints on the acceptable emission measure distribution
since the final function must be single valued at each T_e.
An example of the resulting distribution of $\int_{\Delta h} N_e^2 \, dh$ with
temperature, for Procyon (α CMi, F5 IV-V)
is shown in Figure 1, taken from Brown and Jordan (7). Such
distributions form the starting point of further modelling.

(b) Methods of Determining the Density and Pressure

The emission measure may be written as

$$E_m = \int_{\Delta h} N_e^2 \, dh \tag{8}$$

and expressed in terms of $P_e = N_e T_e$ and dh/dT such that

$$E_m = \int_{\Delta h} \frac{P_e^2}{T_e^2} \cdot \frac{dh}{dT} \cdot dT \tag{9}$$

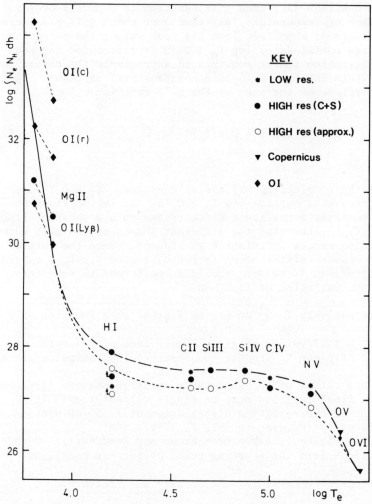

Figure 1

The emission measure distribution for Procyon (7) using fluxes
from IUE spectra obtained at high and low resolution. Some
data from the Copernicus satellite are included (22). The points
for O I refer to excitation by collisions (c), recombination (r)
and via Ly β. The full line is from the model by Ayres et al.
(61).

In order to proceed beyond this point assumptions must be made.
There is a variety of these in the literature, and a full review
will not be given. Since the energy input function is the unknown quantity which is sought assumptions concerning this or
the relative magnitudes of the radiative and conductive fluxes
will be deliberately avoided. (See below also). Without such
assumptions it is essential to know P_e, in order to find the
conductive flux and to compare specific energy input processes
with energy losses deduced from the observations.

The high fluxes observable from solar EUV and X-ray lines
allow both good spatial and spectral resolution to be achieved.Also
ratios of lines involving weak transitions can be measured accurately. Thus methods of measuring N_e through density sensitive line
ratios can be widely applied in solar physics. Several reviews
of these diagnostic techniques have been given recently (8,9,10).
Unfortunately many of the methods cannot be applied to the spectra
of late-type stars now being obtained with the International
Ultraviolet Explorer satellite (IUE) because the lines involved
are too weak to observe at the high resolution required. Some
exceptions are discussed below.

Ideally lines from the same ion, having different dependences
on N_e, and little dependence on T_e, should be used. Pairs of
lines from the ions C III and Si III are in principle suitable(11,13),
depending on the range of N_e, but it is difficult with IUE to
observe both components of the line pairs. A line from one ion
can be compared with a line from a different stage of ionization,
but this is clearly not ideal. Doschek et al. (12) and Cook and
Nicolas (13) have discussed methods involving ratios of C III
to Si III and C III to Si IV, in the context of IUE observations.
However, in stars earlier than around G 8 the strength of the
continuum above $\sim$ 1800 A makes it difficult, or impossible, to
observe the C III and Si III lines at 1909 A and 1892 A. In
giants and supergiants later than around K 2 these lines are
not present, the features observed at low resolution being due
to lines of S I (14). Thus the method seems to be limited to
main sequence stars later than $\sim$ K 0 and perhaps a few giants
around K 0. (Some binary systems may have exceptional spectra,
eg. HR 1099, G5 V, (15)).

For bright objects where high resolution observations are
possible the density sensitive O IV lines around 1400 A could be
resolved from the stronger Si IV doublet, but their absolute
intensity is   usually rather low.

In the late type giants and supergiants, Brown et al. (3)
have pointed out that line ratios within the C II multiplet at
$\sim$ 2335 A can be used to measure N_e in the range 10^7 - 10^9 cm^{-3}.
The necessary transition probabilities have been calculated by

Dankwort and Trefftz (16) and Jackson (17) has calculated the total collision stength. For example application of this method to β Gru (M2 II) gives $N_e \sim 10^8$ cm^{-3}. The long wavelength spectra of late-type stars can be observed at high resolution with moderate exposure times and the method seems potentially useful.

If even long exposures at high resolution do not show the density sensitive lines an alternative way of at least limiting N_e can be used. This method, given below, derives from analyses of limb to disk line ratios in the solar atmosphere (18) and has been applied to Procyon (7,19).

The optical depth at line centre, τ_o, in a Doppler broadened line can be written in terms of the absorption coefficient, leading to

$$\tau_o = 1.2 \times 10^{-14} \, \lambda(A) \, f_{12} \, M_i^{\frac{1}{2}} \, \frac{N_E}{N_H} \int \frac{N_{ion}}{N_E} \frac{N_1}{N_{ion}} N_H \, T_i^{-\frac{1}{2}} \, dh \qquad (10)$$

where f_{12} is the oscillator strength, and M_i the atomic weight of the atom or ion concerned. The ion temperature can be expressed in terms of observed FWHM of the line, $\Delta\lambda$, since

$$T_i^{\frac{1}{2}} = \Delta\lambda \, M_i^{\frac{1}{2}}/7.1 \times 10^{-7}\lambda \qquad (11)$$

Now the expression for τ_o and that for the flux (equation 6) contain many similar quantities, since Ω_{12} can be expressed in terms of f_{12} and $\bar{g}$ the gaunt factor, i.e.

$$\Omega_{12}/\omega_1 = 1.6 \times 10^{-2} \, f_{12} \, \bar{g} \, \lambda(A) \qquad (12)$$

Thus the ratio of the opacity to the surface flux can be written, approximately as

$$\tau_o/F_* = 6.1 \times 10^{-6}\lambda^2(A)T_e^{3/2} \, \exp(W_{12}/kT_e)/\bar{g} \, P_e \, \Delta\lambda \qquad (13)$$

Although in some cases (see below) τ_o can itself be measured through the ratios of lines from a common upper level (20), in main sequence stars only the O I triplet may be suitable. It is also possible to identify which lines have $\tau > 1$ through departures from the ratios of central intensities expected under optically thin conditions, for lines within a multiplet. The

appearance of self-reversals in lines also indicates a high
opacity. Then limits can be placed on P_e from equation (13).
When applied to Procyon this method places quite stringent
limits on P_e (7).

    The opacity can be measured directly for many lines of
neutral atoms and Fe II observed at high resolution in the spectra of late-type giants and supergiants. Although this can lead
to $\int_{\Delta h} N_H \, dh$, which is valuable, the line formation processes
are complex and it is not easy to combine the flux and opacity
measurements (3).

    A method of finding the <u>minimum</u> pressure has been developed
(21,22). It is assumed that the hottest observed line is formed
in an isothermal corona over a height such that

$$d \ln P_e/dh = -0.43/\Delta h = -1/H \qquad (14)$$

where $H = 1.4 \times 10^8 T_{max}/g_*$ is the scale height. Then

$$P_{min} = 1.3 \times 10^{-4} (E_m \, g_* \, T_{max})^{\frac{1}{2}} \qquad (15)$$

is the minimum pressure. (For a collisionally excited line with
flux $\propto N_e^2$ a value a factor $\sqrt{2}$ smaller would be more appropriate.)

    In spectra obtained with IUE the hottest line observed
varies between C II and N V, but emission measures obtained from
X-ray observations can be used in the same way, and are indeed
preferable.

    Upper limits to the pressure in the corona can be found
from models of stellar chromospheres made by obtaining an optimum fit to the fluxes and profiles of lines such as Ca II H and K
and Mg h and k. The electron and gas pressure is calculated as
part of the modelling procedure, and the total pressure at about
8000 K can be used to limit the electron pressure at higher
temperatures. Some models for individual stars are mentioned in
Section 4.

    Even though the pressure may be measurable at one or two
temperatures it is necessary to make assumptions concerning the
pressure variation as a function of height. It is usual to
assume either that $P_e = $ const., (below the corona) on the grounds
that the atmosphere is much less extended than the local isothermal scale height, or that the pressure varies according to
hydrostatic equilibrium. The latter assumption is easy to
include in the formulation and is adopted below on the grounds
that it is better than assuming a constant pressure. In the

presence of strong flows or turbulent motions, this will still
only be an approximation.

Just as the pressure associated with the random motions of
individual electrons is $P_e = N_e k T_e = \frac{1}{2} m_e V_e^2$, so random large
scale turbulent motions can be considered as exerting a pressure.
Thus the pressure associated with the turbulent motions is

$$P_T \simeq \frac{1}{2} \rho \overline{V}_p^2 \qquad\qquad (16)$$

where $\overline{V}_p^2$ is the mean square velocity amplitude.

The total pressure then becomes

$$P_{tot} = P_g + \frac{1}{2} \rho \overline{V}_p^2 \qquad\qquad (17)$$

where $P_g = 1.8 P_e$ is the total gas pressure (for $T_e \gtrsim 2 \times 10^4 K$).

The pressure associated with the non-thermal motions in
the solar atmosphere is not large and causes an increase to P_e
of only $\sim 25\%$. To give an upper limit, if $V_p = C_s$, the sound
velocity, the additional term is only $\sim 80\%$ of P_g. Thus unless
super-sonic turbulence is present the correction to P_e lies
within the uncertainty inherent in the diagnostic methods
discussed above.

In the absence of flows the equation of hydrostatic equilibrium is

$$d P_e/dh = -7.1 \times 10^{-9} P_e g_*/T_e \qquad\qquad (18)$$

Allowing for a flow over a constant cross-section A leads to
the additional term

$$\frac{1}{A} \rho^2 u^2 \frac{d}{dh} (1/\rho A) \qquad\qquad (19)$$

where u is the local flow velocity.

Flows will be excluded from the analysis which follows.

(c)   The Temperature Structure

If P_e is slowly varying then it is a good approximation to
remove it from under the integral in equation (9), which refers
only to the region of line formation, not the whole atmosphere.

The temperature gradient is also, initially, assumed to be constant over Δh. Then

$$d\ T_e/dh = P_e^2/1.4\ E_m T_e \tag{20}$$

where Δh has been taken as the region where $\Delta \log T_e = \log T_m \pm 0.15 = 0.30$. Other authors prefer to remove $T_e^{5/2e} dT_e/dh$ in regions where the conductive flux is almost constant. Then the factor $\sqrt{2}$ in (20) is replaced by 3/2. The assumption of dT_e/dh constant over Δh appears crude but it should be remembered that for the solar atmosphere E_m is known at intervals less than 0.30 in $\log T_e$.

Equation (20) can be combined with equation (18) to give

$$P_e\ d\ P_e/d\ T_e = -7.1 \times 10^{-9}\ g_*\ 1.4\ E_m \tag{21}$$

or $\quad P_e^2 = P_T^2 + 2.0 \times 10^{-8}\ g_*\ \int_{T_e}^{T_T}\ E_m\ dT \tag{22}$

where T refers to the highest T_e at which E_m is   known.

Thus if the pressure and E_m are known at one temperature the variation of P_e and T_e with height can be found.

Figure 2 shows the temperature structure for Procyon for three boundary conditions on P_o, the pressure at $2 \times 10^5 K$(7).

(d)   Effects of Limiting Emitting Area

Suppose that the emission in the EUV lines comes solely from an area $A = X(T_e).4\pi R_*^2$, where $X(T_e) \leq 1$. In deriving the emission measure distribution all values must be <u>increased</u> by a factor $1/X(T_e)$. For the <u>same</u> pressure, assuming these are measured from line ratios, the temperatures gradient would be reduced by the same factor.

The estimate of the minimum pressure from equation (15) will be increased, but by a factor of $1/X(T_e)^{\frac{1}{2}}$. In this case combining P_e with E_m in equation (20) gives no change in dT_e/dh, at the top of the atmosphere. If $X(T_e)$ was constant with T_e there would also be no change in dT_e/dh lower in the atmosphere, but if $X(T_e)$ decreases with decreasing T_e, as is likely, there would be some increase in dT_e/dh at the lower temperatures.

Multi-component models can be made in the same way. Since it is difficult to find the departures from a uniform stellar atmosphere it is pointed out that a comparison between P_e measured from line ratios and P_e measured from opacity arguments can in

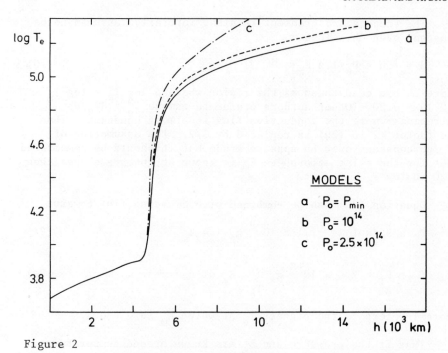

Figure 2

Models of the temperature structure of the atmosphere of Procyon
(7) for three values of the pressure at 2×10^{5}K. At temperatures
below 8000 K the model of Ayres et al. (61) is used.

principle reveal a non-uniform surface emission, since the latter
method depends on the emitting area and will overestimate P_e if
this is assumed to be the whole stellar surface. In practice
the accuracy obtainable at present may be insufficient.

3.   THE ENERGY BALANCE

     It is necessary to model the atmosphere in order to deter-
mine the net conductive flux through particular regions. In
many stars a simple estimate of the net conductive flux shows
that it is unimportant compared with radiation losses. In such
cases the energy input may balance only the radiation losses,
which do not depend on the model, only the emission measure
distribution. However, in order to make a comparison between
this empirical energy input and particular processes a model is
required.

     The energy balance equation can be written as

$$\Delta F_m = \Delta F_R + \Delta F_c + \Delta F_k \tag{23}$$

where ΔF_m is the energy flux required to balance ΔF_R - the radiation losses, ΔF_c the net conductive flux and, ΔF_k the net energy flux from flows. F_k is given by

$$F_k = \tfrac{1}{2}\rho u^3 + 5\rho \frac{k}{M_H} Tu - \rho g_* Ru \tag{24}$$

these terms arising from the kinetic energy flux, the enthalpy flux and the rate of doing work against the gravitational field. (See references (23) and (24) for discussions of energy and momentum flux).

ΔF_R and ΔF_c will now be formulated in terms of the parameters used for modelling. The theory of stellar winds will be treated in later papers in this volume.

(a)   The Radiation Loss

Using a temperature range $\Delta \log T = \log T_m \pm 0.15$, since this is the typical range of T_e over which the emission lines are formed the flux lost by radiation from such a layer is

$$\Delta F_R(T_e) = \int 0.8\, N_e^2 P_{rad}(T_e)\, dh \quad \text{erg cm}^{-2}\text{s}^{-1} \tag{25}$$

(expressions involving $\dfrac{dF}{dT} = \dfrac{0.43}{0.30} \dfrac{1}{T_e}$ can be used if preferred).

The radiative power loss function has been calculated by several authors (25,26,27). It is usual to make a series of straight line fits to the power loss function and there is no large discrepancy between that adopted by different authors, except where a decreasing gradient over the whole range from $\sim 2 \times 10^4$–10^7K is assumed. This should be avoided because the sign of the gradient changes at around 2×10^5K.

Over each region Δh the radiation flux can be written approximately as

$$\Delta F_R(T_e) \simeq 0.8\, E_m\, P_{rad} \tag{26}$$

Thus P_e is not required explicitly, only the emission measure distribution.

The calculations of Post et al. (27) are particularly useful for stellar work since the contribution from each element is given individually. The effects of different stellar abundances can then be taken into account if necessary. If the atmosphere is not static but instead ions are caused to pass rapidly across a steep temperature gradient then enhancement of radiation may occur through the sensitivity of the collisional excitation rate to T_e. For example the emission from helium lines is particularly sensitive to such processes (28,29). The related change in ion populations, for example for the lithium-like ions, may also cause enhanced radiation (4).

As mentioned above the low density in the envelopes of giants and supergiants leads to long ionization and recombination times. The radiation losses under these circumstances bear closer examination.

(b)   The Net Conductive Flux

The conductive flux is given by

$$F_c = -\kappa \, T_e^{5/2} \, dT/dh \tag{27}$$

where $\kappa \simeq 10^{-6}$.
For dT/dh from equation (20)

$$F_c = -\kappa \, P_e^2 \, T_e^{3/2}/\sqrt{2} \, E_m \tag{28}$$

Thus in the region above 2×10^5K, where in the Sun the emission measure gradient is $\simeq 3/2$, an assumption of constant pressure obviously leads to F_c = const. (30). It is not necessary to use P_e = const. and the formulation below continues with P_e varying according to hydrostatic equilibrium. From (34),

$$\frac{dF_c}{dT_e} = \frac{-\kappa}{\sqrt{2}} \, \frac{d}{dT_e} \, \frac{P_e^2 \, T_e^{3/2}}{E_m} \tag{29}$$

$\frac{dP_e}{dT}$ can be found from equation (21) in terms of P_e, E_m and g_*, and after some algebra, the general expression for ΔF_c becomes

$$\Delta F_c = 9.9 \times 10^{-9} \, g_* \kappa \, T_e^{5/2} - 0.49 \, \kappa \, P_o^2 \, \frac{T_e^{3/2}}{E_m} \, (\frac{3}{2} - \frac{d \log E_m}{d \log T_e})$$

$$- 9.9 \times 10^{-9} \, g_* \, \kappa \, \frac{T_e^{3/2}}{E_m} \, (\frac{3}{2} - \frac{d \log E_m}{d \log T_e}) \int_{T_e}^{T_o} E_m \, dT \tag{30}$$

where P_o and T_o refer to $T_e = 2 \times 10^5$K.
This reduces to the form given by Jordan (31) if E_m is replaced by a general power law $E_m = a \, T_e^b$.

It is clear from equation (30) that ΔF_c does not depend only on the emission measure gradient, but also the absolute value of E_m unless $d \log E_m/d \log T_e = 3/2$. Equation (30) then reduces to the simple form (first term) given by Jordan (32), when conductive is a <u>loss</u> term. When $d \log E_m/d \log T_e < 3/2$ conduction becomes an energy <u>deposition</u> term.

At a given temperature the ratio of $\Delta F_R/\Delta F_c$ always depends on E_m, the local absolute value. The variation of $\Delta F_R/\Delta F_c$ with T_e depends on the emission measure gradient and on 'a'.

If the terms of equation (30) are examined using typical solar values or values from the analysis of Procyon it is found that once $d \log E_m/d \log T_e \leq 0$ (which is below $\sim 2 \times 10^5$K) then the second term dominates. For typical giants, where P_e, E_m and g_* are all lower, this may not be so, but the absolute value of ΔF_c may be far smaller than ΔF_R.

Therefore, for near main sequence stars, <u>below</u> $\sim 2 \times 10^5$K, conduction deposits energy whilst radiation loses energy. In the sun it has been known for many years that the energy deposited by conduction at the base of the transition region is far larger than is radiated away in the layer immediately below. It has been suggested that this energy deposition will drive various types of motions until the energy is carried to a region from whence it can be radiated (31,33-36).

The condition for radiation to stably dispose of the conducted energy is $\Delta F_R \geq \Delta F_c$
i.e.

$$0.8 \; E_m^2 \; P_{rad} \geq 0.49\kappa \; P_o^2 \; T_e^{3/2} \; (\frac{3}{2} - \frac{d \log E_m}{d \log T_e})$$ (31)

where $d \log E_m/d \log T_e < 3/2$,
which is independent of gravity for near main-sequence stars.

Alternatively one can say that all energy conducted back from a hot corona at $T_e > 2 \times 10^5$K must eventually be radiated away, but since the majority of the conducted energy is deposited immediately simply choosing a base temperature T_b such that

$$F_c \; (at \; T_o) = F_R \; (T_b \; to \; T_o)$$ (32)

ignores the problem of the stability. To make correct use of this boundary condition the radiation losses in a <u>dynamic</u> atmosphere would need to be considered. Dynamic models will be required for stars which appear to have hot coronae and steep transition region temperature gradients.

The use of either $\Delta F_R = \Delta F_c$ or F_c (at T_o) = $F_R(T_b$ to T_o) to determine P_o should be avoided. If conduction is large, P_o will be underestimated, but if conduction is small and ΔF_R is balanced only by direct deposition of mechanical energy (as in Procyon) P_o will be overestimated.

(c)   The Region Above 2×10^5K

The formulation for ΔF_R and ΔF_c given above is general
and can be applied provided the local emission measure is known.
EUV observations, for example for the IUE satellite, allow the
region up to $T_e \sim 2 \times 10^5$K to be studied. Lines formed at higher
temperatures are observed above 1000 A in the solar spectrum but
they are either weak magnetic dipole transitions or transitions
between excited states (eg. O VII 1s2s ^{3}S-1s2p ^{3}P). X-ray
observations can be used to investigate the region above
$T_e \sim 10^6$K. In order to find the total radiation loss or the
structure of the atmosphere between 2×10^5K and 10^6K some inter-
polation must be adopted. Observations of fluxes in two different
energy bands in the X-ray region can be used to determine an
average temperature and the corresponding emission measure; this
approach has been widely applied to solar observations. Some
measurements of X-ray fluxes, or upper limits to these, are
available from rocket flights, the ANS satellite, HEAO I satellite
and Einstein observatory (37-42).

The principle of finding the volume emission measure from the
X-ray fluxes is essentially the same as that used for individual
lines, but the total contribution of the lines and continuum in
the relevant wavelength range must be known as a function of T_e
(43,44). If fluxes from only one energy band are observed then
$\int_V N_e^2 \, dV$ cannot be found uniquely, but the range of suitable
combinations of $\int_V N_e^2 \, dV$ and T_c can be investigated. The corresponding values of ΔF_R for the coronal region can then be
established.

There is no agreed method of interpolating through the
region between 2×10^5K and 10^6K. Also controversy exists concerning suitable scaling laws for coronal parameters. Rosner
et al. (45) have proposed a scaling law based on constant pressure
and boundary conditions on F_c. Whilst this may adequately fit
small dense closed loop structures in the solar atmosphere the
assumption of constant pressure makes it unsuitable for application to open stellar coronae as a whole. Mullan (46,47) has
predicted coronal parameters by extending Hearn's (48) minimum
energy flux concept.

The present writers prefer to assume that the emission
measure gradient above $T_e \sim 2 \times 10^5$K has a value of $\simeq 3/2$, as
found in the solar atmosphere. The justification is that there
appears to be underlying cause in the stability of the atmosphere
leading to this particular gradient (31). If the emission
measure above 2×10^5K is written as

$$E_m = a \, T_e^{3/2} \qquad\qquad\qquad\qquad (33)$$

then equation (22) for hydrostatic equilibrium becomes

$$P_e^2 = P_o^2 - 8.0 \times 10^{-9} \; a \; g_*(T_e^{5/2} - T_o^{5/2}). \qquad (34)$$

Defining T_c, the 'coronal' temperature as that temperature at which P_e and $dT/dh \rightarrow 0$ gives

$$T_c^{5/2} - T_o^{5/2} = 1.2 \times 10^8 \; P_o^2 / \; a \; g_* \qquad (35)$$

where T_o and P_o are the temperature and pressure at $2 \times 10^5 K$. Thus if E_m and P_o are known at $2 \times 10^5 K$, the coronal temperature can be predicted. This method has the advantages that neither constant pressure nor a particular form of the heating function are assumed, but the hypothesis that the emission measure gradient is constant from star to star is as yet unproven. Further X-ray observations are required before a satisfactory method of treating the coronal parts of the atmospheres can be established.

With the same formulation the radiation losses above T_o can be summed to give

$$F_R(T_o) = 1.6 \times 10^{-16} a \; (T_c^{1/2} - T_o^{1/2}), \qquad (36)$$

where $P_{rad} = 6.0 \times 10^{-17}/T_e$ has been used.

The conductive flux back at T_o, from equations (28) and (35) becomes

$$F_c(T_o) = 5.6 \times 10^{-15} \; g_* \; (T_c^{5/2} - T_o^{5/2}) \qquad (37)$$

(d) Use of Line Widths

If high resolution observations are available then it is possible to use the widths of optically thin lines to examine the non-thermal motions present. The method below has been used by McWhirter et al. (26) and later authors. For Doppler broadening and a turbulent velocity (most probable velocity), ξ_o, the full-width at half maximum (FWHM) $\Delta\lambda$ is given by

$$\frac{\Delta\lambda}{\lambda} = 7.1 \times 10^{-7} \; (\frac{T_e}{M_i} + \xi_o^2 \; \frac{M_p}{2k})^{\frac{1}{2}} \qquad (38)$$

where M_i is the ion to proton mass ratio. The non-thermal energy density, E, can be expressed in terms of the r.m.s. velocity, $<V_T^2> = 3/2 . \xi_o^2$,

$$\text{as } E = \tfrac{1}{2} \; \rho <V_T^2> \qquad (39)$$

If this energy is related to the passage of waves through the atmosphere, the energy flux, ϕ_m, can be written as

$$\phi_m = 2 \ EC$$

where C is the appropriate propagation velocity. The variation of ϕ_m with temperature can then be compared with dissipation rate implied by the values of ΔF_m discussed in the previous section.

Given the uncertainties in present measurements of P_e and line widths the comparisons may not lead to firm conclusions but in the region up to 2 x 10^5K, where ΔF_m is determined mainly by ΔF_R useful constraints may emerge.

The obvious wave processes with which comparisons may be made are sound waves, Alfvén waves and shock waves. The propagation velocities for sound waves and Alfvén waves are, respectively,

$$C_S = (\gamma P/\rho)^{\frac{1}{2}} \simeq 1.5 \ x \ 10^4 \ T_e^{\frac{1}{2}} \qquad cm \ s^{-1}$$

and $\quad C_A = B/(4\pi\rho)^{\frac{1}{2}} \simeq 2 \ x \ 10^{11} \ B \ T_e^{\frac{1}{2}}/P_e^{\frac{1}{2}} \quad cm \ s^{-1}.$

Following Kuperus (49) the flux carried by shock waves can be written as

$$\phi_m = 4\rho \ C_S^3 \ (M^2 - 1)^2/3 \ (\gamma+1)^2 M \tag{40}$$

where M is the Mach number.

This type of approach can be used to show that in the solar atmosphere at $T_e \sim 3 \ x \ 10^4$K a pure acoustic flux cannot carry sufficient energy to heat the corona above. On the other hand fields of only 2 Gauss lead to Alfvén wave fluxes equal to the acoustic flux. Heating mechanisms will not be discussed further in this paper. Discussions of the rôle of acoustic heating in the low chromosphere may be found in the series of papers by Ulmschneider and colleagues (50). Rosner et al. (51) have revived interest in DC heating modes. For stars with magnetic fields and gas pressures not too dissimilar from the sun the succession of MHD wave processes proposed by Osterbrock (52) may well be relevant. In particular the viscous damping of short-period Alfvén waves (in the weak field limit) is difficult to rule out for either the solar atmosphere or that of Procyon (7, 32).

4.   RESULTS

In the long term a detailed study of a wide range of stars will be required in order to investigate the dependence of coronal structure and heating on the stellar parameters. Meanwhile surveys of Mg II fluxes (53,54) have shown distinct differences between

the gravity dependence of Mg II flux and the energy flux predicted from the deposition of acoustic energy. Moreover, the
limited sample of high gravity stars does not show the predicted
flux decrease with decreasing effective temperature. (The flux
is $\propto T_{eff}^4$ rather than T_{eff}^8). These trends can be seen from
Figure 3, taken from ref. (53). The observed fluxes are compared
with calculations of the acoustic flux generated (55) and that
available to the chromosphere (56). These differences are also
apparent in the survey of X-ray fluxes (42).

Some of the methods discussed in the previous sections have
been applied previously to stars for which chromospheric models
and IUE spectra are available (57). Since then the IUE fluxes
have been found to require correction (58). The examples below
are restricted to stars whose emission line fluxes are known to
have been corrected.

(a)   Near Main Sequence Stars

The Sun (G2V). Many of the methods discussed above have
been developed from solar work. The value of solar observations
lies in the high spatial and spectral resolution that can be
achieved, for both strong and weak lines, over a very wide region
of the spectrum. Stellar observations are limited to the integrated fluxes, for the stronger lines, in the region above the
Lyman continuum and below $\sim$ 100 A. EUV fluxes should also be
measurable for a few nearby stars. Thus one must rely on whatever is known about the spatial structure of the solar atmosphere,
and about the rôle of the magnetic field in particular, to obtain
guidance in modelling stellar coronae. However, the 'average'
quiet sun model that would be obtained from integrated fluxes
would give a reasonable idea of the relative importance of the
various total energy loss and deposition terms. For the
purpose of comparisons with other stars the parameters and energy
terms are as follows. The electron pressure at 10^5K and coronal
temperature are 5.6×10^{14} cm^{-3}K and 1.5×10^6K respectively.
The overall radiation losses amount to $\sim 5 \times 10^6$ erg cm^{-2} s^{-1} (23)
with only $\sim 4 \times 10^4$ erg cm^{-2} s^{-1} above $T_e \sim 2 \times 10^5$K. The main
energy loss from the corona is by conduction back to the chromo-
sphere from whence it is radiated, but not apparently in a static
configuration.(31,34) ie. the condition for stable deposition
of the conductive flux (equation 31) is not satisfied. Line
profile measurements show the presence of non-thermal motions, but
interpreted as due to the passage of acoustic waves the flux
carried is insufficient to account for coronal heating (31,59).

Procyon (α CMi, F5 IV-V). The F-stars are important in that
models of the convective zone structure suggest that the acoustic
flux generated is largest around the late F's, dropping rapidly
for earlier type stars (60). Procyon is sufficiently close and

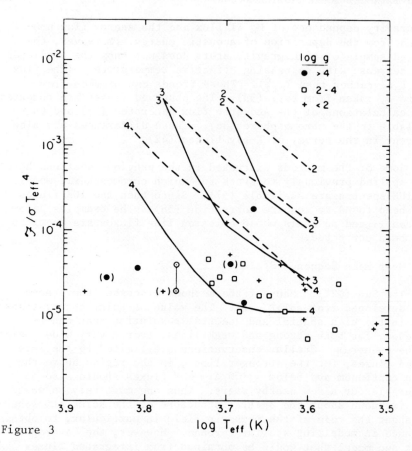

Figure 3

The ratio of Mg II flux to surface luminosity plotted against
the effective temperature. The dashed lines show the theoretical
acoustic flux generated (55) and the full lines show the flux
calculated to be available to the chromosphere and corona (56).
The Figure is from Linsky and Ayres (53).

bright to allow UV spectra to be obtained with IUE at both
low and high resolution, and the methods described above can be
applied (7,19). Also a model of the chromosphere has been
made from Ca II and Mg II profiles and fluxes (61). The emission
measure distribution has been shown in Figure 1. The electron
pressure can be limited through P_{min} and opacity arguments and
the chosen model has $P_e \sim 1.2 \times 10^{14}$ cm^{-3}K at 2×10^5K. The
electron pressure at $\sim 10^4$K is about a factor of three higher
than that in the chromospheric model (61). Figure 2 shows the
temperature structure which results from three pressure boundary
conditions. The 'coronal' temperature in the final model is
$\sim 3 \times 10^5$K. Neither the pressure or temperature agree with the

predictions of heating by the shock dissipation of acoustic flux
(60,62). It is interesting to note that in spite of the low
conductive flux at 2 x 10^5K, ($\lesssim$ 3 x 10^3 erg cm^{-2} s^{-1}) the atmosphere has a steep 'transition region'. Also the large surface
flux of $\sim$ 14 times the solar value, at C IV,does not imply a
correspondingly larger pressure. Line widths have also been
measured and although they show the presence of non-thermal
motions the data are not consistent with the deposition of acoustic
flux. The Mg II flux from Procyon is about three times that in
the sun. The total radiation loss above 10^4K is about a factor
of four greater than from the sun, $\sim$ 2 x 10^6 erg cm^{-2} s^{-1}. Thus
more energy must be deposited, but this appears to be lost through
radiation in a lower pressure, more extended chromosphere than
in the sun. Only upper limits to the X-ray flux from Procyon
are available (38,63) giving L_x < 10^{28} erg s^{-1}. With $T_c \sim$3 x 10^5K
this gives an emission measure less than 3 x 10^{26} cm^{-5}, consistent
with the results shown in Figure 1.

    ε Eri K2 V. ε Eri was observed with IUE in the international
collaborative programme (15). A model of the chromosphere has
been made by Kelch (64). A recent paper by Simon et al. (65)
corrects the previous IUE fluxes and concludes that the electron
pressure must be $\sim$ 3.6 x 10^{15} cm^{-3}K, higher than in the previous
model, which gave 2.3 x 10^{14} cm^{-3}K. X-ray luminosities of
3.8 x 10^{28} erg s^{-1} and 2.0 x 10^{28} s^{-1} have been reported from
HEAO I(41) and the Einstein observatory (42). If the UV surface
flux ratio is combined with the X-ray luminosity through the
'coronal' radiation loss, (equation 26) and the scaling law
given by equation (35), then the values of the transition region
electron pressure and coronal temperature which result are
5.2 x 10^{15} cm^{-3}K and T_c = 3.5 x 10^6K. The conductive flux back
from the corona (from equation 37) is then 4.6 x 10^6 erg cm^{-2} s^{-1},
greater than can be radiated away at around 10^5K (ie. 1.0 x 10^5
erg cm^{-2} s^{-1}), but less than the total which is lost from the
chromosphere. The radiation loss above 2 x 10^5K, from equation(36)
is $\sim$ 9 x 10^5 erg cm^{-2} s^{-1}. The total radiation losses will be
about twice those from the sun, given the larger surface flux of
the Mg II emission and UV emission. This is contrary to the
results expected from the generation of acoustic energy.

(b) Giants and Supergiants

    IUE observations have shown that although evidence for
hot ($\sim$ 10^5K) material is present through C IV emission in giants
earlier than around K 0, even by K2 the lines are barely strong
enough to detect (66,67,68). Supergiants show a similar behaviour,
with C IV being visible in the G2 Ib star, α Aqr, but not in
λ Vel (K5 Ib) (67,68). Linsky and Haisch (69) have suggested
that there is a sharp division in the H-R diagram between stars
which do and do not show C IV emission, the former having hot

coronae like the sun. The position of their dividing line is
similar to that proposed by Stencel and Mullan (70) for the
region where there is evidence of mass out flow from the asymmetry
of the Mg II lines. Other loci relating to evidence of outflow
from Ca II asymmetries (72) and of mass-loss from circumstellar
lines (73) have been proposed.

Although the methods discussed above are in principle
applicable to stars showing predominantly emission from cool
species much of the necessary work on the line excitation
processes and derivation of ion and atom densities has not yet
been carried out.

Two examples of late-type giants are now briefly discussed.

β Gem. KO III. β Gem shows emission from C IV and other
moderately ionized species and is of interest since it lies
close to the dividing line proposed by Linsky and Haisch (69).
A chromospheric model has been calculated from the Ca II and
Mg II fluxes and profiles (74). Some early observations were
made from the Copernicus satellite (75) showing that both H Lyα
and the Mg II lines have profiles with blue rather than red wing
enhancements suggesting that any outflow is not an accelerating
wind. The present authors have obtained a long exposure low
resolution spectrum with the IUE satellite. The surface fluxes
of lines formed between 2×10^4K and 3×10^5K are lower than
from the sun by factors of up to 5.

The chromospheric model (74) gives $P_g \sim 1.3 \times 10^{-2}$ dynes cm^{-2}
at ~ 8000 K, giving an upper limit for P_e of 4.7×10^{13} cm^{-3}K.
The emission measure from the C IV line places a lower limit on
P_e of 7×10^{12} cm^{-3}K. Using these pressures the coronal
temperature is predicted to lie between 1.6×10^6K and
2×10^5K, and the X-ray luminosity to be less than 5×10^{28}
erg s^{-1}. Linsky (private communication) informs us that β Gem
is observed as an X-ray source by the Einstein observatory.

The conductive flux back at 2×10^5K must be $< 1.6 \times 10^4$
erg cm^{-2} s^{-1}, compared with the local radiation loss of
$\sim 4.3 \times 10^3$ erg cm^{-2} s^{-1}. The radiation losses below 2×10^5K
and the Mg II flux are lower than in the sun by about a factor
of four.

Although the results are highly dependent on the pressure,
which is not known to better than a factor of two, a coronal
temperature similar to that of the sun is possible through the
combination of lower pressure, lower emission measure and lower
gravity (see equation 35). The energy requirements are, however,
smaller.

α Tau K5 III. α Tau does not show C IV emission, and the
surface fluxes of lines such as C II and Si II are lower than in
β Gem by around a factor of 15, ie. $\sim 10^{-2}$ the solar values (14).
The spectrum is composed mainly of lines from neutral atoms and
singly charged ions, eg. H Lyα, C I, O I, S I, C II, Si II, Fe II.
The strength of the O I resonance lines is likely to be due to
pumping by H Lyβ as in α Boo (76). This behaviour is typical
of other late-type giants observed with IUE (66,67,77). The O I
lines appear to excite the nearby S I lines, resolved in high
resolution spectra (14). Lines of C I and O I are present which
owe their stength to radiation trapping in lines with a common
upper level (3,14).

The line formation processes require further work but initial
studies of opacity and density sensitive lines suggest $2 \times 10^{19} <$
$\int N_H$ dh $< 8 \times 10^{21}$ cm^{-2}, and $N_e \sim 10^8$ cm^{-3}. The chromospheric
model (74) gives $P_e \sim 6.1 \times 10^{12}$ cm^{-3}K at $\sim$ 7200 K. Assuming
the C II lines to be formed in an isothermal 'corona' at 10^4K gives
$P_{min} \sim 10^{12}$ cm^{-3}K.

The radiation losses from Mg II are lower than in β Gem by
about a factor of five (54). High resolution observations of
H Lyα, Mg II and the O I lines (14,75) reveal strong red wing
enhancements indicating an accelerating outflow. The S I lines
are also asymmetric. Interpreted as a spread in velocities limits
the relative outflow velocity to < 10 km s^{-1}. The energy carried
in a stellar wind then appears to be small compared with the total
radiation losses.

The upper limit on the C IV flux can be used to limit the
temperature gradient at $\sim 10^5$K, and hence estimate the coronal
temperature. If there is really no material above $\sim 10^4$K this is
an artificial procedure and a spuriously high value of T_c would
result. With values of P_e given above T_c would lie between
7.4×10^6K and 1.6×10^6K. Apart from arguments concerning the
escape velocity, α Tau would then emit a substantial X-ray flux
($> 2 \times 10^{28}$ erg s^{-1}), whereas it is not reported as an X-ray
source. A typical upper limt of 6×10^{27} erg s^{-1} for K giants
in the Einstein survey (42) would limit the coronal temperature
to $< 2 \times 10^5$K. Thus the low temperature solution with a real
absense of material above $\sim 2 \times 10^4$K seems appropriate.

The total energy requirement of α Tau appear to be at least
an order of magnitude less than predicted by the deposition of
acoustic flux.

It has become clear over the past few years that neither
the broad properties or detailed structure of late-type
chromospheres and coronae fit the early predictions based on the

generation and propagation of acoustic flux in the sub-photospheric convective zone. These theories must either be substantially revised or the controlling factor sought elsewhere,
for example in the combined efforts of the magnetic field and
convective motions. As the energy requirements for the different
types of stars became available from both UV and X-ray observations other correlations, for example between rotation and
activity (78,79), can be investigated. Even in the cool giants
it may be necessary to invoke magnetic modes (Alfvén waves) to
provide sufficient momentum to drive the mass loss (80). The
change in the structure of the coronae apparent from the decrease
of C IV emission does seem to be closely related to the develop-
me of an accelerating outflow as seen in the Mg II lines (72).
However, the energy input required, as deduced from Mg II fluxes
(see Figure 3 and ref. 54) or IUE spectra, decreases steadily,
rather than suddenly, across the region around K0 III ($\sim$4700 K).
Thus the change in the structure appears to be related more
closely to the large drop in surface gravity for stars evolved
beyond K0 III rather than to a sudden change in the total energy
flux requirements. Cass nelli (81) has recently reviewed the
current state of observations and theories relating to mass loss.
Although the present paper has concentrated on methods of investigating the structure of late type coronae and their relative
conductive fluxes, the momentum flux remains a crucial factor
to be included in any satisfactory theory.

References

1. Pottasch, S.R.: 1964, Space Sci. Rev. $\underline{3}$, p. 816.
2. Seaton, M.J.: 1962, in 'Atomic and Molecular Processes'
       (Ed. D. Bates) Academic Press, p. 375.
3. Brown, A., Ferraz, M.C. de M. & Jordan, C.: 1980, Proc. of
       Symposium 'The Universe at Ultraviolet Wave Lengths:
       The First Two Years of IUE', NASA Special Publication,
       (Ed. R.D. Chapman). In Press.
4. Dupree, A.K., Moore, R.T. & Shapiro, P.R.: 1979, Astrophys.
       J. (Letts.) $\underline{229}$, p. L101.
5. Jordan, C. & Wilson, R.: 1971, in 'Physics of the Solar
       Corona'(Ed. C.J. Macris) D. Reidel Publ. Co., Dordrecht,
       Holland, p. 211.
6. Withbroe, G.L.: 1975, Sol. Phys. $\underline{45}$, p. 301.
7. Brown, A. & Jordan, C.: 1980 , submitted to Mon. Not. R.
       astr. Soc.
8. Dupree, A.K.: 1978, in 'Advances in Atomic and Molecular
       Physics', Vol. 14, Academic Press, New York, p.293.
9. Jordan, C.: 1979, in 'Progress in Atomic Spectroscopy'-
       Part B., (Ed. W. Hanle and H. Kleipoppen), Plenum
       Press, New York & London, p. 1453.
10. Dere, K.P. & Mason, H.E.: 1980, Proc. of NASA Skylab Active
       Region Workshop (Ed. F.Q. Orrall) In Press.
11. Gabriel, A.J. & Jordan, C.: 1973 in 'Case Studies in Atomic
       Collision Physics', Vol. 2 (Eds. E.W. McDaniel and
       M.R.C. McDowell) North-Holland, p. 210.
12. Doschek, G.A., Feldman, U., Mariska, J.T. & Linsky, J.L.:
       1978, Astrophys. J. (Letts.) $\underline{226}$, p. L35.
13. Cook, J.W. & Nicolas, K.R.: 1979, Astrophys. J. $\underline{229}$, p. 1163.
14. Brown, A. & Jordan, C.: 1980, Mon. Not. R. astr. Soc. $\underline{191}$,
       p. 37P.
15. Linsky, J.L., Ayres, T.R., Basri, G.S., Morrison, M.D.,
       Boggess, A., Machetto, F., Wilson, R., Cassatella, A.,
       Heck, A., Holan, A., Stickland, D., Schiffler, F.H.,
       Blanco, C., Dupree, A.K., Jordan, C. & Wing, R.F.:
       1978, Nature $\underline{225}$, p. 19.
16. Dankwort, W. & Trefftz, E.: 1978, Astron. Astrophys. $\underline{65}$,
       p. 93.
17. Jackson, A.R.G.: 1972, J. Phys. B.: Atom. Molec. Phys. $\underline{5}$,
       p. L83.
18. Burton, W.M., Jordan, C., Ridgeley, A. & Wilson, R.: 1971,
       Phil. Trans. Roy. Soc. Lond. $\underline{A270}$, p. 81.
19. Brown, A. & Jordan, C.: 1980, Proc. of Second European IUE
       Conference, Tübingen, ESA SP-157, p. 71.
20. Jordan, C.: 1975, IAU Symposium 68 (Ed. R. Kane) p. 109.
21. Gerola, H., Linsky, J.L., Shine, R., McClintock, W., Henry,
       R.C. & Moos, H.W.: 1974, Astrophys. J. (Letts.) $\underline{193}$,
       p. L107.
22. Evans, R.G., Jordan, C. & Wilson, R.: 1975, Mon. Not. R. astr.
       Soc. $\underline{172}$, p. 585.

23. Athay, R.G.: 1976, 'The Solar Chromosphere and Corona: Quiet
         Sun', D. Reidel Publ. Co.
24. Chiuderi,C. & Kuperus, M.: 1976, in 'The Energy Balance and
         Hydrodynamics of the Solar Chromosphere and Corona'
         IAU Colloquium No. 36 (Ed. R.M. Bonnet & Ph. Delache)
         p. 223.
25. Cox, D.P. & Tucker, W.H.: 1969, Astrophys. J. 157, p. 1157.
26. McWhirter, R.W.P., Thonemann, P.C., Wilson, R.: 1975, Astron.
         Astrophys. 40, p. 63.
27. Post, D.E., Jensen, R.V., Tarter, C.B., Grasberger, W.H.,
         Lokke, W.A.: 1977, Atomic Data and Nuclear Data Tables,
         20, p. 597.
28. Jordan, C.: 1975, Mon. Not. R. astr. Soc. 170, p. 429.
29. Jordan, C.: 1980, Phil. Trans. Roy. Soc. Lond. A297, p. 54.
30. Athay, R.G.: 1966, Astrophys. J. 145, p. 784.
31. Jordan, C.: 1980 , Astron. Astrophys. 86, p. 355.
32. Jordan, C.: 1976, Phil. Trans. Roy. Soc. Lond. A281, p. 391.
33. Giovanelli, R.G.: 1949, Mon. Not. R. astr. Soc. 109, p. 372.
34. Kuperus, M. & Athay, R.G.: 1967, Sol. Phys. 1, p. 361.
35. Kopp, R.A. & Kuperus, M.: 1968, Sol. Phys. 4, p. 212.
36. Moore, R.L. & Fung, P.C.: 1972, Sol. Phys. 23, p. 79.
37. Nugent, J. & Garmire, G.: 1978, Astrophys. J. (Letts.) 226,
         p. L83.
38. Mewe, R., Heise, J. Gronenschild, E.H.B.M., Brinkman, A.C.,
         Schrijver, J. & den Boggende, A.J.F.: 1975, Astrophys.
         J. (Letts.) 202, p. L67.
39. Catura, R.C., Acton, L.W. & Johnson, H.M.: 1975, Astrophys.
         J. (Letts.) 196, p. L47.
40. Cash, W., Bowyer, S., Charles, P.A., Lampton, M. Garmire, G.
         & Riegler, G.: 1978, Astrophys. J. (Letts.) 223, p. L21.
41. Walter, F.M., Linsky, J.L., Bowyer, S. & Garmire, F.: 1980,
         Astrophys. J. (Letts.) 236, p. L137.
42. Vaiana, G.S., Cassinelli, J.P., Fabbiano, G. Giacconi, R.,
         Golub, L., Gorenstein, P., Haisch, B.M., Harnden, F.R.,
         Johnson, H.M., Linsky, J.L., Maxson, W.C., Mewe, R.,
         Rosner, R., Seward, F., Topka, K. & Zwaan, C.: 1980,
         Astrophys. J. In Press.
43. Kato, T.: 1976, Astrophys. J. Suppl. 30, p. 397.
44. Raymond, J.C., Cox, D.P. & Smith, B.W.: 1976, Astrophys. J.
         204, p. 290.
45. Rosner, R., Tucker, W.H., Vaiana, G.S.: 1978a, Astrophys. J.
         220, p. 643.
46. Mullan, D.J.: 1976, Astrophys. J. 209, p. 171.
47. Mullan, D.J.: 1978, Astrophys. J. 226, p. 151.
48. Hearn, A.G.: 1975, Astron. Astrophys. 40, p. 355.
49. Kuperus, M.: 1965, Rech. Astron. Obs. Utrecht, 17, p. 1.
50. Schmitz, F. & Ulmschneider, P.: 1980, Astron. Astrophys. 84,
         p. 191.
51. Rosner, R., Tucker, W.H., Coppi, B. & Vaiana, G.S.: 1978b,
         Astrophys. J. 222, p. 317.

52. Osterbrock, D.E.:   1961, Astrophys. J. 134, p. 347.
53. Linsky, J.L. & Ayres, T.R.: 1978, Astrophys. J. 220, p. 619.
54. Basri, G.S. & Linsky, J.L.: 1979, Astrophys. J. 234, p. 1023.
55. Renzini, A., Cacciari, C., Ulmschneider, P. & Schmitz, F.:
        1977, Astron. & Astrophys. 61, p. 39.
56. Ulmschneider, P., Schmitz, F., Renzini, A., Cacciari, C.,
        Kalkofen, W. & Kurucz, R.L.: 1977, Astron. & Astrophys.
        61, p. 515.
57. Jordan, C.: 1980, in 'Highlights of Astronomy' Vol. 5,
        D. Reidel, Dordrecht, p. 533.
58. Holm, A.: 1979, SRC IUE Newsletter No. 4.
59. Athay, R.G. & White, O.R.: 1978, Astrophys. J. 226, p. 1135.
60. de Loore, C.: 1970, Astrophys. & Space Sc. 6, p. 60.
61. Ayres, T.R., Linsky, J.L. & Shine, R.A.: 1974, Astrophys. J.
        192, p. 93.
62. Kuperus, M.: 1965, Rech. Astron. Obs. Utrecht, 17, p. 1.
63. Cruddace, R., Bowyer, S., Malina, R., Margon, B. &
        Lampton, M.: 1975, Astrophys. J. 202, p. L9.
64. Kelch, W.L.: 1978, Astrophys. J. 232, p. 931.
65. Simon, T., Kelch, W.L. & Linsky, J.L.: 1980, Astrophys. J.
        237, p. 72.
66. Brown, A., Jordan, C. & Wilson, R.: 1979, Proc. of Symposium
        'The First Year of IUE'(Ed. A. Willis), p. 232.
67. Dupree, A.K.: 1980, in 'Highlights of Astronomy' Vol. 5
        D. Reidel, Dordrecht, p. 263.
68. Hartmann, L., Dupree, A.K. & Raymond, J.C.: 1980, Astrophys.
        J. (Letts.) 236, p. L143.
69. Linsky, J.L. & Haisch, B.M.: 1979, Astrophys. J. (Letts.)
        229, p. L27.
70. Stencel, R.E. & Mullan, D.J.: 1980, Astrophys. J. In Press.
71. Linsky, J.L.: in 'Turbulence in Stellar Atmospheres' IAU
        Colloquium 51. In Press.
72. Stencel, R.E.: 1978, Astrophys. J. (Letts.) 223, p. L37.
73. Reimers, D.: 1977, Astron. & Astrophys. 57, p. 395.
74. Kelch, W.K., Linsky, J.L., Basri, G.S., Chiu, H.Y., Chang,
        S.H., Maran, J.P. & Furenlid, I.: 1978, Astrophys. J.
        220, p. 462.
75. McClintock, W., Linsky, J.L., Henry, R.C., Moose, H.W. &
        Gerola, H.: 1975, Astrophys. J. 202, p. 165.
76. Haisch, B.M., Linsky, J.L., Weinstein, A. & Shine, R.A.:
        1977, Astrophys. J. 214, p. 785.
77. Carpenter, K.G. & Wing, R.F.: 1979, Bull Am.astr. Soc. 11,
        p. 419.
78. Skumanich, A.: 1972, Astrophys. J. 171, p. 565.
79. Ayres, T.R. & Linsky, L.J.: 1980, Astrophys J. In Press.
80. Haisch, B.M., Linsky, J.L. & Basri, G.S.: 1980, Astrophys. J.
        235, p. 519.
81. Cassinelli, J.P.: 1979, Ann. Rev. Astron. Astrophys. 17,
        p. 275.

NUMERICAL STUDIES OF THE ENERGY BALANCE IN CORONAL LOOPS

J. H. Underwood*, S. K. Antiochos** and J. F. Vesecky**

*Jet Propulsion Laboratory, Pasadena, California
**Stanford University, Stanford, California

Abstract. In closed magnetic field regions of the solar corona the
plasma is largely confined into loop or arch-like structures of
enhanced density. These loops have a lifetime much longer than
the coronal radiative cooling time, so that a quasi-static model
in which the energy input to each volume element is balanced by
radiative losses, and losses or gains by conduction, is appropriate. Such models are believed to have application to the study
of stellar coronae, and have been treated analytically by other
authors. Our treatment is numerical, and includes the effects of
gravity and the variation of cross-sectional area of a loop along
its length. The programs have been used to examine the effects
of various heat input functions on the distribution of temperature,
density, etc., within the loop. As an example of the use of the
method, we examine the effect of form of the energy input on the
emission measure function, and show that little information regarding the coronal heating mechanism can be gleaned from measurements of spectral line intensities, integrated over a complete
loop or set of loops.

1. INTRODUCTION

In regions of the solar corona outside of coronal holes,
i.e. closed magnetic field regions, the coronal plasma is largely
confined into loop or arch-like structures of higher than average
density. Such features are easily seen on x-ray or extreme ultraviolet (euv) images of the corona. Although loops are often seen
to undergo rapid changes in brightness indicative of a sudden
injection of material or a change in the magnitude of the energy
input, many are observed to last for periods of hours or even

R. M. Bonnet and A. K. Dupree (eds.), Solar Phenomena in Stars and Stellar Systems, 227–233.
Copyright © 1981 by D. Reidel Publishing Company.

days, much longer than the radiative cooling time (∿1000 sec).
For such "quasi-static" loops, it seems appropriate to consider a
steady state model, in which the energy input to each volume element is balanced by losses due to conduction and radiation. Such
models have been treated analytically by several authors (Rosner,
Tucker and Vaiana, 1978; Craig, McClymont and Underwood, 1978;
Jordan, 1980). However, the analytical approach becomes intract-
able when such factors as the variation of loop area along its
length, non-uniform energy input, and the effects of gravity are
considered. For this reason we have developed a numerical model
for quasi-static loops. This numerical method was originally
applied to a study of the simple case in which the energy input
per unit volume of a coronal loop is uniform (Vesecky, Antiochos
and Underwood, 1979, hereafter VAU). In this paper we extend the
method to other energy input functions, with the particular
objective of investigating how the form of the energy input function is reflected in the form of the emission measure versus temperature curve. Since the emission measure function, in effect,
determines the relative strengths of optically thin spectral
lines emitted by the corona and transition zone in the euv and
x-ray regions, this investigation provides an indication of how
much information on the coronal heat input function can be deduced from intensity measurements of such lines.

2. THE MODEL

     As our treatment of the quasi-static loop model is numerical
rather than analytic, we can include: gravity; an energy input
function that is dependent on density, temperature and/or position along the loop; an accurate form for the radiative losses,
and a variable cross-sectional area for the loop geometry.

     Observations in the x-ray region provide the rationale for
taking the area variation into account in the computations. The
assumptions used in the computations are fully discussed by VAU
and need be summarized only briefly here.

2.1 Geometrical assumptions

     The plasma is assumed to be confined in a magnetic loop due
to a line dipole at a depth d below the chromospheric (30,000K)
level (Fig. 1). The coordinate s is measured from the dipole
along the field lines. If the cross-sectional area of the loop
is a at the chromospheric level, it is Γa at the apex; A(s) is the
cross-sectional area at distance s. The length of the loop ℓ is
related to Γ and h, the height of the loop from dipole to apex,
by:

$$\ell = 2h \cos^{-1}(\Gamma^{-\frac{1}{2}}) \qquad\qquad (1)$$

These geometrical assumptions may alternatively be replaced by
others, e.g. those appropriate for a constant area loop.

2.2 Physical assumptions

(i) The energy input by the "coronal heating mechanism" is
locally balanced by radiative and conductive losses, so that:

$$\varepsilon = \nabla.F_c - E_r \tag{2}$$

Here ε and E_r are, respectively, the energy input and radiated
per unit volume, and F_c is the conductive flux along the loop.

(ii) The plasma is optically thin, so that E_r is given by:

$$E_r = n^2 \, \Lambda(t) \tag{3}$$

Where n is the electron density and $\Lambda(t)$ is the radiative loss
function as computed, for example, by Raymond, Cox and Smith
(1976).

(iii) Conduction of energy takes place along the field
lines only, so that (2) may be written in the one-dimensional
form:

$$\frac{1}{A(s)} \quad \frac{d}{ds} \quad \left| A(s) \; K \; \frac{dT}{ds} \right| = n^2 \; \Lambda(T) - \varepsilon \tag{4}$$

Where $K = 10^{-6} \, T^{\frac{5}{2}}$ (Spitzer, 1962)

(iv) The plasma is in hydrostatic equilibrium, so that

$$\frac{dp}{ds} = -\rho \; g_{\parallel} \tag{5}$$

Where ρ = density, p = pressure and $g_{\parallel}$ = component of gravity
along field lines.

2.3 Boundary conditions

Equations (4) and (5) can be rewritten as three first-order
differential equations for dT/ds, dn/ds and dF_c/ds. For their
solution, their boundary conditions must be specified. We have
chosen to specify :

(i) The temperature T_B at the base of the loop. In the
majority of cases, T_B was chosen to be $3 \times 10^4 K$.

(ii) The conductive flux $(F_c)_B$ through the base of the
loop. As VAU have pointed out, there are good reasons to believe

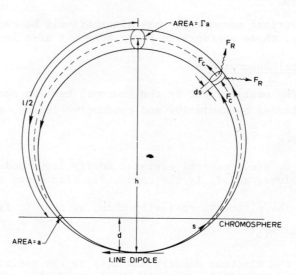

Figure 1.   Geometry of quasi-static loop model.

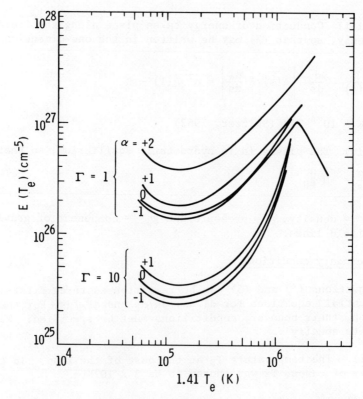

Figure 2.   Emission measure curves for a loop of length 10^{10}cm.

that $(F_c)_B$ is small compared with the conductive flux in the corona, and in the computations presented here, it has been set equal to zero.

(iii) The temperature gradient $(dT/ds)_0$ at the apex of the loop. By symmetry, this is also zero.

Given these boundary conditions, the equations are then solved by a "shooting" technique. A base pressure is assumed, and the equations numerically integrated upward from the base until either a) the top of the loop is reached or b) dT/ds goes to zero. If the boundary condition at the top of the loop is not satisfied, a new pressure is chosen and the integration repeated. This iterative procedure is repeated until $(dT/ds)_0$ is sufficiently close to zero (for details see VAU). As an alternative procedure, one could fix the pressure at some point in the loop (the top, say), and iterate on the total energy input until the boundary conditions were satisfied.

2.4 Definitions

The differential emission measure $\xi(T)$ is defined by:

$$\xi[T(s)] = A(s) \ n^2(s) \ \left|\frac{dT}{ds}\right|^{-1} \tag{6}$$

For comparison with observations of the intensities of optically thin spectral lines in the U.V. and x-ray regions of the spectrum, we define the emission measure $E(T_e)$ by

$$E(T_e) = \int_{0.71Te}^{1.41Te} \xi(T) \ dT \tag{7}$$

Where T_e is the optimum temperature for forming the line (cf. Jordan, 1980). It is evident from these definitions that the absolute value of the emission measure is dependent on the cross-sectional area of the loop.

2.5 Relevance of other assumptions

Equations (4) and (5), together with the appropriate boundary conditions, completely determine the temperature and density structure in a loop. Further assumptions will either be redundant or contradictory. Examples of such assumptions are:

a) $E(T_e) = aT_e^b$

(in particular $b = 3/2$) (Jordan 1980)

(b) "Constant conductive flux"

(c) "Minimum Flux Corona"

$$\frac{d}{dT} (F_r + F_c) = 0 \text{ (Hearn 1975, 1977)}$$

(d) "Constant pressure transition region, isothermal corona."
It is not possible to choose the pressure and temperature scale
heights in the corona and transition region independently. If
the pressure scale height >> temperature scale height in the
transition region, and pressure scale height << temperature scale
height in the corona, there must be an intermediate region where
the two scale heights are comparable.

3. SAMPLE COMPUTATIONS

     In order to examine the conjecture that the emission measure
distribution integrated over an active region loop, obtained by
analysis of spectral lines, can be used to infer properties of
the coronal energy input function such as the scale length for
energy deposition, a theoretical emission measure function was
computed for a number of trial cases. An energy input function
of the type $\varepsilon = \varepsilon_0 (T/10^6)^\alpha$ was adopted. Evidently $\alpha < 0$ is
equivalent to cases in which the energy is deposited primarily
near the base of the loop, $\alpha = 0$ to uniform energy input, and
$\alpha > 0$ to energy deposition primarily near the loop apex.

     The loop length was chosen to be 10^{10}cm. Fig. 2 shows the
emission measure $E(T_e)$ versus temperature variation calculated
for a loop with $\Gamma = 1$, $\varepsilon_0 = 10^{-4}$ ergs/cm^2/sec, and exponents
α of -1, 0, 1 and 2 (top four curves). Similar curves are shown
for $\Gamma = 10$ and $\alpha = -1$, 0 and 1. All curves refer to a loop of
area 1 cm^2 at the apex. It should be noted that after reaching
a peak of $E(T_e)$ at 1.41 T_e = temperature at the top of the loop,
all curves turn over and decrease to zero. This has been shown
only for the $\Gamma = 1$, $\alpha = 0$ case, as we are primarily interested in
the behavior at lower temperatures.

     In the $\Gamma = 1$ cases, we see that the curves vary about a
factor 3 in the absolute value of $E(T_e)$. However, this absolute
value is dependent on the cross-sectional area of the loop, and
all curves could be brought into coincidence (at a given temperature) by adjusting this area. Alternatively, the total power
deposited in the loop could be altered - this would lead to a
change in the loop pressure required to give a static solution.

     If all the curves are brought into coincidence at some
temperature in this manner, it is found that they differ relatively

little throughout the temperature range (for example, it can be seen that the curves for $\alpha = 0$ and $\alpha = -1$ differ by less than 10%). This means that the emission measure integrated over a single loop must be determined with relatively high accuracy even to determine whether the energy is deposited primarily at the top or at the bottom. The situation is somewhat better for the $\Gamma = 10$ cases, although the curves for $\alpha = -1$, 0, 1 differ nowhere by as much as a factor 2. However, it is evident that the variation of Γ has a far bigger effect on the emission measure gradient than the location of the energy input.

REFERENCES

Craig, I. J. D., McClymont, A.N. and Underwood, J. H.; 1978, Astron. Astrophys. 70, 1.

Hearn, A. G.; 1975 Astron. Astrophys. 40, 355.

Hearn, A.G.; 1977 Solar Phys. 51, 159.

Jordan, C.; 1980 Astron. Astrophys. 86, 355.

Raymond, J.C. , Cox, D.P., and Smith, B.W. 1976; Ap. J. 204, 290.

Rosner, R., Tucker, W.H. and Vaiana, G. S. 1978; Ap. J. 220, 643.

Spitzer, L.; 1962 "Physics of Fully Ionized Gases" (New York, Wiley-Interscience) Chap. 5.

Vesecky, J.F., Antiochos, S.K. and Underwood, J.H. 1979; Ap. J. 233, 987.

ENERGY, BALANCE, AND ARMAGNAC

Randolph H. Levine

Harvard-Smithsonian Center for Astrophysics
Cambridge, Massachusetts 02138 USA

An exhaustive study has been made of a newly-discovered object of astronomical interest, Armagnac. Celestial coordinates are not available at this time. In several follow-up studies the position of the object (and of the observers) will be determined as closely as possible.

Observations of Armagnac were carried out on 28 August, 1980, at L'Observatoire de Vic Fezensac. We would like to thank the staff of the institution for their attention to the equipment. During a single 2.5 hour run we were able to obtain several observations in the visible, infrared, and ultraviolet parts of the spectrum. Some members of the observing team attempted speckle observations. Reports of this imaging and interferometry will be reported at a (much) later data, after those observers are located.

A preliminary model for Armagnac is shown in Figure 1. It is apparently not a highly collapsed object, although several are known to exist in its vicinity (and some of these are reported to be degenerate). From spectral analysis we conclude that there is an outer layer which is high in Silicon, and a core region dense enough to be in liquid form. There is some suggestion from our time series of observations that intermittent turbulence or mixing of this core (perhaps associated with rotation of the object) leads to significant mass flow and a strong wind. The infrared observations detect organic molecules, such as simple alcohol, in this wind. The observing team felt that this was a significant finding.

R. M. Bonnet and A. K. Dupree (eds.), Solar Phenomena in Stars and Stellar Systems, 235–237.
Copyright © 1981 by D. Reidel Publishing Company.

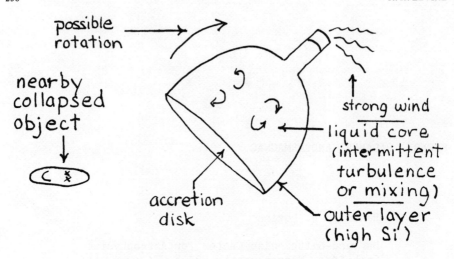

Figure 1. Preliminary schematic model for Armagnac.

Finally, we have retained our energy and our balance and deduced a Q vs. T curve for Armagnac. This is shown in Figure 2. Notice the significant departure from the often-assumed $T^{3/2}$ power law. Our data are reasonably well fit by a $T^{6/5}$ dependence. This is strong evidence that Armagnac is associated with a steep transition region. Finding a corona upon getting higher cannot be ruled out. Shorter wavelength (i.e., higher frequency) observations are necessary to settle this issue.

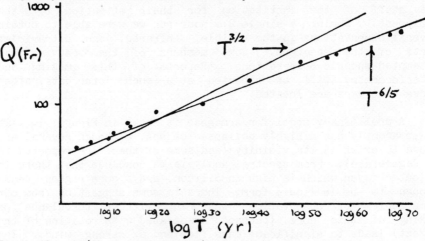

Figure 2. Q (in French francs) vs. T for Armagnac

This work was supported by a NATO Institute for Advanced Study. The investigators are still looking for support. I thank several reoxygenated referees for useful comments on this paper.

Note added in proof:

A parallel study of the Armagnac system has been undertaken by J. Underwood of the Jet Propulsion Laboratory. From close time sampling at different spatial resolutions over various fields of view he finds Q=96 fr for V=0.7 liters and Q=206 fr for V=1.5 liters. This apparently $T=30$ data. The preliminary conclusion is that Q varies as $V^{1.002}$. A joint study is planned in order to verify this effect. Other observing sites are being investigated in order to obtain better seeing, which tended to degrade over the course of our investigations.

THEORIES OF HEATING OF SOLAR AND STELLAR CHROMOSPHERES

P. Ulmschneider

Institut für Theoretische Astrophysik,
Im Neuenheimer Feld 294,
D-6900 Heidelberg
Federal Republic of Germany, formerly

Institut für Astronomie und Astrophysik,
Am Hubland,
D-8700 Würzburg
Federal Republic of Germany

ABSTRACT. In the outer atmosphere of stars a rise of the kinetic temperature to values above T_{eff} is possible only if a large and persistent amount of mechanical heating is present. Constraints derived from empirical chromosphere models allow selection of important heating mechanisms from among a great number of possible processes. It appears that for non-magnetic regions short period acoustic waves and for magnetic regions Alfvén and slow mode magnetohydrodynamic waves are the dominant mechanisms. For non-magnetic cases new acoustic energy generation rates are reported. Non-magnetic theoretical chromosphere models for the sun and 10 other stars are discussed and compared with observations. Chromospheric heating in early type stars is briefly mentioned.

1. ENERGY BALANCE IN STELLAR CHROMOSPHERES

When in 1941 Edlén conclusively demonstrated that the mysterious solar coronal lines were produced by extremely highly ionized metals a firmly established astrophysical world had been shattered. The quarter century before Edlén, due to the work of Bohr, Saha, Milne, Eddington and others had seen the very successful explanation of stellar spectra based on the principle of radiative equilibrium introduced by Schwarzschild. This principle states that in the outer atmosphere of stars energy is exclusively transported by radiation. Applications before Edlén

239

however had shown that radiative equilibrium invariably lead to
an outwardly decreasing temperature distribution which now was in
obvious contradiction to Edlén's discovery of an extremely hot
corona.

In recent years stellar observations of X-ray emission, of
UV lines from highly ionized atoms, of Fe II emission lines and
of the He 10830 Å line have indicated that the hot shell found in
the case of the sun is for stars not an exception but rather a
rule (32, 33, 45, 47). Let us define a hot shell as the stellar
layer adjacent to the photosphere where the temperature increases
outwardly to values higher than T_{eff}. Fig. 1 shows our present
state of knowledge of the existence of hot shells around stars.
It is seen that very likely all stars have hot shells. The inner
parts of hot shells are called chromospheres.

Consider a gas element in a stellar chromosphere. If an
amount of heat dQ enters this element the change of entropy S per
gram is

$$dS = \frac{dQ}{\rho T} .$$ (1)

Here ρ is the density and T the kinetic temperature. The entropy

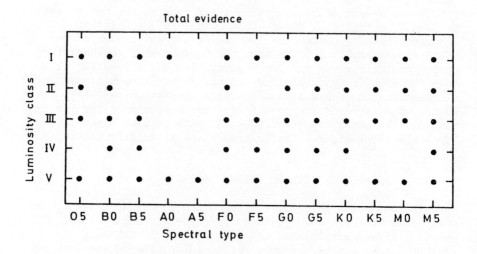

Figure 1. Stellar types (dots) where hot shells have been
           detected

or energy conservation equation valid for stellar chromospheres
can be written

$$\frac{\partial S}{\partial t} + u \frac{\partial S}{\partial x} = \frac{dS}{dt}\bigg|_{Rad} + \frac{dS}{dt}\bigg|_{Visc} + \frac{dS}{dt}\bigg|_{Cond} + \frac{dS}{dt}\bigg|_{Mech} \qquad (2)$$

where x is the geometrical height, t the time and u the stellar
wind velocity. The right hand side of equ. (2) represents the
entropy gain due to radiative-, viscous-, conductive- and mechanical (that is acoustic or magnetic) heating originating from outside the gas element. For the solar chromosphere the time dependence and the influence of the stellar wind can be neglected and
the left hand side of equ. (2) is zero. Furthermore in the low
and middle chromosphere viscous heating and thermal conduction
are very small (42). Thus the energy balance is mainly between
radiative cooling and mechanical heating.

In a grey atmosphere for instance the gas element gains
entropy by absorption of photons proportional to the mean intensity J but at the same time looses photons proportional to the
integrated Planck function B

$$B = \frac{\sigma}{\pi} T^4 \; . \qquad (3)$$

We thus have

$$0 = \frac{4\pi\kappa(J-B)}{T} + \frac{dS}{dt}\bigg|_{Mech} \qquad (4)$$

Here κ is the opacity per gram and σ the Stefan-Boltzmann constant. The sun at chromospheric heights can be crudely considered
as a black body with an effective temperature T_{eff}. Because
photons radiate only into one half space the mean intensity can
roughly be written

$$J = \frac{1}{2} \frac{\sigma}{\pi} T_{eff}^4 \; . \qquad (5)$$

In absence of mechanical heating, that is in radiative equilibrium, it is seen from equ's (3) to (5) that the temperature decreases to a boundary temperature of

$$T = \sqrt[4]{\frac{1}{2}} T_{eff} \simeq 0.8 \; T_{eff} \qquad (6)$$

which for the sun (T_{eff} = 5770 K) is about 4900 K. This was the
state before Edlén.

The observed large temperature increase with $T \gg T_{eff}$ in the stellar chromospheres thus signifies after equ. (4) that in order to satisfy energy conservation a <u>large and persistent amount of mechanical heating is necessary to hold the atmospheres in a steady state</u> configuration. If this mechanical heating were switched off radiative equilibrium would be quickly reestablished. The time constant for reestablishment of radiative equilibrium is the radiative relaxation time

$$t_R = \frac{C_v}{16\kappa\sigma T^3} \tag{7}$$

where C_v is the specific heat per gram. Note that t_R for the solar chromosphere is in the range of minutes.

2. POSSIBLE CHROMOSPHERIC HEATING MECHANISMS

What mechanisms are responsible for the heating of stellar chromospheres? Clearly there are a great number of possible mechanisms which however are not all equally important. Moreover their importance may vary greatly from point to point in the atmosphere. Chromospheres are differentiated e.g. into dense and thin regions, into regions of large and small magnetic field strength, into regions of plane and very special magnetic field geometries. It presently appears that the chromospheric heating mechanisms can be grouped into three radically different types; <u>explosive heating,</u> <u>quasisteady heating</u> and <u>wave heating</u> (45).

The prototypes of the <u>explosive mechanisms</u> are those that generate flares which however occur rather infrequently and in special magnetic configurations. Somewhat more steady types of explosive mechanisms could be those that give rise to microflares, spicules and to the high velocity jets recently discovered by Brückner et al. (8). Aside from direct heating these mechanisms could bring considerable amounts of mass into the corona from where it is observed to flow back into the chromospheric network contributing to the enhanced emission.

A typical example of the second type, the <u>quasi-steady heating mechanisms</u> is the one recently proposed by Rosner et al. (37) for solar active regions. Here magnetic field tubes are twisted by the differential rotation of the sun. This twisting is relaxed by anomalous current dissipation that heats the flux tube. Note that the Rosner et al. mechanism has recently been criticized by Kuperus, Ionson and Spicer (28). Another mechanism of this type is the one of Somov and Syrovatskii (39). These

authors suppose that the formation of active regions is accompanied by the development of quasi-steady current sheets where magnetic field reconnection takes place. Both the explosive and
quasi-steady mechanisms are strongly correlated with the magnetic
field.

A third type of chromospheric heating mechanism is wave heating. A rich spectrum of different types of waves has been observed on the sun owing its existence to the four principal restoring forces. Pressure gives rise to acoustic waves, buoyancy
to both internal gravity waves and convection, magnetic tension
to Alfvén waves, coriolis forces to Rossby waves. Simultaneous
action of more than two restoring forces produces additional wave
forms like e.g. the fast and slow mode magnetohydrodynamic waves.

The long period acoustic modes have recently been reviewed
by Deubner (15). Typical energy fluxes in the 160.0 min oscillation of Kotov et al. (27) are less than $2 \cdot 10^3$ erg/cm^2 s at the
base of the photosphere if these waves were propagating. Presently however it is still debated whether these modes actually exist.
The 5 min oscillations are an outstanding well observed phenomenon on the sun with amplitudes of the order of 500 m/s. Observations however show (13) that these acoustic modes are largely
standing waves with a phase shift of about 90° between velocity
and brightness fluctuations. Canfield and Musman (10) find for
these waves an energy flux of $8 \cdot 10^5$ erg/cm^2 s at 490 km height
and of $2 \cdot 10^4$ erg/cm^2 s at 1000 km.

Short period waves with periods from a few seconds to minutes have been detected by Deubner (14) using lines of C, Fe and
Na in the visible. As these waves have periods less than the
acoustic cut-off period of about 180 s they will propagate and
transport energy. Deubner finds a short period acoustic flux of
between 10^8 and 10^9 erg cm^{-2} s^{-1} which probably is an upper estimate. This flux value has been discussed by Cram (11) and criticized by Durrant (16). Independently Stein (40) has found theoretically a short period acoustic flux of between $7 \cdot 10^6$ and 10^8
erg/cm^2 s with a frequency spectrum that peaks considerably above
the acoustic cut-off frequency (ω_1 = .034 Hz) of the temperature
minimum. In observations of Si II lines with the OSO-8 satellite
in the upper chromosphere Athay and White (2) obtain a flux of
only 10^4 erg cm^{-2} s^{-1}. Because of the low resolution of OSO-8
this probably is an underestimate. As the resolution of the OSO-8
spectrometer used for the Si II observations is 20 arc sec the
measurements from this instrument could very likely lead to serve
horizontal averaging. In addition the wavelength of short period
acoustic waves is usually small compared to the width of the contribution function of the spectral line which results in vertical
averaging. Both difficulties were recognized by Athay and White.
Thus presently the magnitude of the observed short period acoustic

flux is uncertain by a considerable margin even if one takes into account that most of this flux would be dissipated at the height of Si II formation. Note however, that short period acoustic wave energy is clearly seen in Lyα observations of Artzner et al. (1) from OSO-8 and also in radio observations by Butz, Hirth and Fürst (9).

Just recently Brown and Harrison (7) have observed gravity waves on the sun. These waves so far were difficult to detect because they cannot exist in the unstable convection zone and are strongly damped in the radiative damping zone at heights of less than 100 km. Convection the unstable version of gravity waves on the other hand is easily seen as granulation. Here the overshooting of fast rising convective elements will give rise to acoustic waves which however are already included in the above mentioned short period wave observations.

Although magnetohydrodynamic waves must exist on the sun they have so far not been detected except for Alfvén waves above sunspots (5). Recently Stein and Leibacher (43) as well as Stein (41) have computed Alfvén wave fluxes of between 10^8 and $3 \cdot 10^9$ erg/cm^2 s from strong field regions. The reason why such large fluxes have not been observed probably lies in the fact that Alfvén waves are transverse waves with small variations of the gas pressure. Most of the Alfvén wave flux is reflected by the transition layer while an average flux of about $3 \cdot 10^5$ erg/cm^2 s is transmitted into the corona. Alfvén surface waves are proposed as heating mechanism for coronal loops (24). Stein (41) has computed a flux of slow mode mhd waves which is of the same order of magnitude as the Alfvén wave flux due to monopol sound generation in both cases. He finds that the fast mode mhd flux however is much smaller. Yet fast mode mhd waves have been proposed as heating mechanism for coronal loops (20). Finally the horizontally propagating Rossby waves have large periods and appear to carry little energy.

3. EVIDENCE FROM EMPIRICAL MODELS

What parameters can be derived from empirical chromosphere models that allow a selection among the many possible heating mechanisms? The most accurate and elaborate stellar chromosphere models are those for the sun by Vernazza, Avrett, Loeser (48) henceforth called VAL 80. These models are based on a wealth of visible UV, infrared, line and continuum data. VAL 80 have presented a series of models valid for a broad range of solar regions from a dark inner cell point to a very bright network element. In spite of the fact that these models are very sophisticated in that they carry out detailed solutions of the hydrostatic, radiative transfer and statistical equations for a great number of lines and

continua, they suffer from a possibly dangerous restriction. They
largely neglect the dynamical nature of the atmosphere. The VAL
80 models include dynamical effects only in form of a microturbulence distribution and assume a smooth temperature profile. Here
the presence e.g. of large amplitude acoustic waves, through the
nonlinearity of the Planck function and the velocity-temperature
correlation in the wave, may generate important effects (12).

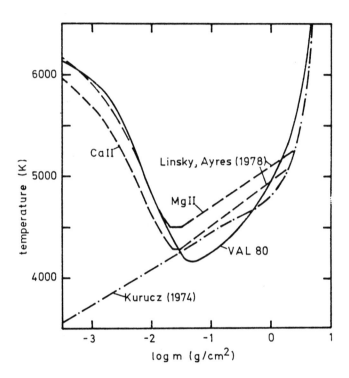

Figure 2. Empirical solar atmosphere models

Using information only from line profiles mainly of Ca II
and Mg II but also of Si II, Si III, C II empirical chromosphere
models for a considerable number of stars have been constructed
(see reviews 32, 33, 45). For stars other than the sun such
models are presently the only available chromosphere models.
Fig. 2 shows for the sun a comparison of Ca II and Mg II line
models of Linsky and Ayres (34) with the VAL 80 average sun model
and a theoretical LTE radiative equilibrium model of Kurucz (29).
There are systematic differences between these models. The Mg II
model is hotter than the Ca II model and in the photosphere both
line models are hotter than the VAL 80 model. This may be partly

due to a different weight given to the bright network areas by
the line models. On the other hand dynamical effects also explain
part of this discrepancy. The empirical VAL 80 model in the
temperature minimum area is much cooler than the theoretical model.
As non-LTE effects (Cayrel mechanism) are expected to further
raise the theoretical temperature it seems highly likely that this
discrepancy is due to the dynamical nature of the atmosphere as
discussed below.

Keeping in mind these uncertainties in the empirical chromosphere models we now list parameters which may be useful for the
selection of heating mechanisms.

I. The total chromospheric radiation loss F_E

As the existence of chromospheres is directly linked to the
availability of mechanical energy the total chromospheric radiation loss F_E is in principle the most powerful selection criterion. The difficulty here is that this parameter is not easily
evaluated from empirical models. E.g. considerable controversy
exists in the literature as to whether H^- loss is a dominant contribution or not (3, 25, 35, VAL 80). Here the difficulties are
the importance of non-LTE effects and how to separate the chromospheric loss from the photospheric radiative equilibrium loss. A
similar, however much less severe situation exists for the Ca II
losses, $F_{Ca\ II}$. Only the chromospheric Mg II losses, $F_{Mg\ II}$, due
to the low photospheric contribution can be measured relatively
unambiguously. However, the Mg II losses represent only about 20
percent of the total losses and this percentage varies from star
to star (34).

II. The height of the temperature minimum m_T

Both in empirical models (VAL 80, Fig. 49) and in theoretical
models (38) the temperature minimum coincides closely with a
height where the mechanical dissipation increases rapidly with
altitude. This height is usually measured either on a geometrical
x_T or on a mass column density scale m_T. Because purely radiative
means exist that raise the kinetic temperature in an atmosphere
(Cayrel mechanism) the energy balance at the temperature minimum
region in empirical models has to be closely checked.

III. The steepness of the chromospheric temperature rise dT/dm

The steepness of the chromospheric temperature rise depends sen-
sitively on the distribution of both mechanical dissipation and
radiative loss rates in the atmosphere.

IV. The variation of F_E, $F_{Ca\ II}$, $F_{Mg\ II}$, m_T, dT/dm with magnetic
    field strength $\vec{B}$

The empirical VAL 80 models of magnetic and nonmagnetic regions
show characteristic differences. While the total flux F_E appears
strongly affected, the temperature minimum heights m_T are not
much changed. This is also observed for stars other than the sun.
Due to a different coverage by plage areas of stars of similar
T_{eff} and gravity the total Ca II and Mg II emission $F_{Ca\ II}$,
$F_{Mg\ II}$ can vary by a factor of ten (4). On the other hand the
Wilson-Bappu-effect which concerns the width of the Ca II emission
core and thus the height of the temperature minimum does not show
a great age or magnetic field dependence (17).

V. The variation of F_E, $F_{Ca\ II}$, $F_{Mg\ II}$, m_T, dT/dm with T_{eff},
   gravity and average field strength $\vec{B}$

The systematic variation of the chromospheric parameters F_E,
$F_{Ca\ II}$, $F_{Mg\ II}$, m_T or dT/dm with T_{eff} and gravity is a powerful
selection criterion which for chromospheric heating mechanisms
is useful even if the absolute magnitude of the produced heating
flux is uncertain. For example see Stein (41), and the acoustic
heating results below.

4. ACOUSTIC HEATING AS THE DOMINANT MECHANISM FOR THE LOW
   CHROMOSPHERE

On basis of the data available from empirical chromosphere models
we now try to identify important heating mechanisms for stellar
chromospheres. In Tab. 1 total chromospheric and coronal radiation
losses given by various authors and summarized by Ulmschneider
(45) are compared with recent determinations from VAL 80. Values
in brackets are taken from (45). In spite of considerable dis-
agreements due to differing views on the H⁻ losses and the Ca II
IRT the total losses do not appear to be greatly in dispute: A
solar chromospheric heating mechanism should provide a mechanical
flux of about $F_E = 6 \cdot 10^6$ erg/cm² s at the base of the chromosphere.

    From this flux value it is immediately obvious that from
the wave mechanisms the long period modes are excluded. The same
can be said about the 5 min oscillations. These waves can be excluded also on other grounds. The long wavelengths of the 5 min
oscillations preclude any appreciable viscous or conductive heating. Radiative damping as a heating mechanism by these waves is
effective only in the lower photosphere. Thus shock dissipation
remains as the only potential heating mechanism. Shock dissipation
is however excluded for the 5 min oscillations because of the
observed 90⁰ phase shift. It is a general property of shock waves

Table 1. Chromospheric radiative loss rates

Source	Ulmschneider (1979) loss [erg/cm^{2}s]	VAL 80 loss [erg/cm^{2}s]
H$^-$	$30 \cdot 10^5$	$4 \cdot 10^5$
Ca II	$8 \cdot 10^5$	$30 \cdot 10^5$
Mg II	$10 \cdot 10^5$	$9 \cdot 10^5$
H$_\alpha$	$5 \cdot 10^5$	–
Mg I, Na I, Ca I, Fe I, Fe II, etc.	$4 \cdot 10^5$	$(4 \cdot 10^5)$
Lα	$2 \cdot 10^5$	$3 \cdot 10^5$
Corona and transition layer	$\underline{3 \cdot 10^5}$ $62 \cdot 10^5$	$\underline{(3 \cdot 10^5)}$ $53 \cdot 10^5$

that velocity and temperature shock simultaneously producing a
0^0 phase shift. Thus the 5 min oscillation cannot produce a
temperature minimum at the observed height.

The explosive and quasi-steady mechanisms primarily apply to
the upper chromosphere and the corona where magnetic effects
dominate. For the lower and middle chromosphere where most of the
chromospheric energy loss originates these mechanisms do not
apply because of the energy requirement. Tab. 1 shows that the
averaged energy requirements for coronal loops are by an order
of magnitude smaller than for the chromosphere.

Thus the only remaining powerful heating mechanisms are
short period acoustic waves and magnetohydrodynamic waves. In
spite of the observational uncertainty it appears that the energy
requirement of $F_E = 6 \cdot 10^6$ erg/cm^2 s is met for short period
acoustic waves. In addition, these waves as shown below, are able
to produce the temperature minimum at the observed height. In network areas calculations of Stein and Leibacher (43) as well as
Stein (41) show that Alfvén and slow mode mhd waves are both
efficiently produced. The magnitude of this mhd wave flux appears
sufficient but is still rather uncertain. A detailed study is
missing. Alfvén waves are difficult to dissipate in the chromosphere (44), but surface waves in the presence of strong gradients
of the Alfvén velocity are thought to dissipate significantly
(24). Yet it seems difficult for Alfvén waves to explain the

height of the temperature minimum. Here the slow mode waves are an attractive companion/alternative. Slow mode waves in regions of strong magnetic fields are essentially acoustic waves that propagate along the magnetic field lines. They would be seen as acoustic waves, being included in Deubners (14) observations. They could form shocks at the temperature minimum area and through their magnetic nature would explain the variability of the stellar Mg II and Ca II emission. Moreover Stein (41) has shown that both slow mode and Alfvén waves can explain the missing gravity dependence of the stellar Mg II emission.

At this point an important fact should be noted. If the computation of the empirical chromospheric radiation flux is not grossly in error the chromosphere cannot accept more energy than F_E. Thus any mechanism which provides F_E must be the dominant one and the other mechanisms should be found to be considerably less energetic. We thus conclude that for nonmagnetic regions the short period acoustic waves very likely are the dominant mechanism while for network areas with strong magnetic fields Alfvén and slow mode waves appear as the main mechanical input for the low and middle chromosphere. Here the acoustic-like slow mode waves seem to be especially important for producing - through the onset of shock dissipation - the required rapidly increasing mechanical heating at the temperature minimum area $x \geq x_T$.

5. ACOUSTIC ENERGY GENERATION

For the generation of non-magnetic acoustic waves a fairly well developed theory exists (31, 40) while for the production of magnetohydrodynamic waves only rough estimates (41, 43) are presently available.

The acoustic energy generation in stars is calculated in two steps. For a star of given T_{eff} and gravity one first constructs a convection zone model where for the mixing length theory one in addition needs the parameter $\alpha = \ell/H$, the ratio of mixing length to pressure scale height. Then Lighthills (31) theory is applied. In its simplest form this latter theory neglects magnetic fields as well as gravity and considers a homogeneous atmospheric layer with density ρ_0 and pressure p_0 as well as a localized field of turbulent velocities $\vec{v}$. The turbulent field generates small density ρ', pressure p' and velocity $\vec{u}$ perturbations. In the turbulent field $|\vec{v}|$ is not assumed small but $|\vec{u}| << |\vec{v}|$. One has

$$\frac{\partial \rho'}{\partial t} + \frac{\partial \rho v_i}{\partial x_i} = 0 \tag{8}$$

$$\frac{\partial \rho v_i}{\partial t} + \frac{\partial \rho v_i v_j}{\partial x_j} + \frac{\partial p'}{\partial x_i} = 0 \tag{9}$$

$$p' = c_o^2 \, \rho' \tag{10}$$

where c_o = const is the sound velocity. From these equations a wave equation is found

$$(\nabla^2 - \frac{1}{c_o^2} \frac{\partial^2}{\partial t^2})\rho' = -\frac{1}{c_o^2} \frac{\partial^2 \rho_o v_i v_j}{\partial x_i \partial x_j} \tag{11}$$

with a quadrupole source term which can be solved for large distances $|\vec{x}| \gg |\vec{x}'|$ from the turbulent field

$$\rho' \simeq \frac{\rho_o}{4\pi c_o^4} \frac{x_i x_j}{|\vec{x}|^3} \int \frac{\partial^2}{\partial t^2} v_i v_j d^3 x'. \tag{12}$$

With the relation $u = \rho' c_o/\rho_o$ valid for acoustic waves the acoustic flux is the time average

$$F_M = \overline{p'u} = \frac{c_o^3}{\rho_o} \overline{\rho'^2} = \frac{\rho_o}{4\pi c_o^5} \frac{x_i x_j x_k x_\ell}{|\vec{x}|^6} \int\int \overline{\frac{\partial^2}{\partial t^2} v_i v_j \frac{\partial^2}{\partial t^2} v_k v_\ell} d^3 x' d^3 x'' \tag{13}$$

For the turbulent velocities assumptions must be made on the spacial and temporal correlations $v(\vec{x}',t')v(\vec{x}'',t'')$. These assumptions together with the mean turbulent velocity v from the convection zone model allow evaluation of equ. (13),

$$F_M = \int 38 \frac{\rho_o v^8}{c_o^5 \ell} \, dx \tag{14}$$

Fig. 3 shows (drawn) the acoustic energy flux F_M computed this way by Renzini et al. (36). It is seen that F_M rises rapidly with increasing T_{eff} and decreasing gravity. As the convection zones become inefficient for early type stars, F_M abruptly decreases at high T_{eff}.

Newest UV and X-ray observations indicate however that especially for late type dwarf stars the homogeneous atmosphere assumption in Lighthills theory breaks down. Recent work of Bohn (6) following Stein (40) aside of improving the treatment of H_2 molecules includes gravity in Lighthills theory. Stein (40) has shown that then the source term at the right hand side of equ. (11) is essentially replaced by

$$A \frac{\partial^2 \rho_o^{\frac{1}{2}} v_i . v_j}{\partial x_i \partial x_j} + B \frac{\omega_1}{c_o} \frac{\partial \rho_o^{\frac{1}{2}} v_j . v_z}{\partial x_j} + C \frac{\omega_1^2}{c_o^2} \rho_o^{\frac{1}{2}} v_z v_z \qquad (15)$$

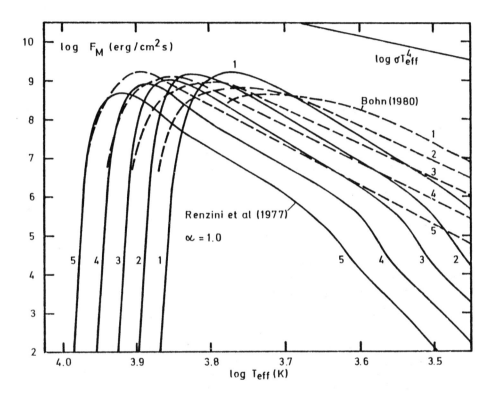

Figure 3.   Non-magnetic acoustic energy generation rates as
function of T_{eff} with log g as parameter

where z is the vertical direction, ω_1 the acoustic cut-off frequency and A, B, C constants. Bohn finds that especially towards
late type dwarf stars the much more efficient dipole and monopole
source terms in equ. (15) dominate as is shown in Fig. 3 (dashed).
For late type stars the acoustic flux is there considerably mag-
nified.

As shown by Stein and Leibacher (43) and Stein (41) the inclusion of strong magnetic fields modifies equ's (9), (10) still

further leading to monopole source terms for Alfvén and slow mode
waves as well as to quadrupole terms for fast mode waves. These
fluxes considerably further enhance Bohn's (6) values in strong
magnetic field areas.

In addition to the total wave flux F_M, Stein (40) and Bohn
(6) also give the monochromatic flux $dF_M/d\nu$. The peak of this
flux spectrum is for the sun roughly at the frequency

$$\nu_{Max} = \frac{1}{P_{Max}} \simeq \frac{10}{2\pi} \omega_1 = \frac{10}{4\pi} \frac{\gamma g}{c} \simeq \frac{c}{H} \qquad (16)$$

where γ is the ratio of specific heats. From a peak given by equ.
(16) the acoustic spectrum falls off to both larger and smaller
frequency. The decay towards larger frequency is produced by the
decrease of velocity with frequency in the turbulence spectrum.
Here different assumed turbulence spectra (Kolomogoroff, Spiegel
or exponential (40)), give different decays of the acoustic flux
with frequency. Towards lower frequency the decrease of the flux
is due to the acoustic cut-off frequency which decreases with
increasing temperature.

6. THEORETICAL SOLAR CHROMOSPHERE MODELS

For a given type of star and given magnetic field structure theoretical chromosphere models can be constructed in principle by
solving the magnetohydrodynamic equations and the radiative transfer equation. At the present time this procedure has been carried
out only for non-magnetic cases and only for rather simplified
circumstances. The specification of two parameters T_{eff} and gravity allows to compute the acoustic flux F_M as shown above. The
third parameter α is usually thought to vary only within the
narrow limits $\alpha = 1.0$ to 1.5 (19). A comparison of recent radial
and nonradial pulsation calculations with observed solar 5 min
oscillations does not disagree with this view (38). Thus for non-
magnetic cases essentially a two parameter set of theoretical
chromosphere models is obtained.

In the most recent calculations summarized by Ulmschneider
(45) a series of severe simplifications are made. Instead of a
spherically propagating acoustic spectrum one assumes a plane
monochromatic wave with frequency ν_{Max} and neglects effects due
to ionization. The radiation transport is evaluated with a grey
LTE two-stream approximation.

Consider a plane stellar atmosphere bounded below by a
piston and at the top by a transmitting, fluid type boundary.
Within this slab shockfronts act as internal boundaries separating

continuous regions. For these regions the hydrodynamic equations can be written

$$\frac{\partial \rho}{\partial t} + \frac{\partial \rho u}{\partial x} = 0 \tag{17}$$

$$\rho\frac{\partial u}{\partial t} + \rho u\frac{\partial u}{\partial x} + \frac{\partial p}{\partial x} + \rho g = 0 \tag{18}$$

$$\frac{\partial S}{\partial t} + u\frac{\partial S}{\partial x} = \frac{dS}{dt}\bigg|_{Rad} \tag{19}$$

while the Hugoniot relations connect across the shocks. With the equations

$$p = \rho\frac{\mathcal{R}T}{\mu}, \quad \rho = \rho_0(\frac{T_0}{T})^{3/2}\, e^{-\frac{(S-S_0)}{\mathcal{R}}\mu}, \quad c^2 = \gamma\frac{\mathcal{R}T}{\mu} \tag{20}$$

valid for neutral ideal gases where $\mathcal{R}$ is the gas constant and c the sound velocity, three of the five thermodynamic variables p, ρ, T, c, S can be eliminated. ρ_0, T_0, S_0 refer to the undisturbed atmosphere. The radiative transfer equation is solved in the LTE two stream approximation

$$\pm\frac{1}{\sqrt{3}}\frac{dI^\pm}{dx} = -\kappa\rho(I^\pm - B) \tag{21}$$

where

$$J = \frac{I^+ + I^-}{2}, \quad B = \frac{\sigma}{\pi}T^4 \tag{22}$$

and with equ. (4)

$$\frac{dS}{dt}\bigg|_{Rad} = \frac{4\pi\kappa}{T}(J - B). \tag{23}$$

Here κ is the grey opacity per gram and $I^\pm$ the specific intensity. The boundary conditions are:

at the piston: $u = -\sqrt{\frac{2F_M}{\rho c}}\sin(2\pi\nu_{Max}t)$, $I^+ = \frac{\sigma}{\pi}T^4 + \frac{\sqrt{3}\sigma}{4\pi}T_{eff}^4$ (24)

at the top: $u = u(x-(c+u)\Delta t, t-\Delta t)$, $I^- = 0$.   (25)

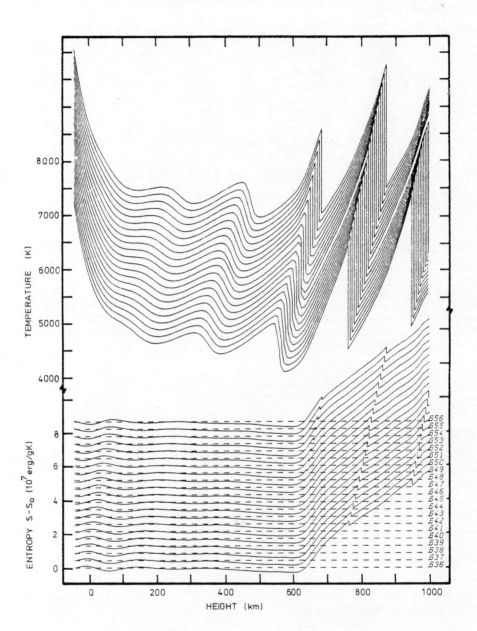

Figure 4.   Temperature and entropy as function of height. Succes-
            sive time steps ($\Delta t = 0.95$ s) are displaced for clari-
            ty. Scales are for the lowest curves

Starting initially with a radiative equilibrium atmosphere labeled T_{RE} in Fig. 5 the temperature and entropy distributions after some time are shown in Fig. 4. Roughly after 20 shocks have been transmitted at the top boundary the mean time averaged quantities approach a steady state. Fig. 5 shows a comparison of thus obtained theoretical chromosphere models (drawn) with empirical models (dashed) for the sun (46). The range of α indicates the remaining freedom of choice for the theoretical models. Relatively good agreement is seen for both the height of the temperature minimum and the chromospheric temperature gradient.

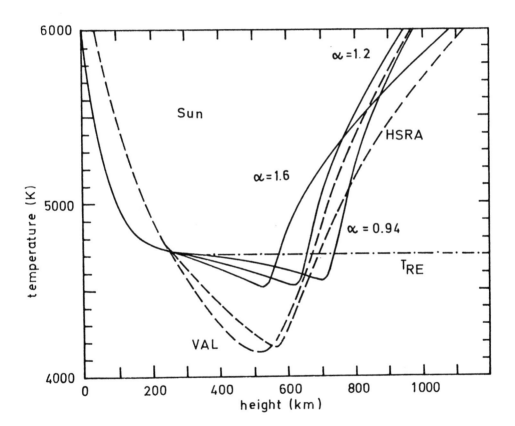

Figure 5.   Time averaged theoretical and empirical solar models

A rather unexpected feature of the theoretical models (c.f. Fig. 5) is the photospheric temperature depression below the radiative equilibrium distribution T_{RE}. This is due to the large amplitude of the acoustic waves and the nonlinearity of the Planck

function. As in steady state the time averaged mechanical dissipation must be equal to the averaged radiation loss we have from
equ. (4)

$$\frac{\overline{\frac{dF_M}{dx}}}{} = \overline{4\pi\kappa\rho(J-B)}.$$ (26)

In the upper photosphere a negligible amount of radiation damping
leads to $dF_M/dx \simeq 0$. The mean intensity J originating from optical depth $\tau \simeq 1$ where acoustic waves have small amplitude is
essentially constant. Thus roughly $J \simeq B$. But the phase $T^4 =
(\overline{T} + \Delta T)^4$ of the wave contributes disproportionately much com-
pared with the phase $T^4 = (\overline{T} - \Delta T)^4$ such that $\overline{T} < T_{RE}$. Note that
this dynamical behaviour explains readily the observed temperature
depression below Kurucz's (29) model (c.f. Fig. 2). The discrepancies between empirical and theoretical models in the photosphere (c.f. Fig. 5) are mainly due to the grey approximation
used for the theoretical models as can be seen by comparing the
T_{RE} distribution of Fig. 5 with Kurucz non-grey model in Fig. 2.

7. EMPIRICAL AND THEORETICAL CHROMOSPHERE MODELS OF LATE TYPE
      STARS

With the same methods as used for the sun theoretical chromosphere
models of other stars can be constructed. Here likewise only
models disregarding magnetic fields are presently available. The
non-magnetic theoretical models are uniquely determined by specifying (in addition to the parameter α) only the two variables
T_{eff} and gravity.

In Fig. 6 theoretical temperature minima for eleven late
type stars are shown (triangles) together with temperature minima
from semiempirical models (dots) based on Ca II line observations
(38). The acoustic energies used for the computation of these
theoretical chromosphere models are calculated in the approximation of Renzini et al. (36) and assuming α = 1.25. A rather
good agreement is seen except for late type dwarf stars where
there is increasing discrepancy towards low T_{eff}. E.g. for 70 OphA
a factor of five and for EQ Vir a factor of 145 more acoustic
energy is needed to bring agreement between theory and observation. This discrepancy however is now largely eliminated due to
Bohn's (6) new values of the acoustic flux (see Fig. 3). For the
active chromosphere star EQ Vir a factor of about three remains
even if Bohn's flux is taken into account. This discrepancy is
very likely due to the neglect of magnetic fields in the computation of the acoustic energy generation of Fig. 3.

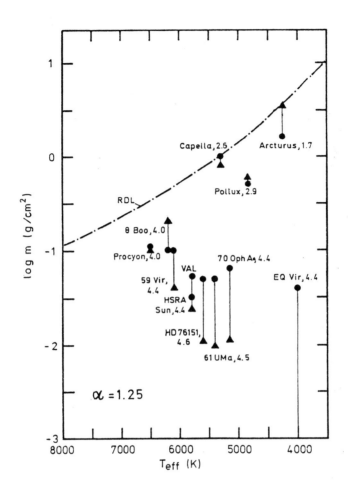

Figure 6. Theoretical and empirical heights of the temperature
           minimum for individual stars

     Fig. 7 after Schmitz and Ulmschneider (38) shows theoretical
mean temperature distributions for various stellar models identified by T_{eff} and log g. Here again α = 1.25. These temperatures
should be compared with semiempirical chromosphere models and
chromospheric temperature gradients shown in Figs. (8) and (9)
based on Ca II K line observations of Kelch, Linsky and Worden
(26). As can be seen by comparing the theoretical models (T_{eff},
log g) = (4000 K, 2), (4000 K, 4) and (6000 K, 4) of Fig. 7 the
chromospheric temperature gradient increases with increasing
gravity and decreasing T_{eff} in agreement with the observations
(Fig. 9).

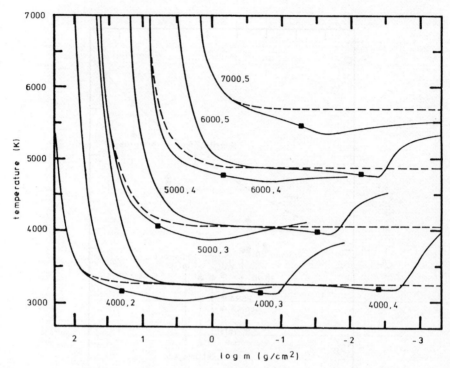

Figure 7. Time averaged theoretical atmosphere models for stars
of indicated T_{eff} and log g. Radiative equilibrium
models are shown dashed

A comparison of Fig. 7 and 8 (dashed) shows considerable
differences in the temperature structure caused by the grey and
non-grey approximations used in the theoretical and semiempirical
models respectively. However even if non-grey radiative transport
were taken into account in the theoretical models the fact that
all these models show photospheric temperature depressions would
not vanish as this is a consequence of the large amplitude of the
waves and the nonlinearity of the Planck function as discussed
for the solar case.

Interestingly however the empirical models invariably show
photospheric temperature enhancements. In Fig. 7 filled squares
show the height of shock formation. It is seen that for stars of
large gravity and low T_{eff} the heights of shock formation are
closely correlated with the temperature minimum positions. These
chromosphere models have been called S-type chromospheres. Stars
with high T_{eff} or low gravity have shock formation heights considerably different from the temperature minimum heights. Such
stellar models have been called R-type chromospheres as there the
process of radiation damping determines the position of the tempe-

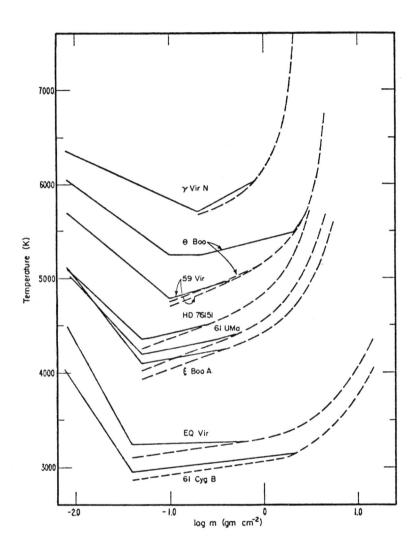

Figure 8. Empirical atmosphere models (Kelch, Linsky, Worden 1979)

rature minimum. R-type chromospheres have rather extensive photospheric temperature depressions. From their values of T_{eff} and g,
stars like θBoo, αAur, αOri or αCMi should have R-type chromospheres. These stars are observed to have extensive photospheric
temperature enhancements. The remaining stars of Fig. 8 are all
S-type chromosphere stars and have small photospheric temperature
enhancements. Thus theoretical computations and observations
complement each other. Where an empirical model shows high tempe-

rature enhancement, the corresponding theoretical model shows
large temperature depression. This behaviour has been explained
(38) by the fact that the Planck function at the frequencies of

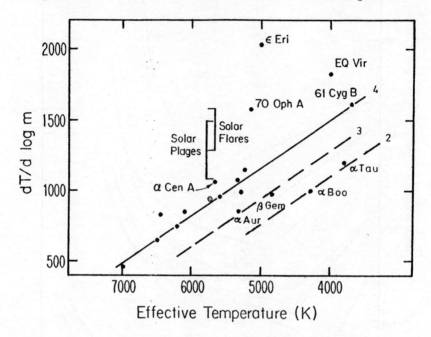

Figure 9. Empirical chromospheric temperature gradients for
           dwarf stars (Kelch, Linsky, Worden 1979) together with
           values for αAur, βGem, αBoo and αTau. Lines are label-
           led by log g

the Ca II K and Mg II h,k lines has a very steep temperature dependence such that only the wave crests of the large amplitude
acoustic waves are seen at these UV frequencies. Thus the infered
empirical models based on Ca II K are considerably hotter than
the time averaged theoretical models. As in turn the temperature
dependence for Mg II k (λ 2793 Å) is much larger than for Ca II K
(3933 Å) we expect by the same effect the Mg II models to be much
hotter than the Ca II models. This has actually been observed
(c.f. Fig. 2).

     As the chromospheric radiation flux F_E should be roughly
equal to the acoustic energy at the temperature minimum F_{MT} if
the acoustic mechanism is the dominant chromospheric heating
mechanism these two quantities have been compared in previous
work. Rough agreement has been found (38). As a considerable controversy has recently arisen in the literature about the importance of the H^- losses these comparisons of the acoustic flux
and the total chromospheric losses do not carry much weight pre-

sently. Fig. 3 of Schmitz and Ulmschneider (38) shows the comparison of the acoustic flux at the temperature minimum of eleven stars with the average Mg II k emission flux given by Basri and Linsky (4). The acoustic fluxes also decrease with decreasing T_{eff} and are roughly a factor of ten above the Basri and Linsky line. This factor of ten agrees with the ratio of total to Mg II k line losses as given above by Tab. 1. It is clear that the variability of the Mg II k line emission in stars of similar T_{eff} and gravity cannot be explained by the present non-magnetic theoretical models. Here slow mode heating models will have to be constructed.

8. CHROMOSPHERIC HEATING IN EARLY TYPE STARS

The recent X-ray observations with the Einstein satellite (47) together with UV observations of O VI, N V and Si IV lines summarized by Ulmschneider (45) and in Fig. 1 conclusively demonstrate the existence of hot shells in early type stars. As these stars do not have efficient convection zones (c.f. Fig. 3) another wave energy generation mechanism must be at work. A very attractive possibility is that observed turbulent gas motions are amplified by the radiation field of the star. Hearn's (21, 22, 23) mechanism considers amplification of isothermal acoustic waves by the κ-mechanism. If primordial magnetic fields are present on early type stars slow mode waves could be amplified by the same process. Rough calculations for non-magnetic cases produce acoustic energies which are in relative good agreement with empirical values derived from photospheric turbulence (30). A detailed investigation is however missing at the present time.

REFERENCES

(1) Artzner, G., Leibacher, J., Vial, J.C., Lemaire, P.,
      Gouttebroze, P.: 1978, Astrophys. J. 224, pp. 83-85.
(2) Athay, R.G., White, O.R.: 1978, Astrophys. J. 226,
      pp. 1135-1139.
(3) Ayres, T.R.: 1980 preprint.
(4) Basri, G.S., Linsky, J.L.: 1979, Astrophys. J. 234,
      pp. 1023-1035.
(5) Beckers, J.M.: 1976, Astrophys. J. 203, pp. 739-752.
(6) Bohn, H.U.: 1980, to be published.
(7) Brown, T.M., Harrison, R.L.: 1980, Astrophys. J. Letters,
      in press.
(8) Brückner, G.E., Bartoe, J.D.F., von Hoosier, M.E.: 1978,
      in E. Hansen and S. Schaffner eds., Proceedings OSO-8 workshop.
(9) Butz, M., Hirth, W., Fürst, E.: 1979, Astron. Astrophys.
      72, pp. 211-214.

(10)   Canfield, R.C., Musmann, S.: 1973, Astrophys. J. 184,
       pp. L131-L136.
(11)   Cram, L.E.: 1977, Astron. Astrophys. 59, pp. 151-159.
(12)   Cram, L.E., Keil, S.L., Ulmschneider, P.: 1979, Astrophys.
       J. 234, pp. 768-774.
(13)   Deubner, F.: 1974, Solar Physics 39, pp. 31-48.
(14)   Deubner, F.: 1976, Astron. Astrophys. 51, pp. 189-194.
(15)   Deubner, F.: 1980, Highlights of astronomy 5, pp. 75-87.
(16)   Durrant, C.J.: 1979, Astron. Astrophys. 73, pp. 137-150.
(17)   Glebocki, R., Stawikowski, A.: 1978, Astron. Astrophys. 68,
       pp. 69-74.
(18)   Golub, L., Maxson, C., Rosner, R., Serio, S., Vaiana, G.S.:
       1978, Astrophys. J., in press.
(19)   Gough, D.O., Weiss, N.O.: 1976, Monthly Not. Roy. Astr. Soc.
       176, pp. 589-607.
(20)   Habbal, S.R., Leer, E., Holzer, T.E.: 1979, Solar Phys. 64,
       pp. 287-301.
(21)   Hearn, A.G.: 1972, Astron. Astrophys. 19, pp. 417-426.
(22)   Hearn, A.G.: 1973, Astron. Astrophys. 23, pp. 97-103.
(23)   Hearn, A.G.: 1976, Physique des mouvements dans les atmo-
       sphères stellaires, Colloque intern. du CNRS 250, R. Cayrel
       and M. Steinberg eds., p. 65.
(24)   Ionson, J.A.: 1978, Astrophys. J. 226, pp. 650-673.
(25)   Kalkofen, W., Ulmschneider, P.: 1979, Astrophys. J. 227,
       pp. 655-663.
(26)   Kelch, W.L., Linsky, J.L., Worden, S.P.: 1979, Astrophys.
       J. 229, pp. 700-712.
(27)   Kotov, V.A., Severny, A.B., Tsap, T.T.: 1979, Solar Oscil-
       lations and the Internal Structure of the Sun, Report USSR
       Academy of Sciences.
(28)   Kuperus, M., Ionson, J.A., Spicer, D.F.: 1980, Annual Rev.
       Astron. Astrophys., in press.
(29)   Kurucz, R.L.: 1974, Solar Physics 34, pp. 17-23.
(30)   Lamers, H.J.G.L.M., de Loore, C.: 1976, Physique des mouve-
       ments dans les atmosphères stellaires, Colloque intern. du
       CNRS 250, R. Cayrel and M. Steinberg eds., pp. 453-458.
(31)   Lighthill, M.J.: 1952, Proc. Roy. Soc. London A 211,
       pp. 564-587.
(32)   Linsky, J.L.: 1980, Proc. of IAU Coll 51, Stellar Turbu-
       lence, D.F. Gray, J.L. Linsky eds., Springer, Heidelberg,
       pp. 248-277.
(33)   Linsky, J.L.: 1980, Annual Rev. Astron. Astrophys., in press.
(34)   Linsky, J.L., Ayres, T.R.: 1978, Astrophys. J. 220,
       pp. 619-628.
(35)   Praderie, F., Thomas, R.N.: 1976, Solar Phys. 50, pp. 333-
       342.
(36)   Renzini, A., Cacciari, C., Ulmschneider, P., Schmitz, F.:
       1977, Astron. Astrophys. 61, pp. 39-45.
(37)   Rosner, R., Golub, L., Coppi, B., Vaiana, G.S.: 1978,
       Astrophys. J. 222, pp. 317-332.

(38) Schmitz, F., Ulmschneider, P.: 1980, Astron. Astrophys.,
     in press.
(39) Somov, B.V., Syrovatskii, S.I.: 1977, Solar Phys. 55,
     pp. 393-399.
(40) Stein, R.F.: 1968, Astrophys. J. 154, pp. 297-306.
(41) Stein, R.F.: 1980, preprint.
(42) Stein, R.F., Leibacher, J.W.: 1974, Annual Rev. Astron.
     Astrophys. 12, pp. 407-435.
(43) Stein, R.F., Leibacher, J.W.: 1980, Proc. IAU Coll. 51,
     Stellar Turbulence, D.F. Gray, J.L. Linsky eds., Springer,
     Heidelberg, pp. 225-247.
(44) Uchida, Y., Kaburaki, O.: 1974, Solar Phys. 35, pp. 451-
     466.
(45) Ulmschneider, P.: 1979, Space Sci. Rev. 24, pp. 71-100.
(46) Ulmschneider, P., Schmitz, F., Kalkofen, W., Bohn, H.U.:
     1978, Astron. Astophys. 70, pp. 487-500.
(47) Vaiana, G.S. et al.: 1980, preprint.
(48) Vernazza, J.E., Avrett, E.H., Loeser, R.: 1980, Astrophys.
     J. Suppl., in press.

MECHANICAL FLUX IN THE SOLAR CHROMOSPHERE

P. Mein

Observatoire de Paris-Meudon

We investigate energy transport by pressure waves, derived from observations of chromospheric line profiles. For a given frequency ω, the flux can be written

$$F_\omega = \mathcal{A} < V \cdot \Delta P^* > \tag{1}$$

where V is the vertical velocity and ΔP the pressure fluctuation. In the adiabatic case, it can be expressed in two different ways, according to the number of available lines :

a) If only 1 line is available, we can use :

$$F_\omega = (1/2) P \left[\gamma/(\gamma -1)\right] \times |\Delta T/\bar{T}| \times |V| \cos \theta \tag{2}$$

P = local pressure, $|\Delta T/\bar{T}|$ = temperature fluctuation amplitude, $|V|$ = velocity amplitude, θ = phase shift between velocity and temperature. V and T can be related to observable quantities (doppler shifts d and intensity fluctuations ΔI) by using weighting functions W_V and W_T, derived from a mean model atmosphere :

$$d = \int W_V (z) V(z) \, dz \tag{3}$$

$$\Delta I/\bar{I} = \int W_T (z) (\Delta T(z) /\bar{T}(z)) \, dz \tag{4}$$

In fact, weighting functions generally show that, for a given line, formation altitudes are not the same for V and T. Since the value of θ is very crucial, the best accuracy is obtained by interpolations between 2 lines, although equation (2) refers only to one altitude.

R. M. Bonnet and A. K. Dupree (eds.), Solar Phenomena in Stars and Stellar Systems, 265–268.
Copyright © 1981 by D. Reidel Publishing Company.

b) If 2 (or more) lines are available, we can use

$$F_\omega = (1/2)\ \rho |v|^2\ V_g = (1/2)\ \rho\ |v|^2\ V_s^2/\ V_\phi \tag{5}$$

ρ = local density, V_g = group velocity, V_s = sound velocity,
V_ϕ = phase velocity. Using weighting functions W_v, it is possible
to derive V_ϕ from doppler shifts in two different lines. It is
very important to note that expression (5) is valid with V_g (or
V_s^2/V_ϕ) and not V_s. Replacing V_g by V_s leads to strong overesti-
mation of F_ω in the solar case.

I will report some results by N. Mein and B. Schmieder. Data
are time sequences of high resolution spectra (Sac Peak), in
5173 mgI (core and wings) and 8498, 8542, 3933 CaII. Cross correlation functions and Fourier transforms provide mechanical fluxes.
Formation altitudes and smoothing by radiative transfer are deduced from weighting functions.

FLUX VERSUS FREQUENCY

As an example, fig. (1) shows the flux deduced from doppler
shifts in 3933 and 8542 CaII. At periods larger than 400s, network convective motions are not relevant to this analysis. At
high frequencies, dashed line corresponds to the flux without
correction of smoothing, and solid line corresponds to the flux
with correction. This result refers roughly to the altitude
1550 km in the HSRA model atmosphere. Computations performed
with the VAL lead to a smaller flux because of a larger distance
between the line formation regions (V_ϕ larger).

The integral over frequencies leads to 4×10^3 erg cm^{-2}s^{-1}.
An attempt has been made in order to extend the frequency range.
Between periods 60 and 30 s, an overestimate has been derived
from doppler shifts in the K line by using expression (5) with
$V_g = V_s$. The total flux reaches 6×10^3 erg cm^{-2}s^{-1}.

FLUX VERSUS HEIGHT

In fig. (2), results quoted V/V refer to expression (5),
and V/I to expression (2). Results derived from (2) are more
uncertain, because of possible non-adiabatic propagation. This
accounts probably for the departure between V/V and V/I (low
temperature fluctuations). Fluxes are plotted in the range 400-
120 s. Extrapolations to 60 and 30 s. are also indicated. As a
comparison, circles show results by Lites and Chipman. Let us
note also that Athay and White derived the value 10^4 at the top
of the chromosphere, from OSO-8 observations.

If we remind that the radiative loss in transition region

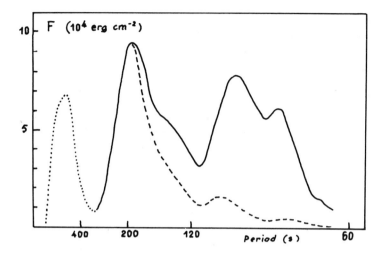

Fig. 1. Flux versus frequency

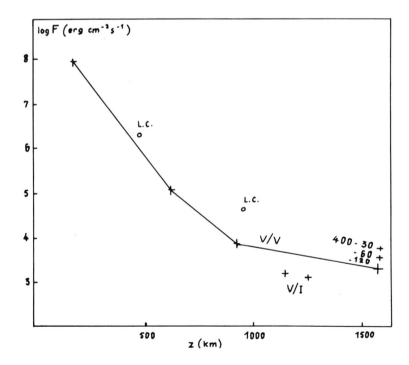

Fig. 2. Flux versus height

and corona is around 3×10^5 erg cm^{-2} s^{-1}, we see that almost
two orders of magnitude are lacking in the mechanical .flux.
Other mechanisms are needed in order to account for the heating
of upper layers.

Let us point out again the importance of phase shift measurements in flux calculations. In case of reflected or trapped
waves, the group velocity is far below the sound velocity. This
should be reminded for example in estimates of energy balance in
stellar coronae.

REFERENCES

Mein, N., Mein, P. : 1980, Astron. Astrophys. 84, 96.

Schmieder, B., Mein, N. : 1980, Astron. Astrophys. 84, 99.

Mein, N., Schmieder, B. : submitted to Astron. Astrophys.

MAGNETIC HEATING IN THE SUN

Claudio Chiuderi

Istituto di Astronomia, Università di Firenze, Italy

Abstract. The observational evidence for magnetic heating in the solar corona is presented. The possible ways of investigati theoretically the nature of the heating processes are critically discussed. Merits and disadvantages of the basic mechanisms so far proposed are reviewed. Finally, a preliminary application of the magnetic heating concept to stellar coronae is presented.

1. INTRODUCTION

In the last few years, comprising what may be called the post-Skylab era, our conception of the structure of the solar corona has dramatically changed. As a result of the observations performed during the Skylab missions and successively during rocket flights, a number of traditional assumptions had to be abandoned. In the following I shall briefly outline the main observational findings and the resulting changes in theoretical ideas about solar coronal heating. For a more detailed description of the new trend in coronal physics, I refer the reader to the excellent review by Vaiana and Rosner (1978).

The main result of the new observations has been the recognition of the extremely inhomogeneous nature of the solar corona. The fact that the solar atmosphere had a structure was of course known before. Sunspots, plages, spicules, prominences and coronal streamers are just a few of the structures familiar to every solar observer. In spite of this, for long time the solar corona has been modeld as an homogeneous medium, by arguing that the observed structures were simply local deviations or "perturbations" of an otherwise undifferentiated medium. It was

269

R. M. Bonnet and A. K. Dupree (eds.), Solar Phenomena in Stars and Stellar Systems, 269–288.
Copyright © 1981 by D. Reidel Publishing Company.

also felt that the homogeneous models were somehow the theoretical counterpart of the observational averaging due to the instruments' finite resolution.

The new observations have forced us to abandon this view.
The structuring present in the solar corona is so extreme, that
it now seems more sensible to consider it rather a collection of
widely different structures, whose properties should be studied
separately. The individual structures share many properties of
the previous homogeneous models and it is in this connection that
the concept of the structures as separate "mini"-atmospheres was
introduced in the literature (Vaiana and Rosner, 1978).

As a result of the observational revolution just described,
the theoretical studies have mostly concentrated on the properties of the two classes of structures that appear to be extreme,
especially when observed in the X-ray range: the dark coronal
holes and the bright coronal loops. Also, due to the fact that
most of the intensity in the EUV and X-ray ranges comes from the
active regions, a great deal of studies on coronal heating has
concentrated on how the active regions are heated. Of course,
the intensity of radiation does not tell the whole of the story,
since there are other losses, e.g. the solar wind, that are more
important outside of the active regions. Thus the local energy
input should show a much less dramatic variation than the intensity as we move from a coronal hole to an active region.

A second important point that emerges from the most recent
observations is the fact that the corona is in a state of continuous dynamical evolution. The spatial and temporal scales of
this evolution or, in brief, activity, cover a very wide range.
It is fair to say that some degree of activity is always observed and our present lower limits on space and time scales seem
more related to the finite resolution of the observing instruments than to physical reality. As we shall see in the following,
there are strong theoretical arguments that suggest that the clue
to the understanding of many important physical processes may
in fact reside in structures whose size is well below the present resolution limits.

If we keep in mind the two basic aspects of the solar corona, structuring and activity, and look for a unified explanation
of the observed facts, we must conclude that there is only one
quantity that appears a likely candidate, namely the magnetic
field. The presence of well-defined structures naturally suggests
that they delineate the general topology of the magnetic field.
Even if a direct measure of coronal fields has proven so far impossible, the extrapolation, under certain assumptions, of the
measured line-of-sight photospheric fields clearly indicate the
close connection of the field lines with the observed structures.

The association of the magnetic field with solar activity is also a well-established fact.

The magnetic field can play a simple passive role, for example by confining the plasma forming the various structures, or an active one, if magnetic heating mechanisms are operating. The two aspects are however strongly correlated, and represent just two facets of the complex manifestation of the presence of a magnetic field. We thus conclude that *the magnetic field appears to be the key factor in shaping, maintaining and evolving the coronal structures.*

Since most of the following discussion will be centered on the properties of coronal loops, one example of which is shown in Figure 1, a brief description seems in place.

Figure 1. A coronal loop observed in Ne VII (Skylab, NRL).

Loops, are bright, arch-like structures, clearly visible at EUV and X-ray wavelengths, indicating a range in temperatures from less than 10^5 K to a few units in 10^6 K. A typical loop may reach a height of 50,000 km and have transverse dimensions of a few thousand km, but smaller or longer loops are by no means uncommon. Loops appear to be relatively stable structures, with lifetimes of the orders of hours or even days. Most of the X-ray luminosity comes from collections of loops forming the solar active regions. For a more detailed description we refer to the monograph on Solar Active Regions (Orrall, 1980).

2. CORONAL LOOPS AND HEATING MECHANISMS

The presence of coronal structures is so intimately connec-

ted with the existence of magnetic fields, and in particular
with magnetic heating mechanisms, that it is quite natural to
analyze the properties of loops and see if they give us some
clue on the nature of heating processes. In the past few years
a large amount of work has been devoted to produce models of
loops and to confront them with observations. The proposed models can be basically divided into two groups:

     i) Thermodynamic (TD) models.
    ii) MHD models.

The TD models are essentially one-dimensional models in
which a "realistic" energy equation is adopted. Examples of this
category are given in the papers by Landini and Monsignori-Fossi
(1975), Rosner et al. (1978), Craig et al. (1979), Hood and
Priest (1979). In the MHD models, the full 3D-structure of the
magnetic field is retained but the use of the energy equation
is avoided, for instance by prescribing a pressure profile, as
done in Van Hoven et al. (1977), Chiuderi et al. (1977), Chiuderi
and Einaudi (1980). Since our purpose here is to investigate the
nature of the heating mechanisms, only the TD models will be
discussed.

The basic assumptions, common to most of the models so far
proposed, are the restriction to *static* situations ($\underset{\sim}{v}$ = 0), the
neglect of gravity ($\underset{\sim}{g}$ = 0) and the adoption of a cylindrical geometry. Each of these assumptions is, at best, questionable and
has been relaxed in subsequent work. Thus, flows have been introduced by Noci (1980) and Cargill and Priest (1980), gravity
has been introduced by Veseky et al. (1979) and Wragg and Priest
(1981). However, if any of these simplifying assumptions is
dropped, even a semi-analytical discussion becomes impossible and
a large amount of computing work is needed. In the following I
shall therefore adopt this over-simplified scheme, keeping in
mind that in fact some of the conclusions could be subject to
revision when a more rigorous approach is followed, as appropria-
tely stressed by Carole Jordan during this Advanced Study Institute.

From the equation of hydrostatic equilibrium, in absence of
gravity,

$$0 = \nabla p + \frac{1}{c} \, (\underset{\sim}{J} \times \underset{\sim}{B}) \qquad\qquad (2.1)$$

we get

$$0 = \nabla p \cdot \underset{\sim}{B} \qquad .$$

Therefore, the pressure is constant along the magnetic field lines and we may limit ourselves to the study of the *equilibrium*

along a single field line. Since thermal conduction is strongly channeled along the magnetic field (the transverse thermal conductivity is suppressed by more than twelve orders of magnitude in the solar case), the energy equation is similarly projected along $\underline{B}$ and reads:

$$\frac{d}{ds}\left[F_c(T)\right] = \frac{d}{ds}\left(\kappa_o\, T^{5/2}\, \frac{dT}{ds}\right) = E_R(T) - E_H(T)\,, \qquad (2.2)$$

where s is the coordinate measured along a field line, F_c the conductive flux, E_R and E_H, the radiative power density and heating power density, respectively. The plasma is assumed to be optically thin, as appropriate for the density and temperature ranges of interest here. $E_R(T)$ can be written in the form

$$E_R(T) = n^2\, \Phi(T)\,, \qquad (2.3)$$

where n is the electron density and $\Phi(T)$ has been computed by many authors. Two examples of $\Phi(T)$ are shown in Figure 2.

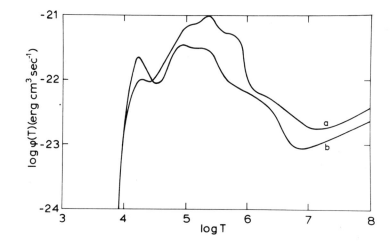

Figure 2. The radiative loss function $\phi(T)$ as given by
a) Cox and Tucker (1969) and
b) Mc Whirter et al. (1975).

It is common practice to represent the $\phi(T)$ in the form

$$\Phi(T) = \Phi_o\, T^{-\alpha} \qquad (2.4)$$

where Φ_o and α are piecewise constants. In the range $8 \times 10^5 K < T < 10^7\,K$, $\alpha \approx 1$. Taking advantage of the assumed constancy of

the pressure we see from Eq. (2.3) that $E_R \sim T^{-(\alpha+2)}$. For lack
of better information the term E_H in the energy equation is
similarly parametrized in the form,

$$E_H = H \ T^\gamma \quad , \qquad (2.5)$$

where p = const has been used again to eliminate dependences
other than on temperature. A number of possible heating mechanisms can indeed be parametrized as in Eq. (2.5), as shown by
Rosner et al. (1978). An unambiguous determination of H and γ
would therefore give us some hint on coronal energy sources.

The energy equation, Eq. (2.2) can easily be integrated
numerically, to produce the temperature profile along a magnetic
field line, T = T(s). A number of parameters enter the energy
equation, but a detailed analysis (Chiuderi et al., 1980) shows
that the *equilibrium* properties of the model depend on p, H
and γ , but only negligibly on the values of the base temperature, T_0 , and base heat flux, F_0 . From the same analysis it is
also possible to show that the numerical results are well represented (within a few percent) by the following expressions (scaling laws):

$$H = (\gamma + 7/2) T_M^{-(\gamma+7/2)} p^2 (a+b \ T_M^{1/2}) \qquad (2.6)$$

and

$$pL = G(\gamma) \ T_M^{7/2} (c+d \ T_M^{1/2})^{-1/2} \qquad (2.7)$$

Here T_M is the maximum temperature of the loop and L its
length. The constants a, b, c, d as well as the function $G(\gamma)$
can be explicitly given. The above scaling laws are valid for
$T_M > 8 \times 10^5$ K, independently of the value of F_0 and thus apply
to both thermally isolated and non-isolated loops. Since p, L
and T_M are measurable quantities, Eq.s (2.6) and (2.7) allow
in principle the determination of H and γ . Before attempting
such a determination, however, we must consider two other aspects,
namely the thermal stability of the proposed loop models and the
effect of the errors in the measured quantities.

The possibility of occurrence of a thermal instability is
readily understood from the fact that $E_R \sim T^{-1}$ for relevant coronal temperatures. Thus, the system reacts to a cooling perturbation by increasing the radiative output, which in turn cools the
system even further. A meaningful comparison of theoretical models
and observations must therefore involve thermally stable configurations and the stability of the equilibrium models must be carefully tested. Stability analyses have been performed by a number
of authors (Antiochos, 1979, Habbal and Rosner, 1980, Hood and
Priest, 1980, Chiuderi et al., 1980) with somewhat conflicting

results. The disagreement among the various authors is most likely due to differences in boundary conditions, to which stability
is exceedingly sensitive. At the present time the discrepancies
are being resolved, and a fair conclusion seems to be that stability *does not impose severe restrictions* on the equilibrium models. It can be shown, in fact (Chiuderi et al. 1980), that for
every value of γ , there is a large range of values of   H   that
produce stable loop models with

$$0.1 \lessgtr p \text{ (dyne cm}^{-2}) \lessgtr 3.0 \quad ,$$
$$1 \times 10^6 \lessgtr T_M(K) \lessgtr 3.5 \times 10^6 \quad ,$$
$$2 \times 10^4 \lessgtr L \text{ (km)} \lessgtr 4 \times 10^5 \quad .$$

The presence of errors, on the other hand, affects to such
an extent the estimate of   H   and γ , that no useful inferences
on heating mechanisms can be made on the basis of the scaling
laws Eq.s (2.6) and (2.7). In fact, let us assume that each of
the measurable quantities p,L   and T_M appearing in Eq. (2.7)
is known with an error as small as 10%. From the explicit expression of $G(\gamma)$ it is then possible to show that the realtive error
on   G, $\Delta G/G \simeq 0.5$. This in turn implies a very high uncertainty
on γ . In fact, if for example the value of   G   turns out to be
$G = 0.14 \pm 0.07$, the estimated γ would be

$$\gamma = 0.0^{-2.2}_{+6.7}$$

No matter what heating mechanism is operating, γ would certainly
be comprised in such a large error bracket and all information on
the heating process is lost.

The use of scaling laws, although promising, seems therefore
very unpredictive. As remarked by various authors (Gabriel, 1976,
Craig et al. 1979, Jordan, 1980) the temperature profile is rather insensitive to the details of the heating mechanism and a
better quantity for a meaningful comparison seems to be the
emission measure. Furthermore, the scaling laws discussed here
are derived from a rather restrictive set of assumptions. In particular, a very crucial ingredient, the magnetic field, plays only a very marginal role. The success of attempts to deduce information on the heating processes by the direct use of observed
quantities will probably depend both on better observations and
better models. From theorists in particular it is required a
much better job in trying to identify the *observable signatures*
of every model.

3. PHYSICS OF HEATING PROCESSES

As we have seen, time is not ripe to get clues on the nature of coronal heating processes directly from observations. But

even if the modeling exercises of the previous Section were successful, we would be a long way from understanding the physics
of the heating mechanisms. Therefore, as an alternative to the
previous approach, we may ask: what does physics, and especially
plasma physics, suggest as possible ways of heating the coronal
structures? In agreement with our previous discussion, we shall
assume that the magnetic field plays a dominant role, and try
to answer the basic questions concerning the source of the energy that is deposited in the corona, the propagation mode and the
dissipation mechanism.

The proposed heating mechanisms are traditionally grouped in
two categories: wave heating and current heating. The first class
assumes that some type of MHD wave propagates through the upper
solar atmosphere, damping part of its energy by means of one or
another dissipative process. In the second class, the attention
is focused on the presence of electrical currents that produce
heat through ohmic dissipation. The distinction is made mostly
for reasons of convenience and actually there are many properties
that are shared by both classes of models.

An MHD wave is a moving disturbance in a pre-existing magnetic field. The magnetic field lines are disturbed locally, plu-
cked or shaken, and this produces a magnetic stress that propagates away from the position where the disturbance was originally
produced. But a stressed magnetic field implies the presence of
electrical currents and our wave can be described in terms of
these currents as well. The ultimate source of the wave (or current) energy resides in the agent that stresses the field. In the
case of the solar corona the accepted suggestion is that the turbulent motions present at photospheric levels are responsible for
the field-stressing process, i.e. for the generation of waves or
alternatively currents. In the photospheric plasma the gas pressure, p, generally exceeds the magnetic pressure, $B^2/8\pi$,
($\beta = 8\pi p/B^2 > 1$) and the high electrical conductivity tightly couples plasma and field together. Thus any material motion is bound
to drag the magnetic field lines and to generate magnetic stresses. In this way free energy is fed from the photosphere into
the system of field lines extending upwards into the coronal regions and the interplanetary space. In the corona, the field lines are still frozen-in, but $\beta < 1$ and the magnetic field has
the possibility of rearranging itself in a lower energy state by
releasing part of the stress. This is alternatively seen as a
damping of the waves or a decrease in the intensity of the currents.

In spite of the many similarities just outlined, a wave is
generally throught as a more orderly motion than a current. For
small amplitude waves in a plasma, there are well-defined wave
modes, obeying specific dispersion relations. The ordered energy

contained in a wave can decrease due to a variety of processes, collisional or collisionless, and may end up in part as disordered motion of the particles, that is heat. In the current heating models, on the other hand, we just concentrate on one mechanism. The direct transfer of the energy contained in the bulk motion of the charges into thermal motion. In a sense this is the most direct and straightforward way of transforming magnetic energy into thermal energy and this explains the interest in current heating theories.

In the following I shall not attempt to cover even a small part of the vast literature on heating mechanisms. I shall rather give a few general concepts and describe the consequences of a couple of models that have been recently proposed. Detailed updated reviews of the subject, containing extensive bibliographies can be found in Hollweg (1980a) and Kuperus et al. (1981).

3.1 Wave heating

Let us consider the possible wave modes induced by material motions in a perfectly conducting plasma. Since the expected motions are relatively slow, we are mostly interested in the low-frequency MHD waves. Even with this restriction there are several possibilities and consideration of the spatial inhomogeneity of the plasma further widens the choice. The high degree of structuring present in the solar corona increases the importance of the modes that are peculiar of non-homogeneous plasmas. One of such modes, the so-called surface Alfvén waves, has been recently proposed as a possible candidate for coronal heating (Ionson, 1978). Therefore I shall briefly outline the properties of this mode, after recalling those of the more familiar magnetic waves of homogeneous plasmas.

Starting from the standard MHD equations (see e.g. Boyd and Sanderson, 1969):

$$\frac{\partial \rho}{\partial t} + \nabla \cdot (\rho \, \underset{\sim}{v}) = 0 \quad ,$$

$$\rho \left[\frac{\partial \underset{\sim}{v}}{\partial t} + (\underset{\sim}{v} \cdot \nabla) \, \underset{\sim}{v} \right] = - \nabla p + \frac{1}{c} \, (J \times \underset{\sim}{B}) =$$

$$= - \nabla p + \frac{1}{4\pi} \, (\nabla \times \underset{\sim}{B}) \times \underset{\sim}{B} \quad , \qquad (3.1)$$

$$\frac{\partial \underset{\sim}{B}}{\partial t} = \nabla \times (\underset{\sim}{v} \times \underset{\sim}{B}) \quad ,$$

$$p \rho^{-\gamma} = \text{const.}$$

and considering the perturbations of a static $(\underset{\sim}{v} = 0)$ equilibrium, we arrive at this set of linearized equations,

$$\frac{\partial \rho_1}{\partial t} + \nabla \cdot (\rho_0 \underset{\sim}{v}_1) = 0 \quad ,$$

$$\rho_0 \frac{\partial \underset{\sim}{v}_1}{\partial t} = - \nabla p_1 + \frac{1}{4\pi} \left[(\nabla \times \underset{\sim}{B}_0) \times \underset{\sim}{B}_1 + (\nabla \times \underset{\sim}{B}_1) \times \underset{\sim}{B}_0 \right] , \qquad (3.2)$$

$$\frac{\partial \underset{\sim}{B}_1}{\partial t} = \nabla \times (\underset{\sim}{v}_1 \times \underset{\sim}{B}_0) \quad ,$$

where the suffices 0 and 1 refer to the equilibrium and perturbation quantities respectively. Let us first consider the case of an unstructured medium. All the unperturbed quantities are then constant and we can perform a Fourier analysis on the system (3.2) by assuming that first-order quantities are proportional to $\exp\left[i(\underset{\sim}{k} \cdot \underset{\sim}{r} - \omega t)\right]$.

The linear differential system (3.2) then becomes a linear homogeneous algebraic system that admits non-trivial solutions when

$$\omega^2 = k^2_{\parallel} \, c^2_A \qquad\qquad (3.3)$$

or

$$\omega^4 - k^2(c^2_A + c^2_S)\,\omega^2 + k^2 \, k^2_{\parallel} \, c^2_A \, c^2_S = 0 \qquad\qquad (3.4)$$

In the preceding equations $k_{\parallel} = (\underset{\sim}{k} \cdot \underset{\sim}{B}_0)/B_0$ and

$$c^2_A = B^2_0/(4\pi \, \rho_0) \quad , \qquad c^2_S = \gamma \, p_0/\rho_0$$

are the squared Alfvén and sound speeds. The dispersion relation (3.3) corresponds to (shear) Alfvén waves and the two solutions of (3.4) to fast and slow magnetosonic modes.

However, if the basic magnetic field, B_0 has a structure, new modes arise. Let us consider the simple one-dimensional case $\underset{\sim}{B}_0 = B_0(x) \, \underset{\sim}{e}_z$. Since the zeroth-order quantities appearing in the linearized system (3.2) now depend on x , we cannot Fourier-analyze along x , and we must assume for each perturbed quantity, $f_1(r , t)$ a form

$$f_1(\underset{\sim}{r} , t) = f_1(x) \exp\left[i(k_y y + k_z z - \omega t)\right]$$

The system (3.2) then becomes a system of ordinary differential

equations, that can be reduced to a single differential equation
for the x-component of the velocity, v_x . The final equation
turns out to be

$$\frac{d}{dx} \left[\frac{\varepsilon(x)}{q(x)} \frac{dv_x}{dx} \right] - \varepsilon(x) \, v_x = 0 \qquad (3.5)$$

where $\varepsilon(x)$ and $q(x)$ are known functions of x. The properties
of the solutions of Eq. (3.5) have been discussed by Wentzel
(1979) and Roberts (1981).

To simplify the discussion let us consider the case of a
sharp discontinuity, located at x = 0. Then all equilibrium
quantities, and therefore ε and q are constant but assume different values for negative or positive x-values. Eq. (3.5) is
easily solved on either side of the discontinuity and the x-depen-
dence of v_x is given by

$$v_x \sim \exp(- |q \, x|) \; .$$

The amplitude of motions decreases exponentially as we move away
from the surface of discontinuity, and we speak therefore of a
"surface" wave. In the astrophysically relevant case in which
$\beta \ll 1$ on either side of the discontinuity, it is easy to show
that $v_z \simeq 0$, so that there are no motions along the magnetic
field. The imposition of the continuity of v_x and total pressure at x = 0 gives the approximate dispersion relation

$$\omega^2 = k_z^2 \frac{B_1^2 + B_2^2}{4 \, (\rho_1 + \rho_2)} \qquad , \qquad (3.6)$$

where the suffices 1 and 2 distinguish the two different regions.
The phase velocity given by Eq. (3.6) is intermediate between
the Alfvén velocities on the two sides of the discontinuity. A
wave obeying the dispersion relation (3.6) is called an Alfvén
surface wave. Such a wave exists also in the more realistic case
of a diffuse discontinuity, along with a continuum of "body" waves,
i.e. shear Alfvén waves. Moreover, the two types of modes can
couple, giving rise to a number of effects that may be of relevance for the heating of coronal loops, as I shall now describe.

Let us consider a single loop, whose physical properties
vary with r , the minor radius. When this loop is shaken by the
turbulent photospheric motions, according to the general scenario previously described, both surface and body waves are generated. The body waves are shear Alfvén waves corresponding to the
oscillatory characteristics of the spatially localized region in
which they are excited. There may be circumstances where a surface wave can resonantly couple to a shear wave so that a considera-

ble fraction of the energy contained in the surface wave is ab-
sorbed by the body wave. In principle this does not seem very
exciting since the continuum of body waves already contains a
certain amount of energy that could be used for heating. However,
in the region where the resonant absorption takes place, which
forms a thin layer   a pure MHD description does not hold and
effects such as the finite electron inertia and finite ion gyroradius must be taken into account. The modified waves resulting
from the inclusion of these effects are known as kinetic Alfvén
waves. The point of interest to heating theories is that kinetic
Alfvén waves have been shown to dissipate much more efficiently
than ordinary Alfvén waves (Hasegawa and Chen, 1976). Thus the
resonant absorption of surface waves by kinetic Alfvén waves and
their subsequent dissipation by-passes one of the main problems
faced when building a heating mechanism based on Alfvén waves,
that dissipate very inefficiently.

Ionson (1978) has used these ideas to construct, to a considerable detail, a complete dynamical model of a coronal loop. He
envisages that the resonant absorption takes place in a thin
sheath that envelops the entire loop. The estimated thickness
of the sheath, where the irreversible heating takes place, is of
the order of a few kilometers. As a result of the intense heating
a large temperature gradient is established between the sheath
and the contiguous plasma, so large, in fact, to overcome the
effect of the small cross-field thermal conductivity. Thus, considerable energy extraction takes place from the sheath in a
neighbouring boundary layer, whose thickness is estimated of the
order of 100 km. The plasma heated in the boundary layer acquires
buoyancy and accelerates upwards along the field lines, being
continuously replaced by cooler plasma coming from below. When
the hot upflowing plasma reaches the loop's summit, an instability mechanism helps him crossing the field lines and entering the
interior of the loop. At this point, being no longer in contact
with the hot boundary layer, the plasma cools by radiation, con-
denses into clumps and falls along the field lines. Thus, a convection pattern develops within the loop, as suggested by the
observation of the so-called "coronal rain" loops. Ionson suggests
that an electrostatic Rayleigh-Taylor instability is responsible
for the field crossing at the top of the loop, where gravity and
density gradient are oppositely directed. The inversion of density profile is in turn caused by the local evacuation of the plasma due to the convective pattern.

Ionson's model is a very articulate one and contains a great
deal of physics. At the present time it is probably the most quantitative model of loop heating. One interesting feature is that
the establishment of a circulation pattern is by no means peculiar
to his particular heating mechanism, but applies equally well to
any process capable of dumping a large amount of energy in a thin

sheath. However, there are problems. The high-temperature layers are well below the present resolution limits and a direct observational check of the model appears impossible. From the available observations one has the impression that heating takes place over more extended regions, and no mechanisms have been found capable of spreading the heat outside of the boundary layer. Also the restriction to the linear regime is open to doubts. Above all one cannot avoid the feeling of an exceeding complexity of the mechanism: each step is reasonable, but too many steps must follow in sequence to make the machine work!

Other wave modes can be considered as candidates for heating: Alfvén shear waves (Hollweg, 1978, 1980) or fast magnetosonic waves (Habbal et al., 1980). The main problem with these modes is connected with our poor understanding of the damping processes. Another general statement to be made on most loop heating models concerns the lack of a stability analysis for the proposed configuration, although such an analysis is available in a few cases (Zweibel, 1980).

3.2 Current heating

The coronal plasma is highly, but not perfectly conductive and the presence of resistivity offers the seemingly simplest way of transforming magnetic energy into heat. To evaluate how viable is the idea of ohmic heating, let us consider a thermally isolated static loop. The integration of the energy equation over the loop's volume gives

$$\int_{V'} (E_H) \, dV = \int_V (E_R) \, dV \tag{3.7}$$

where V' is the fraction of the total loop volume V effectively heated. If heating is due to ohmic dissipation, $E_H = \eta J^2$, where η is the resistivity. The radiative term is again written as $E_R = n^2 \Phi(T)$. We use cylindrical geometry and assume that the current flows in a sheath of thickness $\delta\ell = a$, the minor radius of the loop. All quantities referring to the sheath will be labelled by the subscript s. The condition of stationarity (3.7) then gives

$$J_s^2 = \frac{n^2 \Phi}{2\eta_s} \left(\frac{a}{\delta\ell} \right) \tag{3.8}$$

From Ampère's law: $\nabla \times \underset{\sim}{B} = \left(\frac{4\pi}{c} \right) \underset{\sim}{J}$ we get

$$J_s \simeq \frac{c}{4\pi} \left(\frac{\delta B}{\delta\ell} \right) , \tag{3.9}$$

and combining (3.8) and (3.9),

$$(\delta B)^2 = \frac{8\pi^2}{c^2} \; n^2 \; a^2 (\frac{\phi}{\eta_s}) (\frac{\delta \ell}{a}) \; . \tag{3.10}$$

The resistivity that appears in Eq. (3.10) can be either "classical" (i.e. due to electron-ion collisions) or "anomalous" (when the plasma is turbulent). Let us examine in turn these two possibilities.

i) Classical resistivity. In this case the drift velocity of the electrons, v_D , must be less than v_c , the critical velocity for the onset of plasma turbulence. From $J_s = e \, n_s \, v_D$, $v_D < v_c \simeq (k_B \, T_s/m_i)^{\frac{1}{2}}$, where m_i is the ion's mass, and Eq. (3.10) we deduce two *separate* inequalities for the size $\delta \ell$ of the current channel and the jump of the magnetic field, δB:

$$\frac{\delta \ell}{a} > (\frac{m_i \, \phi_o}{2 \, k_B \, e^2 \eta_c}) \, (\frac{T_s}{T})^{5/2} \; , \tag{3.11}$$

$$(\delta B) > (\frac{4 \, \pi^2}{c^2} \; \frac{m_i \phi_o}{k_B \, e^2 \eta_c^2}) \, a^2 \, n^2 \, T \, (\frac{T_s}{T})^4 \tag{3.12}$$

In Eq.s (3.11) and (3.12) we have assumed $\phi(T) = \phi_o T^{-1/2}$ (McWhirter et al. 1975) and $\eta_s = \eta_c \, T^{-3/2}$ (Spitzer, 1962). To give a numerical estimate of the above inequalities we use the "typical" values for active regions (Vaiana and Rosner, 1978): $T = 2.5 \times 10^6$ K, $n = 5 \times 10^9$ cm^{-3}, $a = 2 \times 10^8$ cm and $\phi_o = 5 \times 10^{-20}$ (c.g.s.) $\eta_c = 2.5 \times 10^{-7}$ (c.g.s.). We thus arrive at

$$\frac{\delta \ell}{a} > 5.2 \times 10^{-3} \, (\frac{T_s}{T})^{5/2} \; ,$$

$$\delta B > 1.5 \times 10^4 \, (\frac{T_s}{T})^4 \; G \; .$$

These two limitations, and especially the second one, give inacceptably high lower limits: classical resistivity does not work.

ii) Anomalous resistivity. When $v_D > v_c$, the plasma becomes turbulent and the relevant collisions are those between the electrons and the dominant plasma excitations. In the case of ion-acoustic turbulence (Rosner et al. 1978b) an approximate expression for η_s is (Papadopoulos, 1977):

$$\eta_s = \eta_a \; n_s^{-1/2} \; (T_e/T_i)_s \quad , \qquad (3.13)$$

with
$$\eta_a \; 10^{-2} \; (\frac{4\pi \; m_e^2}{e^2 m_i})^{1/2} \qquad (3.14)$$

The ratio T_e/T_i in the sheath that appears in (3.13) is a re-
minder of the fact that ion-acoustic turbulence requires
$T_e \gg T_i$. In the following we shall assume $(T_e/T_i)_s = 10$. The
numerical factor in front of the expression (3.14) is somewhat
uncertain: we have adopted the value given by Sagdeev (1967).
Repeating the procedure followed for the classical resistivity
we now find for the same values of the physical parameters already used,

$$\frac{\delta\ell}{a} < 4.5 \times 10^{-8} \; (\frac{T_{es}}{T})^{1/2} \quad , \qquad (3.15)$$

$$\delta B < 13 \quad G \quad . \qquad (3.16)$$

If the electron temperature in the sheath $T_{es} = 10 \; T$ we get
$\delta\ell \approx 15$ cm, still larger than the ion gyroradius. The limitations
given by Eq.s (3.15) and (3.16), although not preposterous, are
at best marginally acceptable. Many arguments can be given to
illustrate the intrinsic weakness of heating mechanisms based on
anomalous ohmic dissipation. For instance, turbulence produces
enhanced diffusion and the sheath tends to thicken, thus viola-
ting the condition given by (3.15). More serious is the consideration of the time that is needed to dissipate a current, whose
characteristic are dictated by the assumption of a turbulent
state plus the maintenance of a stationary state. The dissipative timescale, in fact, is given by

$$\tau = \frac{4\pi \; a^2}{\eta_s \; c^2} \; (\frac{\delta\ell}{a})^2 \quad , $$

and by using (3.15) we easily arrive at

$$\tau < 5 \times 10^{-7} \quad s \quad ! \qquad (3.17)$$

Thus, if the current has the right characteristics, the dissipation time becomes exceedingly short.

Anomalous resistivity seems to pose more problems than it
solves. It is difficult to generate and to maintain, since due
to (3.17), we have to regenerate continuously the conditions for
the onset of turbulence. A possible way out could be the relaxation of the condition of a steady mechanism. Heating would then
be achieved by a series of intense, sporadic events, whose cumu-

lative effects appears to be continuous only because of our poor
time resolution. In this context, the proposal of Levine (1974)
acquires a particular interest. Coronal heating may be due to
numerous mini-reconnections, occurring randomly in space and time. I believe that the original concept of Levine's model, reconnection in neutral sheaths, should be replaced by that of localized tearing modes in current sheaths, that appear to be more
easily realized in the physical world, since they only require
a magnetic field with a high local shear. A quantitative analysis
of this mechanism is badly needed.

4. MAGNETIC HEATING IN STARS

We want now to address the following question: can we apply
our limited knowledge of the physics of solar corona to stars
other than the Sun? The question makes sense after the X-ray
observations performed by the EINSTEIN satellite, that proved
that a great deal of main sequence stars do in fact possess some
sort of corona. We have seen in the previous Sections that little can be said, at a quantitative level, on the heating mechanisms operating in the Sun. One has the impression, however, that
the basic ideas are there. The interaction of turbulent photospheric motions with the magnetic fields extending up in the corona
induces a certain amount of magnetic stress, that eventually
results in irreversible heating of the regions that are sources
of the X-ray emission. If these ideas are correct, we can expect
a certain variation of the X-ray luminosity as we move along the
main sequence. In fact, the variation of the structure of the
convection zone and the rate of rotation along the H-R diagram
produce changes in the surface magnetic flux level. On the other
hand, the strength of the surface turbulence determines the degree of twisting of coronal magnetic fields. From diagram models
we can estimate both the value of the emerging magnetic field
and of the surface turbulence. Thus, we should be able to predict
the X-ray flux as a function of the star's spectral type.

A first attempt in this direction has been recently made by
Belvedere et al. (1980a). The final result is achieved in two
steps. First, the expected X-ray flux is given in terms of the
mean emerging toroidal field and the surface turbulent velocity.
Second, the latter quantities are computed by using kinematical
dynamo theories, calibrated in order to reproduce the known solar 22-years cycle. The first part is straightforward. Starting
from a purely B_z-field , we generate a B_θ component by twisting motions whose velocity is v_θ . From the magnetic induction equation

$$\frac{\partial B_\theta}{\partial t} = B_z \frac{\partial v_\theta}{\partial z} \qquad (4.1)$$

The energy contained in the newly generated B_θ field, $(B_\theta^2/8\pi)$, is the reservoir that provides the fuel for heating. Consider an isolated loop of volume V: the total heat produced per unit time is:

$$\int E_H \, dV \simeq \xi \frac{\partial}{\partial t} (B_\theta^2/8\pi) \, V = (\xi/4\pi) \, B_\theta \, B_z \frac{\partial v_\theta}{\partial z} \, V \simeq$$

$$\simeq (\xi/4\pi) \, (B_\theta/B_z) B_z^2 (v_\theta/L) \, V, \qquad\qquad (4.2)$$

where L is the length of the loop, ξ is the efficiency of the (unknown) process of energy conversion, and use has been made of (4.1). In a stationary state (4.2) also gives the X-ray luminosity of a single loop, since at the temperatures of interest most of the radiative output falls into the X range. The contributions from the single loops (of the form (4.2)), can now be added to give the total X-ray luminosity, L_x, of an active region. It is clear from dimensional reasons that the final expression of L_x will still be given by (4.2), provided that L, B_z and V are replaced by L_0, B_{z0}, V_0 which are measures of the typical dimensions, magnetic field and emitting volume of the active region, respectively. Assuming (Golub et al. 1980) that $B_\theta/B_z \simeq 1$ for all loops, we finally get $L_x \simeq (\xi/4\pi) B_{z0}^2 v_\theta (V_0/L_0)$ or, for the X-ray flux,

$$F_x \simeq (\xi/4\pi) \, B_{z0}^2 \, v_\theta \qquad\qquad (4.3)$$

The second step, the evaluation of B_{z0} and v_θ, by means of stellar dynamo theory, is of course much more involved, and will not be presented here. It suffices to say that the interaction of rotation with convection appears to be the basic mechanism for the generation of both differential rotation and magnetic activity. The variation of these quantities with spectral type along the lower main sequence has been computed in a series of papers by Belvedere et al. (1980 b,c,d,e), who use as a calibration criterion the requirement that the computed activity cycle for the Sun matches the observed 22 years period. This condition fixes the dynamo parameters for the solar case. The same values are then adopted for the other stars. If we identify B_{z0} with the average emerging toroidal field and v_θ with the surface convective velocity deduced from convection zone models, it is possible to predict the X-ray flux on the basis of Eq. (4.3). The results are shown in Fig. 3, for two different values of ξ. The histogram represents the median of the observed values and the vertical bars do not give errors but indicate the spread in the observed values for each spectral type.

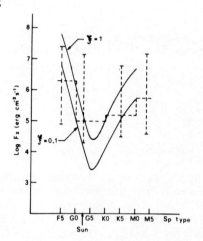

Figure 3. Predicted and observed X-ray fluxes for lower main
sequence stars (Belvedere et al. 1980a).

There is a general agreement between the shape and the numerical values of the computed and observed X-ray fluxes. Considering the extreme simplicity of the model and the fact that the
only adjustable parameter is ξ , I consider Figure 3 positive
evidence of the correctness of the basic scheme of magnetic heating. We are very far from a theory: the main problem, that of
the physical nature of the heating processes is simply bypassed.
Nevertheless the conjecture that the heating of the external
layers of stars, evidenced by their X-ray emission, is the result of an intimate interplay of rotation and convection seems
to be strengthened by this preliminary application. I feel that
the different implications of this conjecture should be thoroughly investigated and the quantitative details worked out to
transform it in a theory.

REFERENCES

Antiochos, S.K.: 1979, Astrophys. J. Lett., 232, L125.
Belvedere, G., Chiuderi, C. and Paternò, L.: 1980a, Astron.
    Astrophys., in press.
Belvedere, G., Paternò, L. and Stix, M.: 1980b, Geophys. Astro-
    phys. Fluid Dynamics, 14, 209.
Belvedere, G., Paternò, L. and Stix, M.: 1980c, Astron. Astrophys.
    86, 40.
Belvedere, G., Paternò, L. and Stix, M.: 1980c, d, Astron. Astro-
    phys., in press.
Boyd, T.I.M. and Sanderson, J.J.: 1969, "Plasma Dynamics", Barnes
    and Noble, New York.

Cargill, P.J. and Priest, E.R.: 1980, Solar Phys., 65, 251.
Chiuderi, C., Giachetti, R. and Van Hoven, G.: 1977, Solar Phys. 54, 107.
Chiuderi, C. and Einaudi, G.: 1981, Solar Phys., in press.
Chiuderi, C., Einaudi, G. and Torricelli-Ciamponi, G.: 1980, Astron. Astrophys., submitted.
Cox, D. and Tucker, W.H.: 1969, Astrophys. J., 156, 87.
Craig, I.J.D., Mc Clymont, A.N. and Underwood, J.H.: 1978, Astron. Astrophys., 70, 1.
Gabriel, A.H.: 1976, in "Proceedings of Colloquium N. 36 of I.A.U.". (R.M. Bonnet and Ph. Delache Ed.s), G. de Bussac, Clermont-Ferraud, 375).
Habbal, S.R. and Rosner, R.: 1979, Astrophys. J. 234, 1113.
Habbal, S.R., Leer, E. and Holzer, T.E.: 1979, Solar Phys. 64, 207.
Hasegawa, A. and Chen, L.: 1976, Phys. Fluids, 19, 1924.
Hollweg, J.V.: 1978, Solar Phys.,56, 305.
Hollweg, J.V.: 1980a, in "Skylab Workshop on Active Regions" (F.Q. Orrall, Ed.), in press.
Hollweg, J.V.: 1980b, Solar Phys., submitted.
Hood, A.W. and Priest, E.R.: 1979, Astron. Astrophys., 77, 233.
Hood, A.W. and Priest, E.R.: 1980, Astron. Astrophys., 87, 126.
Ionson, J.A.: 1978, Astrophys. J., 226, 650.
Jordan, C.: 1980, Astron. Astrophys., 86, 355.
Kuperus, M., Ionson, J.A. and Spicer, D.S.: 1981, Ann. Rev. Astron. Astrophys., to appear.
Landini, M. and Monsignori Fossi, B.C.: 1975, Astron. Astrophys., 42, 413.
Levine, R.H.: 1974, Astrophys. J., 190, 457.
Mc Whirter, R.W.P., Thonemann, P.C. and Wilson, R.: 1975, Astron. Astrophys., 40, 63.
Noci, G.: 1980, Solar Phys., in press.
Orrall, F.Q. (Ed.): 1980, "Skylab Workshop on Active Regions", in press.
Papadopoulos, K.: 1977, Rev. Geophys. Space Sci., 15, 113.
Roberts, B.: 1981, Solar Phys., 69, 27, 39.
Rosner, R., Tucker, W.H. and Vaiana, G.S.: 1978, Astrophys. J. 220, 643.
Rosner, R., Golub, L., Coppi, B. and Vaiana, G.S.: 1978b, Astrophys. J., 222, 317.
Sagdeev, R.Z.: 1967, Proc. Symp. Appl. Math., 18, 281.
Spitzer, L.: 1962, "Physics of Fully Ionized Gases", Interscience, New York.
Vaiana, G.S.: 1980, in Proc. AAS/SPD Workshop "Cool Stars, Stellar Systems and the Sun" (A.K.D. Dupree, Ed.) Smiths. Astrophys. Obs. Special Report 389, 195.
Vaiana, G.S. and Rosner, R.: 1978, Ann. Rev. Astron. Astrophys., 16, 393.
Van Hoven, G., Chiuderi, C. and Giachetti, R.: 1977, Astrophys. J., 213, 869.

Vesecky, J.F., Antiochos, S.K. and Underwood, J.H.: 1979,
    Astrophys. J., 233, 987.
Wentzel, D.G.: 1979, Astrophys. J., 227, 319.
Wragg, M.A. and Priest, E.R.: 1981, Solar Phys., 69, 257.
Zweibel, E.: 1980, Solar Phys., 66, 305.

MAGNETOHYDRODYNAMICS OF THIN FLUX TUBES

H. C. Spruit

Max-Planck-Institut für Astrophysik
Garching bei München, W. Germany

Abstract

In this contribution we present a formalism for studying
the magnetohydrodynamics of a magnetic field consisting of
isolated flux tubes. Applications are given for the following
problems: a) the value of the field strength in magnetic
elements at the stellar surface, b) the transfer of energy from
the convection zone to the chromosphere via the magnetic field,
(c) the instability of toroidal magnetic fields in a convection
zone and its effect on the dynamo process.

I. INTRODUCTION

The magnetic field at the solar surface is observed to be
quite inhomogeneous (e.g., Harvey, 1977). It consists of
patches of strong field (1800 G at τ = 1), separated by
essentially field free regions. For historical reasons these
patches have different names, depending on their size. The
biggest ones (up to 50 000 km) are spots, the smallest ones (200
km or less) are called filigree elements or facular points. All
these structures are probably magnetic flux tubes. They are
oriented almost vertically (due to buoyancy, Meyer et. al., 1979)
in the top of the convection zone and rapidly fan out in the first
1500 km above the photosphere (see Figure 1).

R. M. Bonnet and A. K. Dupree (eds.), Solar Phenomena in Stars and Stellar Systems, 289–300.
Copyright © 1981 by D. Reidel Publishing Company.

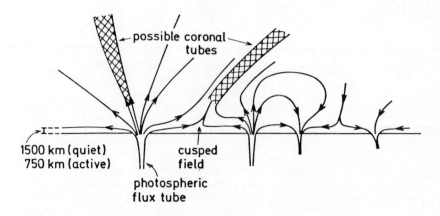

Fig. 1 Magnetic field structure near the solar photosphere.
Merging occurs within a height of 1500 km. (schematic).

There are theoretical reasons to believe that a division of
the solar magnetic field into ropes or sheets may extend deeper
down into the convection zone as well. For example, Peckover and
Weiss (1978), Galloway et. al. (1977), Parker (1979, Ch. 16) have
shown that convective eddies have a strong tendency to expell
magnetic flux tubes. Thus, the field would preferentially be located
on the boundaries between convective eddies. It would seem
important to take into account this essential inhomogeneity of
the field structure, when studying magnetohydrodynamic processes
like dynamo action or chromospheric heating by MHD waves generated
in the convection zone. In the traditional treatment of these
processes, a background magnetic field, varying slowly in space
has generally been used however.

    In this contribution, we describe a formalism which
approaches the same processes from the opposite point of view, by
regarding the magnetic field as consisting of isolated flux
tubes, and studying the dynamics of such tubes. The simplifying
assumptions which make this approach possible are:

i) the electrical conductivity is infinite (frozen-in field)

ii) the tube is *thin* compared with all relevant length scales.

In addition, we assume that the tubes are *untwisted* and that the perturbations are *adiabatic*. These two assumptions are not essential however.

Since the tube is thin, it is at all times in pressure equilibrium with its surroundings:

$$P + B^2/8\pi = P_e \quad ,$$ (1)

where P and B are the gas pressure and field strength at some point inside the tube and p_e is the external gas pressure. This is because the time required for perturbations to propagate across the tube radius can be made arbitrarily small by making the tube thin.

The tube can only be in pressure equilibrium with its surroundings if the internal pressure is less than the external pressure. Even if this condition is satisfied everywhere along the tube at a given instant, during its subsequent evolution the pressure distribution may change such that Equation (1) cannot be satisfied everywhere. Thus, we should expect that in general the thin tube approximation as defined above can only be applied during a limited time interval or in a limited area in space.

In the following, we describe some of the results obtained so far for thin tubes. This is first done for the simpler case of longitudinal motions along a vertical tube (section 2), then a more general formalism is given in section 3. The following astrophysical applications are discussed: a) instability of vertical tubes in a convection zone, value of the field strength at the solar surface (section 2); b) transversal oscillations in vertical tubes, transport of energy by such waves from the convection zone to the chromosphere (section 4); c) instability of toroidal flux tubes in the deeper parts of the convection zone, consequences of this instability for the operation of stellar dynamo's (section 5).

2. LONGITUDINAL MOTION IN VERTICAL TUBES

Consider a thin magnetic flux tube oriented vertically in a stratified atmosphere. Assume that the field strength B and the flow velocity v are uniform across the tube (but variable with depth). For flows *along* the tube, the equations of motion and continuity are then:

$$\rho(\frac{\partial v}{\partial t} + v \frac{\partial v}{\partial z}) = -\frac{\partial P}{\partial z} + \rho g \quad ,$$ (2)

$$\frac{\partial}{\partial t}\frac{\rho}{B} + \frac{\partial}{\partial z}\left(v\frac{\rho}{B}\right) = 0 \quad ,$$ (3)

(Defouw, 1976). Here the depth z is taken positive in the
direction of $\vec{g}$. Equation (2) is just the equation of motion of a
fluid in one dimension. The magnetic forces do not enter because
they are perpendicular to $\vec{B}$. Equation (3) expresses the conservation of mass, in terms of the mass per unit length of the tube,
ρ/B. To solve these equations, the lateral equilibrium condition
(1) and an energy equation have to be added. The formal derivation
of (2) and (3) from the basic MHD equations has been given by
Roberts and Webb (1978).

For a tube to be in static equilibrium we must have $dP/dz =$
ρg. The precise form of the function $P(z)$ and $\rho(z)$ in this equilibrium depends on the temperature distribution $T(z)$ in the tube.
A simple case is $T(z) = T_e(z)$, the tube is then in temperature
equilibrium with its surroundings. Defining

$$\beta = \frac{8\pi P}{B^2} \quad , \tag{4}$$

we find for such a tube that β = const. If we consider *linear*,
adiabatic perturbations of this equilibrium, the thin tube
equations can be combined into a single equation for the velocity
amplitude v_l (Defouw, 1976, Roberts and Webb, 1978, see also Spruit
and Zweibel, 1979):

$$\frac{\partial^2 v}{\partial z^2} + \frac{1}{2H}\frac{\partial v}{\partial z} + \frac{1}{2H^2}\left[\omega^2 \frac{H}{g}\left(\frac{2}{\gamma}+\beta\right) + (1+\beta)\delta\right] = 0, \tag{5}$$

where H is the pressure scale height, ω the frequency of the
perturbation and δ the superadiabaticity, $\delta = d\ln T/d\ln P - (1-1/\gamma)$.

For an isothermal atmosphere with $\gamma = 5/3$ we have $\delta = -0.4$
and the solutions of (5) are waves. They are similar to the waves
associated with pressure fluctuations in an elastic tube filled
with fluid (varicose waves in a blood vessel for example). Due to
the effects of gravity, the waves have a cut off frequency below
which they are evanescent. This cut off frequency is very close to
the acoustic cut off frequency of the atmosphere. The asymptotic
(high frequency) propagation speed v_∞ of the waves is

$$v_\infty = c\, v_A/(c^2 + v_A^2)^{1/2} \quad . \tag{6}$$

This speed is less than both the sound speed c and the Alfvén
speed v_A inside the tube. We call this wave the longitudinal tube
mode (as opposed to the transversal mode discussed in Section 4).

If the stratification is not isothermal, the perturbations
may behave as instabilities rather than waves. This may be seen by
inspection of Equation (5), which suggests that for given boundary
conditions solutions with $\omega^2 < 0$ exist if δ is made sufficiently
large, i.e. if the stratification is sufficiently superadiabatic.

Since it is driven by the superadiabaticity, we call this instability convective. The instability has been studied in detail by Webb and Roberts (1979), Spruit and Zweibel (1979), Spruit (1979) and Unno and Ando (1979). A numerical solution of (5) for the stratification of the solar atmosphere and convection zone (Spruit and Zweibel, 1979) shows that for flux tubes near the solar surface to be stable to this process, they must have $\beta < 1.8$. This corresponds to surface field strengths (defined as the field strength measured at the level $\tau_{5000} = 1$ inside the tube) $B_s > 1300$ G. For magnetic field measurements at an effective height of, say, 150 km above $\tau = 1$, this corresponds to observed field strengths $B_{obs} > 800$ G. The observed field strengths on the sun are above this value. This is reassuring since the time scale of the instability (a few minutes) is so short that the chances of observing an unstable tube would be small.

The instability described above may also be the cause of the particular value of the field strength observed in magnetic elements (about 1800 G at $\tau = 1$). Consider a flux tube brought to the surface, during the emergence of an active region, with a field strength less than that required for stability. The field strength could for example be about $B_s = 700$ G, corresponding to equipartition of the magnetic energy density with the kinetic energy of the convective turbulence. How would the tube evolve due to the instability? When the tube emerges at the surface, a down-flow is initiated by radiation cooling. This flow is accelerated by the instability, thereby emptying the tube (internal pressure decreases). This increases the field strength. A high field strength however, favors stability, so we guess that the instability is *self-limiting*, i.e. after a finite displacement the flow stops, leaving the tube in a new equilibrium state. Such equilibria do indeed exist (Spruit, 1979). In the case of the sun, this process of "convective collapse" of a tube will transform a tube with a surface field strength of 700 G into one with 1800 G, which is just the observed value.

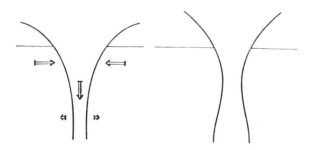

"convective collapse" of a flux tube

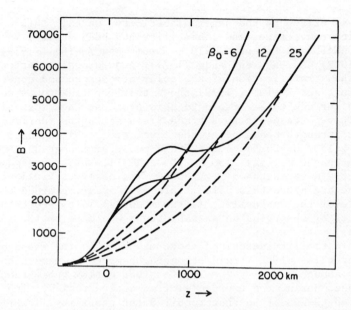

Fig. 2 Field strength as a function of depth for three flux
tubes of different strength. Broken: field strength in the
original constant β - tube, solid: field strength in the collapsed
state. β_o: value of β in the original state.

The effects of the collapse, in terms of the field strength
as a function of depth, are illustrated in Figure 2. It is seen
that the effects are significant only in the top 1000 km of the
convection zone, where the superadiabaticity δ is significant. The
particular value of the surface field strength would thus be
determined by a surface effect operating in the top of the
convection zone.

3. GENERAL EQUATIONS FOR THIN TUBES

Defouw's thin tube equations (2, 3) are valid only for
longitudinal flow in a vertical tube. More general equations can
be derived which are valid for the motion of the fluid in a tube
which follows an arbitrary path through a stratified fluid (Spruit,
1981a).

Let $\hat{l}$ be a unit vector parallel to the axis of the tube, and
let l be the arc length measured along the tube. Because the tube
is thin and untwisted, we can write its magnetic field vector as

$$\vec{B} = B(\vec{r}_l, l)\,\hat{l} \quad , \tag{7}$$

where r denotes the coordinates in the plane perpendicular to axis. The general equation of motion for an arbitrary field is

$$\rho \frac{d\vec{v}}{dt} = -\vec{\nabla} P + \frac{1}{4\pi} (\vec{\nabla} \wedge \vec{B}) \wedge \vec{B} + \rho \vec{g} \; . \tag{8}$$

We substitute (7) into (8), and the average over $\vec{r}_\perp$, i.e. over the cross section of the tube. For details of the derivation we refer to Spruit (1981a). The resulting equation of motion is

$$\frac{d\vec{v}}{dt} = -\frac{1}{\rho} \frac{\partial P}{\partial l} \hat{l} + \vec{g} \cdot \hat{l} \, \hat{l} + \frac{\rho}{\rho + \rho_e} V_A^2 \vec{k} + \frac{\rho - \rho_e}{\rho + \rho_e} (\hat{l} \wedge \vec{g}) \wedge \hat{l} , \tag{9}$$

where $\vec{k} = \partial \hat{l} / \partial l$ is the curvature of the path of the tube. All quantities in (9) are functions of time and a suitable coordinate along the tube (the arc length l , or a Lagrangian coordinate). The first and second terms in (9) give the parallel part of the acceleration of the fluid, due to the pressure gradient along the tube and the component of gravity along the tube. The third and fourth terms are perpendicular to the tube and are due to the tension in the field lines and to buoyancy, respectively.

For a perfectly conducting fluid, the equations of continuity and induction can be combined into (e.g. Roberts, 1967):

$$\frac{d}{dt} \frac{\vec{B}}{\rho} = \frac{\vec{B}}{\rho} \cdot \vec{\nabla} \vec{v} \; . \tag{10}$$

For the magnetic field of a thin tube (Equation 7) this yields

$$\frac{d}{dt} \frac{\rho}{B} = -\frac{\rho}{B} \hat{l} \cdot \frac{\partial \vec{v}}{\partial l} \; . \tag{11}$$

This is the generalisation of Equation 3 and expresses again the conservation of mass in terms of the mass per unit length of the tube.

If the tube is almost vertical and the horizontal motions are small (i.e. we linearize with respect to the horizontal motion but not the vertical motion), Equation (11) reduces to (3), and the vertical component of (9) reduces to (2), i.e. we recover Defouw's equations. In addition however, the horizontal component of (9) yields

$$\frac{\partial^2 \xi}{\partial t^2} = -g \frac{\rho - \rho_e}{\rho + \rho_e} \frac{\partial \xi}{\partial z} + \frac{\rho}{\rho + \rho_e} V_A^2 \frac{\partial^2 \xi}{\partial z^2} \; , \tag{12}$$

where ξ is the horizontal displacement of the tube. This equation governs the transversal oscillations of a vertical tube.

4. THE MODES OF A THIN TUBE

We call the solutions of Equation (12) the transversal tube mode since the motion is purely perpendicular to the tube (also called "taut wire mode" by Wilson, 1979). By contrast, the solution of Defouw's equations is a longitudinal mode. These are the two modes of a thin, untwisted tube of circular cross section. If we would include twist, a third mode would appear, which is the torsional Alfvén wave, propagating at the Alfvén speed V_A. Together, the three modes are analogous to the transversal, longitudinal and torsional modes of an elastic wire under tension. The analogy is not quite exact, especially due to the presence of gravity. The longitudinal mode, for example, can become unstable (section 2). Secondly, the separation into a purely longitudinal and a purely transversal mode occurs only in a *vertical* tube. At other angles with respect to gravity there are still two modes (not counting the Alfvén wave) but the corresponding fluid motions have both longitudinal and transversal components (cf. Section 5).

4.1 Transversal tube waves

The longitudinal tube wave has been discussed in section 2 so we concentrate on the transversal mode in this section.

In the absence of gravity, the first term on the RHS of (12) vanishes, and the solutions are simple nondispersive waves with propagation speed

$$V_p = (\frac{\rho}{\rho + \rho_e})^{1/2} V_A .$$
(13)

This speed differs from the Alfvén speed because the effective inertia of the tube is increased by a factor $\rho/(\rho + \rho_e)$, due to the fact that a certain amount of external fluid moves with the tube during its motion.

More interesting for stellar applications is the case when gravity is not neglected. Suppose that the tube is in temperature equilibrium with its surroundings (the same situation as discussed in section 2) and that the atmosphere is *isothermal* with scale height H. Then the wave equation (12) becomes (Spruit 1981a)

$$\frac{1}{g}(2\beta + 1)\frac{\partial^2 \xi}{\partial t^2} = \frac{\partial \xi}{\partial z} + 2H\frac{\partial^2 \xi}{\partial z^2} .$$
(14)

The solutions of (14) are of the form

$$\xi = exp(i\omega t + ikz - \frac{1}{4}z/H) ,$$
(15)

where $kH = \pm \frac{1}{4}(\omega^2/\omega_c^2 - 1)^{1/2} ,$
(16)

and $\qquad \omega_c^2 = \dfrac{g}{8H} \dfrac{1}{2\beta + 1}$. $\qquad\qquad\qquad$ (17)

Below the *cut off frequency* ω_c the waves are evanescent (k imaginary); wave propagation is possible only for $\omega > \omega_c$. The cut off frequency is always less than the acoustic cut off frequency, $\omega_a = c/(2H)$, of the same atmosphere. This has important consequences for the propagation of tube waves in the solar atmosphere, as we will discuss presently.

Transversal tube waves are easily excited by the turbulent flow in the convection zone. In the vertical tubes near the solar surface for example, they are induced by the vertical gradient in the horizontal component of the flow in granules. In such tubes, β is of the order 1, so that from (17), $\omega_c \approx 9\ 10^{-3}\ \text{s}^{-1}$, whereas $\omega_a \approx 3\ 10^{-2}\ \text{s}^{-1}$. The dominant frequency associated with the granulation is of the order $1.3\ 10^{-2}\ \text{s}^{-1}$, *below* cut off frequency. This is probably the reason why propagating acoustic waves have not been found in the solar atmosphere. It is *above* the tube cut off, however. This implies that a significant fraction of the motion induced in the tube by convection can propagate into the chromosphere without significant reflection. Thus, transversal tube waves are available as a candidate for chromospheric and coronal heating. In Spruit (1981a) we estimated that the horizontal motions of the order of 1 km s^{-1} present in the granulation would lead to amplitudes of 10 km s^{-1} in the chromosphere. Such amplitudes are indeed observed as transversal displacements of H_α fibrils. If the surface density of tubes is such that the average field strength on the solar surface is 5 G, the waves carry an average energy flux of about $3\ 10^6\ \text{erg s}^{-1}\ \text{cm}^{-2}$, enough to heat the chromosphere.

What happens to the waves after they reach the chromosphere can not be described with the present formalism. At this level the tubes have fanned out and merged with each other (see Figure 1) so that the thin tube approximation does not hold any more. At this height, the Alfvén speed also starts increasing rapidly, so that the waves will probably be reflected significantly. At the same time, the amplitudes are so large (near the Alfvén speed) that non linear effects will become important and dissipation through shock formation may occur.

5. STABILITY OF HORIZONTAL TUBES

During the solar cycle, the differential rotation continually produces toroidal flux tubes in the convection zone from which sections rise to the surface, producing the observed bipolar active regions. For the dynamo process to operate properly, this toroidal field must remain in the convection zone for a significant fraction of the duration of the cycle, and only occasionally may (part of)

a tube rise to the surface. This poses theoretical problems
because of two processes which affect a toroidal field in the
convection zone. If the temperature in the tube is the same as in
its surroundings, the internal density is less, and the tube
floats to the surface. It thereby leaves the region where the
dynamo process is thought to occur (lower half of the convection
zone). This effect has been studied by Parker (1975), Unno and
Ribes (1976), and Schüßler (1979). Schüßler's detailed calculations
show that the effect is probably not as serious as had been thought
previously. In addition a tube may be *unstable* however, forming
waves which destroy the original toroidal field. In the following
we discuss this instability, with the aid of the formalism of
section (3).

We approximate the convection zone by a plane parallel layer.
In this layer we study the stability of a horizontal flux tube
which is in equilibrium, such that $\rho_i = \rho_e$. Assume a cartesian
coordinate system with z in the direction of $\vec{g}$, and x along the
direction of the tube. Then after linearization the equation of
motion (9) reduces to (Spruit, 1981b)

$$\frac{\partial^2 \xi}{\partial t^2} = -\frac{1}{\rho}\frac{\partial P_1}{\partial x} + g\frac{\partial \zeta}{\partial x} , \qquad (18)$$

$$\frac{\partial^2 \eta}{\partial t^2} = \frac{1}{2} V_A^2 \frac{\partial^2 \eta}{\partial x^2} , \qquad (19)$$

$$\frac{\partial^2 \zeta}{\partial t^2} = \frac{1}{2} V_A^2 \frac{\partial^2 \zeta}{\partial x^2} + \frac{1}{2}g\left(\frac{\rho_1}{\rho} - \frac{\zeta}{H_\rho}\right), \qquad (20)$$

where ξ, η and ζ are the displacements of the fluid in x, y and z
directions. The perturbations in ρ and P are ρ_1 and P_1, the
unindexed quantities refer to the equilibrium state, and H_ρ is the
density scale height. Equation (19) for the y motion is uncoupled
to the rest and yields simple transversal waves. The equations for
ξ and ζ are coupled. In both solutions of these equations the flow
has a ξ as well as a ζ component, in contrast with the case of a
vertical tube.

Assume that the perturbations are adiabatic (this is a very
good approximation in the solar convection zone). The dispersion
relation for this case, which follows from (18), (20) and the
linearized forms of (1) and (11), is given in Spruit (1981b). It
shows that if the stratification is superadiabatic, there are
always wavelengths for which the tube becomes unstable, no matter
what the field strength is. The character of the instability can
be of two slightly different kinds. If the field is sufficiently
weak, the growth rate has its maximum for infinite wavelength
(Figure 3). This instability is identical in nature to the
convective instability of the medium surrounding the tube. If the
field is strong enough, the growth rate goes to zero for infinite

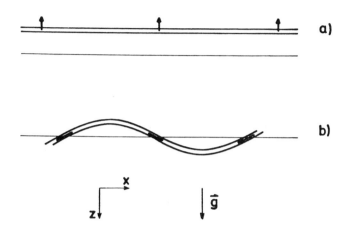

Fig. 3 Types of instability of a horizontal flux tube in a convection zone. a) : the tube as a whole is unstable to vertical displacement, b) : the instability is mediated by a flow from the crests to the troughs of a wave.

wavelength. The transition between the two is determined by the dimensionless number

$$M = \beta \delta ,$$ (21)

where δ is again the logarithmic superadiabaticity. For $M < 0.12$ the instability is of the Parker type, for $M > 0.36$ it is of the convective type (values for $\gamma = 5/3$). There is a continuous transition in between.

Since the wavelength required for instability can be long, it is of importance to study the instability in a spherical rather than a plane parallel geometry. It turns out (Spruit, 1981b) that the instability is stronger in this case. The typical growth time (for not too strong fields) is of the same order as the convective turnover time. Thus we must conclude that the instability is serious because it tends to destroy the toroidal field before it has had time to get wound up sufficiently. We propose that this implies that the toroidal field of the solar dynamo occurs at the interface between the convection zone and the radiative interior, rather than in the convection zone itself.

If the effect of Coriolis forces due to the solar rotation
is included, the growth rates of the instability are reduced
(Spruit, 1981c). This has been found earlier by Acheson (1979)
for a somewhat different physical situation. For the case of the
sun the stabilisation factor, for the most unstable wave, is not
large enough to keep toroidal fields within the convection zone
for a significant fraction of the cycle. In more rapidly rotating
stars however, the effect may be more important. This suggests
that in such stars the distribution of the field through the
convection zone could be qualitatively different from that in the
sun.

REFERENCES

Acheson, D.J., 1979, Solar Physics 62, 23.
Defouw, R.J., 1976, Astrophys. J. 209, 266.
Galloway, D.J., Proctor, M.R.E., Weiss, N.O., 1977, Nature 266, 686.
Harvey, J.W., 1977, Highlights of Astronomy 4, 223.
Meyer, F., Schmidt, H.U., Simon, G.W., Weiss, N.O., 1978, Astron.
    Astrophys. 76, 35.
Parker, E.N., 1975, Astrophys. J. 198, 205.
Parker, E.N., 1979, Cosmic Magnetic Fields, Clarendon Press, Oxford.
Peckover, R.S., and Weiss, N.O., 1978, Mon. Not. Roy. Astron. Soc.,
    182, 189.
Roberts, B., and Webb, A.R., 1978, Solar Phys. 56, 5.
Roberts, P.H., 1967. An Introduction to Magnetohydrodynamics,
    Longmans, London.
Schüßler, M., 1979, Astron. Astrophys. 71, 79.
Spruit, H.C., 1979, Solar Phys. 61, 363.
Spruit, H.C., 1981a, Astron. Astrophys. in press.
Spruit, H.C., 1981b, in preparation.
Spruit, H.C., 1981c, in preparation.
Spruit, H.C., and Zweibel, E.G., 1979, Solar Phys. 62, 15.
Unno, W., and Ando, H., 1979, Geophys. Astrophys. Fluid Dyn.
    12, 107.
Unno, W. and Ribes, E., 1976, Astrophys. J. 208, 222.
Webb, A.R., and Roberts, B., 1978, Solar Phys. 59, 249.
Wilson, P.R., 1979, Astron. Astrophys. 71, 9.

CORONAL HOLES AND SOLAR MASS LOSS

Jack B. Zirker

Sacramento Peak Observatory*

During the declining phase of solar activity, and possibly at other phases, solar mass loss occurs in the form of high speed streams in the solar wind. These streams originate in low density, low temperature coronal regions, called "holes." The physical properties of the wind streams and the holes will be discussed. Preliminary models for the three-dimensional structure of the wind, at different phases of the solar activity cycle, will be sketched. Remaining problems in developing a self-consistent theory of solar wind acceleration will be reviewed.

Parker, E. N.: 1963, "Interplanetary Dynamical Processes," J. Wiley and Sons.
White, O. R. (ed.): 1977, "The Solar Output and its Variation," Colorado Associated University Press.
Zirker, J. B. (ed.): 1977, "Coronal Holes and High Speed Wind Streams," Colorado Associated University Press.

During past years we have learned a great deal about the origins of the solar wind. In particular, there has been an explosion of knowledge about the influence of the structure of the solar corona and its large-scale magnetic field on the spatial and time variations of the wind. At the present time we are looking forward to the flight of the International Solar

*Operated by the Association of Universities for Research in Astronomy, Inc. under contract with the National Science Foundation.

R. M. Bonnet and A. K. Dupree (eds.), Solar Phenomena in Stars and Stellar Systems, 301–318.
Copyright © 1981 by D. Reidel Publishing Company.

Polar Mission, sometime in the early 1990s, to refine and
extend the knowledge that we've gained in the past decade.   It
is an appropriate time in which to review what we know and
where we are headed.   The understanding that we gain from the
Sun will hopefully serve as a guide toward understanding solar
mass loss in other stars.

        This lecture has three parts.   In Part I, I will summarize
some of the salient features of the solar wind, particularly,
properties of the high speed wind streams that seem to be among
the most important and most easily understood aspects of the
wind.   In Part II, I will review the properties of the coronal
holes, where all the high speed wind and perhaps all the wind,
originates.   In Part III, I will try to sketch some of the most
important questions that remain to be answered and indulge in a
little speculation.   The reader who is interested in more
details on this whole subject should look into the references
listed at the end of this chapter (1).

I.   THE SOLAR WIND
     A.   The Average Wind

        We must always keep in mind that nearly all the *in situ*
measurements we have of the solar wind were taken in or near
the ecliptic.   Some interplanetary probes, such as the Helios
experiment, and the Mars and the Jupiter probes, have sampled
the wind at large distances from the earth but still within the
ecliptic plane.   Most of the heliosphere remains as unknown
territory.

        The solar wind group at the Los Alamos Scientific
Laboratory has collected one of the longest and most
homogeneous sets of direct measurements of the wind presently
available.   They find that the speed of the solar wind near the
earth fluctuates with timescales of minutes to days.   Figure 1,
drawn from their data (2), shows a histogram for the wind speed
for each year between 1962 and 1974.   The wind speed ranges
between 250 and 800 km/s.   In years of declining and near
minimum solar activity, such as 1962, 1964, 1973 and 1974, very
high wind speed (above 650 km/s), is observed more
frequently.   As we shall see a little later on, these high
speeds occur in discrete streams in the solar wind. Except for
the appearance of this high speed tail, there doesn't seem to
be a very systematic variation of the average wind speed
throughout the solar cycle.   The Los Alamos group has examined
the properties of wind with high, average and low mean
speeds.   "Low" is defined as anything below 350 km/s and "high"
is anything above 650 km/s.   Table 1 summarizes their results
(3).   The first point to notice is that the average proton flux

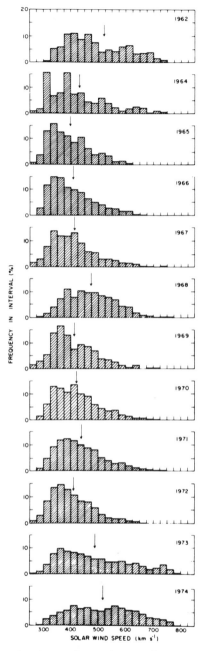

Figure 1. Distribution of wind speed over a solar cycle. The arrows indicate annual means. Reprinted courtesy J. Gosling *et al*. and Journal of Geophys. Res., © 1977 The American Geophysical Union.

TABLE 1*

Plasma Characteristics of Various Types of Solar Wind Flows

PARAMETER	AVERAGE			LOW SPEED			HIGH SPEED		
	MEAN	σ	% VAR	MEAN	σ	% VAR	MEAN	σ	% VAR
N (cm^{-3})	8.7	6.6	76	11.9	4.5	38	3.9	0.6	15
V (km s^{-1})	468	116	25*	327	15	5*	702	32	5*
NV (cm^{-2}s^{-1})	3.8×10^8	2.4×10^8	63	3.9×10^8	1.5×10^8	38	2.7×10^8	0.4×10^8	15
ϕ_v (degrees)**	-0.6	2.6	430	$+1.6$	1.5	94	-1.3	0.4	31
T_p (^{0}K)	1.2×10^5	0.9×10^5	75	0.34×10^5	0.15×10^5	44	2.3×10^5	0.3×10^5	13
T_ε (^{0}K)	1.4×10^5	0.4×10^5	29	1.3×10^5	0.3×10^5	20	1.0×10^5	0.1×10^5	8

*Reprinted with permission of W. C. Feldman and the Colorado Associated University Press.
**Angle between wind vector and earth-sun line. ϕ > 0 means corotation with sun.

near the earth is about the same within a factor of two for all conditions of wind speed. Secondly, the percentage variation about the mean for all quantities, with the possible exception of wind speed, is the lowest for high speed wind. In other words, the high speed wind represents the most stable, uniform and structureless type of wind known and in this sense, is easiest to interpret. Near the Earth, and indeed at distances greater than some tens of solar radii from the Sun, the wind is a collisionless plasma so that the electron and proton temperatures are not equal. The number density of helium ions (alpha particles) represents 4 to 5% of the total, independent of the value of the maxium speed. This table doesn't show a rather curious property at low wind speed, however: namely, that the helium abundance can very radically from one low speed interval to another and even during a particular interval. The full range covers two orders of magnitude.

We can now estimate the average mass loss from the Sun, assuming isotropic flow. Using the product of nv from Table 1, we find a value of 2×10^{-14} solar masses per year, an amount which is utterly trivial for the evolution of the Sun. A more interesting quantity is the angular momentum loss of the Sun due to the wind. The wind flow is not quite radial from the Sun but appears to flow from a direction making an angle between 1 and 2 degrees either east or west of the Sun. Feldman *et al.* (3) caution us about relying too much on their measurements because they contain systematic uncertainties greater than the measured deviation from zero and because all the measurements relate to the ecliptic and there may be significant variations in wind direction with solar latitude. Nevertheless, even a rough estimate suffices to illustrate the possible importance of this effect. The rate of angular momentum loss is $nvmv\phi_v r$, where m is the proton mass and r is the astronomical unit. If we select values from Table 1 corresponding to the average wind, assume that the flow direction is *one degree east* of the radial direction and integrate over a $60°$ belt of latitude centered on the solar equator, we find that the wind exerts a braking torque on the Sun of 3×10^{30} dyne cm. Since the angular momentum of the Sun is of the order of 2×10^{48} dyne cm sec, the wind torque is sufficient to brake the solar rotation in about 7×10^{17} s or 2×10^{10} years. This is only a few times longer than the age of the Sun.

As we said before, this calculation is subject to many uncertainties, but it does suggest that the wind could have been a significant factor in spinning down the Sun.

B. High Speed Wind Streams

As we have seen above, the solar wind in the ecliptic reaches speeds of 800 km/s a small fraction of the time. The flight of Mariner 2 to Venus in 1962 provided data that showed that these high speed events are organized in streams flowing nearly radially from the Sun and rotating with the Sun with the equatorial rotation period of 27.1 days. Moreover, the rotation of these streams past the Earth was highly correlated with the reccurrence of geomagnetic storms. Feldman *et al.* (4) have summarized the physical properties of 19 high speed streams observed between 1971 and 1974 by plasma analyzers aboard IMP 6, 7 and 8. They defined high speed as anything above 650 km/s and found the following properties.

First, the streams have an average full width at half maximum of 90° in solar longitude. In other words, the most common pattern they saw was a system of four quadrants with alternating high and low speed wind. Within a stream, the interplanetary magnetic field is unipolar, directed either predominantly inward or outward from the Sun. These two results remind one strongly of the sector structure of interplanetary magnetic fields, first described by Wilcox and Ness (5). The high speed streams overlap but do exactly coincide with the magnetic sectors. Streams last anywhere from one to 18 rotations and are most stable and persistent during the declining phase of the solar cycle, namely 1973 to 1974.

Geophysicists have been aware since 1934 that geomagnetic storms occur with a period of 27 days, which implies a solar influence. Bartels (6) gave the name "M-region" to the source on the Sun of this influence. The Mariner 2 results showed that high speed streams are the connecting link between the Sun and the Earth. But the question remained open during the early 1970s as to the source of the streams. Statistical studies pointed to areas between active regions on the Sun. A final convincing identification was made by Krieger, Timothy and Roelof in 1973 (7). They identified a coronal hole as the source of a high speed wind stream.

Thus, coronal holes would be interesting for no other reason than that they are the source of high speed wind. But in addition, they display a number of other intrinsic properties which are interesting and challenging in the context of stellar physics. We turn now to a discussion of the holes themselves.

II. CORONAL HOLES

Figure 2 shows a large coronal hole observed from Skylab

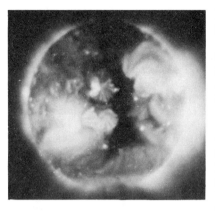

Figure 2. The corona photographed in X-rays from Skylab. Coronal hole No. 1 is the black north-south lane. Reprinted courtesy A. Krieger of American Science and Engineering.

with an X-ray camera built by the American Science and Engineering, Inc. The camera records X-rays in wavelength bands 44 to 60 Å, which are produced by thermal bremsstrahlung at temperatures between 1 and 3 million degrees. Although Skylab produced perhaps the best observations ever made of coronal holes, their existence was first recognized by Waldmeier (8) around 1957, who noticed persistent and recurring gaps in the monochromatic emission line corona recorded with his coronagraph. Around 1968, the Harvard experiment on OSO-4 recorded large dim areas in the corona as observed in the Mg X line at 625 Å (9), and later in OSO-7 the same regions appeared on the disk in the Fe XV line 284 Å (10). Synoptic observations of the white light corona made by the High Altitude Observatory with a K-coronometer over a decade, clearly show the evolution and birth of coronal holes (11). Thus a long history of observations that hinted at the existence of large dim regions in the corona preceded Skylab but the opportunity to observe these regions continuously on the disk for nine months from Skylab made all the difference in understanding their properties.

A. Empirical Properties

The Skylab data confirmed the assertion of Krieger, Timothy and Roelof (7) that *solar streams originate in coronal holes*. Of 69 central meridian passages of coronal holes between solar latitudes of ± 30°, 75% were found to be associated with streams near Earth, 9% possibly associated and 16% not associated. The *lifetime* of coronal holes is also consistent with the hypothesis that they are the coronal roots of high speed wind streams. Of the nine coronal holes present during Skylab, two had lifetimes shorter than three rotations, four lasted more than five rotations and three had lifetimes of ten rotations or more. The solar *poles* were covered by coronal holes throughout Skylab. The area of these polar holes shrank and nearly disappeared with the rise of solar activity toward the maximum in 1980. Coronal holes that extend in a north-south direction, such as the one shown in Figure 2, seem to be

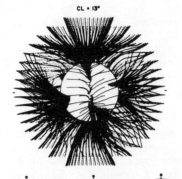

Figure 3. Potential magnetic field lines in the corona for Carrington rotation 1602/1603. Reprinted courtesy R. H. Levine, M. D. Altschuler, J. W. Harvey and B. V. Jackson: 1977, Astrophysical Journal 215, p. 636, published by the University of Chicago Press; © 1977, The American Astronomical Society.

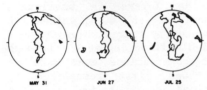

fairly uncommon. With the rise of solar activity since 1974, new coronal holes have appeared at mid-latitudes, are small and last at most a few rotations.

Coronal holes contain a *unipolar magnetic field* which diverges rapidly with increasing height. This conclusion is based upon calculations rather than measurements of the coronal magnetic field. Coronal magnetic field measurements are still in their infancy, although a group from the High Altitude Observatory, resident at the Sacramento Peak Observatory, is attempting to measure them throughout the flight of the Solar Max Mission. In order to calculate magnetic field lines in the corona, one begins with global measurements of the line-of-sight component of the magnetic field over the photosphere and assumes that no electrical currents flow in the corona. It is then possible to solve Laplace's equation for a potential field between the photosphere and an imaginary spherical potential surface commonly placed at about 2 solar radii from the Sun's center. The tangential component of the magnetic field is assumed to vanish on this outer boundary; the measured field is taken as a condition on the inner boundary. Figure 3 illustrates some typical results of this technique. Only field lines that cross the outer spherical surface have been drawn in this figure, i.e., all field lines that loop back to the photosphere inside the spherical surface have been suppressed. This figure suggests that all field lines that extend far out into interplanetary space across the outer spherical boundary, are rooted in coronal holes. (This may be an artifact of the calculation, however.) The corona can thus be divided into closed magnetic field regions from which no wind escapes and open magnetic field regions rooted in coronal holes from which *all* the wind escapes. The accuracy and validity of these potential magnetic fields is open to question, but they do seem to outline at least the large-scale coronal structures that can

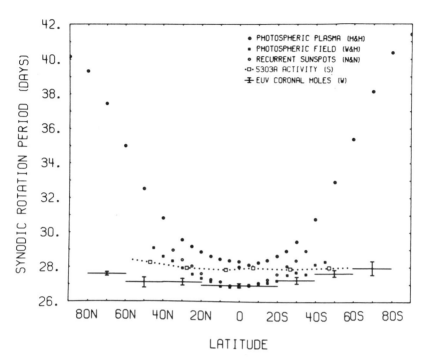

Figure 4. Rotation periods of photosphere, sunspots, emission line corona and coronal holes. Reprinted courtesy A. Krieger and the Colorado Associated University Press.

be seen in white light during eclipses, for example.

Note the rapid divergence of the field lines. Within the inner corona, the magnetic energy density is larger than the gas energy density so that the magnetic field strength is sufficient to confine and channel the gas flow. The diverging geometry provides a kind of Laval nozzle whose shape, in combination with the radial variation of energy and momentum deposition, controls the acceleration of the wind. Such diverging geometries tend to decrease the height of the critical point in solar wind solutions, as Parker pointed out in the early 1960s, and as Kopp and Holzer have confirmed more recently (12). Finally, coronal holes do not seem to partake of the differential rotation of the photosphere. They rotate instead nearly as *rigid bodies* as Figure 4 illustrates. This property has raised all sorts of questions regarding the existence of a deep-seated rigidly rotating source of magnetic flux (13).

The solar *chromosphere* underlying coronal holes differs

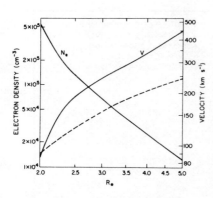

Figure 5. Density and speed
along the axis of a polar
hole. Reprinted courtesy R.
Munro, B. V. Jackson and the
Astrophysical Journal, published
by the University of Chicago
Press; © 1977 The American
Astronomical Society.

very little from that under
normal quiet coronal regions.
The only perceptible difference
is a slight reduction in con-
trast of the chromospheric network as seen in the He I line at
10830 Å.

B. Derived Properties

The High Altitude Observatory operated an externally
occulted coronagraph aboard Skylab that recorded pictures of
the white light corona continuously. Since the white light
originates from photospheric light scattered by free coronal
electrons, the distribution of electrons can be determined from
an analysis of the intensity and polarization of the white
light corona. Munro and Jackson (14) carried out such an
analysis for the hole at the north pole of the Sun. By
comparing a series of photographs taken during several
rotations, they were able to determine the geometry of the
boundaries of the hole. The hole has the shape of an expanding
funnel centered over the polar axis. Assuming cylindrical
symmetry, they were then able to derive a three-dimensional
model for the electron density distribution within the hole.
Their results for the axes of the hole are shown in Figure 5.
Next, assuming that the solar wind stream with a typical proton
flux of 3×10^8 cm^{-2} s^{-1} escapes from the hole, and using the
equation of continuity, they determined the velocity profile
along the hole's axis, also shown in Figure 5. The velocity
increases from about 80 to 450 km/s between 2 and 5 radii.
Since the velocity of sound is only about 150 km s^{-1}, the
critical point of this flow lies within this distance range,
i.e. the principle acceleration of the solar wind in the hole
occurs close to the solar surface. Without further
assumptions, they could not determine the temperature profile
in the hole, but the fact that the maximum acceleration of the
wind occurs below 2.5 solar radii suggests that energy and/or
momentum is being deposited to at least this height.

The spectroscopic data derived during Skylab have been
thoroughly analyzed in an attempt to determine the run of

temperature and pressure and the energy balance throughout a
hole. These attempts have been partially successful in
understanding some of the major attributes of the hole. For
example, the hole is dim at all wavelengths, primarily because
the electron density is less by a factor of three than in a
normal region at the same height. The spectra also imply a
much lower temperature at a given height in a hole than in its
neighborhood. At a height of .1 R_θ for example, Mariska finds
a temperature of 1 to 1.1 million degrees, lower by as much as
600,000 degrees than the neighborhood.

The low temperature at a given height implies a smaller
temperature gradient in the hole and as a result the heat
conducted downward through the transition zone is lower by a
factor of four than in the neighboring quiet regions. Table 2
shows this result as well as other terms involved in the energy
budget of a hole and a normal closed field, coronal region.
Notice that the solar wind loss (extrapolated to the Sun from
measurements near Earth) greatly exceeds downward heat
conduction in the hole. Notice also that the energy loss
associated with downward enthalpy flux may be the principle
loss in both kinds of regions (15). At this time we have no
reason to believe that the total energy supply to a coronal
hole is any different than that to its neighborhood but the
distribution of losses among the different mechanisms may
differ in the hole and in its neighborhood.

The spectroscopic data from Skylab were insufficient to
establish the maximum temperature in a hole. This quantity is
very important if we are to understand the mechanisms that
accelerate the solar wind. In the classical Parker theory, it
is the gradient of the gas pressure that accelerates the
wind. The wind speed near Earth rises monotonically with the
maximum coronal temperature and with the distance over which
maximum temperature prevails. In the most favorable case, an
isothermal wind, a temperature of about 2.5 million degrees
would be needed to produce a speed as high as 800 km/sec.
While we are not sure of the maximum temperature in a coronal
hole, we can be sure that the stream is not isothermal, so that
a temperature in excess of 2.5 million would be needed.

Recently, the High Altitude Observatory and Center for
Astrophysics collaborated in an experiment to measure the
proton temperature in a coronal hole from 1.5 to 3 solar radii
(16). The proton temperature was derived from the profile of
the Lyman alpha radiation that originates in the chromosphere
and is scattered from neutral hydrogen atoms in the corona.
Despite uncertainties due to the presence of turbulent motions
and systematic outflow, the most probable proton temperature
lies between 1.4 and 1.8 million degrees at a distance of 2

TABLE 2

Coronal Energy Losses (erg cm^{-2} s^{-1})

Loss	Coronal Holes (Open B)	Quiet Corona (Closed B)
Heat Conduction	6×10^4	2×10^5
Radiation	10^4	10^5
Wind*	7×10^5	$< 10^5$
Subtotal	8×10^5	3×10^5
Enthalpy (Transition Zone)	$2.5 \times 10^5 - 2.5 \times 10^6$	$2.5 \times 10^5 - 2.5 \times 10^6$
Total	$10^{6.0} - 10^{6.5}$	$10^{5.7} - 10^{6.4}$

*Gravitational, kinetic and enthalphy energy losses.

radii from the Sun's center. This recent result reinforces a suspicion that arose during analysis of the Skylab data, namely that the temperature in a coronal hole is too low to produce the observed acceleration at low heights and the final wind speed near Earth. There is now reason to think that *momentum*, in addition to energy, is being deposited throughout a coronal hole and helps to accelerate the wind stream. This momentum deposition might occur, for example, by a gradient in wave pressure due to acoustic or MHD waves that also deposit energy at these heights. These ideas have been explored theoretically by a number of theorists but a consistent model of wind stream heating and acceleration by waves has not yet appeared.

One of the complications in building such a model is that the classical theory of heat conduction does not apply to a low density plasma like the wind. A diffusive conduction theory is valid where the mean free path of electrons is small compared to the distance over which the temperature changes appreciably. This condition breaks down within 2 or 3 solar radii of the Sun's center in a wind stream. There is no generally accepted theory of heat transport to replace the classical theory, however.

C. A Model for the Heliosphere Near the Minimum of Solar Activity

As we have seen, the solar wind is dominated by long-lived, stable wind streams from a relatively small number of holes during the declining phases of solar activity. During this time, the polar holes have reached their maximum areal extent and some of the largest holes seen at mid-latitudes, are simply equatorial extensions of these polar holes. These simplifying conditions have led to the construction of the schematic model to describe the three-dimensional wind just before solar minimum.

The basic insight derives from an investigation by Wagner (17). He correlated the sign of the interplanetary magnetic field with the appearance of recurrent coronal holes at different latitudes, as observed in the Fe XV line from OSO-7. Wagner found that the best correlation existed between the interplanetary magnetic field in the ecliptic and coronal holes at solar latitudes between 40 and 80°. In short, the magnetic field observed near the Earth seemed to be rooted near the polar regions of the Sun. Since the wind streams follow the diverging field lines near the Sun, Wagner's result implies that the wind near the Earth is dominated by polar flows from the Sun.

This result was extended by Hundhausen, who examined the

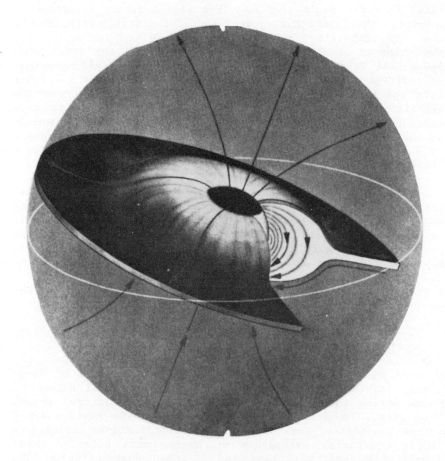

Figure 6.   Tipped dipole model of coronal magnetic fields.
Reprinted courtesy A. Hundhausen and the Colorado Associated
University Press.

correlation between wind speeds near the Earth and the
recurrence of high latitude polar holes that are extensions of
the polar holes.   Once again, a high correlation was found.
Hundhausen was led to the highly simplified model of the
heliosphere shown in Figure 6.   Just before solar minimum, the
large-scale magnetic field of the Sun can be represented as a
modified dipole whose axis is tipped with respect to the solar
rotation axis.   Large coronal holes cover both of the solar
poles.   Magnetic field lines do not loop directly from one
solar hemisphere to the other but approach each other
asymptotically along a neutral current sheet that lies along
the equator of the magnetic dipole.   Near the solar surface

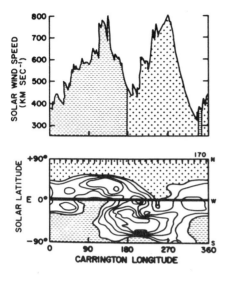

Figure 7. A comparison of coronal brightness (proportional to Ne), coronal magnetic polarity and the properties of wind streams near Earth. Polar holes influence wind flow in the ecliptic. Reprinted courtesy of A. Hundhausen and the Colorado Associated University Press.

this current sheet extends from a belt of closed magnetic field that runs around the Sun near the solar equator. Because the field lines are closed, the electron density in the belt is high and it appears bright in white light. Figure 7 is a map of the coronal white light brightness for a full rotation toward the end of the Skylab mission. It shows this closed belt very well. Because the belt and its extension, the magnetic neutral sheet, are tipped with respect to the ecliptic, solar rotation sweeps alternately portions of the northern and southern polar coronal holes into the line of sight from the Earth. As a result, we detect at Earth, first a high-speed stream with a magnetic polarity characteristic of the northern hemisphere, then slow wind as the closed magnetic belt comes into the line of sight, and then a high speed wind stream, with a characteristic southern hemisphere magnetic polarity. These relations are shown in Figure 7, and are compared there with the actual wind streams and magnetic polarity observed at Earth during this rotation. The model seems to characterize the heliosphere fairly well under these extremely simplified circumstances prevalent around solar minimum. Naturally, the situation gets much more complicated as solar activity builds up and large areas of the Sun are covered with active regions at mid-latitudes. We hope that the international Solar-Polar Mission will be able to test this simplified model and that it will fly near solar minimum when interpretation is easiest.

III. SOME QUESTIONS FOR THE FUTURE

    At this stage, we are left with a long list of questions regarding the origin of the solar wind and the structure and evolution of coronal holes. In this last section I'll discuss a few of these.

Much of the discussion above has concerned the high speed wind, partly because it is steadier and more uniform in its properties and partly because it is more challenging to the theorist. To make further progress in modeling the high speed streams, we obviously need more observations to guide the theory. We would like to know the profiles of temperature and velocity near the critical point between 2 and 10 radii. As we have seen, a start has been made to get this information, using the scattered resonance line of hydrogen. Mechanical waves are probably important in accelerating the wind. We need to get an observational handle on them. What is the wave flux as a function of height? What are the characteristic wave modes? What is the height dependence of the deposition of energy in momentum? To get better stream models, we need to incorporate some new physics, such as a refined theory of electron heat transport in diffuse gases.

With all this emphasis on the high speed wind, we must not forget the slow wind. At this point I think it's fair to say that we don't have a clear idea where on the Sun the slow wind originates. One possibility is near the edges of the high speed streams. Nolte and his colleagues (18) have shown that the peak speed within a wind stream increases with the area of a coronal hole associated with it. This finding is consistent with MHD calculations that show that the field lines are more closely parallel in the middle of a hole, and diverge rapidly toward the edges (19). The large increase of cross-sectional area with height that such field line divergence implies, would tend to slow down the wind near the edge of a hole. But this is probably not the whole story. There are unquestionably other places on the Sun than coronal holes where open field lines appear and there are opportunities for the flow to be modified with increasing distance from the Sun. We don't really understand how this occurs and how it may bear on the origin of the slow wind.

We are just beginning to piece together information on the evolution of coronal holes throughout the solar cycle. Broussard and his colleagues collected all the X-ray and XUV solar images that they could find throughout the period 1963 to 1974 (20). They found that during solar minimum, polar coronal holes are prominent and decrease as activity increases. At solar maximum, coronal holes occurred poleward of the sunspot belts and in the equatorial region between them. The holes were small and lasted only a few rotations. Information on the development in coronal holes during the present solar cycle, Cycle 21, is building up (21).

We obviously need more information about the morphology of coronal hole development before we can tackle the larger

question of the origin and the pattern of development of
holes. To quote Harvey and Sheeley (22), "Non-polar holes form
whenever the magnetic flux from bipolar magnetic regions
interacts to produce a large region of locally unbalanced
flux. The term 'locally balanced' is not precisely defined and
the burden to explain the origin of holes is shifted to
explaining why magnetic fields erupt, interact and dissipate in
the way that they do." This statement accurately represents
our state of ignorance at the present time. We have only
phenomenological models to explain both the pattern of
development in latitude and the possible organization in
longitude that coronal holes seem to display throughout the
solar cycle. For example, Svalgaard and Wilcox (13) have
observed magnetic patterns that last for more than one sunspot
cycle and suggest that they arise from a long-lived subsurface
structure, possibly certain modes of the solar dynamo.

What we have learned so far about the structure of the
heliosphere and its evolution throughout the activity cycle
emphasizes the essential role of the large-scale magnetic field
in channeling the gas flow and the energy fluxes. Is it
possible that all the complications I've described in this
chapter, would disappear in an early type star, where magnetic
fields are presumably absent or weak? I doubt it. Even such
stars show rapid fluctuations in chromospheric lines
like Hα (23) and possibly also in X-rays. I'd be surprised if
their winds did not fluctuate in space and time. Possibly some
analog of a coronal hole (i.e. large-scale inhomogeneities in
the lower atmosphere) exists.

References

(1) General references:
     Zirker, J. B. (ed.): 1977, Coronal Holes and High Speed
     Wind Streams (Colorado Univ. Press, Boulder).
     Hundhausen, A. J.: 1972, Coronal Expansion and Solar
        Wind (Springer-Verlag, New York).
     Withbroe, G. L. and Noyes, R. W.: 1978, Ann. Rev. Astro.
        Ap. 16, p. 363.
(2) Gosling, J. T., Asbridge, J. R., Bame, S. J. and Feldman,
        W. C.: 1976, J. Geophys. Res. 81, p. 5061.
(3) Feldman, W. C., Asbridge, J. R., Bame, S. J. and Gosling,
        J. T.: 1977, in "The Solar Output and its
        Variation," O. R. White (ed.), p. 351 (Colorado
        Association Univ. Press, Boulder).
(4) Feldman, W. C., Asbridge, J. R., Bame, S. J. and Gosling,
        J. T.: 1976, J. Geophys. Res. 81, p. 5054.
(5) Wilcox, J. M. and Ness, N. F.: 1965, J. Geophys. Res.
        70, p. 5793.

(6)   Bartels, J.:   1934, J. Geophys. Res. 39, p. 201.

(7)   Krieger, A. S., Timothy, A. F. and Roelof, E. C.:   1973,
      Solar Phys. 29, p. 505.

(8)   Waldmeier, M.:   1957, Die Sonnenkorona, 2 (Verlag
      Birkhauser, Basel).

(9)   Munro, R. and Withbroe, G. L.:   1972, Astrophys. J. 176,
      p. 511.

(10) Neupert, W. M. and Pizzo, V.:   1974, J. Geophys. Res. 79,
      3701.

(11) Hansen, R. and Hansen, S., quoted in Hundhausen, A. J.:
      1977, Coronal Holes and High Speed Wind Stream, J. B.
      Zirker (ed.), p. 225 (Colorado Associated Univ.
      Press, Boulder).

(12) Kopp, R. A. and Holzer, T. E.:   1976, Solar Phys. 49, p.
      43.

(13) Svalgaard, L. and Wilcox, J. M.:   1975, Solar Phys. 41,
      p. 461.

(14) Munro, R. H. and Jackson, B. V.:   1977, Astrophys. J.
      213, p. 874.

(15) Pneuman, G. W. and Kopp, R. A.:   1978, Solar Phys. 57, p.
      49.

(16) Kohl, J. L., Weiser, H., Withbroe, G. L., Noyes, R. W.,
      Parkinson, W. H., Reeves, E. M., Munro, R. H. and
      MacQueen, R. M.:   1980, Astrophys. J., submitted.

(17) Wagner, W. J.:   1976, Astrophys. J. 206, p. 583.

(18) Nolte, J. T., Krieger, A. S., Timothy, A. F., Gold, R.
      E., Roelof, E. C., Vaiana, G., Lazarus, A. J. and
      Sullivan, J. D.:   1976, Solar Phys. 46, p. 303.

(19) Pneuman, G. W. and Kopp, R. A.:   1971, Solar Phys. 18, p.
      258.

(20) Broussard, R. M., Sheeley, Jr., N. R. Tousey, R.,
      Underwood, J. H.:   1978, Solar Phys. 56, p. 161.

(21) Sheeley, Jr. N. and Harvey, J. W.:   1978, Solar Phys. 59,
      p. 159.

(22) Harvey, J. W. and Sheeley, Jr., N. R.:   1979, Space Sci.
      Rev. 23, p. 139.

(23) Conti, P. S. and Vaiana, V. S.:   1976, Astrophys. J. 209,
      p. L37.

THE SOLAR WIND: MACROSCOPIC FEATURES AND MICROSCOPIC PROBLEMS

Steven J. Schwartz

Department of Applied Mathematics
Queen Mary College
Mile End Road, London E1 4NS   U.K.

ABSTRACT

   This paper presents a brief review of the theory and observations of the solar wind. The success of simple fluid models is
contrasted with the lack of understanding of the detailed micro-
physics on which such fluid theories are, or should be, based.

1. INTRODUCTION

   The sun is observed to lose mass in the form of a large scale
supersonic wind, which blows throughout the interplanetary space.
While this gross aspect of the solar atmosphere has been understood
theoretically since 1958 (36), many of the detailed observations
currently made by satellite experiments are not easily explained
by, or accommodated in, a simple hydrodynamic picture. This paper
provides a brief review of our current levels of understanding
concerning the persistent expansion of the solar corona.

   From the fluid point of view, pressure gradients associated
with the hot corona drive an outward flow of coronal material.
The condition that the solar atmosphere must ultimately come into
pressure balance with the relatively low pressure of the galactic
medium reduces the class of possible hydrodynamic solutions to a
single critical solution which varies continuously from low flow
velocities near the sun (consistent with observations) through a
sonic point to supersonic flow at large heliocentric distances.
Models based on this fluid-like behaviour of the solar wind are
reasonably successful at explaining the gross features of the

319

R. M. Bonnet and A. K. Dupree (eds.), Solar Phenomena in Stars and Stellar Systems, 319–330.
Copyright © 1981 by D. Reidel Publishing Company.

wind as observed by satellites from 0.3 AU to beyond the earth's orbit at 1 AU.

However, many observed details are not in agreement, even qualitatively, with the predictions of fluid models. This is perhaps not so surprising, since the implicit assumption of frequent particle collisions inherent in the fluid approximation is clearly naive. For example, in the vicinity of the earth's orbit, collisional mean free paths are of the order of 1 AU. The resulting non-Maxwellian nature of the particle distributions has consequences for transport processes (thermal conduction, viscosity, etc.), isotropy and other features of the interplanetary medium. Also included in these "microprocesses" are the beams of minor ions, mainly helium, observed in the solar wind plasma.

The plan of this paper is as follows. Section 2 reviews the fluid aspects of the solar wind, while section 3 is devoted to the microstructure. Section 4 presents some concluding remarks. The interested reader will find more complete discussions on these and related topics in several review articles (4,7,11,12,24,26, 29,31,40,46) and in the literature at large.

2. THE MACROSCOPIC SOLAR WIND

2.1 Basic Models

The simplest model of the solar wind assumes it to be a radially symmetric steady expansion of a single (electron-proton) fluid. Mathematically, this fluid is governed by the following set of equations:

$$\frac{d}{dr} (\rho u r^2) = 0 \quad , \tag{2.1}$$

$$u\frac{du}{dr} + \frac{1}{\rho} \frac{dp}{dr} + \frac{GM_o}{r^2} = 0 \quad , \tag{2.2}$$

and

$$\frac{d}{dr} (p\rho^{-\gamma}) = 0 \quad , \tag{2.3}$$

expressing respectively mass continuity, momentum conservation and an adiabatic equation of state. Here, ρ is the total mass density, u the radial expansion velocity and p the total (electron + proton) pressure. The solution to (2.1)-(2.3) is elucidated by noting that an energy integral exists, namely

$$E = \frac{1}{2}u^2 + \frac{\gamma}{\gamma-1}\frac{p}{\rho} - \frac{GM_\odot}{r} = \text{const.} \qquad (2.4)$$

It is also illuminating to combine (2.1)-(2.3) into a single equation for the mach number $M \equiv (\rho u^2/\gamma p)^{\frac{1}{2}}$:

$$\frac{M^2 - 1}{2M}\frac{dM}{dr} = F(M,E,r) \qquad (2.5)$$

where F is a known function of E, M and r.

The function F has only one zero which defines a critical radius, r_c . Inspection of the left hand side of (2.5) then reveals that at r_c the solution either makes a transition between sub-and supersonic behaviour, or else reaches a local maximum or minimum. Thus the solution topologies look like those in Fig. 1. Of these solutions, only the bold one satisfies the boundary conditions that a) the flow is sub-sonic near the sun, consistent with observations, and b) the pressure goes to zero as r tends to infinity, so that the solution can be in equilibrium with the interstellar medium. Physically, the existence of a hot corona implies that the energy constant (2.4) is positive and sufficiently large that the subsonic solutions with u→ 0 correspond to pressures in excess of the

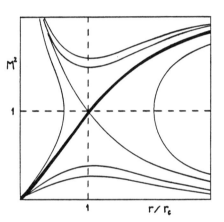

Figure 1. Solar wind solution topologies.

interstellar pressure. Whence the corona must expand and that expansion becomes supersonic. (The real solar wind probably terminates in a shock, the heliopause, at r ∿ 100 AU.)

Typically, $r_c \sim 5$ solar radii ($R_\odot$) and at 1 AU (=215 $R_\odot$) flow velocities average u ∿ 400 km/s corresponding to $M^2 \sim 50$, with particle number densities ∿ 5 cm^{-3}. (A complete statistical survey of solar wind parameters can be found in (12)) Also typically, one fluid models produce flow velocities which are too low and densities which are too high, both by a factor ∿ 2.

This simple one fluid description of the solar wind can be, and has been, made increasingly more sophisticated. Additional forces, such as mechanical and/or magnetohydrodynamical (MHD)

wave pressure gradients (1,22), centrifugal, magnetic and
frictional or viscous forces (48), can be placed on the right hand
side of (2.2). Equation (2.3) can be replaced by a thermal energy
equation involving thermal conduction, and heating due to wave
dissipation and friction. In the stellar problem the forces and
heating due to radiation may also be important. Since the
collisional coupling between protons and electrons is actually
weak, a better approximation would also involve two complete
(coupled) sets of fluid equations, one for each species (20).
And more recently, the effects of non-spherical symmetry have
been investigated (cf. subsection 2.3 below).

        Testing solar wind models by comparing them with observations
is a tricky business. The interplanetary medium is never as
steady, uniform, etc. as reflected in the models. And physical
processes are often investigated theoretically by adjusting free
parameters (e.g. ad hoc conduction laws, heating rates, etc.).
Some surveys of parameter ranges and their general consequences
have been carried out recently (28,30,33). Basically, if one is
interested in describing the gross macroscopic fluid-like aspects
of the solar wind reasonably well by astrophysical standards
(i.e. to within factors of 2 - 10) then basic fluid models of the
solar wind are sufficient and successful.

2.2 Magnetic fields from the sun

        The sun's magnetic field is basically that of a dipole
stretched out into the interplanetary medium by the solar wind.
Fig. 2 shows this configuration viewed from the ecliptic plane,
while Fig. 3 shows the magnetic field in the ecliptic plane itself
as viewed from a northern solar latitude. The spiral pattern in
Fig. 3 is caused by the sun's rotation coupled with the freezing-
in of the field with the solar wind plasma. The field lines
connect solar wind material with the material's point of origin
on the solar surface. In practice, the surface separating
northern hemispheric field from that originating in the southern
hemisphere is not coincident with the ecliptic, and probably not
planar. Thus the earth passes through regions of more-or-less
uniform field polarity followed by regions of opposite polarity
(known as magnetic sectors).

        These gross aspects of the interplanetary magnetic field are
important for a couple of reasons. Near the sun, the field is
sufficiently strong to dominate and/or control the solar wind
flow (cf. subsections 2.3 and 2.4 below). And particle
cyclotron frequencies are large, in the sense that transport and
thermal properties may differ in directions parallel and perpendicular to the background field.

        Superimposed on the average magnetic field is a spectrum of

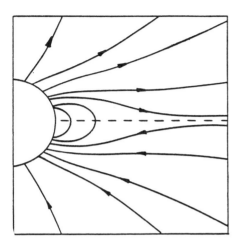   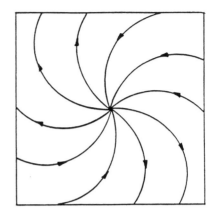

Figure 2. Solar magnetic field   Figure 3. Solar magnetic field
                                         in the ecliptic plane

smaller scale fluctuations. These fluctuations are characterized
by their large amplitude, their energy density being comparable
to the average field energy density, and the fact that they
correspond mainly to fluctuations in field direction rather than
magnitude. This latter property has led to the conclusion that
they are predominantly Alfven (MHD) waves (5). They appear to be
travelling outward and are therefore believed to be mainly of
solar origin. This spectrum of magnetic fluctuations extends
from very low frequencies ($\lesssim 10^{-4}$ Hz) up to frequencies ~ 10 Hz,
with a reasonable power law spectrum, the bulk of the energy
residing at the low frequency end. The waves have been proposed
as major sources of solar wind acceleration (22) and as a factor
influencing plasma microprocesses and transport processes (39;
cf. section 3 below).

2.3 Coronal Holes and High Speed Solar Wind Streams

      Current thinking suggests that the solar magnetic field
strongly controls the solar wind. Near the solar equator, the
predominance of closed loops of field lines inhibits or prevents
the expansion of coronal material into interplanetary space.
Associated with these closed field regions, then, are periods of
highly irregular, relatively slow speed (~ 300 km/s), dense (~ 10
part./cm^3) solar wind. By contrast, the wind eminating from the
regions of open magnetic field associated with polar coronal holes
(sometimes extending into equatorial latitudes) is more uniform
and structureless (3), high speed (~ 700 km/s) and less dense
(~ 4 part./cm^3). The survival of coronal holes for many solar
rotations implies that these regions of high speed solar wind

will be observed to reappear regularly, in close association with
magnetic sectors, and are often referred to as "streams".

    This interpretation of observations has serious implications
for solar wind modelling, since theories of steady coronal
expansion obviously more closely correspond to the high speed
structureless solar wind streams. Among the salient differences,
the high speed streams have proton temperatures in excess of
electron temperatures,anisotropic proton distributions with
higher temperatures perpendicular to the background field than
parallel to it (cf. subsection 3.3 below). Additionally, solar
wind from coronal holes suffers a more rapid expansion than a
simple spherical wind would. This alters estimates of the
necessary energy supply/unit area at the solar surface. It also
complicates the solution topologies of Fig. 1 by the possible
introduction of more critical points (28). Further solar wind
modelling will obviously involve a careful look at the inter-
dependence of the magnetic field geometry and the wind
itself (10,38).

2.4 Angular Momentum and Mass Loss

    The sun loses mass via the solar wind at a rate of 3×10^{-14}
$M_\odot$/year, corresponding to a timescale $\sim 3 \times 10^{13}$ years, which is
long compared to the sun's lifetime ($\sim 5 \times 10^9$ years), although it
is possible that the solar wind was much greater during the sun's
early history (21). Thus it is unlikely that this mass loss will
affect the sun's present evolution.

    By contrast, the angular momentum carried away by the solar
wind may be significant. Because of the strong solar magnetic
field, the corona and solar wind co-rotates with the sun out to a
distance r_A where the kinetic energy density in the wind ($\frac{1}{2}\rho u^2$)
is comparable to the magnetic energy density ($B^2/8\pi$); i.e. where
the flow velocity is comparable to the Alfven speed $v_A = B/(4\pi\rho)^{\frac{1}{2}}$.
Thus the angular momentum loss of the sun is enhanced by this
larger effective moment arm, and is roughly given by

$$\frac{dL_\odot}{dt} \sim \Omega_\odot \, r_A^2 \, \frac{dM_\odot}{dt} \qquad\qquad (2.6)$$

where $\Omega_\odot$ is the angular frequency of the sun's rotation (i.e.
like that angular momentum imparted to particles at a radius r_A
by the co-rotation). Thus the solar angular momentum changes
over a timescale

$$\tau_L \sim \frac{L_\odot}{dL_\odot/dt} = \frac{M_\odot \Omega_\odot R_\odot^2}{(dM_\odot/dt)_{\Omega_\odot} r_A^2} = \frac{(R_\odot)^2}{(r_A)} \tau_M \qquad (2.7)$$

where $\tau_M \sim 3 \times 10^{13}$ y is the mass loss timescale. Now $r_A \sim 50 R_\odot$ so that

$$\tau_L \sim 10^{10} \text{ years} \qquad (2.8)$$

Inclusion of the angular momentum carried away in the field increases the estimated angular momentum loss (2.6)(31), while more detailed calculations tend to produce less angular momentum loss (47,48). These estimates are consistent with the losses inferred from observations of the solar wind flow velocity in the azimuthal direction, but the observations are also consistent with zero azimuthal flow (and therefore no angular momentum loss) to within experimental error (31,12).

Thus it remains possible, though not certain, that the solar wind can effectively brake the solar rotation and in turn influence the sun's evolution. Further observations, particularly measurements out of the ecliptic plane, could provide a more definitive answer to the actual angular momentum lost by the sun via the solar wind.

3. THE MICROSCOPIC SOLAR WIND

3.1 Why a Plasma Description is Needed

The models presented in the previous section considered the solar wind as a fluid. Thus they implicitly assume that individual particles suffer many collisions while traversing a typical scale height. However, a consideration of the parameter B_T, which measures the ratio of the electron mean free path to the temperature scale height in the solar wind, yields

$$|B_T| \sim \lambda_{mfp} \mid \nabla \ln T_e \mid \sim 1 \qquad (3.1)$$

at 1 AU, so that the assumption that the plasma is collision-dominated, and hence fluid-like in behaviour, is clearly inappropriate.

From the particles' point of view, B_T measures the deviation of the particle distribution functions from Maxwellian. $|B_T| \ll 1$ implies small deviations and is the approximation involved in calculating classical transport coefficients (e.g. thermal conduction (45); cf. subsection 3.2 below) used in fluid theories. $|B_T| > 1$ corresponds to a nearly collisionless plasma, in which

particle distributions can be highly non-Maxwellian and for which
the transport theory is not well understood. Since the solar wind
involves essentially a conversion of thermal energy into bulk flow,
these transport problems have non-trivial consequences for
accurate solar wind modelling.

For the sake of simplicity, I shall regard microscopic as
referring to any process requiring consideration of particle
behaviour or involving lengths and times very much less than
those of the bulk flow (e.g. short wavelength MHD waves). Also,
I shall restrict my attention to the high speed solar wind since,
as previously argued, it best represents a structureless steady-
state medium. Finally, it is worth noting that the solar wind
plasma is a very interesting medium as far as plasmas are concerned,
with many small scale features which can be routinely investigated.

3.2 Heat Conduction

The $|B_T| << 1$ approximation described above is used to derive
the standard electron conduction law

$$\underline{q}_e = - \kappa_o T_e^{5/2} \nabla T_e \qquad (3.2)$$

where $\underline{q}_e$ is the electron heat flux density and $\kappa_o \sim 8 \times 10^{-7}$
ergs cm^{-1} s^{-1} $deg^{-7/2}$ (44) (the proton conduction is smaller by
$(m_e/m_p)^{\frac{1}{2}}$). $|B_T|$ generally increases with radial distance in the
solar wind. As it does so, (3.2) becomes inappropriate and the
plasma deviates significantly from Maxwellian, the electron
distribution taking on a highly asymmetric shape. The skewness
of this distribution, based on collision-dominated theory, may
eventually drive a collisionless plasma microinstability (i.e.
one which grows rapidly compared to a collision frequency) of
some sort (15,37). The resulting microturbulence can, in
principle, scatter the electrons, bringing the distribution back
closer to Maxwellian equilibrium and thereby regulating the
transport of heat. Singer and Roxburgh (43) tested a variety of
published solar wind fluid models and found that all of them
were micro-unstable within 11 $R_\odot$ from the sun. Unfortunately,
in order to determine the transport processes, a nonlinear theory
of plasma microturbulence is required. No firm theory yet exists,
despite considerable effort. A recent detailed review of solar
wind heat conduction driven instabilities appears elsewhere (40).
Other attempts at modelling interplanetary heat conduction have
considered the global behaviour of individual electron trajectories
and the resulting distributions (23,37,42).

The actual electron distribution measured at 1 AU consists
of two near-Maxwellian components: a cool "core" containing most
of the electrons and a hot "halo" drifting with respect to the
core (13). The halo electrons carry most of the heat flux,

which is much less than the theoretical predictions of fluid models
using (3.2). In the lack of a rigorous consistent theory of heat
conduction, several fluid models have incorporated ad hoc
reductions of the conduction (48) or collisionless behaviour (25).

3.3 Thermal Anisotropies

The collisionless nature of the solar wind plasma also has
consequences for the isotropy of the proton distribution.
Conservation of the two collisionless adiabatic invariants
(the first being the particle's magnetic moment)

$$c_{\perp} = T_{\perp} / B_o \tag{3.3}$$

and

$$c_{\parallel} = T_{\parallel} B_o^2 / n^2 \tag{3.4}$$

(where $T_{\perp,\parallel}$ are the temperatures perpendicular and parallel to
the magnetic field of strength B_o) (6) implies that the protons
should become highly anisotropic with $T_{\parallel} \gg T_{\perp}$. These
distributions are also expected to drive microinstabilities (40)
although again a nonlinear theory of plasma turbulence is
desperately needed to determine their consequences.

In fact, the observed situation is even worse, as the high
speed solar wind protons are anisotropic in the opposite sense,
$T_{\perp} / T_{\parallel} \sim 2.5$ between 0.5 AU and 1 AU (2,34). Some in situ
source of perpendicular heating is obviously required. One
strong possibility is wave heating, although it appears that the
interplanetary spectrum of solar produced waves are insufficient
and some local wave source is required (41; cf. subsection 3.4
below).

3.4 Minor Ions

Although the solar wind is primarily composed of protons and
electrons, a significant population of other ions, notably helium,
is also present ($\sim$ 4% by number). Of particular interest is
the observation that, in the high speed solar wind, the heavier
helium ions flow faster than the bulk of the protons (and are
also hotter). The relative He-p streaming speed is on the order
of the Alfven speed, v_A, and follows the wide variations in v_A ,
increasing with decreasing heliocentric distance (35). This
relationship suggests that the streaming is controlled by some
wave-related phenomena. Previous work has attempted to explain
the acceleration of minor ions in terms of resonant and non-
resonant wave-particle interactions (27).

A recent proposal (41) relies on the decrease of v_A with

increasing radial distance to argue that waves must actually be
slowing down the helium beams, implying a transfer of energy from
helium to the waves via the growth of an instability. The subsequent damping of these waves could provide sufficient thermal
energy to explain the observations of $T_{p\perp} > T_{p\parallel}$ discussed in
subsection 3.3, although the initial source of the beams is not
addressed in this work.

4. CONCLUSIONS

I have briefly reviewed present day solar wind theory.
In some ways this review is notable for its many omissions.
A wide variety of other interesting and fundamental problems also
receive considerable attention. Among them are solar wind stream-
stream interactions and flare produced disturbances (8,9,32,16);
high frequency electrostatic waves and associated wave-particle
interactions (17,18,19); energetic particles (14); as well
as a whole host of phenomena occurring in the magnetosphere and/
or associated with the earth's bow shock and other planetary
magnetospheres now observed directly.

In summary, general solar wind research needs answers in
(at least) three broad fields:

1) The mechanism of coronal heating and energy balance
2) The interaction of the solar magnetic field and the solar
wind flow
3) A detailed understanding of the microprocesses governing
the internal structure of the solar wind plasma

Additionally, the solar wind is an excellent laboratory for
studying plasma physics, and may prove to be the best testing
ground for future theories of plasma turbulence.

REFERENCES

(1) Alazraki, G. and P. Couturier: 1971, Astron. Astrophys. 13,
     p. 380.
(2) Bame, S.J., J.R. Asbridge, W.C. Feldman, S.P. Gary and
     M.D. Montgomery: 1975, Geophys. Res. Lett. 2, p. 373.
(3) Bame, S.J., J.R. Asbridge, W.C. Feldman and J.T. Gosling: 1977,
     J. Geophys. Res. 82, p. 1487.
(4) Barnes, A.: 1979, in "Solar System Plasma Physics,
     Twentieth Anniversary Review", Vol 1, E.N. Parker, C.F.
     Kennel and L.J. Lanzerotti eds., North-Holland Publ. Co.,
     Amsterdam, p. 249.
(5) Belcher, J.W. and L. Davis, Jr.: 1971, J. Geophys. Res. 76,
     p. 3534.

(6) Chew, G.F., M.L. Goldberger and F.E. Low: 1956, Proc. Roy.
     Soc. A236, p. 112.
(7) Cuperman, S.: 1980, Sp. Sci. Rev. 26, p. 277.
(8) Dryer, M.: 1974, Sp. Sci. Rev. 15, p. 403.
(9) Dryer, M. and R.S. Steinolfson: 1976, J. Geophys. Res. 81,
     p. 5413.
(10)Durney, B.R. and G.W. Pneuman: 1975, Solar Phys. 40, p. 461.
(11)Feldman, W.C.: 1979, in "Solar System Plasma Physics,
     Twentieth Anniversary Review", E.N. Parker, C.F. Kennel
     and L.J. Lanzerotti eds., North-Holland Publ. Co.,
     Amsterdam.
(12)Feldman, W.C., J.R. Asbridge, S.J. Bame and J.T. Gosling:
     1977, in "The Solar Output and Its Variations", O.R. White
     ed., Colo. Univ. Press, Boulder, Colo.
(13)Feldman, W.C., J.R. Asbridge, S.J. Bame, M.D. Montgomery
     and S.P. Gary: 1975, J. Geophys. Res. 80, p. 4181.
(14)Fisk, L.A.: 1979, in "Solar System Plasma Physics,
     Twentieth Anniversary Review", E.N. Parker, C.F. Kennel and
     L.J. Lanzerotti eds., North-Holland Publ. Co., Amsterdam.
(15)Forslund, D.W.: 1970, J. Geophys. Res. 75, p. 17.
(16)Gosling, J.T.: 1975, Rev. Geophys. Sp. Phys. 13, p. 1053.
(17)Gurnett, D.A. and R.R. Anderson: 1977, J. Geophys. Res. 82,
     p. 632.
(18)Gurnett, D.A. and L.A. Frank: 1978, J. Geophys. Res. 83, p.58.
(19)Gurnett, D.A., E. Marsch, W. Pilipp, R. Schwenn and
     H. Rosenbauer: 1979, J. Geophys. Res. 84, p. 2029.
(20)Hartle, R.E. and P.A. Sturrock: 1968, Astrophys. J. 151,
     p. 1155.
(21)Heymann, D.: 1977, in "The Solar Output and Its Variations",
     O.R. White ed., Colo. Univ. Press, Boulder, Colo., p. 405.
(22)Hollweg, J.V.: 1973, Astrophys. J. 181, p. 547.
(23)Hollweg, J.V.: 1974, J. Geophys. Res. 79, p. 3845.
(24)Hollweg, J.V.: 1975, Rev. Geophys. Sp. Phys. 13, p. 263.
(25)Hollweg, J.V.: 1976, J. Geophys. Res. 81, p. 1649.
(26)Hollweg, J.V.: 1978, Rev. Geophys. Sp. Phys. 16, p. 689.
(27)Hollweg, J.V. and J.M. Turner:1978, J. Geophys. Res. 83,
     p. 97.
(28)Holzer, T.E.: 1977, J. Geophys. Res. 82, p. 23.
(29)Holzer, T.E.: 1979, in "Solar System Plasma Physics,
     Twentieth Anniversary Review", Vol. I, E.N. Parker,
     C.F. Kennel and L.J. Lanzerotti eds., North-Holland Publ.
     Co., Amsterdam, p. 101.
(30)Holzer, T.E. and E. Leer: 1979, J. Geophys. Res. 84.
(31)Hundhausen, A.J.: 1972, "Coronal Expansion and Solar Wind",
     Springer-Verlag, New York.
(32)Hundhausen, A.J.: 1973, J. Geophys. Res. 78, p. 7996.
(33)Leer, E. and T.E. Holzer: 1979, J. Geophys. Res. 84.
(34)Marsch, E., W.G. Pilipp, H. Rosenbauer, R. Schwenn and
     K-H. Muhlhauser: 1980, in "Lecture Notes on Physics,
     Proc. Solar Wind 4", Springer-Verlag, Berlin, in press.

(35) Marsch, E., H. Rosenbauer, W. Pilipp, K-H. Muhlhauser and
     R. Schwenn: 1980, in "Lecture Notes on Physics, Proc. Solar
     Wind 4", Springer-Verlag, Berlin, in press.
(36) Parker, E.N.: 1958, Astrophys. J. 128, p. 664.
(37) Perkins, F.: 1973, Astrophys. J. 179, p. 637.
(38) Pneuman, G.W. and R.A. Kopp: 1971, Solar Phys., 18, p. 258.
(39) Scarf, F.L.: 1970, Sp. Sci. Rev. 11, p. 234.
(40) Schwartz, S.J.: 1980, Rev. Geophys. Sp. Phys. 18, p. 313.
(41) Schwartz, S.J., W.C. Feldman and S.P. Gary: 1980, J.
     Geophys. Res., in press.
(42) Scudder, J.D. and S. Olbert: 1979, J. Geophys. Res. 84, p. 2755.
(43) Singer, C. and I.W. Roxburgh: 1977, J. Geophys. Res. 82,
     p. 2677.
(44) Spitzer, L. Jr.: 1962, "Physics of Fully Ionized Gases",
     Wiley (Interscience), New York.
(45) Spitzer, L. Jr. and R. Harm: 1953, Phys. Rev. 89, p. 977.
(46) Volk, H.J.: 1975, Sp. Sci. Rev. 17, p. 255.
(47) Weber, E.J. and L.J. Davis Jr.: 1970, J. Geophys. Res. 75,
     p. 2419.
(48) Wolff, C.L., J.C. Brandt and R.G. Southwick: 1971,
     Astrophys. J. 165, p. 181.

OBSERVATIONS AND THEORY OF MASS LOSS IN LATE-TYPE STARS

L. Hartmann

Harvard-Smithsonian Center for Astrophysics

ABSTRACT

The properties of the winds from luminous late-type stars are discussed. The features of high mass loss rates, cold outflows, and terminal velocities smaller than the surface escape velocities are very different from the comparable characteristics of the solar wind. Previous wind theories therefore have suggested mechanisms completely removed from solar behavior. Preliminary space-based observations also suggested a clear-cut dichotomy between hot, low mass loss winds and cold, massive outflows. However, the discovery of "hybrid" atmosphere stars has shown that escaping cool circumstellar shells can exist even when the acceleration region of the wind is hot ($\sim 10^5$K). This recent result indicates that the mass loss from giants and supergiants may be a manifestation of solar-type activity after all. A theory developed by Hartmann and MacGregor, which investigates the effects of wave modes known to exist in the solar wind propagating through low-gravity atmospheres, has shown that dissipating magnetic waves can account for mass loss rates in accord with observations. A natural byproduct of the Alfven wave dissipation is heating of the outer atmospheres. Gas temperatures decrease with decreasing gravity, because the consequent increase of density in the outflow enhances radiative cooling. For a constant energy flux per unit area, one finds that the wave dissipation results in coronal (10^6K) heating for log g $\gtrsim$ 2, hybrid atmospheres and warm (10^5K) coronae for log g $\sim$ 1-2, and extended warm chromospheres ($\sim 10^4$K) in the inner regions of M supergiant winds, in reasonable agreement with observation. Thus further explorations of the relationships between solar winds and luminous stellar winds, and of the possible connection between coronal heating and mass loss even in low gravity stars, appear to be the most promising avenues of investigation at present.

R. M. Bonnet and A. K. Dupree (eds.), Solar Phenomena in Stars and Stellar Systems, 331–347.
Copyright © 1981 by D. Reidel Publishing Company.

I. INTRODUCTION

·The study of mass loss from late-type stars has primarily concentrated on winds from the coolest, most luminous stars. This is so
not only because the winds of such stars have the highest mass loss rates,
but also because their cold circumstellar shells can be studied optically.
In recent years the advent of space-based observations, particularly
with the International Ultraviolet Explorer satellite, have enabled astronomers to bridge the gap between supergiant winds and the solar wind by
studying mass loss in higher temperature regimes. The most promising
approach at present suggests that the same momentum and energy fluxes
are present in all cool stellar winds, but changing atmospheric responses
as a function of the stellar surface gravity alter the observed properties
of the outflows.

Although certain stars may lose mass primarily via pulsational
waves, and all cool stars may lose the bulk of their material in planetary
nebula or supernova phases, this review will concentrate on the ubiquitous
mass loss that occurs during most of a star's existence as a cool giant
or supergiant.

II.   OBSERVATIONS OF MASS LOSS

a) Wind Components and Kinematics

Adams and McCormack (1935) first discovered blueshifted absorption components in the strong resonance lines of low-ionization species
in the spectra of late-type giants. The circumstellar shells responsible
for these features exhibit expansion velocities significantly smaller than
the surface escape velocities of the associated stars. It remained for
Deutsch (1956, 1960) to demonstrate that gas was actually escaping by
observing the circumstellar components in the spectra of companion stars
widely separated from their mass-losing primaries. Using modern techniques it has proved possible to resolve the circumstellar shell of α Ori
on the sky in the K I circumstellar line out to about $600\,R_*$ (Bernat and
Lambert 1976; Bernat et al. 1978).

The region of the HR diagram in which optical CS lines are seen
is indicated in Fig. 1. In general, the coolest and most luminous stars
posess the highest mass loss rates. There is also a general trend of
decreasing wind velocities with decreasing gravity, also indicated in Fig.
1 (Reimers 1977), although there are regions in which stars of nearly
identical spectral types have very different wind velocities. In all cases
the wind terminal velocities are substantially below the stellar surface
escape velocities.

Although the boundary of observed circumstellar lines in the HR
diagram has been interpreted as a mass loss boundary, below which winds

become tenuous as well as hot (Mullan 1978), it seems likely that ionization effects can cause the circumstellar lines to vary much more than the
actual mass loss rate (Dupree and Hartmann 1980).

Fig. 2 shows Deutsch's (1960) correlation between the equivalent
width of the Ca II K circumstellar absorption component for M giants.
Of particular interest is the fact that three pulsating M giant variables,
δ Oph, ν Vir, and β And are found to vary their CS equivalent width
with velocity along the same relation as defined by the other (constant)
stars. This fundamental relationship has been explained by Reimers
(1974) in terms of the basic ionization structure of cool winds, as ela-
borated in the following section.

Although a great deal of effort has been expended on determining
mass loss rates (cf. Reimers 1977; Bernat 1977), at present they must
be regarded as order-of-magnitude estimates. The principal difficulty
in turning a column density into a mass loss rate is our lack of understanding of the ionization structure. The inner portions of late-type
winds appear to be in a sufficiently high ionization state that optical CS
lines disappear (Weymann 1962); this can be demonstrated from the
spatially resolved K I line emission in α Ori (Bernat and Lambert 1976).
Uncertainties in the inner radius of the optically thick CS region yield
large errors in $\dot{M}$. In addition, possible velocity gradients (Reimers
1975) and wind turbulence or terminal velocity variations (Hartmann,
Dupree and Raymond 1980b) can affect the derived mass loss rate. The
net result is that almost every scaling law of mass loss based on stellar
parameters that has been developed fits the data equally well (or poorly;
see Goldberg 1979).

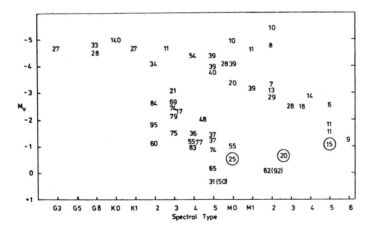

Fig. 1. Stars with observed optical circumstellar lines, labeled by the
observed velocity shifts in km s^{-1}. The circled values are means for
M giants. From Reimers (1975).

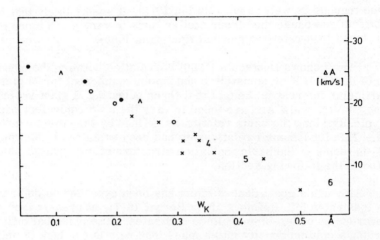

Fig. 2. Deutsch's line strength-velocity correlation of circumstellar
Ca II equivalent widths. Crosses are individual stars, 4, 5, and 6 are
means for M4, M5, and M6, dots represent δ Oph, open circles ν Vir,
and $\wedge$ represents β And; the last three stars are observed at different
phases. From Reimers (1975).

     Dust particles can condense in the outer atmospheres of some
cool stars (Gillette et al. 1968; Woolf and Ney 1969; Gilman 1969; Gehrz
and Woolf 1971). However, the processes by which dust is formed are
not well understood. Salpeter (1976a, b) has discussed dust formation in
which high densities and low temperatures are required. However,
spatially resolved infrared studies of α Ori indicate that less than 20%
of the dust emission occurs at distances $< 12R_*$ (Sutton et al. 1977).
Recent work has cast doubt on whether any correlation between mass
loss rates and infrared emission exists (Hagen 1978).

     Wind variability is mostly observed in the stars near the boundary
line of the circumstellar line region in the HR diagram (Reimers 1977;
Dupree and Baliunas 1979). This is not surprising, given that these
stars have the highest wind velocities and smallest radii, hence the shortest flow timescales. In addition, stars near the CS boundary may have
generally higher wind temperatures (Hartmann, Dupree, and Raymond
1980a, b) and so changing ionization effects may contribute to CS line
variability.

     Multiple CS shells have been seen in a few late-type giants, which
also may indicate variable mass loss. Although the optical CS line profiles in α Ori have remained remarkably constant over decades (Weymann
1962; Goldberg 1979), observations of α Ori in CO have revealed the existence of two CS shells, one at the optical CS velocity of -11 km s^{-1} with
a temperature $\sim 200K$ extending over $\sim 30R_*$, and a second at -17 km
s^{-1} at $\sim 70K$ extending over $\sim 1000R_*$ (Bernat et al. 1979). The two

shells can also be seen in K I (Goldberg et al. 1975). Thus it is beginning to look as if variability may be a common feature of late-type winds.

Observational evidence for chaotic velocity fields in cool winds
has also accumulated. Indicators of components with large velocity
shifts, possibly even supersonic, in the upper atmospheres of K supergiants have come from observations of eclipsing binaries (cf. Wilson and
Abt 1954). In a study of Fe II emission in α Ori, Boesgaard (1979) has
found evidence for large turbulent velocities. In addition, she found
redshifted Fe II emission; if this truly represents infall, the mass motions
in the chromosphere of α Ori must be very complex indeed.

b) Temperature Structure of Cool Winds

It has become increasingly clear that a knowledge of the temperature structure in late-type stellar winds is important to an understanding
of the underlying mass loss mechanism. The circumstellar shells observed in optical transitions must have temperatures of the order of a
few thousand degrees or less; measurements in CO of the distant shells
of α Ori indicate temperatures of $\sim$ 100K (Bernat et al. 1979).

However, the temperatures of material in the distant circumstellar
shell tell us relatively little about the nature of the outflow. Observations
of temperatures in the inner few stellar radii of the wind, where the gas
is accelerated, represent the most crucial information. Studies of many
stars clearly indicate that an extended warm chromosphere exists at the
base of a typical late-type stellar wind. Wilson and Abt (1954) were able
to use eclipse measurements to trace an increase in the excitation temperature from $\sim$ 3000 to $\sim$ 7000K in the atmosphere of ζ Aur out to about
0.17R$_*$. However, the analysis is complicated by the possible excitation
effects of the B star companion. Weymann (1962) analyzed Ca II IR triplet
data to show that an inner Ca III region must exist close to the surface in
α Ori; he also showed that the gas temperature in this region was probably below $\sim 8 \times 10^4$K. Reimers (1974) explained Deutsch's (1960)
equivalent width vs. velocity relation as the effort of a varying chromospheric flux on Ca II ionization at the inner shell boundary, changing the
line strength and velocity in the differentially expanding wind. Alternatively, an extended warm chromosphere varying in size could accomplish
the same thing.

The clearest picture of the temperature and velocity structure of
a wind from a luminous late-type star has been assembled for the best-
studied object, α Ori (Goldberg 1979). The Hα equivalent width is larger
than expected for the star's effective temperature; this has been interpreted to imply a sizeable chromospheric column density of material at
$\sim$ 5000K exists, strengthening the line. The absorption core is blue-
shifted by about half the wind terminal velocity of 11 km s^{-1}. Furthermore, the absorption velocity does not share in the photospheric pulsation
of 5 km s^{-1} with a period $\sim$ 2100 days (Fig. 3). Similar results hold for
the Ca II IR triplet. Thus the chromosphere of α Ori must be spatially
distinct from the photosphere, and probably extends over several stellar radii.

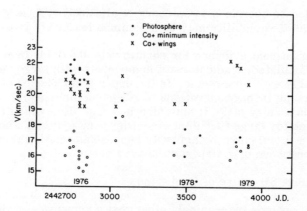

Fig. 3. Photospheric and Ca II infrared triplet velocities as a function
of time in α Ori. The Ca II lines are presumably formed in an extended
chromosphere, as they do not follow the photospheric pulsation. From
Goldberg (1979).

Further confirmation of this picture has come from radio observations. Altenhoff, Oster, and Wendker (1979; see also Bowers and
Kundu 1979) have detected radio emission at frequencies below 20 GHz
that are in excess of the expected photospheric contribution. Altenhoff
et al. show that this result may be explained by invoking free-free radiation from a region of a few stellar radii near the star. Since free-free
emission is formed in LTE, an angular diameter measurement coupled
with a flux observation yields the electron temperature from the surface
brightness. In principle it should be possible to perform these measurements at a couple of different frequencies, probing different radii due to
the variation of free-free opacity, and building up a picture of the temperature stratification of the extended chromosphere. At present all
that can be said is that the ionized fraction of the gas is $\sim 1\%$, and the
temperature is probably $\sim$ 4000-5000K over a region of $\sim$ 2-3 stellar
radii (Altenhoff, Oster, and Wendker 1979).

The physical extension of chromospheres in cool giants is probably related to the absence of gas near solar coronal temperatures
suggested on both observational and theoretical grounds (Weymann 1962;
Parker 1963). The situation has become much clearer with the arrival
of ultraviolet emission-line observations made with IUE. The general
notion that high atmospheric temperatures occur only on high-gravity
stars was reinforced by Linsky and Haisch (1979), who observed solar-
type transition region lines like C IV and N V, formed at temperatures
$\sim 10^5$K. These authors found a sharp division in the HR diagram between stars with 10^5K emitting gas and those without. This dividing line
occurred near the CS boundary of Reimers (1977) and near the "STL,"
a boundary calculated by Mullan (1978) as defining the region where

stellar coronae should vanish, based on minimum-flux coronal theory.
Linsky and Haisch therefore suggested that a rapid increase in mass
loss at the STL prevented coronal temperatures from being achieved
through wind expansion cooling.

The suggested sharp division has been substantially blurred by
subsequent work. Dupree and Hartmann (1980) found that stars could
not be neatly divided into "solar" or "non-solar" types (i. e., with 10^5K
emission or without) purely on the basis of position in the HR diagram.
The suggested sharp boundary in mass-loss was not found by Stencel
(1978) and Stencel and Mullan (1980), who interpreted Mg II and Ca II
emission asymmetries in terms of mass loss. Rough HR diagram boundaries of opposite asymmetries can be drawn, although there is a large
amount of overlap. As can be seen in Fig. 4, the Mg II, Ca II, and CS
boundaries are not in identical positions. The impression is that of mass
loss rates gradually increasing with increasing luminosities, becoming
visible in successively weaker lines. The rough boundary of C IV emission (at limits detectable with IUE) found by Dupree and Hartmann (1980)
is near to, but not identical with, the other boundaries.

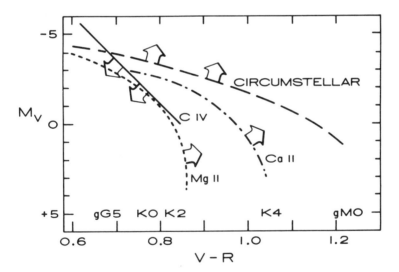

Fig. 4. Boundaries of temperature structure and mass loss in the HR
diagram. Stars with C IV emission at levels detectable with IUE are
below the line labeled "C IV" (Dupree and Hartmann 1980). The upper-
most line marks the region in which optical circumstellar features are
seen (Reimers 1977). The "Mg II" and "Ca II" lines mark dividing lines,
on either side of which the ratio of the violet emission peak to the red
emission peak (V/R) of the respective line changes from greater than to
less than unity in the majority of stars. V/R values < 1 are interpreted
to suggest blueshifted absorption from escaping material (Stencel 1978;
Stencel and Mullan 1980). From Dupree and Hartmann (1980).

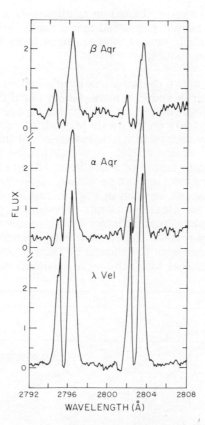

Fig. 5. IUE Mg II line profiles in the stars β Aqr (G0Ib), α Aqr (G2Ib), and λ Vel (K5Ib). The G supergiants exhibit high-velocity circumstellar shells at - 60 km s^{-1} (β Aqr) and - 120 km s^{-1} (α Aqr). From Hartmann, Dupree, and Raymond (1980a).

The most important alteration of the findings of Linsky and Haisch was the discovery of "hybrid atmosphere" stars, combining aspects of both the "solar" and "non-solar" types (Hartmann, Dupree, and Raymond 1980a). In Fig. 5 we show that the hybrid stars possess perfectly normal cool wind absorption components in Mg II (and also Ca II), expanding at relatively high velocities ~ 100 km s^{-1}. However, as Fig. 6 demonstrates, N V, C IV, and Si IV emission is also present, with surface fluxes of these lines the same order of magnitude as in the Sun. Thus the hybrid stars have physical conditions in their outer layers which are intermediate between solar and cool supergiant conditions, indicating a more gradual transition between "solar" and "non-solar" types. Most important, the observations of hybrid stars show that it is possible to have both high-temperature gas (presumably near the star) and a cold circumstellar shell (at large radial distances). Coupled with the failure

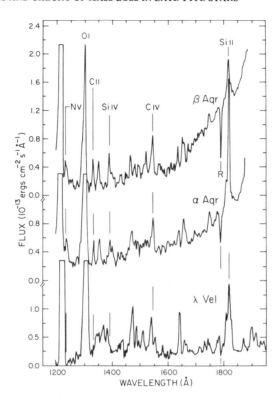

Fig. 6. IUE short wavelength observations of β Aqr, α Aqr, and λ Vel. Despite having cold circumstellar shells, α Aqr and β Aqr also exhibit emission from solar-type transition region lines such as N V, C IV, and Si IV, indicating gas at temperatures of at least 2×10^5K with surface emission measures comparable to the Sun's. They also possess emission from lower excitation lines, such as S I λ1475, λ 1296, typical of cool wind stars like λ Vel. From Hartmann, Dupree, and Raymond (1980a).

of past theories of cool supergiant mass loss, this result stands in contrast to past supposition that the winds of luminous late-type stars have nothing to do with solar-type activity.

III. THEORIES OF MASS LOSS

From the limitations on the presence of high-temperature gas in cool, luminous stars, plus the evidence that expansion has already begun in low temperature chromospheric regions (Wilson 1960; Weymann 1962), it is clear that the mass loss cannot be thermally driven. This fact, coupled with the low terminal velocities observed, has led investigators to suggest mass loss mechanisms in giants completely different from processes operating in the solar wind. However, the previously suggested mechanisms have difficulties in explaining the general phenomenon of out-flow in late-type stars, as detailed below.

a) Radiation Pressure on Dust

The fact that mass loss rates are greatest for the coolest stars which also form dust in their expanding envelopes has prompted several authors to investigate dust-driven mass loss (Gehrz and Woolf 1971; Kwok 1975). Because dust opacity per gram is much larger than for gas, radiation pressure on the dust can transfer a substantial
p. 10, l. 27: (Sutton et al. 1977).
p. 11, l. 28:

A major difficulty with this process can be seen in the wind models constructed by Kwok (1975). The gas in these wind models does not cool to grain-forming temperatures until it escapes out to ~ 1.5 R$_*$. The question arises as to just what process forces the gas out to 1.5 R$_*$, and why it should not be considered the principal mass loss mechanism. This difficulty does not even take into account the evidence for extended chromospheric regions. This almost certainly means that grain-forming gas temperatures are not reached until a more distant radius than Kwok assumed. Also, as previously mentioned, spatially resolved infrared measurements indicate that most of the emitting dust is at radii > 12 R$_*$ (Sutton et al. 1977).

The principal difficulty with the dust-driven wind theory is that most of the stars shown in Fig. 1 are too hot to be expected to form dust in their near environs, or are insufficiently luminous to generate an appreciable radiative pressure. In particular, the hybrid stars, with an inner region of temperatures reaching at least 2×10^5 K, are inauspicious objects from which to infer the importance of dust to the mass loss pro-cess. There is no reason to suppose that the mass loss from the majority of stars in the upper righthand corner of the HR diagram occurs for a qualitatively different reason than the mass loss from the extreme α Ori-type stars.

b) Radiation Pressure in Lyman α

Recently Haisch, Linsky, and Basri (1980) investigated a sug-gestion originally due to Wilson (1960), that chromospheric radiation from Ly α can drive mass loss. This possibility is easily dismissed by noting that the observed Ly α emission in cool stars contains about three to four orders of magnitude smaller momentum flux than is present in the winds of such stars at infinity. Haisch et al. find that Ly α pressure is important dynamically only over a very small region in the wind, possibly leading to supersonic flow. This conclusion is questionable, since no correct integration through the critical point was accomplished, and the adopted temperature distribution, which has important effects on the force calculations, is not uniquely determined from observations.

Haisch et al. make the much more fruitful suggestion that Alfvén waves drive the mass loss.

c) Magnetic Wave-Driven Winds

Consider a late-type giant below the CS-STL boundary line. The evidence from IUE in transition-region lines (Linsky and Haisch 1979; Dupree and Hartmann 1980) is that such a star has a solar-type transition region and corona. Presumably the star possesses magnetic field and wind behavior that is also solar-like in character. Now suppose the star evolves slightly, through the CS-STL boundary. Although we do not clearly understand the origins of solar activity, there is no obvious reason why the field strengths, mechanical energy fluxes, etc., should vary dramatically. This leads naturally to the idea that perhaps it is not the energy and momentum fluxes needed to heat the outer layers and drive the wind that are changing, but rather the response of the atmosphere to such energy and momentum fluxes, a response which changes as the stellar surface gravity decreases. The existence of hybrid stars, combining "solar" and "non-solar" characteristics in their outer atmospheres, suggests that we should look to the Sun for a mass loss mechanism which can operate in low-gravity stars as well.

Now it has become apparent that the solar wind is not only driven by thermal heating at its base, but that a significant source of heat or momentum is being added to the wind at large distances from the Sun's surface as well (cf. Munro and Jackson 1977). With the identification of Alfvén waves in the solar wind (Belcher and Davis 1971), it has become plausible that these waves are the source of the extra momentum deposition, and that they account for the high-speed streams in the solar wind (Belcher 1971; Hollweg 1973; Jacques 1977, 1978).

These considerations led Keith MacGregor and myself to investigate the effects of Alfvén wave fluxes of the same order of magnitude in surface flux as those invoked to explain solar high-speed streams (Hartmann and MacGregor 1980). The theory is schematic at this stage, because one cannot predict or observe magnetic field strengths or wave surface fluxes. However, it is attractive in that it adopts a phenomenon known to exist in the solar wind, calculating the differing responses of atmospheres with differing surface gravities.

Assuming a radial magnetic field and a polytrope relation $P \propto \rho^{\gamma}$, the wind equation of motion is derived to be

$$
\left[u^2 - a^2 - \frac{\epsilon}{4\rho} \left(\frac{1 + 3M_A}{1 + M_A} \right) \right] \frac{du}{dr}
$$
$$
= \frac{2u}{r} \left[a^2 - \frac{GM}{2r} + \frac{\epsilon}{4\rho} \left(\frac{1 + 3M_A}{1 + M_A} \right) + \frac{\epsilon}{4\rho} \frac{r}{\lambda} \right], \tag{1}
$$

(Hartmann and MacGregor 1980), where u is the wind velocity,
$a = (\gamma P/\rho)^{1/2}$ is the sound speed, ρ is the gas density, $\epsilon = \delta B^2/8\pi$
is the wave energy density, and M_A is the Alfvénic mach number = u/V_A.
The parameter λ is the length scale for dissipation of the waves. This
wind equation has critical points to those arising in solar wind theory
(Kopp and Holzer 1976; Jacques 1977, 1978). The results showed that
quite reasonable mass loss rates can be achieved with Alfven wave fluxes
on the order of 10^6 erg cm^{-2} s^{-1} (see Table 1). In this connection it is
interesting to note that for typical empirically determined mass loss
rates and wind velocities, the observed wind kinetic energy losses in
terms of surface flux are of the same order of magnitude as the surface
fluxes in the solar wind (Hartmann 1980).

Table 1

Wind Models with Wave Damping

$M(M_\odot)$	$R(R_\odot)$	T_o (°K)	F_o (erg cm^{-2}s^{-1})	B_o (gauss)	$\dot{M}(M_\odot \text{yr}^{-1})$	$V(\text{km s}^{-1})$
16	400	10^4	4.20×10^5	5	3.14×10^{-8}	47.7
16	400	10^4	3.36×10^6	10	4.51×10^{-7}	51.0
1.33	27	10^4	3.36×10^6	10	1.46×10^{-9}	56.3
10	1330	5000	2.43×10^6	10	3.08×10^{-5}	22.5

From Hartmann and MacGregor (1980)

In Fig. 7 we show wind models calculated using two different
assumptions about wave damping. The advantage of Alfvén waves is that
they are not as easily damped as sound waves, so that they can survive
to greater distances in the wind and thus accelerate the flow more efficiently. The disadvantage of Alfvén waves is precisely the same property.
Undamped waves in solar wind models, give rise to high-speed streams
(Hollweg 1973; Jacques 1977, 1978); similarly, undamped waves result
in impossibly high stellar wind velocities, as demonstrated in Fig. 7.
Therefore we were forced to assume some damping mechanism was operating. Acceptable agreement with observed terminal velocities can be obtained if the wave damping length λ is of the order of a stellar radius
(Fig. 7). MacGregor and I were able to identify several possible mechanisms for wave damping, such as conversion to fast modes and ion-neutral
frictional damping; and by assuming a certain wave period were able to
calculate a reasonable value of λ for frictional damping. However, in
the absence of real predictions of wave periods, propagation patterns,
and magnetic field geometries, we were unable to predict appropriate
damping lengths.

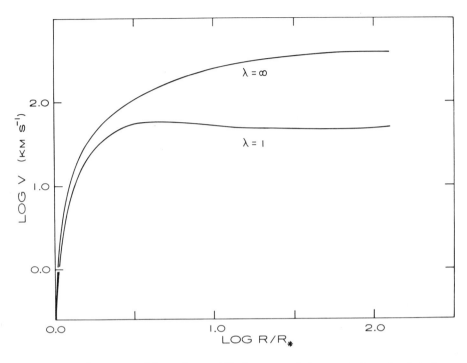

Fig. 7. Velocity profiles of two Alfvén wave-driven winds calculated
for a star of 16 $M_\odot$ and 400 $R_\odot$. If the Alfvén waves are undamped
($\lambda = \infty$), the terminal velocity of the wind is unacceptably high. The
waves must damp within $\sim 1\ R_*$ ($\lambda = 1$) in order to obtain velocities
in reasonable agreement with observations. From Hartmann and
MacGregor (1980).

What, then, can such a schematic theory predict? Interesting
consequences arise from considering the energy deposition of the waves
as well as their associated momentum. The wave damping necessary to
achieve reasonable terminal velocities necessarily heats the wind. In
fact, the Alfvén wave dissipation provides a natural explanation for the
observed extended chromospheres. For most of the wind structure the
energy equation can be approximated by

$$\frac{F}{\lambda} \cong \Lambda(T) N_e N_H,\qquad\qquad (2)$$

where F is the wave energy flux, N_e and N_H are the electron and hydrogen number densities, and $\Lambda(T)$ is the radiative cooling function. Given
a wave flux F sufficient to account for mass loss, and a damping length
λ that yields reasonable terminal velocities, the temperature structure
can be calculated if $\Lambda(T)$ is known. In general $\Lambda(T)$ depends on optical

depth effects, but estimates can be made which give a reasonable indication of the temperature structure. For most winds we can make a further
simplifying approximation, that thermal pressure is not dynamically
important; this permits us to use the density distributions derived by
assuming cool isothermal winds with $T=T_0$ (Table 1) in equation (2).

A model with parameters chosen to approximate the wind of α Ori
predicts an amount of wave heating which results in a chromospheric
temperature of about 5000K extending out to $\sim$ 3R$_*$. This agrees fairly
well with the radio flux measurements. Angular diameter measurements
would be extremely useful in testing this model.

Another interesting consequence of the wave energy balance is that
the calculated wind temperatures rise with increasing surface gravity.
For models with $\log g \gtrsim 2$, one cannot obtain low velocity, cool winds.
If the waves damp, the gas temperature rises above 10^5K and a tenuous
coronal wind results. With no damping, wind velocities are far too high.

Since the CS-STL-transition region emission boundaries occur in
the HR diagram near $\log g \sim 1$ -2, this naturally suggests that the same
wave energy used to drive the wind can account for coronal heating as
well. The disappearance of coronae for $\log g < 2$ is then the result of
increasing wind density and dissipation lengths, which permit the gas to
cool radiatively to lower temperatures. In this regard, the real solar
analog of the hypothesized stellar Alfven wave effects may not be high-
speed solar wind streaming, but heating in coronal holes. The theory
suggests that the radiative energy losses in extended chromospheres
per unit surface area are of the same order of magnitude as the surface
energy fluxes involved in solar coronal heating, a prediction which can
be tested by future observations.

We therefore arrive at a picture of stellar winds which is closely
tied to solar behavior. This model is qualitatively different from the
dust-driven wind theory, in which the base of the wind is cool (T $\lesssim$ 2000K).
The wave-driven winds all have warm inner regions. The changing
responses of stellar atmospheres as a function of gravity result in substantial variations in wind properties without requiring large changes in
the available energy and momentum fluxes.

Two other qualitative predictions of this theory may have future
importance. One would expect that a mechanism sensitive to surface
magnetic field structures would result in time-dependent winds. The
wind models also show that the waves may have sizable amplitudes, and
so might account for the observed turbulent broadening of lines (cf.
Boesgaard 1979). Much more work remains to be done before such suggestions can be made reasonably quantitative.

We have obtained apparent observational confirmation of some of
these ideas from a high-dispersion IUE study of the hybrid star α TrA
(K4II)(Hartmann, Dupree, and Raymond 1980b). We have detected

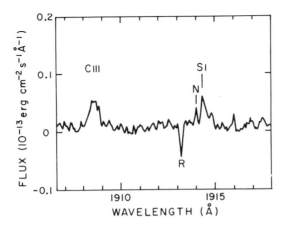

Fig. 8. High dispersion line profiles of S I and C III emission in α TrA. R marks a geometric fiducial mark, and N denotes a radiation noise spike. The S I emission, formed at the base of the wind, has a line width close to that of the instrumental profile; the C III emission, with a FWHM $\sim$ 100 km s^{-1}, is broadened by wind expansion. From Hartmann, Dupree, and Raymond (1980b).

line broadening in the "transition-region" lines, which increases with increasing excitation. The FWHMs of Si III and C III are $\sim$ 100 km s^{-1} (Fig. 8); the FWHMs of the C IV doublet lines are $\sim$ 150-200 km s^{-1}. Since the terminal velocity of the wind from the Mg II CS component is 84 km s^{-1}, we interpret this broadening to reflect wind expansion, with C IV formed as the wind nearly reaches its terminal velocity (see also Dupree 1976 for similar inferences).

There are several important implications of these observations. The significant expansion inferred at Si III and C III temperatures indicates that the wind has been accelerated significantly at temperatures far too low for thermal expansion to be important; but the hot regions are so extensive that it is hard to see how dust can be dynamically effective. Simple emission-measure modelling shows that continuing mechanical heating must occur to account for the high temperature emission. If these regions are really expanding at high velocities, it seems inevitable that the hot gas is occurring in open magnetic field configurations. This in turn suggests that coronal heating mechanisms employing closed magnetic loops are not relevant to hybrid stars; wave damping is an attractive alternative. (Note that this has a solar analogy; Munro and Withbroe (1972) showed that the corona cannot balance the transition region losses in coronal holes, from which most of the solar wind originates. Another heating mechanism must be invoked to balance the radiative losses from these open field regions.)

This suggestion would link the cool supergiant stars with their low-velocity circumstellar shells, the hybrid atmosphere stars, and the Sun in a continuous progression of increasing wind temperatures with increasing stellar gravity.

In conclusion, the study of mass loss from late-type stars appears to be entering a promising new phase, in which we have begun to consider the behavior of cool giants and supergiants from a solar perspective, a perspective which contains important implications concerning the nature of solar activity.

This work was supported in part by NASA grant NAG5-5.

REFERENCES

Adams, W.S., and McCormack, E., 1935, Astrophys. Journ. 81, p. 119.
Altenhoff, W.J., Oster, L., and Wendker, H.J., 1979, Astr. Astrophys.
      73, p. L21.
Belcher, J.W., 1971, Astrophys. Journ. 168, p. 509.
Belcher, J.W., and Davis, R.L., 1971, Journ. Geophys. Res. 76, p. 3534
Bernat, A.P., 1977, Astrophys. Journ. 213, p. 756.
Bernat, A.P., Hall, D.N.B., Hinkle, K.H., and Ridgway, S.T., 1979,
      Astrophys. Journ. (Letters) 233, p. L135.
Bernat, A.P., Honeycutt, R.K., Kephart, J.E., Gow, C.E., Sanford,
      M.T. II, and Lambert, D.L., 1978, Astrophys. Journ. 219, p. 532.
Bernat, A.P., and Lambert, D.L., 1976, Astrophys. Journ. 204, p. 830.
Boesgaard, A.M., 1979, Astrophys. Journ. 232, p. 485.
Bowers, P.F., and Kundu, M.R., 1979, Astron. Journ. 84, p. 791.
Deutsch, A.J., 1956, Astrophys. Journ. 123, p. 210.
_____, 1960, in "Stellar Atmospheres," ed. J.L. Greenstein
      (Chicago: Univ. of Chicago Press), p. 543.
Dupree, A.K., 1976, in "Phys. des Mouvements dans les Atmos.
      Stellaires," ed. R. Cayrel and M. Steinberg, Editions du C.N.R.S.,
      p. 439.
Dupree, A.K., and Baliunas, S.L., 1979, IAUC, no. 3435.
Dupree, A.K., and Hartmann, L., 1980, in "Stellar Turbulence, I.A.U.
      Colloq. 51," ed. D.F. Gray and J.L. Linsky (New York: Springer-
      Verlag), p. 279.
Gehrz, R.D., and Woolf, N.J., 1971, Astrophys. Journ. 165, p. 285.
Gillette, F.C., Low, F.J., and Stein, W.A., 1968, Astrophys. Journ.
      154, p. 677.
Gilman, R.C., 1969, Astrophys. Journ. (Letters), 155, p. L185.
Goldberg, L., 1979, Quart. Journ. Royal Astr. Soc. 20, p. 361.
Goldberg, L., Ramsey, L.W., Testerman, L., and Carbon, D., 1975,
      Astrophys. Journ. 199, p. 427.
Hagen, W., 1978, Astrophys. Journ. (Suppl.) 38, p. 1.
Haisch, B.M., Linsky, J.L., and Basri, G., 1980, Astrophys. Journ.
      235, p. 519.
Hartmann, L., 1980, in proceedings of special workshop on "Cool Stars,
      Stellar Systems, and the Sun," SAO Special Report, in press.

Hartmann, L., Dupree, A.K., and Raymond, J.C., 1980a, Astrophys. Journ. (Letters) 236, p. L143.
_____, 1980b, submitted to Astrophys. Journ.
Hartmann, L., and MacGregor, K.B., 1980, Astrophys. Journ., in press.
Hollweg, J.V., 1973, Astrophys. Journ. 181, p. 547.
Jacques, S.A., 1977, Astrophys. Journ. 215, p. 942.
_____, 1978, Astrophys. Journ. 226, p. 632.
Kopp, R.A., Holzer, T.E., 1976, Solar Phys. 49, p. 43.
Kwok, S., 1975, Astrophys. Journ. 198, p. 583.
Linsky, J.L., and Haisch, B.M., 1979, Astrophys. Journ. (Letters), 229, p. L27.
Mullan, D.J., 1978, Astrophys. Journ. 226, p. 151.
Munro, R.H., and Jackson, B.V., 1977, Astrophys. Journ. 213, p. 874.
Munro, R.H., and Withbroe, G.L., 1972, Astrophys. Journ. 176, p. 511.
Parker, E.N., 1963, "Interplanetary Dynamical Processes," (New York: John Wiley and Sons), p. 256.
Reimers, D., 1974, Mem. Soc. Roy. Liege, 6e Serie, 8, p. 369.
_____, 1975, in "Problems in Stellar Atmospheres and Envelopes," ed. B. Baschek, W.H. Kegel, and G. Traving, (Berlin: Springer-Verlag), p. 229.
_____, 1977, Astron. Astrophys. 57, p. 395.
Salpeter, E.E., 1974a, Astrophys. Journ. 193, p. 579.
_____, 1974b, Astrophys. Journ. 193, p. 585.
Stencel, R.E., 1978, Astrophys. Journ. (Letters), 223, p. L37.
Stencel, R.E., and Mullan, D.J., 1980, Astrophys. Journ. 238, p. 221.
Sutton, E.C., Storey, J.W.V., Betz, A.L., Townes, C.H., and Spears, D.L., 1977, Astrophys. Journ. (Letters) 217, p. L97.
Weymann, R., 1962, Astrophys. Journ. 136, p. 844.
Wilson, O.C., 1960, Astrophys. Journ. 131, p. 75.
Wilson, O.C., and Mot, H.A., 1954, Astrophys. Journ. (Suppl.) 1, p. 1.
Woolf, N.J., and Ney, E.P., 1969, Astrophys. Journ. (Letters) 155, p. 677.

ASPECTS OF LONG-TERM VARIABILITY IN SUN AND STARS

A. Skumanich and John A. Eddy

High Altitude Observatory
National Center for Atmospheric Research
Boulder, Colorado USA 80307

Abstract. Evidence of long-term solar variability is reviewed, including historical data and the tree-ring record of radiocarbon. Epochs of suppressed activity like the Maunder Minimum are shown to be frequent occurences of the last several thousand years, but without no obvious period of recurrence. Weak evidence exists for the 11-year cycle as early as Medieval times, although with insufficient accuracy to establish long-term phase stability.

The evidence for secular and evolutionary changes in the dynamo state of solar-like stars is reviewed. It is found that Ca+ emission in main sequence stars is age dependent and decays as the square-root of main sequence age. Chromospheric emission and hence magnetic flux is linearly related to rotation rate in main sequence stars ; this is believed to represent a fundamental dynamo relation. Angular momentum loss by magnetic braking, produced by erupted magnetic flux that 'escapes' past the Alfvenic point in coronal winds may produce the observed secular changes. It is suggested that the dynamo increases in strength in hotter stars and that the break in the angular momentum distribution is due to a 'closing off' of the coronae of early F-type stars.

The amplitude of the dynamo cycle is similar in old and young stars and is color independent. Chaotic fluctuation appears to increase as one moves 'up' the main sequence to hotter stars. A color and age independence of dynamo periods argues that the convective zone thickness is the determining factor. The recent suggestion that the dynamic state of the atmospheres of solar-like stars is age dependent is not found to be supported by the observational data. Extant non-linear calculations of the dynamo

349

R. M. Bonnet and A. K. Dupree (eds.), Solar Phenomena in Stars and Stellar Systems, 349–397.
Copyright © 1981 by D. Reidel Publishing Company.

yield a scaling law that can be tested against stellar observations.

1. INTRODUCTION

More and more, modern astrophysics has been concerned with the inconstancy of the universe, and on all time scales. The fixed stars that patterned earlier concepts of the cosmos have been replaced in modern thought by burning masses of gas that slowly evolve and more rapidly change : all stars, including the Sun, are surely variable, when examined closely and for sufficient time.

In this review, we consider several aspects of long-term, or secular variability in the Sun and other stars. We are concerned most specifically with variability of solar and stellar dynamos, and the observable manifestation of these phenomena on time scales longer than about 25 years -- an arbitrary limit chosen to de-emphasize the well-studied 11- and 22-year periods of solar variability, which we summarize briefly here for completeness.

The 11-year cycle of solar behaviour is most commonly illustrated in the well-known plot of annual-averaged sunspot numbers (Figure 1). It is better described, however, when the added dimension of solar latitude is included (Figure 2). The familiar butterfly diagram shows the temporal and spatial variability of magnetic active regions in the photosphere and chromosphere, for which sunspots serve as the most easily observed tracer. Thus the patterns shown for sunspots in Figures 1 and 2 apply equally well to chromospheric plages, active prominences, flares, and solar radio bursts, and coronal transient events.

It is not yet known whether or by how much the total luminosity of the Sun (or solar constant) varies in the course of the 11-year solar cycle, although the most recent and precise measurements from the Solar Maximum Mission (1) would imply a probable solar-cycle variation at the level of a few tenths of one percent -- enhanced at times of low activity and suppressed when solar activity increases. This extrapolation, based on an apparent anti-correlation with projected sunspot area found in the first months of SMM data, is an admittedely bold extrapolation

It is well known that the flux of solar radiation that lies outside the 6000K envelope of blackbody radiation (i.e., the ultra-violet, x-ray and radio radiation from the upper chromosphere and corona) is positively correlated with observable solar activity, with the greatest sensitivity at the shortest wavelengths and in the longest radio waves (2). The same is true of variation with the solar cycle, although in this case the real

range of variability is less well established. These irradiance
changes can be readily explained as the result of local, activity-
related increases in the temperature and density of the chromosphere and corona, where short (and long) wave radiation originates. The sign of these changes in irradiance would seem to oppose the total flux changes in the solar constant, if the solar
constant changes are simply anticorrelated with sunspots as the
preliminary SMM data suggests.

Features of solar behaviour for which an 11-year periodicity
is less certain, or unknown, include the occurrence of coronal
holes, the flux and velocity of solar wind particles (as measured
in the ecliptic plane), the character of solar rotation, and the
total magnetic flux. The occurrence of concentrated magnetic regions called "bright points " in the corona and transition region
is thought to be anticorrelated with sunspot number ; moreover,
these features are found over the entire solar surface rather
than restricted to the equatorial zones of activity shown in
Figure 2 (3).

More basic to solar behaviour, but far less easily detected,
is the fundamental, magnetic cycle of 22 years (4). This double
sunspot cycle is classically observed in the polarities of leading spots of sunspot groups and in the aggregate polarity of the
solar poles ; in each case the dominant polarity in each hemisphere is observed to reverse in succeeding 11-year cycles. In
the present solar cycle, for example (≠ 21, whose maximum was
attained in early 1980) the magnetic polarity of leading sunspots in the northern hemisphere is positive, and trailing spots
are negative. In the preceding cycle (≠ 20), the opposite was
true. Poleward migration of equatorial, large-scale magnetic
regions gives the solar poles the appearance of a composite, dipole field of dominant sign -- opposite in the two hemispheres
-- that also reverses each 11-years, though not always simultaneously at the two poles. This clear reversal of the composite
field of the Sun suggests a way that solar activity could modulate terrestrial responses, through a weak interaction with the
more permanent field of the earth ; indeed, evidence has been
cited for possible terrestrial effects of similar period that
might be linked to 22-year magnetic cycle of the Sun. Such a
connection, if real, would be expected to be of minor import,
however, because of the small energies involved. All of the more
energetic features of solar variability obey the simpler half-
cycle of 11-years, if they follow cyclic variability at all.

Evidence is sometimes cited for a 22-year period in the amplitude of annual sunspot number, in the sense that successive
cycles often alternate between higher and lower peak values.
Possible examples can be found in the run of sunspot numbers
shown in Figure , as can an equal number of exceptions. If the

effect is real it may be diluted by possible longer term trends
in the envelope of the curve.

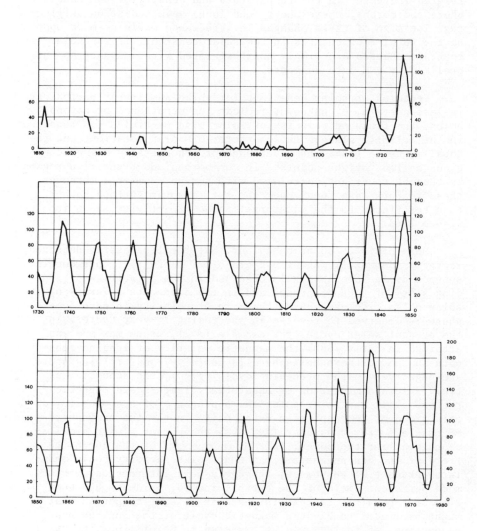

Figure 1.   Annual mean sunspot number,
1610-1979 from Waldmeier (5) and Eddy (6)
with more recent data supplied by NOAA.

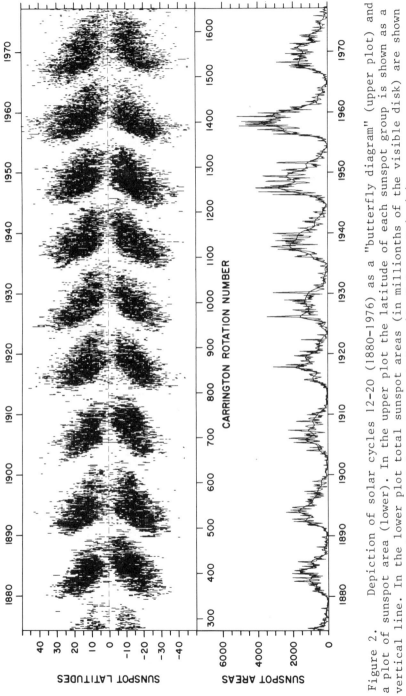

Figure 2.   Depiction of solar cycles 12-20 (1880-1976) as a "butterfly diagram" (upper plot) and a plot of sunspot area (lower). In the upper plot the latitude of each sunspot group is shown as a vertical line. In the lower plot total sunspot areas (in millionths of the visible disk) are shown as monthly averages. (Courtesy Science Research Council of Great Britain).

2. EVIDENCE FOR LONG-TERM VARIABILITY OF THE SUN

2.1. Telescopic Sunspot Record

What is known best of solar variability are day-to-day changes and the 11- and 22-year cycles. Yet it would seem unlikely that an object as massive and complex as the Sun would be constrained to vary significantly only in this narrow, high-frequency range. Our ignorance of longer-term variation probably only reflects the limited period over which we have made direct physical measurements. The longest series of direct observations of the Sun is that of sunspot number, for which we now have a more or less continuous record 370 years long (Figure 1). It is far from a uniform data set. Sunspot numbers become qualitatively accurate only after about 1850, when real-time observations began and corrections could be made for observing conditions (5, 6). Nevertheless, in even this short and limited sample of the best data we find a number of signs of secular change.

Most obvious are changes in the overall level of activity, after the 11-year cycle is removed, as seen in the envelope of annual sunspot numbers (Figure 1). Between the turn of the present century and about 1960, for example, there was a steady rise in the amplitude of successive sunspot cycles, as measured in the peak numbers attained. Since the middle of the 20th century and through all the modern age of space observations we have studied the Sun at a broad peak of this long-term envelope, with probable bias in what we have learned. Another clear feature of the historical sunspot record is the marked, 30-year depression in sunspot amplitude between about 1800 and 1830. Although this earlier depression is deduced from historically reconstructed sunspot data that are less reliable, it is confirmed, as we shall see, in the independent history of auroral incidence and in the wholly objective record of tree-ring radiocarbon.

The most emphatic case of recent secular change in the level of solar activity is the Maunder Minimum in sunspot numbers, a period between about 1645 and 1715 that has been subjected to especial scrutiny (6-13). As in the case of the shorter period of solar depression between 1800 and 1830, the reality of the Maunder Minimum rests as much on independent, indirect and proxy data as it does on direct sunspots counts ; its clear presence in the tree-ring radiocarbon record (14) leaves no doubt that it describes a period of solar behaviour quite unlike the subsequent behaviour of the Sun. Yet periods like the Maunder Minimum are not uncommon in the longer view. As we shall see, two similar features of solar behaviour have been clearly identified in the well-analyzed radiocarbon record between AD 1000 and 1645 (14), and a larger number in the coarser record from even earlier time (7, 15). What we know of their occurrence

suggests : (1) that periods of prolonged depression like the
Maunder Minimum are relatively common features in the secular
behaviour of the Sun perhaps as typical as the last 250 years
of "normal" solar behaviour; and (2) that there is no simple
secular cycle, analogous to the 11-year cycle, that yet seems
obvious in their pattern of recurrence. That is, we cannot yet
predict when the next Maunder Minimum will occur, other than
watching, in a regressive way, the trend of successive 11-year
peaks, as one might watch for winter, or the coming of spring.
Indeed, it now seems possible that these prolonged events in the
life of the Sun may well be aperiodic, and unpredictable, for
their known characteristics can be successfully modelled by passing random noise through a narrow 22-year filter (16).

2.2. Auroral Data

   The historically reconstructed record of sunspot numbers
now reaches back to the introduction of the telescope, or about
1610. Other, direct and indirect data from historical sources
allow us to extend the study of solar behaviour much earlier in
time. By far the best of these are historical accounts of the
Aurora Borealis and Australis, catalogs of which have been com-
piled by a number of workers (17). The earliest auroral reports
are from the 5th century, B.C., however, reports do not become
frequent enough to be of much use until medieval times, about
A.D. 500.

   Even the most complete records of the aurorae require critical interpretation before they can be substituted as an indirect measure of solar activity. Auroral visibility is a strong
function of geomagnetic latitude, and this needs to be taken
into account. Auroral occurrence has commonly been taken as a
direct index of sunspot-associated flare activity, since they
are caused by particle precipitation that is induced by energetic particles from the Sun. In fact, the most complete, modern
auroral data show a sometimes poor correlation with sunspot
number, in the sense that aurorae are often reported when the
Sun is quiet. Moreover, although aurorae show a weak 11 year
variability, peak occurrence lags the corresponding maxima in
sunspot number by several years. With the modern discovery of
coronal holes and their role in modulating the flux of solar
wind particles at the earth (18) we have come to understand
some of the distinction and to look at auroral data more critically as a solar index. It is not surprising, for example, that
some aurorae were reported during the Maunder Minimum. Aurorae
can be induced by sunspot- related flare ejections or by suns-
pot-unrelated coronal holes, and these two causal mechanisms can
produce quite opposite, or even competing, end effects. The two
classes of aurorae may be distinguished in terms of the energy
spectrum of incident particles from the Sun, and a resultant

difference in the lower latitude bound of the auroral oval (19).

Feynmann and Silverman (20) have recently demonstrated that
the secular suppression in sunspot numbers during the 1800-1830
period (sometimes called the "Little Maunder Minimum") is supported by two coincident changes in auroral reports during the time:
a significant drop in the total number of aurorae reported and
a poleward shift in the latitude at which aurorae were seen. The
second effect is consistent with a reduction in the number of
solar-activity-related aurorae during the time and a shrinkage
of the auroral oval to a size characteristic of coronal-hole-
caused aurorae of weaker particle energies. In a more comprehensive study of historical auroral reports, Siscoe (17) has shown
that the Maunder Minimum and the preceding Sporer Minimum in
sunspot activity are clear features of the historical auroral
record, documented both as a dearth of real-time reports and in
contemporary, retrospective accounts. Observers in the New World
and the Old, for example, expressed wonder at the dramatic increase in aurorae, particularly at lower latitudes, beginning about
1715, at the end of the Maunder Minimum.

2.3. Pre-Telescopic Sunspot Reports

Weaker evidence of secular changes in solar activity may be
found in the thinner history of pre-telescopic reports of sunspots (21). The richest sources are those from China, Japan and
Korea, where for centuries occasional sightings of dark features
on the Sun were recorded in local or dynastic histories (21-26).
The dynastic histories of China are a particularly valuable source of naked-eye sunspot data, and in recent years much work has
gone into their reanalysis to search for evidence of the pre-
telescopic sunspot cycle or of secular trends. The earliest reports begin in the first century, B.C., but until about 300 B.C.
they are very sparse. On the average for the 1500 years of records about one sunspot per decade was noted. The record is far
from uniform, however.

Large sunspots can be seen with the naked eye under suitable viewing conditions, as at sunrise or sunset or through thick,
atmospheric haze. Thus it is an interesting fact of history that
through nearly two millenia, when hundreds of such objects were
seen in the Orient, only three were reported in Europe (21). The
dramatic difference is sometimes attributed to climate (27) :
the meteorology of eastern china is characterized by a persistent
continental haze from the Gobi desert that may have made viewing
the Sun easier or less painful. On the other hand the difference
may be one of culture and philosophy : the concept of the sky in
the Orient may well have been less inhibited and prejudiced than
that of early Christian Europe.

In pre-telescopic data from China, Japan, and Korea one finds prolonged periods when sunspot reports were unusually frequent or reduced in number. Thirty-two were reported, for example, in the 12th century in China. Such intensifications, or gaps, are possible evidence of secular changes in the level of solar activity, but such conclusions must be made with care. The possibility of encouragement, suppression, or manipulation of sunspot reports is real, particularly since dark spots on the Sun were taken as portents in court astrology. Stephenson and Clark (21, 22) have been especially careful in interpreting possible trends in these earliest sunspot data ; among tests they apply is whether suspected trends were common to more than one country.

The Maunder Minimum is a distinct feature of all compilations of Oriental sunspot reports, including the recently uncovered data from local (as opposed to dynastic) histories in China (25,11). Weaker evidence of an earlier period of sunspot depression -- the Sporer Minimum, between about AD 1400 and 1550) -- has also been noted in compilations of naked-eye sunspot reports, as has evidence of a prolonged period of higher than normal reports in the 13th century. Both of these earlier features also appear in auroral histories and in the record of tree-ring radiocarbon.

2.4. Tree-Ring Radiocarbon

By far our most useful and reliable source of information on the earlier behaviour of the Sun is the quantitative measure of radiocarbon retained in trees. Carbon-14 is produced from nitrogen atoms in the upper atmosphere of the earth by the impact of high-energy galactic cosmic rays (28, 29) which, in turn, are modulated by solar activity. When the Sun is more active, solar magnetic field irregularities scatter incoming galactic cosmic rays, and the production of radiocarbon is diminished. When the Sun is less active, as during the Maunder or Sporer minima, more galactic cosmic rays impinge on the atmosphere and radiocarbon production is enhanced. Through this mechanism, cosmic rays at the earth are modulated in the course of the 11 year solar cycle by about 20 % and radiocarbon production by about 10 %,as determined by measurements of the cosmic ray neutron flux (30) and by radiocarbon production models (14).

Trees preserve a history of the amount of radiocarbon in the atmosphere in the past, and hence provide an indirect records of the level of past solar activity (31). Atmospheric radiocarbon, in the form of CO_2, enters trees through photosynthesis, and is there locked in datable rings of annual growth. A critical test of tree-rings as a diary of past radiocarbon variability is found in the clear indication, in tree-rings of

the modern era, of times of atmospheric detonation of nuclear
bombs, with an abrupt increase in radiocarbon.

Continuous tree-ring chronologies now extend about 8000
years into the past (32), offering a proxy record of the level
of solar activity extending to the time of the late stone age.
A number of geochemical laboratories have collaborated to recons-
truct this long record and it is under continual improvement and
refinement. The solar history derived from tree-ring radiocarbon
will be blurred by the process of diffusion and CO_2 reservoirs
in atmosphere and oceans, and confused by other effects that
also modulate cosmic ray flux or perturb the natural C-14:C-12
ratio. Strongest of these other competing processes is that of
secular change in the strength of the earth'magnetic moment ;
during the available 8000 year span of tree-ring data the strength
of the terrestrial field has slowly increased to a maximum and
fallen again, following a quasi-sinusoidal modulation of about
the same period. The field strength has varied nearly a factor
two in amplitude, as reconstructed from paleomagnetic studies
(33). These slow, secular changes are readily identified in the
tree ring record of radiocarbon, and are easily removed.

What remains is clear evidence of secular changes in the
overall level of solar activity, on time scales of 100 to 1000
years, that punctuate the full 8000-year record. Among them is
a dramatic increase in the C-14:C-12 ratio at the time of the
Maunder Minimum in sunspot activity, corresponding to a modelled
increase in C-14 production of about 20 % at the time. Also
found in recent, high-precision analyses of tree ring radiocarbon (14,33) are a substantiation of the earlier-inferred Sporer
Minimum,in solar activity (AD 1410-1530), a similar, "Wolf
Minimum" between about AD 1280 and 1340, and an apparent period
of high solar activity, possibly higher than the modern level,
between about AD 1100 and 1250.

Corresponding features are found in auroral records and
with less certainty in naked-eye sunspot counts. The most recent
radiocarbon analyses of Stuiver and Quay (14) clearly demonstrate
a 1:1 correspondence of radiocarbon fluctuations with the envelope of sunspot number, throughout the period of overlap of the
two records. For example, the "Little Maunder Minimum", noted
in sunspot reconstructions and auroral behaviour between about
1800 and 1830, is a definite feature of radiocarbon in tree-rings.

What has not yet been identified in the proxy record of
tree-ring radiocarbon is the 11-year solar activity cycle, which
is severely depressed by the high-frequency filtering of the
global CO_2 reservoir. For the present this remains a difficult
challenge, although there is hope that with improved analysis
technique the weakened signal of the 11-year sunspot cycle may

yet be detected. Were this achieved, valuable information on the
long-term phase stability (or instability) of the solar dynamo
would come to light (34). This in turn could extend the temporal
baseline over which we have reliable information on dynamo phase
stability by almost a factor 30.

A quantitative analysis of the radiocarbon record suggests
that at the time of the Maunder minimum, Sporer minimum, and
Wolf minimum the solar modulation of galactic rays was considerably less than at the times of the lowest sunspot numbers today
(14, 33). This important finding was noted earlier in weaker
evidence by Forman, Schaeffer, and Schaeffer (35) in a study of
cosmic-ray-produced Argon-37 and Argon-39 in recent meteorite
falls. Argon-39 has a half-life of about 270 years and was anomalously abundant in every one of the 30 samples studied by Forman, Schaeffer, and Schaeffer. To explain the high abundance they
were driven to postulate a very low level of cosmic-ray modulation averaged over the last 500 years or so : the Maunder and
Sporer minima were insufficient to explain so low an averaged
modulation unless these epochs described periods of solar activity more quiet than at the minima of the modern 11-year cycle.
Such findings suggest that (1) the historical sunspot number is
at best a limited and non-linear index of solar activity, particularly at the low-activity end of the scale, (2) the zero of the
sunspot number index may lie above the lowest levels of solar
activity, measured in solar wind conditions and cosmic ray modulation, and (3) the limits of solar activity encountered during
the last century or two of modern study of the Sun most probably
sample but a part of the more extended range through which the
Sun has varied in the past millenium.

2.5. The Question of the "Gleissberg Cycle"

Gleissberg (36) and much earlier Wolf called attention to
an apparent period of 50 to 80 years in the amplitude or envelope of annual sunspot number. A period of this approximate length
indeed appears as a weak signal in modern, power spectrum analyses of the sunspot number series, though its reality can be
questioned because of possible aliasing and the serious amplitude uncertainty in all but the most recent 130 years of record
(5). This so-called "Gleissberg Cycle", can be detected by a
casual examination of the sunspot record (Figure 1) ; two enhanced minima appear, roughly 80 years apart, in about 1810 and
1890. The 20th century rise in the level of solar activity, noted
earlier may indeed be a feature of such a secular period. Siscoe
(17) finds evidence of a roughly 80-year period in auroral reports, traceable to the time of the Middle Ages. Stuiver and
Grootes (33) on the other hand, fail to find any similar period
in their power-spectrum analysis of the last 1000 years of precision radiocarbon data which, one might think, should be the

better and more objective index of secular, solar trends. Thus
we are left with a question of whether the period is real or not,
and, if real, whether it is felt in all indices of solar activity, including the solar wind and the mechanism of modulation of
cosmic rays.

The current sunspot cycle (≠ 21 ; see Figure 1) is off possible interest in this regard, for it now seems to challenge the
notion of a simple, 80-year cycle in sunspot numbers. The preceding cycle (≠ 20, with a maximum in 1969) was lower than the one
before it ; in the context of the six or seven preceding cycles,
cycle ≠ 20 seemed to confirm the prediction of a smooth Gleissberg envelope that would span the canonical set of 7 or 8 eleven-
year sunspot cycles. The higher maximum of the present cycle
(with a probable peak sunspot number of about 160) disrupts the
smooth trend toward a lower envelope. If this interruption is a
reflection of an overriding 22-year alternation in cycle amplitudes, and the level of solar activity continues toward a Gleissberg law in succeeding cycles, then the current Gleissberg cycle,
minimum to minimum, will be more than 100 years long. Wolf's original estimate of the length of the cycle was 50 years ; Gleissberg and others proposed 70 to 80 years, based on a longer sample. Proponents of an 80 year period in sunspot numbers will probably tolerate cycles of variable length, justified on the basis
of a corresponding variability of the length of the conventional
sunspot cycle of 11-years.

On the other hand, the apparent elasticity of the cycle and
our limited knowledge of its prior existence makes it possible
that what we deal with is an accidental phenomenon, as suggested
by the random noise model of Barnes, Sargent and Tryon (16),
rather than any physical property of the sun itself.

2.6. Pre-Telescopic Evidence of the 11-Year Cycle of Solar
Activity

As noted earlier, the temporal baseline over which we have
records of the 11-year cycle of solar activity is important to
questions of the physics of the solar dynamo. Ideally, we should
like precise specification of the phase of the cycle for as long
a period into the past as possible. Yet, until very recently,
our only unequivocal evidence for the persistence of the modern
11-year cycle was the limited, direct record of telescopic sunspot observations. In this record (Figure 1), the cycle can be
traced with certainty only to about 1715. In analyses of earlier
data, including historical telescopic observations, naked-eye
reports, and auroral records, an 11-year cycle was commonly presumed and imposed, rather than demonstrated. Wolf (37), for
example, forced sunspot data during periods of apparent dearth
(such as the 17th century) to fit a model with nine maxima in a

century, as has Schove (38) in his extensive reconstructions of past sunspot and auroral records.

It has been proposed in contrast (39), that we could as well defend an opposite view, namely that the 11-year cycle of solar activity did not exist before 1715, and that until proven guilty, the Sun could be presumed innocent of such behaviour. Several more recent studies of historical data have effectively challenged this possibility, and in the process have put the continuous, 11-year cycle on a much firmer footing.

Clark and Stephenson (22) in a critical study of naked-eye sunspot reports from China, Japan, and Korea make a strong case for a likely 11-year cycle in pre-telescopic sunspot data. While the number of reports is inadequate to establish an unequivocal, continued cycle, it is recognizable at the times of richest (most frequent) records, specifically in the early 12th century, when there are 25 oriental reports in a 55-year span.

A weaker, more statistical case has been made by Wittmann (24) and a group from the Yunnan Observatory in China (26) for the persistence of the 11-year cycle through much or all of the entire, 2000-year record of naked-eye accounts. Added evidence for a persistent, 11-year cycle of solar activity has come from a recent and more thorough re-examination of the 2000-year auroral record by Siscoe (17), who finds evidence of an 11-year cycle at the times of richest auroral data from the Middle Ages. In both auroral data and naked-eye sunspots the cycle becomes perceptible only during isolated epochs of higher counts. This limitation, compounded by the variable length of the sunspot cycle in the modern records makes it impossible, at present, to use these early pre-telescopic records for extended tests of phase continuity.

Tantalizing evidence of a possible pre-Cambrian sunspot cycle has recently been found by Williams (40) in a geological analysis of layered sandstone deposits. Williams finds distinct, repetitive patterns in sediment layers laid down in South Central Australia about 670 million years ago. Patterns in the color (deposit content) and width of individual horizontal layers repeat in a mean cycle of 11.3 bands for the best sample of several hundred bands. Williams contends that individual bands, each about one millimeter wide, represent annual sediment which results from summer melting and runoff in pre-Cambrian times. An even stronger, alternating pattern of 22 bands (11 narrow, 11 wide) is also evident. If this is indeed the mark of the Sun, the connection between solar activity and the weather that is implied is much stronger than any effect noted today. What is in doubt, however, is whether the Australian sediment bands are really annual deposits. This is a hard question to answer.

3. EVIDENCE FOR SECULAR AND EVOLUTIONARY CHANGES IN THE DYNAMO
FOR SOLAR-LIKE STARS

3.1. Magnetic Activity (Mean)-Age Relationship

It has been known for some time that magnetic areas on the
Sun are associated with emission reversals in the core of the H
and K resonance lines of Ca II. Indeed, according to Skumanich
et al., (41), it would appear that the strength of the emission
reversal as measured in a 1 Å band centered on the absorption
line, the so-called K-index, varies linearly with magnetic flux
without regard to sign. Consequently a measurement of the K-
index from the entire visible disk of the Sun (star), yields,
after substraction of a zero point, a measure of the total magnetic flux erupted through the solar (stellar) photosphere. In this
lecture we are concerned with the analysis of K-index measurements in solar-type stars and what can be learned about magnetic
dynamo activity in stars of different ages and colors.

The first stellar evidence for secular changes in the mean
dynamo state of solar-like stars came from the observation by
Delhaye (42) of a kinematic difference between dwarf M stars
with Ca II emission and those without. Vyssotsky and Dyer (43)
put this difference on a quantitative basis and showed that emission dwarfs were younger (i.e., had smaller space velocity dis-
persions which was due to their condensation from a galactic gas
with a decayed level of turbulence, refer Spitzer and Schwarzschild (44), Larson (45)) than non-emission dwarfs. This age effect in the Ca II emission strength for solar-like stars was
confirmed and extended by O.C. Wilson and his collaborators, for
references see Wilson and Wooley (46). Because of its interest,
we present a brief review of the analysis that lead to Skumanich's
result.

The basic data consisted of photo-electric K-index measurements of field stars by O.C. Wilson (48). These are illustrated
in Figure 3. The quantity plotted is the K-index of each
star measured in units of a 'neighbouring' continuum but uncor-
rected for continuum bandwith. Being a ratio of counts, it is
dimensionless but when calibrated has the dimension of Angstroms.
To obtain a quantity which represents the total magnetic flux
(of both signs) we define a HK-excess, ΔF, as indicated on the
figure, for each star. The lower envelope is thus assumed to
represent the zero magnetic flux (no chromosphere) state. This
may not be quite correct (49) but is of consequence only for
stars significantly older than the Sun.

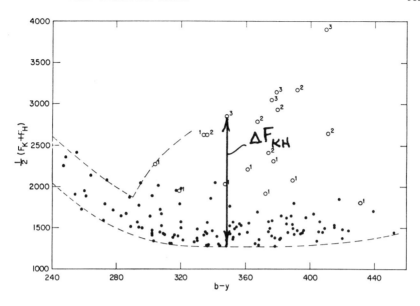

Figure 3. Mean instrumental H-K flux plotted against b - y. Open circles are stars for which emission was seen on 10 A mm^{-1} plates, and numbers beside them are eye-estimated emission intensities. The definition of excess flux is given by the arrow. (After Wilson (48)).

In order to eliminate the color dependence introduced by the use of the neighbouring continuum as a photometric standard we can apply a relative correction factor that represents the variation of the neigbouring continuum with color. We use data as determined for the Hyades main-sequence stars by Wilson (50). This is shown by the ℓ.4 curve in Figure 4. Multiplying Wilson's values by ℓ.4 one obtains $\Delta LCa^{+}/L_{0}(B-V=0.6)$, the ratio of the chromospheric Calcium luminosity in units of the neighbouring continuum luminosity the star would have if at B-V=0.6. This quantity represents a direct measure of the magnetic flux (in relative units) and yields the <u>intrinsic</u> dependence of this quantity on color or effective temperature. In order to use an accurate temperature parameter for each star, the Hβ photometric index (51) was used in conjunction with Stromgren 4-color photometry wherever possible.

The result of this analysis is shown in Figure 5 where Wilson's data are plotted. The location of the Sun is indicated by the cross. Indicated by the solid line is the mean Hyades relation, Wilson (50). This line is an isochrone and represents

the intrinsic variation of the magnetic flux along the main-
sequence. It represents an observational constraint on the
variation of the mean erupted dynamo field with effective temperature (or mass). We note that there is no evidence for a
turnover or disappearance of chromospheric activity to at least
β = 2.66 (b-y=.29, B-V=.45 or F5V) if not earlier, cf. Dravins
(52).

Additional isochrones (broken lines) are also indicated and
were obtained from the Hyades by multiplication by a constant
factor as indicated near each curve. Thus we see that the Sun
falls on an isochrone which is 1/4 the intensity of the Hyades.
Another group of stars that include the Ursa Major nuclear and
stream stars are fitted by an isochrone which is 7/4 more intense than the Hyades. Knowing the age of the Sun, the Hyades, and
the Ursa Major stars permits one to determine a quantitative
Ca II luminosity (magnetic flux) -age relation. The result is a
square-root of age relation.

We note from this figure that the stars seem to separate
into three age groups. One associated with the Ursa-Major cluster,
one with the Hyades (cf. The suggestions of a Hyades (Sirius)
group by Eggin (53)) and one with stars similar to the Sun in age.
This result is similar to that found more recently with more
complete statistics by Vaughn and Preston (54). No explanation
is available. It is unlikely that it indicates three separate
phases of star formation, cf. Wilson (55).

When Skumanich discussed the above results in 1970 with Bob
Kraft, who was in Boulder on a sabbatical, Kraft suggested that
his observations with Greenstein of the Ca II emission equivalent
widths in late-type Pleiades and Hyades would allow one to add
the Pleiades to the relation. His estimate of 2.7 as the intensity ratio of Pleiades to Hyades coupled with the age of the
Pleiades confirmed the square root relation.

Skumanich confirmed Kraft's estimate in the way as was done
in Figure 5. This also allows one to obtain a calibration of
Wilson's flux values (at least for the late-type stars). Figure
III.4. shows a plot of the emission equivalent widths (in A) for
the Pleiades and Hyades versus corrected B-V colors. Also plotted
is the Hyades flux isochrone with different calibration factors
as indicated by the number on the right of each isochrone. 61
Cyg A and B are represented by the lowest triangles on the figure. When scaled by 3.5 (upper triangles) they yield a smooth
continuation of the late (cool) portion of the Hyades isochrone.
This procedure not only defines the continuation of the isochrone, Wilson (50), but also yields the result that 61 Cyg has an
intensity relative to the Hyades of 1/3.5, i.e., the system is
similar in age to the Sun.

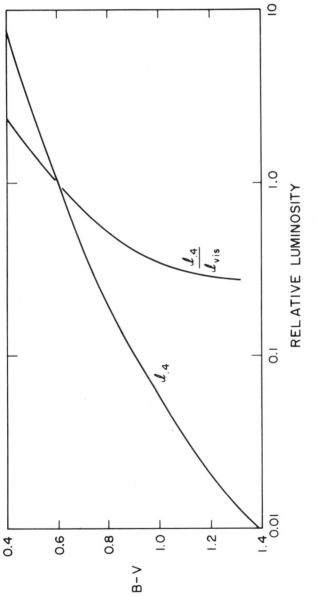

Figure 4. Reference continuum luminosity for main-sequence stars for HK flux measurements and its relative to visual luminosity.

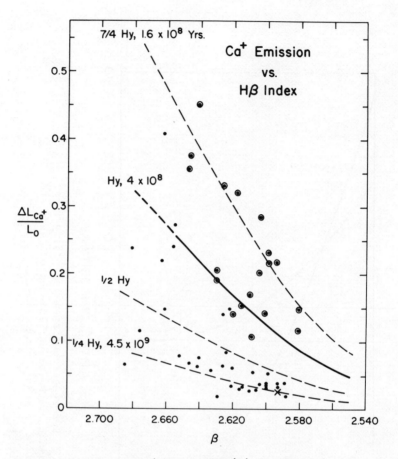

Figure 5. Excess (chromospheric) Ca II luminosity (in arbitrary units, L_0) versus Hβ index. The large circles indicates emission was seen on 10Amm^{-1} plates. x indicates the Sun.

It would appear that the curve 2xHyades gives an 'eyeball least squares' fit to the emission equivalent widths of the Pleiades while 0.6 x Hyades fits the emission equivalent widths of the Hyades (note the large possible errors in B-V estimates from spectral type). This yields a Pleiades to Hyades ratio of 2/0.6 = 3.3, somewhat larger than Kraft's original estimate.

This value coupled with the previous results from Figure 5 yielded the Ca II intensity data indicated in the familiar Figure 7 (47). This shows that the chromospheric Ca II luminosity and, hence, erupted dynamo magnetic flux decays as the inverse square-root of time as the star "cooks" on the main-sequence.

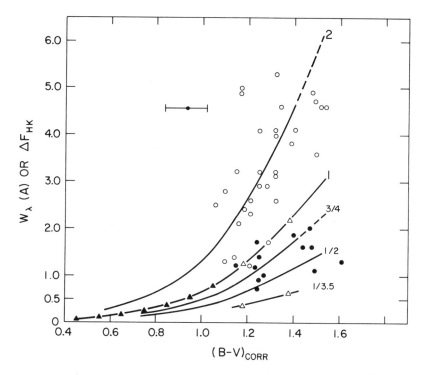

Figure 6. Emission equivalent widths versus color
(corrected for extinction) for Pleiades (open circles)
and Hyades (filled circles). Also plotted to the same
numerical units is the excess HK flux for the Hyades
(filled triangles) as well as constant multiples of
this curve. Open triangles represent 61Cyg. The error
in estimates of B-V is indicated.

This relation has been confirmed by Blanco et al., (56).
However, they claim that the Sun (based on data other than
Wilson's) fits the relation only if the Sun were covered by
plages (which Blanco claims are a factor 40 times brighter than
the quiet Sun). That this is in error is evident from the extended observations of the Sun by Wilson (49) which show at
most only a 50 % rise in chromospheric flux (i.e., in ΔF) from
minimum to maximum activity, refer also to Dravins (57) and
White and Livingston (75).

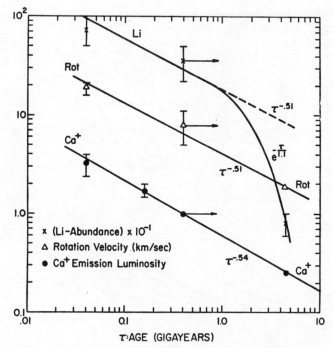

Figure 7.   Ca II excess emission, rotation, and
lithium abundance versus stellar age (Skumanich (47)).

3.2. Magnetic Activity (Mean) - Rotation Relationship

During the '60's the picture gradually emerged that chromospheric activity and rotational braking by magnetic forces
were related. Schatzman (58) advanced the idea that magnetic
braking was responsible for the strong drop in angular momentum
along the main sequence as one enters the region where stars have
deep convective zones and display an apparent onset of chromospheric and hence dynamo activity, Wilson (59). However, as mentioned above, there appears to be no direct evidence, as yet,
for a turnover or disappearance of chromospheric activity in the
F stars. It may very well be that the sharp rise in rotational
velocities near spectral types F5, Kraft (60), may be due to the
fact that magnetic braking becomes <u>inefficient</u> in the F stars
and not because of a decrease of dynamo activity. Rather it is
that the amount of surface magnetic flux which 'escapes' with
the stellar wind becomes a small fraction of the total because
of a high $\beta (= NkT/(B^2/2\mu))$ condition over most of the stellar
surface, refer Pneuman and Kopp (61), in other words the dynamo
and the erupted flux may be strong enough to contain most of the
corona. Whether this is consistent with the X-ray variation along
the main sequence, Vaiana et al., (62), requires knowledge of the

energization mechanism of the corona and its variation along the
main sequence. Nonetheless the detection of strong X-ray corona
as early as B-V=0.30 in the Hyades main sequence, Stern et al.
(63), confirms the lack of turnover in the chromospheric data.

The case for braking in late-type stars received significant confirmation by the observations of Kraft (60) who showed
that stars of the same spectral type but with stronger Ca II
emission had higher rotational velocities. This implies a magnetic flux (Ca II excess)-rotation relation. Using the rotation
data $(4/\pi)$ x $(V_R \sin i)$, for G-stars quoted by Conti (64), Skumanich found that the relationship was linear as is indicated
in Figure 7. Both quantities were found to decay proportiona-
tely. This result represents another observational constraint
on the dynamo mechanism.

Our current picture is as follows : The interaction of rotation and convection generates surface magnetic flux which 'produces' chromospheric activity and Ca II emission and couples to
a corona and solar wind to produce magnetic braking. Magnetic
braking by that fraction of the surface flux which 'escapes' past
the Alfvenic point causes rotation to be age-dependent which
causes the generated magnetic flux -- and hence the dynamo strength
-- to be age-dependent. Evidence for this scenario is found in
the fact that the solar wind does carry a net flux of angular
momentum (Brandt (65), Weber and Davis (66) and Pizzo et al.(67)).
In addition, a theoretical argument which uses the linear field-
rotation relation and leads to the square-root braking curve as
shown in Figure 5 was discussed by Durney and Stenflo (68)
among others, and is given in this volume by Roxburgh (see also
Section 4.1).

The significance of lithium abundance in late type main-
sequence stars with regard to rotational braking was first pointed out by Goldreich and Schubert (69). They argue that magnetic
braking would slow the convective zone differentially with respect to the radiative core. At some time this differential rotation would become unstable and the resulting turbulence generated
at and below the convection zone would carry lithium to sufficiently high temperatures for nuclear burning to deplete it.
Thus quantitative estimates of the lithium depletion time or its
temporal evolution would contribute to an understanding of the
Goldreich-Schubert instability and to the rotational (and hence
dynamo) evolution of solar-like stars. Yet to be addressed is
the effect of a 'burst' of such turbulence on the dynamo itself.
Would it inhibit the dynamo and give rise to a Maunder-like minimum ? Would it allow or suppress additional magnetic modes e.g.
the bipolar mode represented by X-ray Bright Points, Golub (3) ?

Conti (64) and Van Den Heuvel and Conti (70), among others, discussed the lithium depletion rate and argued that the decay proceeded exponentially starting from the Pleiades, using the Van Den Heuvel age for the Hyades -- end of arrow in Figure III.5. Skumanich's analysis of the same Li data is shown in this figure. It would indicate that an exponential burn, i.e., turbulent mixing, turns on sometime after the star passes the Hyades age. This would imply that the Sun may still be experiencing Goldreich-Schubert (G-S) turbulence. However this is hard to reconcile with the Ca II emission which appears normal at this epoch unless the turbulence is intermittent, i.e., bursty. Is it possible, then, that periods of abnormally low dynamo activity in the recent past -- the various minima mentioned above -- are related to bursts of G-S turbulence ? Recent measurements of the rotation velocity of main sequence field stars, Smith (71) and Soderblom (72), allow an update in the magnetic flux (Ca II excess)-rotation relation. In Figure III.6 we investigate this relation with the additional rotation data for G-stars (F8V or GOV except where noted) measured or quoted by Smith. Rather than the chromospheric flux, F, as ordinate we use the square of the inverse ratio of the chromospheric flux of the star (Wilson, various publications) to that of the Hyades at the same color,

$$(\Delta L_{Hy}/\Delta L)^2 \quad \Xi (\Delta F_{Hy}(B-V)/\Delta F_{star}(B-V))^2$$

$$\Xi (\Delta L_{Ca^+}(Hy)/\Delta L_{Ca^+}(star))^2$$

This also serves the role as an age parameter which, if the square-root law holds, represents the ratio of the age of the star to that of the Hyades.

For the clusters we plot $(4/\pi)$ <$V_R \sin i$> (= <V_R>), (cf. Conti (64)) ; for individual stars, $V_R \sin i$; for αCen, (Boesgaard and Hagen (73)) and the Sun, V_R. For late type-stars, see legend, we also identify by the symbol x the value the star would have if, preserving angular momentum, its radius were that of a GO star. With these values one can consider the ordinate as representing angular velocity.

The solid line shows the previous linear relation between magnetic flux (Ca II excess emission) and angular velocity. We note that the Ursa Major star, π'UMa, falls near the predicted position for this cluster. For field stars whose chromospheres are weaker (older) than the sun, the velocity errors are of the order of 1 km/sec, Smith (71). It would appear that the bulk of the observed field stars do not confirm the linear relation that now includes, besides the Sun, αCen, and 61 CygA (and possibly Groombridge 1830). Unfortunately, all these stars are near the detection limit of 2.5 km/sec, Smith (71), and may be affected by some systematic methodological and/or instrumental effect. The vertical line at $(\Delta L_{Hy}/\Delta L)^2 = 25$ represents the approximate 'age' of the universe. Errors in the zero point, F°_{HK} (= lower envelope in Figure 3), or evolutionary effects may allow stars to fall to the right of this line.

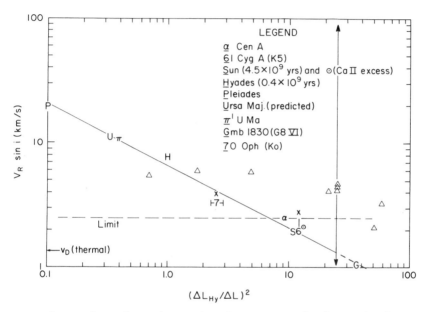

Figure 8. Rotation – Ca II excess emission relation. The solid line is a linear relation. The legend gives the meaning of each symbol used. For a square-root Ca II decay law the ordinate is log (time). S is the Sun plotted at its age, while Θ represents the actual Ca II value.

It would appear, if the data are correct, either that most of the older field stars are anomalously fast rotators (some may be evolved early types) or that the Sun, Smith (71)), as well as α Cen and 61 Cyg A are anomalously slow rotators. Unlike Smith, we feel that the Sun is normal. The more comprehensive data set of Soderblom should elucidate this issue.

In summary we claim that the observational data indicates that the total erupted mean magnetic flux $|\phi|$ (of both signs) for the dynamo in main-sequence solar-like stars, varies such that

$$|\phi| \; \alpha \; \omega \; (T_{eff}, t) \; T_{eff}^{m-n}$$

and

$$\omega \; (T_{eff}, t) \simeq \omega_1 \; T_{eff}^{n} \; / \; \sqrt{t} \; , \quad t \geq t_1$$

where ω is the angular velocity. The quantities ω_1, n, and t_1 have yet to be determined, observationnaly or otherwise, while m can be obtained from Figure 4 once the Hβ-index is converted into T_{eff}.

3.3. Magnetic Activity (Variance) – Age Relationship

In addition to the mean state, variations occur and their period, amplitude, and where observable, their spatial scale are of interest to an understanding of the dynamo mechanism. It is only recently that K-index data have become available, Wilson (49), Vaughn (74), that address the issue of age and color (T_{eff}) effects in dynamo cycles. Wilson (49) as well as White and Livingston (75) have also made a variance study of the Sun.

Vaughn (74) has classified stars whose K-index variance have been followed for more than a decade into two age groups on the basis of their location in a (HK-flux)-color diagram. The variance properties of these two groups are shown in Figures 9 and 10. We have moved the 61 Cyg system to the 'old' star group on the basis of its HK-flux excess (cf. Figure 6). Vaughn (74) gives additional justifications. The HK-flux is plotted versus time with each star arranged in an increasing color sequence, numbers indicated in each column are B-V colors. We have identified the stars that are plotted in the rotation diagram, Figure 8, with the parameter $(\Delta L_{Hy}/\Delta L)^2$ indicated in parentheses.

Vaughn (74) concludes from these figures that the principal difference between the two groups is the presence in the young group of chaotic behaviour even in those rare cases where cyclic-like behaviour appears (refer, 70 OphA in Figure 4. The older stars are quieter and show smooth cycles similar to the solar cycle particularly as one proceeds to the later types. The variance appears to be larger in the young group and also appears to increase to the later types. However, if one takes Wilson's seasonal fluctuations (averaged over essentially all observing seasons) and converts them to percentage fluctuations of the excess (or chromospheric) flux, ΔF, i.e., $(\sigma_{\Delta F}/\Delta F)$ where $\sigma_{\Delta F}$ is the fluctuation expected for a single observation, one finds that the low flux stars (either because of old age or early spectral type) are factors of three or more times more erratic.

One finds, for example, the Sun has $\sigma_{\Delta F}/\Delta F$ = 19 % compared to the 6 % for late-type stars. Thus one has to be careful about zero point effects. It would appear that one finds an increase in dynamo "activity" as one moves 'up' the main sequence to early types. More study is needed to discuss age effects but it is clear that impressions from Figures 9 and 10 can be misleading.

When one considers the ratios of chromospheric activity-cycle range (max-min) to mean flux and range to excess flux (ΔF) one obtains the results as given in the following table.

	RANGE	RANGE
OLD :	MEAN	EXCESS
Sun (G2)	.10	.50
HD81809 (G2)	.20	1.0
61 Cyg A (K5)	.34	.54
YOUNG :		
70 Oph A (K0)	.38	.64

It would appear that the range to excess ratios are similar throughout. It is evident from HD 81809, however, which is quite similar to the Sun as far as mean Ca II flux is concerned, that the data sample is inadequate and that the extant data certainly requires more careful analysis. Nonetheless the above table indicates that the range in the dynamo magnetic flux cycle to mean magnetic flux (range to excess) is less than a factor of two and may be independent of age and spectral type for stars later than G2. We note that Golub et al.(76) present evidence that for the Sun the total average rate of magnetic flux emergence varies by less than a factor or two throughout the cycle, cf. also Howard (77).

A color and age independent result seems to be indicated for dynamo periods also. The period of cycles in young stars, cf. 70 OphA, would appear to be of the same order as in the old stars and, as Vaughn noted, they are also independent of spectral type. All this would be understandable if the thickness of the convection zone were to be determinate for the dynamo period, Stix (78). The thickness of the convective zone changes only slightly with spectral type for stars later than G0 and should be independent of age for stars on the main sequence. The age independence is corroborated by the suggestion by G.E. Williams (see Section 2) that the Sun had an 11-year and 22-year cycle

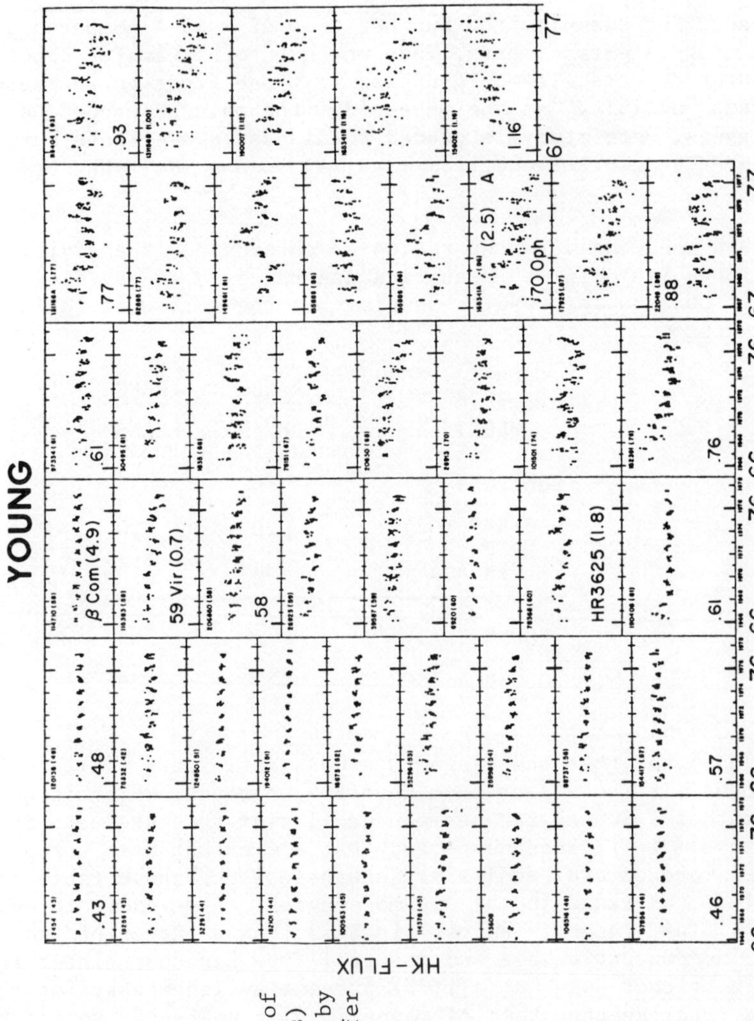

Figure 9. The variation of HK-flux versus time (years) four young stars arranged by ncreasing color index, refer values in lower or upper left (after Vaughn (74)).

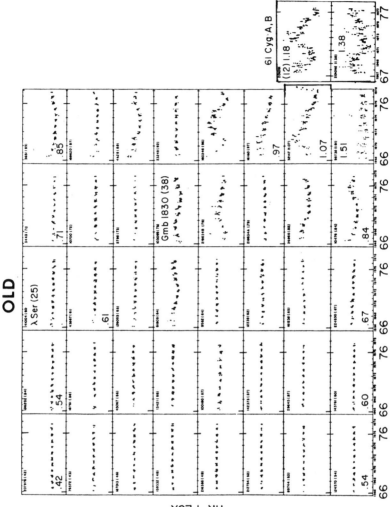

Figure 10. The variation of HK-flux versus time (years) for old stars arranged by color, refer values in lower or upper left (after Vaughn (74)).

$(2/3) \times 10^9$ years ago, even though this is still a long way from
the Hyades phase of 4.0×10^9 years ago.

The fact that cyclic behaviour is not universally observed
raises the issue of the relative contribution of different magnetic features such as network, plages, and spots in the integrated
Ca II emission. In addition, other modes of flux eruption besides
spots, such as X-ray bright points (XBP) may play a more important role in the magnetic state of young stars. The fact that the
number of XBP's are 180° out-of-phase, Golub (3), with spot number may help to mask spot-plage cycles in integrated Ca II emission. Finally, Maunder-minima like states must also play a role.
Many questions remain and only now is the data bank, e.g., observations of X-ray coronae, of rotational modulation of Ca II emission etc., becoming available to help one address these issues.

3.4. Other Age/Dynamo Related Effects

Let us assume that the macroturbulent broadening parameter
derived from stellar spectral lines, Gray (79), is a signature
of the dynamical state of stellar photospheres. As such it bears
information on the nature of motions driven by the convection
zone. Thus secular changes in these motions are of interest as
they imply changes in the convective zone and hence of the energization of the dynamo. In addition if these motions (or some
sub-spectrum) energize the chromosphere and hence Ca II emission,
then secular changes in these motions would confuse secular magnetic flux changes due to rotational braking.

Smith (71) claims to find a decay of macroturbulence with
time. We examine this result using $(\Delta L_{Hy} / \Delta L)^2$ as an age parameter, refer Section 3.1. Figure 11 presents a plat of Smith's
macroturbulence values -- the stars are symbolized as in Figure 8. The turbulence parameter for the Sun observed at the
center (limb) of the disk (80) is indicated by S_1 (S_0) at its
observed age. The result of integrated disk (flux) observations,
reduced by Gray (81), is indicated by the solar symbol at its
observed flux excess value. The reference line is drawn to exaggerate a dependence.

If, following Smith, "we put little weight on the large macroturbulence derived for π'Uma", then the argument for an age
dependence is marginal at best. Soderblom (72) confirms a lower
value for π'Uma. We feel that no dependence exists. In any respect the conclusion "that macroturbulence can be used as an
intermediate-scale age indicator for G stars which are too old
to be dated on the basis of their Li content, K-emission, or
rotational velocities", Smith (71), is not supported by Figure 11.

We mention for completeness the suggestion by Thomas (82)
that variations in the Sun's diameter and surface temperature
may be due to volumetric variations induced by variations in
magnetic buoyancy produced by changes in total magnetic flux.
Other suggestions, Foukal (83), Parker (84), couple changes of
solar (stellar) luminosity to the fluctuations in properties of
the convection zone and the dynamo state. These are problems
that are only now being attacked both observationally and theoretically.

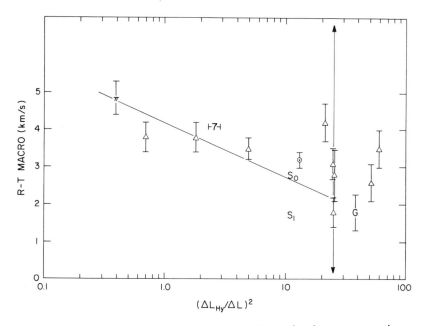

Figure 11. Macroturbulence - Ca II emission regression
            Symbols as in Figure 8.

3.5. Chromospheric-Coronal Activity and the Dynamo State in Evolved Stars

In the course of his study of stellar chromospheres Wilson
(85) noted that the spectroscopic binaries (SB) HR1099 (V711 Tau
= ADS2644A = HD22468, P = 2.8 days) and ξUMaB (ADS8119B = HD
98230, P = 4 days) found as companions in visual binary systems
had enhanced Ca II emission compared to that predicted for their
spectral type from the intensity seen in their coeval companion.
He also noted that these SB's had significantly widened lines,
presumably by <u>rotation,</u> for their spectral type. Wilson suggested
that the enhanced rotation might play a role and, possibly, the
presence of the <u>close</u> companion.

    In a seminal paper dealing with the chromospheres of SB
stars, Young and Koniges (86) showed that evolutionary effects
appear to play a role in the strength of chromospheric activity
in the sense that, for a given orbital period, the more evolved
systems were likely to have enhanced Ca II emission. This is
shown in Figure 12 which shows the HR diagram for SB's with
orbital periods of the order or less than 200 days (the RS Canum
Venaticorum (RS CVn) class of SB's, D. Hall(87)). The higher the
period (inscribed numbers) the farther from the main sequence
before enhanced emission (filled circles) occurs.

    The existence of an evolutionary effect in enhanced emission
stars is further supported by the apparent correlation between
their absolute magnitude and period, Young and Koniges (86). This
correlation, as given by Young and Born (88), is shown in Figure
13. The   least squares fit of $L_v \, \alpha p^{1.3}$ is indicated by the
line. Young and Koniges argue that a Roche lobe effect where the
evolved star fills some critical fraction ($\ll 1$) of the Roche
radius would yield a 4/3 power law.

    However, no such evolutionary effect was discernable in the
Young and Koniges emission equivalent width data for Ca II. We
have plotted this data in Figure III.12. We note that the detection threshold was quite high, at about Wilson's I = 3, so that
only enhanced emission states, i.e., only the upper envelope of
chromospheric activity, are represented here.

    Since more evolved systems have later spectral types it is
imperative that spectral type effects in the background continuum be removed before SB's of different periods are compared.
Skumanich (89) has made a preliminary effort in this direction
by taking the local (Ca II HK) background continuum luminosity
to visual luminosity ratio, $\ell.4/\ell_{vis}$, for the Hyades main-
sequence stars, refer Figure 4, as a correction factor. After
multiplication by this factor, one obtains the L_{CaII}/L_v ratio
for the SB system. The result is shown in Figure 15. A power-
law relation, $p^{-2/3}$, is indicated by the straight line and
appears to represent the upper envelope of activity -- i.e., the
'saturated' state -- of SB systems. This relation (Skumanich
(90)) differs from that of Skumanich (89) in that the Ca II
emission from both components was included.

    Also plotted are four RS CVn stars for which no quantitati-
ve Ca II data was available but which had Mg II data, Basri and
Linsky (91). These are indicated by triangles and were normalized so that the SB HR1099 (P = 2.8 days) fell on the indicated
relation. Wilson assigned I = 5 to this G5 type SB, i.e., the
emission was above the highest point in the continuum between H
and K. Note that the RS CVn star λ And (HD222107, P = 20.5 days)
is represented by three points, the Ca II emission of Young and

Koniges is indicated by the open circle, the Mg II emission, by
a triangle while the lowest point is Ca II data derived from
estimates by Young and Born (88). We note that this star, as
well as others (e.g., HR 1099) in our sample, undergoes flaring
episodes so that the scatter may be due to this effect or there
may be an error in the Young and Koniges value. Nonetheless it
is obvious that additional data for SB's in the quiescent state
is needed to confirm the $P^{-2/3}$ relation. We further note that O
Dra (HD175306, P = 138 days) and αAur (HD34029 , P = 104 days)
define the relation at the high period end.

For this latter SB there is some controversy about the major
source of the activity -- the more evolved primary or the rapidly rotating secondary -- both have Mg II emission, refer Ayres
and Linsky (92). Indeed the issue of which component, if not
both as in σCrB (HD146361, P = 1.1 days), is the active member
introduces a degree of uncertainty in our analysis of the SB
systems. An evolutionary (Roche lobe) effect, Popper and Ulrich
(93), would require the more evolved component (or both in a
equal mass system like σCrB) to be the active source while for
a strict rotational effect (as in main sequence stars) it would
be the faster rotator. The fact that some double line objects
(HD13480A = ιTri, P = 14.7 days as well as αAur) have what
appears to be a dominant Ca II emission feature that comes from
the weaker line (presumably due to higher rotation) component
may indicate the latter effect as argued by Ayres and Linsky (92).

Since essentially all the RS CVn type of SB's seem to be in
synchronous rotation (refer D. Hall 'Activity in Binary Systems
-- this volume) the role of the orbital period is indeterminate,
It may represent an evolutionary effect or a pure rotation effect
(or both). Only in the case of λAnd is the rotation asynchronous
(P_{rot}= 54 days) with the orbital period (P = 20.5 days). We note
that this value of P_{rot} would cause λAnd to fail to satisfy the
Ca II intensity-rotation relation suggested by Figure III.13. A
similar result would occur for αAur if Ayres and Linsky (92)
conclusions hold. It may be that these two systems have just
arrived at the 'saturated' state and have yet to adjust their
angular (orbital + rotational) momentum. In this same vein we may
ask if this is also the case for ξUMaB (P = 4 days), ζAnd (HD
4502, P = 17.8 days) and εUMi (HD153751, P = 39.5 days) which
may be "approaching" or have "left" the 'saturated' state.

With the result that L_V α$p^{1.3}$, we have that the Ca II
chromospheric emission for SB's on the upper envelope varies
such that

$$L_{CaII} \propto P^{2/3}.$$

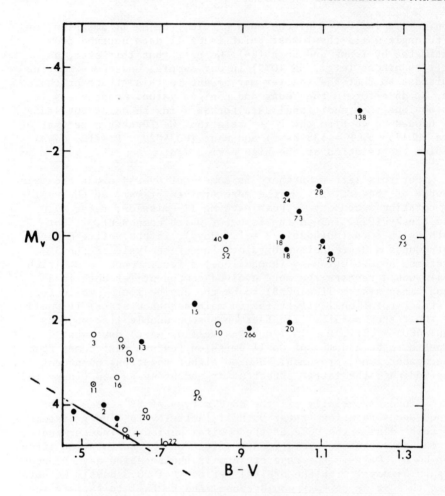

Figure 12.   Color-magnitude location of SB's with
P $\lesssim$ 100 days. Periods (days) are inscribed near each
point. Filled circles have enhanced emission. The main
sequence is shown by the line, + represents the Sun.
(After Young and Koniges (86)).

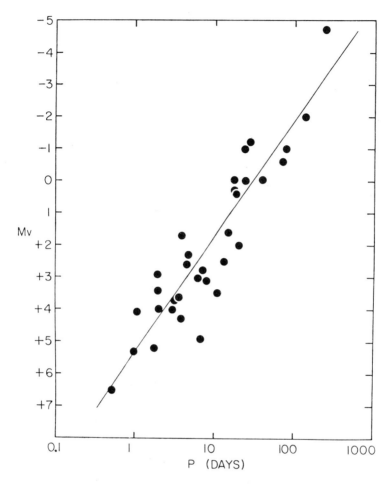

Figure 13. Absolute magnitude-period correlation for SB's
wit enhanced Ca II emission (After young and Born (88)).

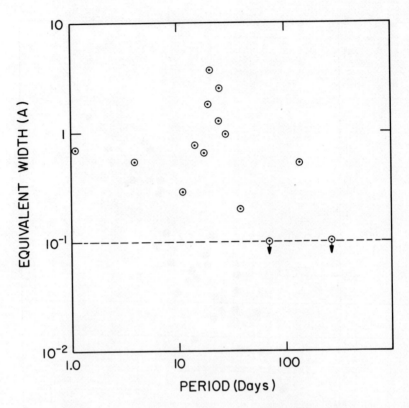

Figure 14. Equivalent width versus period for enhanced emission SB stars. The approximate threshold is indicated by the dashed line.

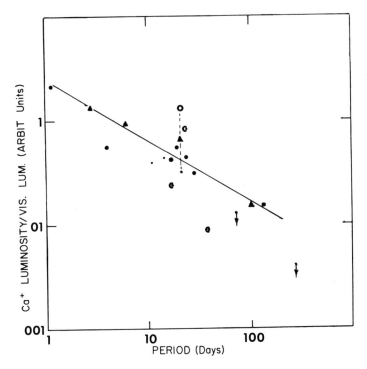

Figure 15. Ca II to visual luminosity ratio (arbitrary units) versus period for enhanced emission SB stars.

If in these stars Ca II emission is a measure of the photospheric magnetic flux, $|\phi|$, and if the component stars are in synchronous rotation so that the rotation rate $\omega(P,t) = P^{-1}$ and is time independent, the above relation implies that the magnetic flux varies such that

$$|\phi| \propto \omega^{-2/3},$$

a dynamo relation which, if true, is quite different from that for main sequence stars. This should not be surprising since the P-sequence in Figures 12 ans 13 is also an internal structure and hence a convective zone sequence. This effect must appear in the above relation. If we argue that a linear ω dependence should hold as in main sequence stars then we would have a structure factor (the T_{eff} term) varying as $P^{5/3}$. Does the convection zone factor in the dynamo increase in 'strength' along the P-sequence by such a law for these stars? The presence of other solar-like phenomena in these stars, e.g., spottedness (photometric wave), X-ray coronae, flares (optical, radio, X-ray) supports the idea of a dynamo induced activity.

We note here that the above results are sensitive to the $\ell.4/_{vis}$ correction factor. It is unlikely that the main-sequence relation holds exactly for the giants along the P-sequence because of increased opacities in the blue. If this ratio were to drop as fast as the $\ell.4$ curve one would find that L_{CaII}/L_{vis} would be better fit by a P^{-1}, however the Mg II data appears to rule out such a steep slope.

Since X-ray coronae measure another aspect of a stars magnetic flux, for example, its spatial extent (via the emission measure) and thermal energy content, it is of interest to consider the relation between X-ray luminosities and Ca II emission, Skumanich (89).

From the available X-ray data (Walter et al. (94), Walter (95)), we form the X-ray luminosity to visual luminosity (combined components) ratio for the systems studied in Figure III.13. The plot of this ratio against period is shown in Figure III.14. as well as a $P^{-2/3}$ law passing through HR1099(P = 2.8 days). A few systems without chromospheric data are included for completeness and are indicated by small dots. The Spectroscopic Binaries systems are identified as in the Ca II diagram, Figure III. 13. Note that ζAnd (P= 17.8 days), and εUMi (P = 39.5 days) fall below the relation as in the Ca II data.

The $P^{-2/3}$ relation as an upper envelope of X-ray activity seems to be ruled out. A P^{-1} relation (lower curve), as proposed by Skumanich (90), seems more appropriate. A similar result has been derived by Walter (96). We note that αAur is included in such a fit. The suggestion by Skumanich (90) that O Dra would prove to be an X-ray source has been confirmed, Walter (95). However its X-ray to visual luminosity ratio falls a factor 5 below the P^{-1} relation and would appear to be anomalous, although we note that this detection was the weakest for this class of stars.

Making use of the L_V (P) relation we find that

$$L_{XR} \propto P^{1/3} \propto (L_{CaII})^{1/2} \propto |\phi|^{1/2},$$

i.e., a square root relation between coronal and chromospheric states. We note that the use of X-ray luminosities derived directly from the flux data with trigonometric estimates of distances can introduce appreciable errors in the X-ray luminosity-period relation, Skumanich (90). It is for this reason that we use luminosity ratios. We further note that one would have a linear X-ray to Ca II emission relation if $L_{CaII}/L_V \propto P^{-1}$, or a quadratic relation if $L_{CaII}/L_V \propto P^{-7/6}$, although these laws are are unlikely (see above).

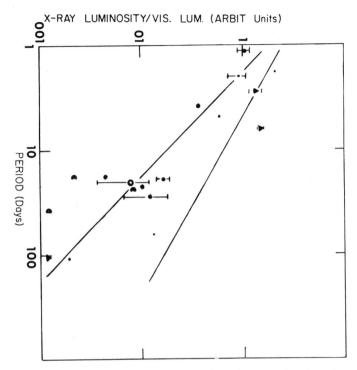

Figure 16. X-ray to visual luminosity ratio (arbitrary units) versus period for enhanced Ca II emission Spectroscopic Binaries stars. The small dots indicate stars without Ca II (or Mg II) observations. Symbols as in Figure III.13.

As an example of the interest in such a relation consider the following. If we argue that the coronal temperature in closed field regions is fixed by the radiation efficiency of the plasma to some common value then the X-ray luminosity is fixed by the emission measure, i.e.,

$$L_{XR} \; \alpha \; \overline{N}_e^2 \; V,$$

where V is the total volume of the X-ray corona. Now Burl et al. (97) have given some evidence that for active regions on the Sun

$$N_e^2 \; V \; \alpha \; (\overline{B}^2 \; V)^{1/2} = B_{rms} \; V^{1/2}$$

On the other hand for the total flux (absolute value) we have $|\phi| \alpha |B_n| A$ where B_n is the photospheric field and A is the area covered by these fields. If we take (B_n/B_{rms}) to be a constant, then we have that

$$L_{XR} \ \alpha \ \overline{N}_e^{-2} \ V \ \alpha \ |\phi| \ V^{1/2} \ A$$

a linear X-ray-CaII relation if the geometric factor -- a measure of the mean extent of the active regions -- is a constant. In the case of a non-linear relation, one will have to allow some of the terms taken constant, e.g., the geometry and/or the coronal temperature, to vary with flux. With regard to the possible scaling of the geometry of active regions a preliminary argument has been presented by Shore and Hall (98).

On the other hand, the magnetic heating argument by Golub et al. (99), see also Belvedere et al. (100), leads to $L_{XR} \ \alpha$ $|\phi|^2$ /A. The current state of the Ca II (or Mg II) data does not allow us to make definite conclusions. In any respects the above arguments illustrate our contention that the X-ray-Ca II relation contains information on the geometric scale and thermodynamic role of magnetic fields in dynamo driven stellar activity.

The issue of how Spectroscopic Binaries evolve to or away from the 'saturated' state or of the possible 'asymptotic' states dealing with the upper or lower period regimes are interesting questions that we shall not discuss here except to note that current evidence from IUE, Dupree (101), does seem to indicate that the W Ursa Major (type W) binaries represent an asymptotic state for SB's with periods less than 1 day.

4. CONCLUSIONS AND OBSERVATIONAL IMPLICATIONS

4.1. On the Nature of Magnetic Braking

In extant discussions, refer Durney (102), of the square-root braking law of G-type stars use is made of a spherically symmetric model of the coronal wind with a radial (monopole) magnetic field. That this is a gross simplification is evident from both X-ray and white light photographs of the corona. Here one sees closed field regions that do not participate in the coronal wind expansion. Thus the presence of such inhomogeneities raises doubts about the derived consistency between the square-root braking law and the linear magnetic flux-rotation relation.

A recent example of a non-radial geometry is the MHD calculation by Pneuman and Kopp (61). They modeled a non-radial coronal wind interacting self-consistently with a dipole magnetic field. They determined the properties of the wind expanding from an open polar region whose fractional area was determined by the

coronal plasma $\beta(=NkT/(B^2/2\mu))$. Pneuman and Kopp took $\beta = 1$ at
the pole with a consequent value (for a dipole) of 1/4 at the
equator. The angular momentum loss from such a configuration
has been calculated by Priest and Pneuman (103). We consider
these two calculations here with the point of view of elucidating the changes that occur from the spherically symmetric model.

The major effect of the equatorial constriction of the coronal wind expansion was an increase in the outflow velocity at
the surface on the open field lines. However by the time the
Alfvenic point was reached along each open flux tube the velocity profile had approached that of the spherically symmetric
(radial) solution for the same thermodynamic surface conditions.
In addition the open magnetic field lines had assumed a nearly
radial distribution that filled 4π steradians. Our own calculations also show that the mass loss rate was unchanged to within
5 %. In other words the constriction of the area of the wind by
the equatorial fields was balanced (to 5 %) by an increase in
the outflow velocity. Thus to a distant observer at or past the
Alfvenic point, the symmetric (radial) model yielded essentially
the same results as the non-symmetric dipole model. We note that
this type of smoothing is seen in the Helios data for the Sun,
Pizzo et al. (67). Presumably, as the coronal β decreases (i.e.,
for $\beta << 1$), so that only a narrow region about the poles is left
for expansion, the invariance of the symmetric representation
must break down and the mass loss rate must drop drastically.

This invariance of flows has important consequences for the
angular momentum loss. The equation for the angular moment loss,
in the absence of 'diffusion' of momentum from the core, has
the form

$$\frac{dJ}{dt} = I \frac{d\omega}{dt} = - \omega r_a^2 \sin^2 \Theta_a \dot{dm}$$

where I is the moment of inertia of the braking region (the convective zone ?), ω is the star rotation rate, and (r_a, Θ_a) the
position of the Alfvenic point for each flux tube containing the
differential mass loss dm. In the symmetric (monopole) case one
has that, since $<\sin^2\Theta> = 2/3$,

$$I \frac{d\omega}{dt} = - \frac{2}{3} \omega r_a^2 \dot{m}$$

or, making use of the definition of the Alfvenic radius,

$$I \frac{d\omega}{dt} = - \frac{2}{3} \frac{\omega}{ua} \frac{|\phi|^2}{4\pi\mu}$$

where u_a is the (radial) wind velocity at the Alfvenic radius, $|\phi|$ the _total_ magnetic flux, $|\phi| = 4\pi R^2 |B|$, and R the star radius, μ the magnetic permeability.

We calculate the amount of flux $|\phi_e|$ which 'escapes'with the stellar wind, i.e. the 'number' of open field lines, for the two cases calculated by Priest and Pneuman. Namely, the non-radial dipole case of Pneuman and Kopp with $\beta(\text{pole}) = 1$ and the radial dipole or $\beta >> 1$ case (here the field is weak enough that the solar wind produces a radial field with $B_n \propto \cos\theta$), -- both with the same distribution of surface magnetic flux. If is an insensitive function, as we claim above, then the ratio of the braking rates should be in proportion to $|\phi_e|^2$; i.e., in the ratio

$$|\phi|^2_{\beta=1} : |\phi|^2_{\beta>>} = 0.33:1.0$$

Priest and Pneuman find

$$\left(\frac{dJ}{dt}\right)_{\beta=1} : \left(\frac{dJ}{dt}\right)_{\beta>>1} = 0.28:1.0$$

The agreement is good and indicates that the radially symmetric flux-form of the braking law holds as long as one uses the 'escaped' magnetic flux $|\phi_e|$ rather than the total magnetic flux $|\phi|$. If the thermodynamics of the corona is affected by the value of β(e.g., magnetic heating effects) then u_a may also vary with $|\phi_e|$ and the simple quadratic flux dependence may break down. We note that in the isotropic or monopole case (where r_a is a constant) containing the _same total_ flux as the non-isotropic but radial dipole case the ratio of braking rates is

$$\left(\frac{dJ}{dt}\right)_{\beta>>1} : \left(\frac{dJ}{dt}\right)_{\text{monopole}} = 0.91:1.0$$

and measures the extent to which $<u_a^{-1}\sin^2\theta_a>$ differs in the two cases. We also note that Priest and Pneuman use the value $r_a = 7.7R_\odot$ instead of the correct value $r_a = 9.9R_\odot$ for their monopole case.

We conclude that the square-root braking law should hold even under conditions of partially closed coronas as long as the 'escaped' flux, $|\phi_e|$, is in constant ratio to the total flux $|\phi|$ as measured by the Ca II emission. However it may very well be that as the dynamo increases in strength as one moves up the main sequence through the F stars, the ratio $|\phi_e| / |\phi|$ drops significantly causing a sharp drop in the braking rate.

4.2. On the Nature of the $\alpha\omega$ -Dynamo

It is not our intent here to review the state of non-linear dynamo models but to discuss the confrontation of one of these models, the $\alpha\omega$ -dynamo, with stellar observations. So far the activity cycle of the Sun has been used to test and specify free parameters in such models. We have indicated above the observational constraints that appear to apply when one considers the dynamo as a function of different states of rotation and convection. If we argue that the Ca II luminosity, ΔL_{Ca^+}, measures the total <u>poloidal</u> flux, $|\phi_p| = 2\pi |B_p| R^2$, then the observations indicate that the mean poloidal field $|B_p| \alpha \omega T_{eff}^{m-n} R^{-2}$ (T_{eff}) , where the constant of proportionality is known from the Ca II emission-magnetic flux calibration (40).

The kinematic $\alpha\omega$-dynamo is described in terms of two specified flows : (1) a shear flow with a mesoscale, L, due to differential rotation, $\Delta\omega L$, characterized by a magnetic Reynolds number $R_{\Delta\omega}$ = (forcing rate/dissipation rate) = $\Delta\omega_L L^2 /\eta$, where η is the magnetic diffusivity, and (2) a small scale, ℓ, helical (due to coriolis effects) convective turbulence modeled by a 'velocity' factor $\alpha (=\omega\ell^2 /\mathcal{K}$, where $\mathcal{K}$ is the convective zone scale height) and characterized by $R_\alpha = \alpha L/\eta$. (We have also taken L as the scale for the mean magnetic field in these Reynolds numbers). The quantity $D_M = R_\alpha R_{\Delta\omega}$ is called the dynamo number. It has a critical value for the existence of steady (but possibly oscillatory) solutions which is obtained by a linear stability analysis of the induction equation. For dynamo numbers exceeding the critical, the magnetic field amplitude grows to some limiting value given by non-linear interactions on the velocity fields that drive the dynamo. It is this state that is of interest to us.

The first discussion of a scaling law for the field amplitudes is that of Roberts (104) whose heuristic argument leads to a poloidal field, B_p, given by

$$B_p \simeq \omega (\mu\rho\ell^2)^{1/2}$$

Where ρ is a characteristic density in the convection zone and μ, the magnetic permeability. Knowledge of the variation of ρ, ℓ with T_{eff} would allow a comparison with observations.

An improvement on Roberts' argument is possible by consideration of the simplest non-linear model, Stix (105). In this model a finite amplitude magnetic field is obtained by the ad-hoc cut-off of α when the field exceeds a critical value B_c. This value represents a free parameter whose scaling in terms of convective and rotational properties has to be known.

Stix's numerical resuts for the toroidal field amplitude, B_T, can be fitted by the expression

$$B_T = 0.13 \left| R_\alpha R_{\Delta\omega} \right|^{1/2} B_c \tag{1}$$

as a function of the dynamo number $\left| R_\alpha R_{\Delta\omega} \right| \geq D_M$ (critical) $=89$. The poloidal field amplitudes were not given, except for a single case where it would appear that either

$$B_P = \left| R_{\Delta\omega} \right|^{-1/2} B_T \tag{2a}$$

or

$$B_P = \left| R_\alpha / R_{\Delta\omega} \right|^{1/2} B_T \tag{2b}$$

Both expressions represent the numerics equally well to the precision of the available numbers. Finally, Stix argues that the critical field B_c must be such that its Lorentz stresses are of the same magnitude as the coriolis forces so as to suppress the helicity effect of the latter. This yields

$$B_c = (2\mu\rho u \ell\omega)^{1/2} \tag{3}$$

where u is the velocity of the convective turbulence.

If the scaling given by equation (2a) holds then we find that the poloidal field scales as

$$B_P = 0.18\omega \left(\frac{\mu\rho\ell^2 \varkappa^{-1} L}{(\eta/u\ell)} \right)^{1/2} \tag{4a}$$

i.e., we obtain a linear field-rotation relation. Further if we take $\eta = \eta_{turb} = 0.1u\ell$, cf. Roberts (104), and $=\ell$ then we obtain the final scaling law,

$$B_P = 0.6 \left(\mu\rho\ell L \right)^{1/2} \tag{4b}$$

We note that, if one takes (without justification) L $=\ell$, one obtains Roberts' result.

On the other hand, the scaling given by (2b) indicates that B_P $\alpha\omega^{3/2}$ which we feel is suspect. This scaling followed from a dimensional analysis by Steinbeck and Krause (106) who showed that

$$\frac{B_p}{B_T} = \frac{a(R_\alpha, R_{\Delta\omega})}{|R_{\Delta\omega}|}$$

where $a(R_\alpha, R_{\Delta\omega})$ is the ratio of the (maximum) amplitudes of the poloidal to toroidal eigenfunctions. Steinbeck and Krause argued that the numerical coincidence, for one of their marginally stable cases, of a_{crit} and $1/2~D_M^{1/2}$ (crit) implied that $a = 1/2~D_M^{1/2}$, or, that $B_p/B_T = 1/2|R_\alpha/R_{\Delta\omega}|^{1/2}$. We note further that Roberts (104) argues for the relation $B_p = |R_{\Delta\omega}|^{-1}B_T$. It is clear that, in both the marginally stable and non-linear problem, a fuller parameter study in the $(R_\alpha, R_{\Delta\omega})$ plane is needed.

The following heuristic derivation of Stix's result may lend some confidence in the square-root scaling law given by equation (2a). We make the <u>ansatz</u> that the effect of a flow on a magnetic field is via the square-root of its magnetic Reynolds number, R_m. In support, we cite the case of a homogeneous turbulent state which produces a field $B_{turb} = R_m^{1/2}~B$ from an ambient mean field B, Steinbeck and Krause (106). Weiss (107) found a similar result for the expulsion and concentration into flux tubes of a uniform field by two-dimensional convective rolls. (We note however that this argument is somewhat weakened by the result of Galloway and Moore (108) who found an $R_m^{1.0}$ scaling for cylindrically symmetric convection in the weak field (i.e., kinematic) regime <u>but</u> a $R_m^{2/3}$ scaling in the non-linear regime).

Thus the conversion of poloidal field to toroidal field by the differential rotation is given by ,

$$B_p \overset{\Delta\omega}{\to} B_T = |R_{\Delta\omega}|^{1/2} B_p \qquad (5a)$$

likewise, the helical turbulence acts so that,

$$B_T \overset{\alpha}{\to} B_p = \frac{1}{9}~|R_\alpha|^{1/2}~B_T \qquad (5b)$$

where the numerical coefficients are selected to fit the calculations and may be weak functions of $(R_\alpha, R_{\Delta\omega})$. For a self-sustaining state one must have, following Roberts, that,

$$(1 - \frac{1}{9}~|R_\alpha R_{\Delta\omega}|^{1/2})~B_T = 0 \qquad (6)$$

so that, in the marginally stable state with $B_T \neq 0$ one requires that the coefficient vanishes or that

$$D_{crit} = R_\alpha R_{\Delta\omega crit} = 9^2 \tag{7a}$$

which is Stix's result (to 10 %). Now for $D > D_{crit}$ and $R_\alpha < 1$, so that $B_p < B_T$, one can take $B_T \simeq B_C$ as a first approximation to the non-linear state. In this case the helical turbulence gives

$$B_p = \frac{1}{9} |R_\alpha|^{1/2} B_c \tag{7b}$$

while the differential rotation effect, equation (5a), gives us a better estimate for B_T , namely,

$$B_T = |R_{\Delta\omega}|^{1/2} B_p = \frac{1}{9} |R_\alpha R_{\Delta\omega}|^{1/2} B_c \tag{7c}$$

which is Stix's result (to 15 %). As far as we can judge (7b) gives a good estimate for Stix's B_p.

In summary, the scenario for the observational test of the $\alpha\omega$-dynamo is as follows. We must determine :

(a) $\omega_1 T_{eff}^n$, the rotation law, by (in the absence of observations) requiring $D(\omega) \geq D_{crit} (T_{eff})$,

and,

(b) $B_p (T_{eff})$ from the "magnetic scaling law", equation (4b).

This requires calculations of convective zone properties along the main-sequence as well as for the RSCVn stars. We note that analysis (a) has been made by Durney and Latour (109), and also Gilman (110), who introduce a role for the Rossby number, $R_0 = (u \ell^{-1}/\omega)$. However, it is unclear if the Durney and Latour analysis is consistent with ours. First they confuse the scales (L, ℓ) and second they take $|R_{\Delta\omega}| = |R_\alpha| = R_0 = O(1)$ whereas we require $|R_\alpha| \leq 1 \leq |R_{\Delta\omega}|$. In our view taking the Rossby number of $O(1)$ imposes a constraint on B that is unwarranted.

Acknowledgements

We wish to thank Drs. T.M. Brown, P.A. Gilman, and R.G. Athay for their careful reading of the manuscript.

REFERENCES

(1)   Willson, R.C., Gulkis, S., Janssen, M., Hudson, H.S., and
      Chapman, G.A., 1980, Science, in press.
(2)   White, O.R., Ed : 1977, "The Solar Output and its Variation"
      Colo. Assoc. Univ. Press, Boulder, pp; 1-48.
(3)   Golub, L. : 1980, Phil Trans. Roy. Soc. A 297, p. 595.
(4)   Howard, R. : 1977, Annual Reviews of Astronomy and Astro-
      physics, pp, 153-173.
(5)   Waldmeier, M. : 1961, "The Sunspot-Activity in the Years
      1610-1960," Schulthess, Zurich,
(6)   Eddy, J.A. : 1976, Science, 192, pp. 1189-1202.
(7)   Eddy, J.A. : 1977, Scientific American 236, pp. 80-92.
(8)   Link, F. : 1977, Astron. Astrophys. 54, pp. 857-861.
(9)   Gleissberg, W. : 1977, Sterne und Weltraum 7-8, pp. 229-233.
(10) Vitinsky, V.I. : 1978, Solar Phys. 57, pp. 475-478.
(11) Cullen, C. : 1980, Nature 283, pp. 427-428.
(12) Landsberg, H. : 1980, Archiv. fur Meteorologie. Geophysik
      und Bioclimatologie B. 28, pp. 181-191.
(13) Eddy, J.A. : 1980, in "Proceedings of a Conference on the
      Ancient Sun", Pepin, R.O., Eddy J.A., and Merrill, R.B.,
      Eds. Pergammon Press, in press.
(14) Stuiver, M., and Quay, P.D. : 1980, Science 207, pp. 11-19.
(15) Eddy, J.A. : 1977, Climatic Change 1, pp. 173-190.
(16) Barnes, J.A., Sargent, H.H., and Tryon, P.V. : 1980, in
      "Proceedings of a Conference on the Ancient Sun", Pepin,
      R.O., Eddy, J.A., and Merril, R.B., Eds. Pergamon Press,
      in press.
(17) Siscoe, G.L. : 1980, Reviews of Geophysics and Space Phy-
      sics 18, pp. 647-658.
(18) Zirker, J.B., Editor : 1977, "Coronal Holes and High Speed
      Wind Streams," Colo. Assoc. Univ. Press, Boulder.
(19) Sheeley, N.R., Jr. : 1978, Solar Phys. 58, pp. 405-422.
(20) Feynman, J. and Silverman, S.M. : 1980, J. Geophys. Res.
      85, pp. 2991-2997.
(21) Stephenson, F.R. and Clark, D.H., 1978, "Applications of
      Early Astronomical Records", Adam Hilger, London.
(22) Clark, D.H., and Stephenson, F.R. : 1978, Quart. J. Roy.
      Astron. Soc. 19, pp. 387-410,
(23) Kanda, S. : 1933, Proc. Imperial Acad. (Japan), 9, pp.
      293-296.
(24) Wittmann, A. : 1978, Astron. Astrophys. 66, pp. 93-97.
(25) Xu Zhentao and Jiang Yaotiao : 1979, Nanjing Daxue Xuebao
      (Ziran Kexue Ban) No. 2, pp. 31-38.
(26) Yunnan Observatory, Ancient Sunspot Records Group ; 1977,
      Chinese Astronomy 1, pp. 347-359.
(27) Willis, D.M., Easterbrook, M.G. and Stephenson, F.R. :
      1980, Nature, 287, pp. 617-619.
(28) Lingenfelter, R.E., 1963, Rev. Geophys. 1, pp. 35-55.

(29) Lingenfelter, R.E. and Ramaty, R. : 1970, in "Radiocarbon
     Variations and Absolute Chronology", J.V. Olsson, Ed.,
     Wiley, New York, pp. 513-537.
(30) Pomerantz, M.A., and Duggal, S.P. : 1974, Reviews of Geo-
     Phys. and Space Physics, 12, pp. 343-361.
(31) Eddy, J.A. : 1978, in "The New Solar Physics", Eddy, J.A.
     Ed., Westview Press, Boulder, pp. 11-34.
(32) Damon, P.E. : 1977, in "The Solar Output and its Variation"
     White, O.R., Ed., Colo. Assoc. Univ. Press, Boulder, pp.
     429-445.
(33) Stuiver, M. and Grootes, P.M. : 1980, in "Proceedings of a
     Conf. On the Ancient Sun", Pepin, R.O., Eddy, J.A., and
     Merrill, R.B., Eds. Pergamon Press, in press.
(34) Dicke, R.H. : 1978, Nature 276,pp. 676-680.
(35) Forman, M.A., Schaeffer, O., and Schaeffer, G.A. : 1978,
     Geophys. Res. Lett. 5, pp. 219-222.
(36) Gleissberg, W. : 1944, Terr. Magnetism and Atmos. Electri-
     city 49, pp. 243-244.
(37) Wolf, R. : 1956, Astron. Mitt. Zurich 1, p. viii.
(38) Schove, D.J., 1955, J. Geophys. Res. 60, pp. 127-146.
(39) Eddy, J.A. : 1977, in "The Solar Output and its Variation"
     White, O.R., Ed.,Colo. Assoc. Univ. Press, Boulder, pp.
     51-71.
(40) Williams, G.E. : 1980, Submitted to Science.
(41) Skumanich, A., Smythe, C., and Frazier, E.N. : 1975, Astro-
     phys. J., 200, 747.
(42) Delhaye, J. : 1953, Compt. Rend. 237, 294.
(43) Vyssotsky, A.N., and Dyer, E.R., Jr. : 1957, Astrophy. J.
     125, p. 297.
(44) Spitzer, L., and Schwarzschild, M. : 1953, Astrophys. J.
     118, p. 106.
(45) Larson, R.B. : 1979, M.N.R.A.S., 186, p. 479.
(46) Wilson, O.C., and Wooley, R.v.d.R. : 1970, M.N.R.A.S., 148,
     p. 463.
(47) Skumanich, A., : 1972, Astrophys. J., 171, p. 565.
(48) Wilson, O.C. : 1968, Astrophys. J., 153, p.221.
(49) Wilson, O.C. : 1978, Astrophys. J., 226, p.379.
(50) Wilson, O.C. : 1970, Astrophys. J., 160, p.225.
(51) Crawford, D.L., and Stromgren, B. : 1966, "Comparison of
     Hyades, Coma and Pleiades Clusters Based on Photoelectric
     u, v, b, y and Hβ Photometry" in Vistas in Astronomy
     (ed. A. Beer) Vol. 8, Pergamon Press (N.J.)
     p. 156, see also Crawford, D.L., and Mander, J.V. :
     1966, Astron. J., p. 114.
(52) Dravins, D. : 1976, "Chromospheric Activity in F and G
     Stars", in Basic Mechanisms of Solar Activity, IAU Symp.
     No. 71 (ed. by V. Bumba and J. Kleczek), D. Reidel Publ.
     Co., Dordrecht, p. 469.
(53) Eggin, O.J. : 1960, M.N.R.A.S., 120, p. 563,
(54) Vaughn, A.H., and Preston, G.W. : 1980, Pub. Astron. Soc.
     Pacific 92, p. 385, see also Vaughn (74).

(55) Wilson, O.C. : 1966, Science, 151, p. 1487.
(56) Blanco, C., Catalano, S., Marilli, E., and Rodono, M. :
     1974, Astron. Astrophys., 33, p. 257.
(57) Dravins, D.,: 1978, "Comments on Solar Chromospheric
     Activity Compared to that of Other Stars" in Proc. JOSO
     Workshop -- Future Solar Optical Observations, Needs and
     Constraints, (ed. G. Godoli, G. Noci, N. Righini)
     Osservazioni e Memorie dell Osservat. Astrofis. di Arcetri,
     Firenze, p. 266.
(58) Schatzman, E. : 1962, Ann. Astrophys. 25, p. 18.
(59) Wilson, O.C. : 1966, Astrophys. J., 144, p. 695.
(60) Kraft, R.P. : 1967, Astrophys. J., 150, p. 551.
(61) Pneuman, G.W. and Kopp, R.A. : 1971, Solar Phys., 18, p.258.
(62) Vaiana, G.S., Cassinelli, J.P., Fabbiano, G., Giacconi, R.,
     Golub, L., Gorenstein, P., Haisch, B.M., Harnden, F.R., Jr.,
     Johnson, H.M., Linsky, J.L., Maxson, C.W., Mewe, R., Rosner,
     R., Seward, F., Topka, K., and Zwaan, C. : 1980, Astrophys.
     J. (in press).
(63) Stern, R., Underwood, J.H., Zolcinski, M.C. and Antiochos,
     S., : "The Einstein Central Hyades Survey : A Progress
     Report" in Proceedings of Americain Astron. Soc./Solar
     Phys. Div. Workshop on Cool Stars, Stellar Systems and the
     Sun, 31 Jan. 1980, Cambridge, Mass. (see also this volume).
(64) Conti, P.S. : 1968, Astrophys. J., 152, p. 657.
(65) Brandt, J. : 1966, Astrophys. J., 144, p. 1221.
(66) Weber, E.J., and Davis, L., Jr., : 1967, Astrophys. J.,
     148, p. 217.
(67) Pizzo, V., Schwenn, R., and Rosenbauer, H. : 1981, J. Geo-
     phys. Res., (in press).
(68) Durney, B.R. and Stenflo, J.O. : 1972, Astrophys. Space
     Sci., 15, p. 307.
(69) Goldreich, P. and Schubert, G. : 1967, Astrophys. J., 150,
     p. 571.
(70) Van Den Heuvel, E.P.J., and Conti, P.S. : 1971, Science,
     171, p. 895.
(71) Smith, M. : 1979, Publ. Astr. Soc. Pacific, 91, 737.
(72) Soderblom, D. : 1980, Ph.D. Thesis, U. of California at
     Davis (as presented at this school).
(73) Boesgaard, A.M. and Hagen, W. : 1974, Astrophys. J., 189,
     p. 85.
(74) Vaughn, A.H. : 1980, Publ. Astron. Soc. Pacific, 92, p.392.
(75) White, O.R. and Livingston, W. : 1978, Astrophys. J., 226,
     p. 679.
(76) Golub, L., Davis, J.M., and Krieger, A.S. : 1979, Astrophys.
     J. (Letters), 229, p. L145.
(77) Howard, R. : 1976, Solar Phys., 47, p. 575.
(78) Stix, M., 1978, "The Solar Dynamo" in Pleins Feux sur la
     Physique Solaire", Deuxième Assemblée Européenne de
     Physique Solaire (Structure Interne du Soleil), Toulouse 8-
     10 Mars 1978, (ed.) S. Dumont and J. Rosch, CNRS Paris,p.37.

(79) Gray, D.F., 1976 : "The Observations and Analysis of Stellar Photospheres" (N.Y., Wiley), p. 426.

(80) Allen, C.W. : 1963, Astrophysical Quantities (2nd ed.), Athlone Press, London.

(81) Gray, D.F. : 1977, Astrophys. J., 218, p. 530.

(82) Thomas, J.H. : 1979, Nature, 280, p. 662.

(83) Foukal, P.V. : 1979, Bull. Amer. Astro. Soc., 11, p. 423.

(84) Parker, E.N. : 1979, Bull. Amer. Astro. Soc., 11, p. 423.

(85) Wilson, O.C. : 1963, Astrophys. J., 138, p. 832.

(86) Young, A. and Koniges, A. : 1977, Astrophys. J., 211, p.836.

(87) Hall, D.S. : 1976, I.A.U. Colloquium No. 29, Part I, p. 287.

(88) Young, A. and Born, B. : 1980, (unpublished).

(89) Skumanich, A. : 1979, Bull. Amer. Astr. Soc., 11, p. 624.

(90) Skumanich, A. : 1980, Presented at 155th Meeting of the American Astronomical Soc., San Francisco, 13-16 Jan.1980.

(91) Basri, G.S., and Linsky, J.W. : 1979, Astrophys. J., 234, p. 1023.

(92) Ayres, T.R., and Linsky, J.W. : 1980, Astrophys. J., 240, (in press).

(93) Popper, D.M., and Ulrich, R.K. : 1977, Astrophys. J. (Letters), 212, p. L131.

(94) Walter, F.M., Cash, W., Charles, P.A., and Bowyer, C.S. : 1979, Astrophys. J., 236, p.212.

(95) Walter, F.M. : 1980, private communication.

(96) Walter, F.M. : 1980, "Rotation and Coronal Activity in G and K Stars" -- presented at this school.

(97) Burl, J.B., Teske, R.G., and Mayfield, E.B. : 1979, Solar Phys. 63, p. 157.

(98) Shore, S.N. and Hall, D.S. : 1980, "A Starspot Model For the RS CVn Stars", presented I.A.U. Symposium No. 73, Montreal.

(99) Golub, L., Maxon, C., Rosner, R., Serio, S., and Vaiana, G.S., 1980, Astrophys. J., in press.

(100) Belvedere, G., Chiuderi, C., and Paterno, L. : 1981, submitted to Astr. and Astrophys. Refer also to C. Chiuderi "Magnetic Heating in the Sun" -- this volume.

(101) Dupree, A.K., Black, J.H., Davis, R.J., Hartman, L., and Raymond, J.C. : 1979, "Chromospheres and Coronae of Late Type Stars", in Proc. Symp. The First Year of IUE, in press (see also Dupree, A.K., -- this volume).

(102) Durney, B.R. : 1976, "On Theories of Solar Rotation" in Basic Mechanisms of Solar Activity, IAU Symposium No. 71 (ed. by V. Bumba and J. Kleczek), D. Reidel Publ. Co., Dordrecht, p. 243.

(103) Priest, E.R., and Pneuman, G.W. : 1974, Solar Phys., 34, P. 231.

(104) Roberts, P.H. : 1974, "Stellar Winds", in Solar Wind Three (ed. C.T. Russell), Inst. of Geophys. and Planet. Phys.,U. of Calif.

(105) Stix, M. : 1972, Astron. Astrophys., 20, p. 9.

(106) Steinbeck, M., and Krause, F. : 1969, Astron. Nachr., 291, p. 49.

(107) Weiss, N.O. : 1966, Proc. Roy. Soc., A293, p. 310, See also Mestel, L. and Weiss, N.O. : 1974, in Magnetohydrodynamics Advanced Course, Swiss Soc. Astrophys., Geneva Obs., Switzerland.

(108) Galloway, D., and Moore, D.R. : 1979, Geophys. Astrophys, Fluid Dynamics, 12, p. 73.

(109) Durney, B.R., and Latour, J. : 1979, Geophys. Astrophys. Fluid Dynamics, 9, p. 241.

(110) Gilman, P.A. : 1980, "Global Circulation of the Sun : Where Are We and Where Are We Going ?" in Highlights of Astronomy, Vol. 5 (ed.P.A. Wayman) p. 91.

THE SOLAR NEUTRINO PROBLEM

Ian W. Roxburgh

Department of Applied Mathematics
Queen Mary College, University of London

Abstract. The problems posed by the low flux of neutrinos from the sun detected by Davis and coworkers are reviewed. Several proposals have been advanced to resolve these problems and the more reasonable (in the author's opinion) are presented. Recent claims that the neutrino may have finite mass are also considered.

I. THE SOLAR NEUTRINO PROBLEM

The energy source for the sun and stars is thought to be the fusion of light elements to produce heavier elements. In the sun it is the fusion of hydrogen to helium, and this reaction must be of the form

$$4 \, {}^1H \rightarrow {}^4He + 2e^+ + 2\nu + \gamma \, ,$$

two neutrinos being produced with the two positrons. The neutrinos escape from the centre without significant interaction with the rest of the solar material, the photons provide the radiative flux that eventually leaves the surface. The flux of neutrinos is readily calculated, since the fusion of four protons releases about $E = 6.10^{18}$ erg/gm, and as the sun radiates at a rate $L_\odot = 3.86 \, 10^{33}$ ergs/sec, this determines the number of protons combining per unit time and hence the rate of production of neutrinos :

$$N = \frac{1}{2} \frac{L}{EM_H} = 2.10^{38} \text{ Neutrinos per sec.}$$

R. M. Bonnet and A. K. Dupree (eds.), Solar Phenomena in Stars and Stellar Systems, 399–406.
Copyright © 1981 by D. Reidel Publishing Company.

The flux of solar neutrinos on the earth's surface is therefore $6.10^{10}/cm^2/sec$.

The detailed reactions in the proton-proton chain that leads to the fusion of hydrogen to helium are shown in Figure 1. There are several ways in which the chain can be completed and the fraction of all reactions completing along one branch rather than the others depends on density, temperature and the abundance of ^4He. In solar conditions about 86 % of reactions complete the chain by the $^3He + {}^3He \rightarrow {}^44He + 2{}^1H$ reaction and therefore the neutrinos produced have energies less than 0.42 MeV. Almost all the other reactions complete through the $Be^7 - {}^7Li$ branch thus producing neutrinos with energies up to 0.86 MeV. Finally, a very small fraction complete on the $^7Be - {}^8B$ branch producing neutrinos with energies up to 14.06 MeV. It is these energetic neutrinos that can be detected with present techniques. In a standard solar model some 0.02 % of reactions complete along this $^7Be - {}^8B$ branch.

FIGURE 1

	Reaction	%	Maximum neutrino energy(MeV)
	$p+p \rightarrow {}^2H + e^+ + \nu$	(99.75)	0.420
	or		
	$p + e^- +p \rightarrow {}^2H + \nu$	(0.25)	1.44 (monoenergetic)
	$^2H + p \rightarrow {}^3He + \gamma$	(100)	
I.	$^3H + {}^3He \rightarrow {}^4He + 2p$	(86)	
	or		
	$^3He + {}^4He \rightarrow {}^7Be + \gamma$		
II.	$^7Be + e^- \rightarrow {}^7Li + \nu$		0.861 (90 %), 0.383 (10 %)
	$^7Li + p \rightarrow 2{}^4He$	(14)	(both monoenergetic)
	or		
III.	$^7Be + p \rightarrow {}^8B + \gamma$		
	$^8B \rightarrow {}^8Be^* + e^+ + \nu$		14.06
	$^8Be^* \quad 2{}^4He$	(0.02)	

The detection technique used by Ray Davis and his collaborators (4) uses the reaction :

$$^{37}Cl + \nu \rightarrow {}^{37}Ar + e^{-} \; .$$

The Chlorine is in the form of $C_2Cl_{4\,4}$ which is readily available and cheap (it is used for dry cleaning), and a tank containing 610 tons is located in Homestake Gold Mine at a depth
of 4400 meters of water equivalent, thus producing a shield
against the effect of Cosmic Rays. The number of Argon atoms
produced in the tank is then counted. The unit of measurement is
SNU (solar neutrino unit) and 1 SNU is one capture per 10^{36} Chlorine atoms per second. The experiment has been running for 10
years and the measured capture rate is 2.2 ± 0.4 SNU.

The theoretical predications depend on an accurate modelling
of the sun, an accurate knowledge of the different reactions
(and their cross sections) that produce neutrinos, an accurate
knowledge of the $^{37}Cl + \nu$ reaction, and estimates of the expected
neutrino flux have varied due to modifications in all these areas.

The predictions of a standard solar model are obtained by
taking the initial sun to be homogeneous, to have a heavy element
abundance $Z = 0.027X$ where X is the abundance of hydrogen, and
adjusting the value of the initial helium abundance Y so that an
evolved model of the sun has the observed luminosity at the present age. This gives a initial helium abundance of 0.24, and a
neutrino flux of 7.2 SNU (1),(2). The factor of 3 difference betweer
prediction and observation constitutes the solar neutrino problem.

II. GENERAL REMARKS ON ATTEMPTS TO RESOLVE THE PROBLEM

At first glance, it might appear that the problem is trivial. Since the fraction of hydrogen burning reactions that
complete on the ^{8}B branch is a sensitive function of temperature,
the solar neutrino measurements are just telling us that the
temperature is a little cooler than our model predictions. Surely this only requires a minor adjustment in those models . This
is not the case however. A decrease in the central temperature
of the sun would result in a decrease in energy production and
the resulting model would no longer produce the energy the sun
is currently radiating. It is this constraint that the nuclear
reactions produce the observed luminosity, that renders it difficult to lower the neutrino flux.

There are, however, several ways round this difficulty. One
is to somehow arrange that the central hydrogen abundance is
higher (and therefore helium abundance lower) than in the standard models. Since the energy production rate is :

$$E \propto X^2 \rho T^{4 \cdot 5},$$

an increase in X requires a lower T_c ; this together with the
concommitant lower value of ^{4}He abundance gives a lower fraction
of reactions completing on the ^{8}B branch and a lower neutrino
flux. Two ways of achieving this have been proposed, one is that
the internal abundance of heavy elements Z is much lower than
the surface abundance, the second is that the inner regions of
the sun are (or were) kept chemically homogeneous by some mixing
process.

The second method of meeting the constraint that the nuclear
reactions produce the observed luminosity but at a lower central
temperature appeals to some other contribution to the support
of the stars in addition to thermal pressure, the most likely
candidates being magnetic fields and rotation. Provided that such
extra support produces a slower decrease of temperature with mass
than in the standard model the total energy production

$$L = \int E_o \, \rho X^2 \, T^{4.5} \, dM$$

can equal $L_\odot$ with a lower central temperature T_c, the regions
away from centre making a larger contribution to the integral
than in the standard model.

Many variations of these themes have been tried by many
authors (myself included), none of them have been widely accepted.
These and other proposed resolutions of the solar neutrino problem have been reviewed by Rood [11], so I will confine my discussion to two models - the low Z 'dirty sun' model and the
mixed model. The magnetic - rotational models require a solar
quadropole movement that is too large compared to current upper
limits.

III. THE DIRTY SUN MODELS

Elementary order of magnitude estimates of the luminosity
of a star [12] yield

$$L \propto \frac{\mu^6 M^{4.33}}{Z} \quad \text{where} \quad \mu = \frac{4}{3 + 5X} \quad .$$

Hence for a given value of L the smaller the value of Z
the smaller the value of μ and hence the larger the hydrogen
abundance X. The same order of magnitude estimates yield a central temperature

$$T_c \propto \frac{M}{R} \propto \left[\frac{LM\mu^3}{X^2}\right]^{2/15} \propto \left[\frac{1}{X^2(3+5\ X)^3}\right]^{2/15}$$

and hence a lower value Z implies a lower central temperature
and a consequential lower neutrino flux. Detailed models of an
evolved sun give the same results but to obtain a model with a
measured neutrino flux of 2.2 SNU requires

$$Z = 0.002, \qquad X = 0.89, \qquad Y = 0.108.$$

There are several criticisms that can be levelled at this
model. The helium abundance is very low, lower than the "universal abundance " left as a relic from the early fireball phase of
the Universe (Y = 0.2). How could the sun have accreted its dirty surface layers with Z $\simeq$ 0.02 ? Even if it passed through a
dense interstellar cloud the solar wind would have effectively
shielded the solar surface from cloud material. Would any accreted material stay in the surface layers only or would it not be
mixed due both to convection in the surface layers and instabilities between layers of different Z ?

While such a dirty model can predict a low neutrino flux
it is difficult to find an acceptable cosmogonical history that
would produce such a model.

IV. MIXED MODELS

In the standard solar model, the central hydrogen abundance
has decreased from its initial value of 0.74 to a value of about
0.38 in the present sun. Since the rate of energy generation is

$$E \propto X^2\rho T^{4\cdot 5}$$

and if the value of X can be kept high near the centre, a lower
temperature will be needed to produce the observed luminosity.
If we assume that an inner fraction q of the total mass is kept
chemically homogenous by some mixing process, we can again construct models that have the observed luminosity at the present
age and calculate the neutrino fluxes. These are given below.

q	0	0.3	0.5	0.8	1
L_ν (SNU)	7.4	5.5	3.5	2.4	1.9

The problem is not therefore to construct a model of the
present sun with low neutrino flux, but to ask whether there is
any good reason to assume the sun is mixed and if so, why this is

apparently not the case for most stars since stellar evolution
theory has had reasonable success in explaining the observations
of   without assuming any substantial mixing. There are,
I think, two serious contendors for the stirring agency, one is
the ^{3}He instability first pointed out by Dilke and Gough (7),
the second is the slowing down of the sun by angular momentum
loss in the solar wind causing mild turbulent diffusion driven
by shear or rotational instabilities (6).

The reactions in the proton-proton chain have different
dependence on temperature and in particular the various methods
of completing the chain do not take place at any appreciable rate
at some distance from the centre of the sun. In the central regions the chain goes to completion with the production of ^{4}He,
further out it leads to a build up of ^{3}He. Just because of the
higher temperature sensitivity of ^{3}He burning an element of ^{3}He
rich material displaced to higher temperatures will burn more
vigourously than the surrounding matter thus providing a means
of exciting an oscillation. Detailed stability calculations by
Christensen-Daalsgard and Gough (5) and by Boury et al (3) confirmed the existence of this instability and results from Boury et al
are given below.

Age (ys)	L/L_0	X_c	Period (secs)	Growth Rate (ys)
10^8	0.72	0.74	6.10^3	-10^7
9.10^9	0.76	0.68	6.10^3	$+10^7$
$2.4.10^9$	0.85	0.56	5.510^3	$+3.10^6$
$4.5.10^9$	1	0.39	4.10^3	-10^6

An initially homogeneous model becomes unstable to a
g , $\ell = 1$ mode after 3.10^8 ys. If this instability leads to mixing then the sun will stay approximately uniform in composition
so reducing the neutrino flux.

Whether or not, this instability is violent as initially
proposed by Dilke and Gough or leads to steady mixing - or
indeed is suppressed by growing chemical inhomogeneity - is an
open question and one urgently in need of further study.

The second possibility is that the solar wind slows down
the surface rotation of the sun leading to a differential rotation beneath the solar convective zone. This differential rotation is then unstable resulting in mild turbulence which both
transports angular momentum outwards and leads to some mixing of
chemical elements in the interior. This mixing can lead to a
steady increase in the surface abundance of ^{3}He and a steady
decrease in the surface abundance of Lithium, the one mechanism

explaining a number of observations. Recent calculations of a
model with slow turbulent diffusion by Maeder and Schatzman (8)
does yield reasonable agreement with these observations and the
low solar neutrino flux, but the consequences of this hypothesis
for stellar evolution theory has yet to be explored.

One problem is the possibility of magnetic coupling between
the central regions and the layers just beneath the outer convective zone. Such a field could be an effective means of transporting angular momentum without turbulent diffusion. Again the
detailed fluid mechanics of such instabilities is far from solved.

V. IS THE PROBLEM ONE OF PARTICLE PHYSICS ?

There has been much discussion recently about the possibility of neutrino oscillations and a finite mass for the neutrinos.
This was first discussed by Pontecorvo (9) who suggested that an
electron neutrino ν_e could change its state to a muon neutrino
ν_μ. If we include in this the possibility of a change of state
to the tauon neutrino ν_τ it is possible to conjecture that of
the electron neutrinos emitted by the sun, only 1/3 on average
are electron neutrinos by the time they reach the earth, the
others being ν_μ and ν_τ. Since only the ν_e' would be detectable
in the C_2Cl_4 experiment the observed flux of ν_e of 2.2 ± 0.4 is
to be compared to 1/3 of the 7.4 of the model predictions i.e.
2.5. SNU.

Grand unified field thoeries relate the oscillation length
(the distance travelled before a change of state) to the difference between the neutrino masses, this length being infinite
if the neutrinos have zero mass. There are suggestions from the
work of Reines et al (10) that have detected these oscillations,
and claims from a group in the USSR that they have a finite mass,
for the electron neutrino from the end point of the β decay
spectrum of Tritium.

This is a subject that will, indeed is, receiving considerable attention and it is too early to pass any opinion. But it
may be that the solar neutrino problem was not a solar problem
at all !

REFERENCES

1. Bahcall, J.N., 1979, Space Science Reviews, vol. 24, p. 227
2. Bahcall, J.N., 1980, Private communication
3. Boury, A., Gabriel, M., Noels, A., Sculflaire, R., and
   Ledoux, P., 1975, Astron. & Astrophys.,vol. 25, p. 455

4. Davis, R., 1978, Proceedings of the Brookhaven Conference on Status and future of Solar Neutrino Research, p. 1, U.S. Dept. of Energy
5. Christensen-Dalsgaard, J., Dilke, F., and Gough, D.O., 1974 Monthly Notices Roy. Astr. Soc., 169, p. 429
6. Dicke, R.H., 1972, Astrophys. J., 171, p. 131
7. Dilke, F., and Gough, D.O., 1972, Nature, 240, p. 262
8. Maeder, A., and Schatzmann, E., C.R. Ac. Sc., série B, 291, 15.09.80, p. 81
9. Pontecorvo, B.C., 1968, Sov. Phys. - JETP, 26, p. 984
10. Reines, F., Sobel, H., and Pasierb, 1980, Phys. Rev. Letters, in press
11. Rood, R., 1978, Proceedings of the Brookhaven Conference on Status and future of Solar Neutrino Research, p. 175, U.S. Dept. of Energy
12. Roxburgh, I.W., 1981, this volume.

CHROMOSPHERES AND CORONAE IN BINARY SYSTEMS

A. K. Dupree

Harvard-Smithsonian Center for Astrophysics
Cambridge, Massachusetts 02138 USA

Abstract. Recent observations and spectral analysis of chromospheres and coronae are discussed for various types of binary systems containing cool stars. These include the W UMa-type systems, the RS CVn variables, and selected widely separated binary systems such as VV Cephei and the giant stars in the Hyades cluster. In many cases evidence for extreme examples of solar activity are found. Radiative losses in ultraviolet resonance lines can be at least one or two orders of magnitude larger than those found in the quiet Sun. In λ And, a RS CVn type system, there is evidence for inhomogeneous structure in the chromosphere and transition region. Rotation appears to be a determinant in the energy losses of these atmospheres although not the only one.

1. INTRODUCTION

Binary systems containing cool stars offer a unique opportunity to investigate the determinants of chromospheres and coronae. Such systems contain cool stars ranging in spectral type from F to M and all luminosity classes. The rotational velocities, usually known because of binary membership or periodic optical variability allow a connection to be made between the energy losses, the energy requirements, and physical parameters of the system.

It has been known since Wilson's (1) studies of the Ca II line that binary systems show enhanced chromospheric activity as compared to most single stars of the same effective temperature and luminosity. The surface fluxes in the Ca II lines are

R. M. Bonnet and A. K. Dupree (eds.), Solar Phenomena in Stars and Stellar Systems, 407–430.
Copyright © 1981 by D. Reidel Publishing Company.

higher. Also the line profiles typically lack a central reversal found in the quiet Sun spectrum (2), and the stars do not obey the Wilson-Bappu relation (3) relating Ca K emission core widths to the absolute visual luminosity of the star. Most recently, the wealth of ultraviolet and X-ray observations shows the ubiquitous nature of these enhanced outer atmospheres in many types of binaries. It appears that there is a correlation of enhanced chromospheric losses associated with rotation of the individual components. This could be expected to arise from the increased concentration of magnetic fields leading to increased heating and radiation - by analogy with the enhanced emission found in the Sun over regions of magnetic field concentration. A few semi-empirical calculations suggest that the pressures are higher than in the quiet Sun and are more typical of those found in solar active regions. Binaries of the RS CVn type show a quite different emission measure structure from that of the Sun. One star of this type also gives evidence for inhomogenous chromospheric structure; another system of the W UMa class shows a chromosphere closely confined to the stellar surface underneath a greatly extended corona apparently encompassing both stars.

In this volume, Linsky (4) has reviewed the presence of chromospheres and coronae in cool stars with particular attention to single stars. In binary systems we shall find an enhanced version of solar activity and explore further the determinants of chromospheres and coronae in the cool stars.

In this contribution, we shall focus first on the contact binary systems - the W Ursae Majoris type. In addition we consider the RS Canum Venaticorum systems that show solar-like activity, the widely separated systems such as those in the Hyades and the massive eclipsing systems such as VV Cephei.

2. THE W URSAE MAJORIS TYPE SYSTEMS

These systems contain cool dwarf stars (spectral type usually A - K) which are spectroscopically similar. Their orbital period is less than one day - generally 6 to 14 hours - making them eclipsing systems. Analysis of the optical light curves indicates that they are contact systems and that the mass ratio of the components differs from unity. The individual stars follow a relation such that $L \alpha M$. Theoretical models suggest that the components share a common convective envelope, that the luminosity is generated principally in one component, and that energy and perhaps mass transfer may occur between the stars (5, 6).

a.) Ultraviolet Measurements

The ultraviolet spectra from such systems (7) shown in

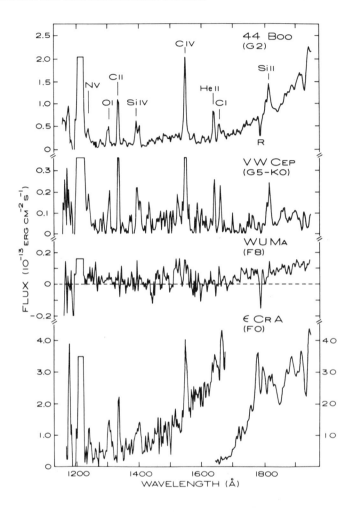

Figure 1: Ultraviolet spectra of four W Ursae Majoris type systems (7) showing spectra very similar to those found in other cool dwarf stars like the Sun. The continuum apparent at longer wavelengths in ε CrA is a consequence of the hotter photosphere of this system; however in the spectrum of 44 Boo the continuum arises from a third unresolved star in the system. Note also the strong presence of He II at λ1640 which can be excited in part by X-rays from the stellar corona (9).

Figure 1 are typical of those found in other cool dwarf stars (8, 9) showing species of C II, Si IV, C IV, and N V. The latter ions indicate the presence of temperatures on the order of 2 x

10^5 K since they are formed under conditions of collisional equilibrium with the electron population. These spectra, when they can be taken frequently enough to resolve the system at various phases as in the case of VW Cephei and 44 Bootis (10), generally show phase modulation of the long wavelength continuum as well as Mg II (λ2800) and the lines of Si II (λ1808) and C IV (λ1550).

A most critical measurement is the flux through the stellar surface in these various species, for it determines a minimum estimate of the required heating and enables the thermal structure of the atmosphere to be determined. Surveys of single cool stars (8, 11) have shown the wide range of surface fluxes that are present (see Figure 2).

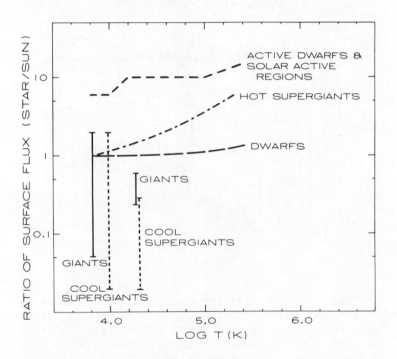

Figure 2:   Typical values of the surface flux in strong ultraviolet emission lines for various types of single cool stars (8). The temperature is the temperature of the maximum contribution of the emission in a collisionally dominated plasma. The stellar surface flux is given relative to that in the quiet solar spectrum.

Generally the dwarf stars show structure similar to the Sun
or to specific features on the Sun - namely active regions. The
more luminous stars - the giants (III) and supergiants (I, II) -
display a wide variation in surface flux which is usually
comparable to or less than that found in the dwarf stars.   The
hot supergiants defined by Hartmann, Dupree, and Raymond (12) as
those showing high temperature species and a massive wind
maintain surface fluxes larger than the quiet Sun.

It is important to remember that the observed radiative
losses define a minimum energy deposition necessary to maintain
the thermal structure of the atmosphere. In many of the luminous
stars there is spectroscopic evidence for a massive stellar wind
which will require momentum and energy deposition as well.

Binary systems generate substantially more radiative losses
than single stars (see Figure 3). The surface fluxes show a
general progression of increasing enhancement with increasing
temperature   much   like that found for a solar active region (13)
and for active dwarf stars (14). We emphasize that these fluxes
are evaluated by assuming a homogeneous distribution of flux over
the stellar surface; in the Sun we know that the surface
distribution is far from homogeneous. Hence our derived stellar
fluxes are merely lower limits to the actual fluxes that can be
present.

There appears to be a dependence of the ultraviolet surface
flux upon binary period (11) as later hypothesized for X-ray
emission too (15, 16). The W UMa stars 44 Boo and VW Cep have
periods of 0.7 day; the RS CVn stars have longer orbital periods:
HR 1099 (2.8 days), λ And (20.5 days), HR 4665 (64.4 days).   The
orbital period of the λ And system is not the appropriate one to
use since the optical light modulation of the system indicates a
rotation period of 54 days. The results in Figure 3 suggest that
the flux is enhanced with decreasing orbital period. Using the
Sun as a guide, we find increased ultraviolet flux over regions
where the photospheric magnetic field is strongest.   It is
plausible that the orbital motion enhances the field
concentration, and hence the ultraviolet emission - possibly by
enforcing rapid rotation as has been suggested for dwarfs (17,
18).

That rapid rotation is present can be directly verified by
measurement of line profiles. The strength of ultraviolet
chromospheric lines makes them especially suitable for this.   In
the system 44 Boo, the Mg II λ2800 transition can be sampled at
various phases in its orbital period (see Figure 4).   The broad
emission core arising from the spectroscopic binary is apparent
in the spectrum which also contains the continuum and narrow
photospheric lines originating from the third member of the

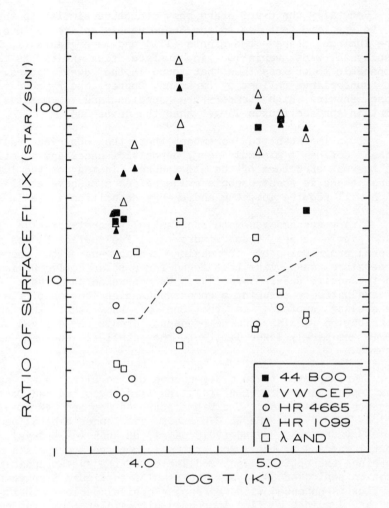

Figure 3: Ratio of stellar emission-line surface flux to that of the Sun for late-type binary stars. The broken line indicates the variations found in a solar active region. Again the pattern of enhancement with increasing temperature is found. It is apparent that the W UMa systems (44 Boo and VW Cep) and the short period RS CVn system (HR 1099) show the largest homogeneous surface fluxes. From Dupree et al. (11).

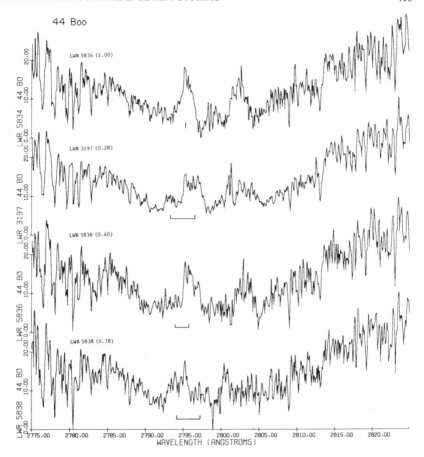

Figure 4: High dispersion IUE spectra of the
Mg II region in 44 Boo as a function of orbital
phase (ref 7). Phases 0.0 and 0.5 represent conjunction. The broad width of the Mg II emission
features corresponds to the expected rotational
velocity of the components if they are synchronously rotating. The relative strength in the
components at elongation is in harmony with the
projected surface area of the two stars. The
solid line indicates the position of each star
predicted from the radial velocity curve. The
spectroscopic binary - the W UMa type system - is
itself a component of a long period visual binary
system containing an F dwarf primary which provides the continuum and narrow lined photospheric
spectrum in this Figure.

system - a single F4 V star. The sample spectra at four phases
show the variation of individual profiles with phase; the central
emission reversal follows the orbital motion of the components of
the W UMa system with an amplitude in agreement with the optical
spectroscopic orbit (19).   Assuming then that all stars of the
W UMa type are synchronously rotating, we can evaluate the
surface flux as a function of orbital velocity of the system
(Figure 5).

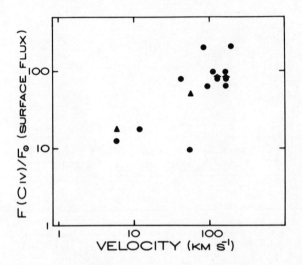

Figure 5:   The surface flux in the C IV (λ1550)
transition as a function of velocity of the components
for several binary systems (7). Synchronous rotation
was assumed for those systems with orbital period less
than 5 days. The 4 W UMa systems have rotational velocities greater than 100 km s^{-1}. Upper limits for the
components of W UMa itself are shown. Several RS CVn
systems are also included. The triangles indicate the
position for Capella with attribution of the total flux
to the secondary (15) or to the primary star.

The surface flux of the C IV transition (λ 1550) is a
convenient index of enhancement and radiative losses in the
stellar atmosphere at temperatures of 2 x 10^5 K.   Inspection of
the   empirical   relation between emergent flux and stellar
rotational velocity shows a clear correlation (7).   The four W
UMa systems have velocities comparable to or in excess of 100 km
s^{-1} - and exhibit the highest surface fluxes in the C IV line.
Both the velocities and fluxes are higher than those of the RS
CVn systems included for comparison.

The correlation suggests a continuity in the effect of rotation upon the radiative losses; however there are still substantial variations in fluxes at a given rotational velocity. In the small sample available to date, there is not a strong correlation among the W UMa type systems themselves between flux and rotational velocity. Comparing VW Cep and ϵ CrA, we find a factor of more than 20 in the luminosity of C IV is accompanied by a sixteen percent increase in the rotational velocity of the primary star. Some of the ultraviolet emission may be due to activity on the stellar surface similar to that found for λ And (20) and HR 1099 (21). The extreme values found for ϵ CrA may result from the larger mass ratio between the components (22) or the earlier spectral type of the system (FO V). However, it is generally thought that the fractional depth of the convective zone is smaller in F stars than in the cooler dwarfs. As we note below the behavior of the X-ray flux is quite different.

b.) X-ray Measurements

Additional information on the coronal structure comes from X-ray observations. The first detection of X-rays from VW Cephei was made by Carroll et al. (23) using HEAO-1. This has been followed by extensive observations with the Einstein satellite by Cruddace and Dupree (24). X-ray emission has been detected in many systems with luminosities between 10^{29} and 10^{30} ergs s^{-1} in the 1-4 keV band. Such luminosities are at the upper end of values of the X-ray luminosities of predominantly single stars found in the Center for Astrophysics stellar survey carried out by the Einstein satellite (25). For the binary systems there is again an indication of a correlation of X-ray luminosity with orbital period of the system (Figure 6).

The W UMa stars appear to have a spread and perhaps a dichotomy in their X-ray luminosities depending upon the type of the system as defined by Binnendijk (27, 28). The so-called A-type systems (not to be confused with spectral class) may have lower luminosities than the W-type stars. In these close binary systems, the more massive star is the larger, brighter component, but whereas in the A-type systems this is also the hotter star, in the W-type systems it is the cooler. For the most part, A-type binaries contain hotter (earlier spectral class) stars and have longer periods than the W-type stars. Moreover the period of the A systems is either constant or changing whereas the W-type systems have orbital periods which are constantly changing (29).

The dependence of X-ray emission on period is unlike that found for the RS CVn stars by Walter and Bowyer (26). The W UMa systems suggest L_x/L_{bol} varies as Ω^3 whereas for the RS CVn systems, the L_x/L_{bol} varies as Ω^1. This is a difference that

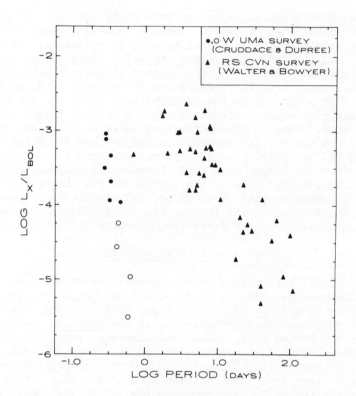

Figure 6: The ratio of L_X/L_{bol} as a function of orbital period for W UMa type stars (24) and RS CVn stars (26). The X-ray flux has been measured by the Einstein satellite in the 1-4 keV band. The filled and open circles denote the W-type and A-type W UMa systems respectively (see text).

may reflect a fundamental dichotomy in the formation of chromospheres and coronae.

Among the W UMa stars themselves there is another puzzle. The X-ray luminosities appear to be roughly constant but the C IV luminosities vary by a factor of 20. In particular in ϵ CrA the highest C IV luminosity (4×10^{29} erg s^{-1}) is associated with the lowest X-ray flux (1.4×10^{29} erg s^{-1}). Such behavior is contrary to that found in solar active regions where enhancement at all energies generally occurs. Tapia and Whelan (22) suggest that the primary star may be evolved off the zero age main sequence although still in a hydrogen burning phase. It is generally thought that a chromosphere and corona decays with age, but such a scenario does not appear to be readily compatible with the high surface flux that we find in the C IV lines.

For VW Cephei a W UMa-type system, the variation of the flux with orbital phase has been measured using the Einstein satellite (30). Surprisingly the modulation does not correspond to the variation of the optical and ultraviolet emissions (see Figure 7). The X-rays show variability of a factor of 2 but they are not in phase with the optical variation. In fact, at the same phase there appears to be variability in sequential epochs.

Apparently the corona in this system is of a scale larger than each component so that the phase modulation found in the photospheric and chromospheric emissions is absent. There may be additional surface activity as well. Hall (31) noted that W UMa type systems may show surface activity as do the "spotted" stars. Moreover, if these stars are in a state of thermal instability as suggested by Flannery (6), there may be mass transfer from one component to the other. A consequence of this could be enhanced emission at the splashdown point; excessive variability at elongation (phase 0.25 and 0.75) would result.

3. RS CANUM VENATICORUM STARS

RS CVn variables are detached binary systems with orbital periods from 1-14 days. Hall (31) has described the class which consists of a hotter component on or near the main sequence with spectral type F to G. In addition there may be long-period systems of the same kind in which the hotter component may be a giant or subgiant star. Capella (α Aurigae) is an example of this long-period group.

These stars are of interest to the study of chromospheres and coronae for they apparently show greatly enhanced versions of solar activity. In particular optical light and color variations have been attributed (31, 32, 33) to the presence of cool spots on the stellar surface. A more detailed discussion of the optical phenomena is given by Hall in this volume (34).

a.) The Ultraviolet Spectrum

The emission spectrum as observed in the ultraviolet is similar to that of the W UMa stars and the Sun (see Figure 8). The surface fluxes can be substantially higher than single stars as we have seen from Figure 3 where the fluxes in strong emission lines for 3 RS CVn type systems HR 4665, HR 1099, and λ And are shown. Enhancement of the radiative losses with rotational velocity of the components is found as discussed in the previous section concerning the W UMa stars. A particularly interesting example of rotation effects may be present in the long-period RS CVn system Capella, in which the primary star (G6 III) appears to

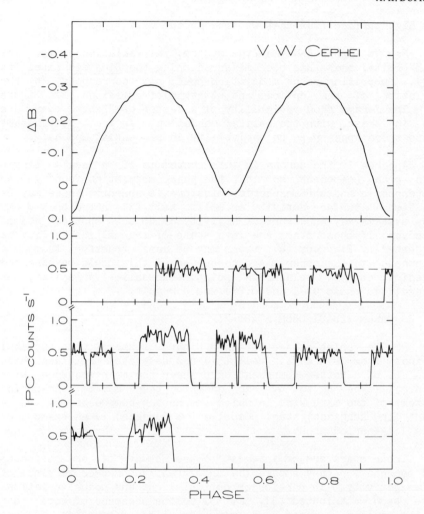

Figure 7: The optical (ΔB) and X-ray variations
of the flux from VW Cephei (30) - a W UMa type system
with a 0.7 day period. The minima and maxima of the
X-rays do not correspond to the optical light curve.
The X-ray measurements represent a continuous 13 hour
observation with the IPC on the Einstein satellite.
The times of zero X-ray flux correspond to earth occul-
tations of the source or data dropouts.

be rotating more slowly than the secondary star (spectral type F9
III). Based on IUE spectra, Ayres and Linsky (15) have
conjectured that the high temperature ultraviolet emission arises
principally from the secondary. This would be in harmony with
the rotation-activity connection found in other systems.

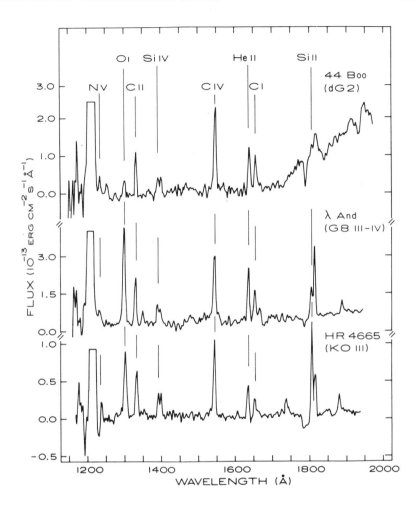

Figure 8: IUE spectra of 2 RS CVn type systems, λ And and HR 4665, and a W UMa-type system 44 Boo.

b.) X-Ray Measurements

Capella was the first RS CVn system to be detected in X-rays by Catura, Acton, and Johnson (35). This was followed by the detection of others with <u>Copernicus</u> (36) and HEAO-1 (37). The launch of the <u>Einstein</u> satellite has enabled a substantial number of these systems to be surveyed both with the Imaging Proportional Counter (16, 38) and the Solid State Spectrometer (39).

Analysis of the X-ray spectra by Swank and White (39) indicates the presence of 2 temperature components (see Figure 9) for a number of these systems.

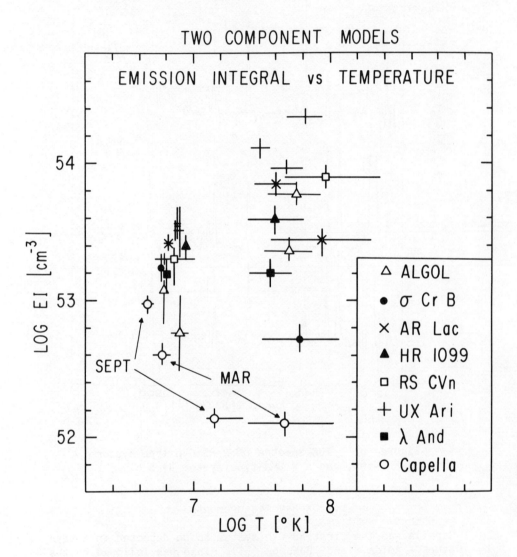

Figure 9: Emission measure as a function of temperature obtained from X-ray spectra of several binary systems. From Swank and White (39).

The low temperature components cluster around T ~ 3-7 x 10^6 K and have similar emission measures for all of the systems. The high temperature component at T $\sim$ 15-100 x 10^6 K shows a larger range of emission measure. The temperatures found on a regular basis in these systems are much higher than those observed in the Sun where the maximum emission measure occurs at 1.5 x 10^6 K. Such a wide spread in temperature at constant or increasing emission measure is also anomalous when compared to observations of solar loop structures in X-rays which typically show a narrow range in temperature. The presence of such hot material is difficult to understand if the corona is not confined in some way. This was realized from the first observations of Capella (40) where a simple scaling of the solar case to the mass and radius of the stars, indicated that the corona would evaporate if the emitting material were not confined to the stellar analogues of active regions. Such an interpretation is also consistent with the high ultraviolet surface fluxes of that system.

In these binary systems, we should not overlook the possibility of an additional source of excitation other than a traditional chromosphere and corona driven by the convection zone and enhanced by rotation. This source could result from the interaction of stellar winds in the binary system causing a standing shock between the stars. Such a model might also account for the variability of the high temperature emission component.

c.) Variability

The optical light shows modulation in many of these systems, a modulation which has been attributed to spots on the stellar surface. Weiler and colleagues (41, 42) have found that RS CVn binaries have shown an increase in the relative emission intensity of the chromospheric lines Ca H and K, Hα, and Lyα at spot minimum.

One of the RS CVn systems, λ And has been studied spectroscopically in the visual and ultraviolet spectral region and evidence is found (43) for modulation of the chromospheric emission with phase of the optical light wave. Figure 10 shows the behavior of the Ca II emission core with optical phase. This system is particularly interesting because the orbital period of 20.5 days differs from the optical light period of 54 days. There is no variation of the Ca II emission as a function of orbital phase (46). The ultraviolet spectrum of λ And shows a modulation consistent with a 54-day period of the Ca II K spectrum (43). Namely the flux in transition region lines increases by a factor of 1.5-2 at spot maximum (light minimum) as compared to its value at spot minimum (light maximum). There also appears to be a change in the asymmetry of the Mg II lines with phase as well.

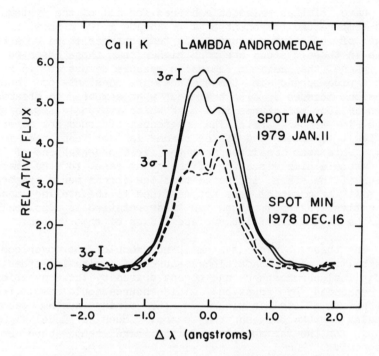

Figure 10:   The Ca II K-emission core of λ And at
spot minimum (broken line) and spot maximum (solid
line). From Baliunas and Dupree (43).

A semi-empirical model of the atmosphere of λ And (44) based
on chromospheric line profiles indicates that the pressure in the
low chromosphere is comparable to that found in solar active
regions.   In this volume, Avrett (45) discusses these models as
related to the energy balance. Thus there is clear evidence for
surface inhomogeneities in λ And. One "side" (associated with
optical light minimum) gives evidence for similarities with
active region phenomena by enhanced Ca II and ultraviolet line
fluxes; the other "side" (associated with optical light maximum)
shows a lowered flux in the ultraviolet and indications of line
asymmetries in Ca II and Mg II flux that can result from mass
outflow.

4. WIDELY SEPARATED BINARIES

For these stars, recent studies have indicated that enhanced
chromospheric activity cannot be easily attributed to the binary
nature of the systems and in addition there are substantial
variations in emissivity among stars of the same luminosity and
effective temperature.

a.) The Hyades Giants

A particularly good example is the giant stars in the Hyades: 77 Tau, γ Tau, δ Tau, and ϵ Tau. Two of the systems 77 Tau and δ Tau are known to be widely separated binaries. These four stars are extremely similar K0 giants judged by all studies to date. Cluster membership indicates that their ages, chemical compositions, and evolutionary history are all alike. Their photospheric similarity is born out by detailed analyses of spectroscopic and photometric observations (47, 48).

Baliunas, Hartmann, and Dupree (49) have measured both the optical Ca K and ultraviolet emissions from these stars. The Ca K lines show a range of a factor of 2 in the emission cores, with 77 Tau and γ Tau showing stronger emission than δ Tau and ϵ Tau. A similar behavior is found in the Mg II surface fluxes. The low dispersion IUE spectra shown in Figure 11 demonstrate that the transition region lines (i.e. C IV, N V) are visible in 77 Tau and γ Tau only. The X-ray fluxes measured for 3 of the 4 giants (50, 51) show a similar variation (see Table 1).

Table 1

Comparison of Two Hyades Giants*

	77 Tau/δ Tau
Ca K normalized emission	2.3
Mg II (h + k) flux	1.7
C IV (λ 1550) flux	6
X-ray luminosity†	10

* Baliunas, Hartmann, and Dupree (49)

† Stern et al. (50, 51)

Both stars, 77 Tau and δ Tau are widely separated binary systems, and differ by an order of magnitude in their ultraviolet and X-ray emissions. It is not possible to assign this discrepancy to the presence of a wind in one or the other star. The Hyades giants are located in a region of the H-R diagram in the vicinity of the onset of mass loss indicators. Optically thick chromospheric lines such as Ca II and Mg II can show asymmetry in the sense of the blue peak weakening relative to the red peak of the line in the presence of a differentially expanding atmosphere. For 77 Tau and γ Tau, the Mg II lines show profiles indicative of mass outflow whereas the Ca II profiles show the opposite asymmetry (49). The observations were not made simultaneously and this may provide the explanation; however

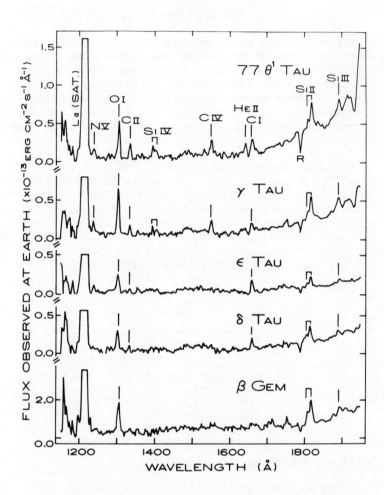

Figure 11: IUE short wavelength low resolution spectra
of the four Hyades giants and the field star β Gem
(49). 77 Tau and δ Tau are known to be widely separated binaries.

repeated measurements of these objects have indicated the same
asymmetry.

Thus we are left with the fact that stars with very similar
physical properties show diverse chromospheric and coronal
emissions, indicating that the atmospheric structure is different
and perhaps the heating mechanism also may reflect the same range
of variation. It may be that accurate determinations of the
rotation rates for these giants, or a measurement of the magnetic

field will provide a clue to their discrepant spectra, but at
present these are puzzling.

b.) Cool Supergiants with a Hot Companion

A laboratory for probing stellar atmospheres is provided by
a binary system composed of a cool supergiant with a hot
companion. Particularly fortuitous was the egress from eclipse
of the secondary of VV Cephei during the first months after the
launch of the International Ultraviolet Explorer in January 1978.
This eclipsing binary system is composed of a supergiant
classified as M2 Iabep and a hot companion - most probably a B
star. Monitoring of the system (52, 53) showed that egress in
the ultraviolet lagged the visible light egress by several months
due to the higher opacity of the M star atmosphere in the
ultraviolet. Fluorescent emission lines dominated the spectrum -
a result of the excitation of species in the atmosphere of the M
supergiant by the ultraviolet continuum of the hot companion.
Later the hot continuum becomes blanketed by absorption lines
arising in the extended atmosphere of the supergiant. The
stratification of ionic species in the atmosphere is marked by
the weakening or disappearance of Fe I absorption features.
Apparently, neutral iron does not extend as far out in the
envelope as Fe II. The Fe II emission profiles also change
dramatically with time. For example, in a period of 2 months,
a sudden enhancement of the blue emission peaks relative to the
red peaks of these lines was observed (54) although the total
fluxes of the lines remained constant. The surface fluxes of the
chromospheric emission lines, Mg II and Fe II are substantially
enhanced over the values found for single stars of comparable
spectral type.

5. CONCLUSIONS

Binary systems exhibit a wide range of stellar parameters
and characteristics. Their binary nature in many cases allows us
to determine physical parameters not easily accessible in single
stars. This selective review has emphasized the following
aspects of chromospheres and coronae in binary systems:

● There are enhanced radiative losses in close binaries
as compared to single stars suggesting that the heating
mechanism is stronger and more effective in these
systems than in the Sun.

● Rotation appears to be a dominant factor - but not a
unique one - in determining the flux of chromospheric
and coronal emissions.

● There are great similarities between the structures
  inferred in stellar atmospheres and the solar
  atmosphere. In particular, inhomogeneities are present
  that may result from the stellar equivalent of solar
  active regions and coronal holes. In many cases there
  are much more vigorous forms of activity on these stars
  than on single stars alone.

● There are some dissimilarities, in particular the
  presence of greatly extended coronae enveloping binary
  systems. Also more energetic activity in the form of
  higher temperatures, is inferred for some binary
  systems as compared to the Sun. There exists the
  possibility of enhanced (variable) emissions resulting
  from the interaction of stellar winds in binary
  systems.

I am grateful to S. Baliunas, L. Hartmann, J. Raymond and F. Walter for discussions and data in advance of publication which aided in the preparation of the manuscript. This research was supported in part by NASA Grants to the Smithsonian Astrophysical Observatory.

REFERENCES

1. Wilson, O.C.: 1966, Science, **151**, 1487.

2. Baliunas, S.L., and Dupree, A.K.: 1979, Astrophys. Journ., **227**, 870.

3. Engvold, O., and Rygh, B.O.: 1978, Astron. Ap., **70**, 399.

4. Linsky, J.L.: 1981, this volume.

5. Lucy, L.B.: 1976, Astrophys. Journ., **205**, 208.

6. Flannery, B.P.: 1976, Astrophys. Journ., **205**, 208.

7. Dupree, A.K., and Preston, S.: 1980, in "The Universe at Ultraviolet Wavelengths", (ed. R. Chapman), NASA Special Publication, in press.

8. Dupree, A.K.: 1980, in "Highlights of Astronomy", (ed. P.A. Wayman), D. Reidel, pp. 263-276.

9. Hartmann, L., Davis, R., Dupree, A.K., Raymond, J., Schmidtke, P.C., and Wing, R.F.: 1979, Astrophys. Journ. (Lett.), **233**, L69.

10. Dupree, A.K., Hartmann, L., and Preston, S.B.: 1979, Bull. Amer. Astron. Soc., **11**, 721.

11. Dupree, A.K., Black, J.H., Davis, R., Hartmann, L., and Raymond, J.C.: 1979, in "The First Year of IUE", (ed. A.J. Willis), University College London, pp. 217-231.

12. Hartmann, L., Dupree, A.K., and Raymond, J.C.: 1980, Astrophys. Journ. (Lett.), **236**, L143.

13. Dupree, A.K., Huber, M.C.E., Noyes, R.W., Parkinson, W.H., Reeves, E.M., and Withbroe, G.L.: 1973, Astrophys. Journ., **182**, 321.

14. Hartmann, L., Davis, R., Dupree, A.K., Raymond, J.C., Schmidtke, P.C., and Wing, R.F.: 1979, Astrophys. Journ. (Lett.), **233**, L69.

15. Ayres, T., and Linsky, J.L.: 1981, Astrophys. Journ., in press.

16. Walter, F., and Bowyer, S.: 1981, Astrophys. Journ., in press.

17. Skumanich, A.: 1972, Astrophys. Journ., **171**, 565.

18. Kraft, R.P.: 1967, Astrophys. Journ., **150**, 551.

19. Popper, D.M.: 1943, Astrophys. Journ., **97**, 394.

20. Baliunas, S.L., and Dupree, A.K.: 1980, in "Cool Stars, Stellar Systems and the Sun", (ed. A.K. Dupree), SAO Special Report No. 389, pp. 101-105.

21. Ramsey, L.W., and Nations, H.L.: 1980, in "Cool Stars, Stellar Systems and the Sun", (ed. A.K. Dupree), SAO Special Report No. 389, pp. 97-99.

22. Tapia, S., and Whelan, J.: 1975, Astrophys. Journ., **200**, 98.

23. Carroll, R., Cruddace, R.G., Friedman, H., Byram, T., Wood, K., Meekins, J., Yentis, D., Share, G., and Chubb, T.: 1980, Astrophys. Journ. (Lett.), **235**, L77.

24. Cruddace, R.G., and Dupree, A.K.: 1980, in preparation.

25. Vaiana, G.S.: 1980, in "Cool Stars, Stellar Systems and the Sun", (ed. A.K. Dupree), SAO Special Report No. 389, pp. 195-215.

26. Walter, F.C., and Bowyer, S.: 1980, preprint.

27. Binnendijk, L.: 1970, in "Vistas in Astronomy", Pergamon Press, **12**, 217.

28. Binnendijk, L.: 1977, in "Vistas in Astronomy", Pergamon Press, **21**, 359.

29. Rucinski, S.M.: 1973, Acta Astronomica, **23**, 79.

30. Dupree, A.K., and Cruddace, R.: 1980, in preparation.

31. Hall, D.: 1976, in "Multiple Periodic Variable Stars", ed. W.S. Fitch, (Reidel: Dordrecht), p. 287.

32. Bopp, B.S., and Evans, D.S.: 1971, Mon. Not. Roy. Astron. Soc., **164**, 343.

33. Vogt, S.S.: 1975, Astrophys. Journ., **199**, 418.

34. Hall, D.S.: 1981, this volume.

35. Catura, R.C., Acton, L.W., and Johnson, H.: 1975, Astrophys. Journ. (Lett.), **196**, L47.

36. White, N.E., Sanford, P.W., and Weiler, E.J.: 1978, Nature, **274**, 1569.

37. Walter, F., Charles, P., and Bowyer, S.: 1978, Astrophys. Journ. (Lett.), **225**, L119.

38. Walter, F., Charles, P., and Bowyer, S.: 1980, in "Cool Stars, Stellar Systems and the Sun", (ed. A.K. Dupree), SAO Special Report No. 389, pp. 35-45.

39. Swank, J.H., and White, N.E.: 1980, in "Cool Stars, Stellar Systems and the Sun", (ed. A.K. Dupree), SAO Special Report No. 389, pp. 47-64.

40. Dupree, A.K.: 1975, Astrophys. Journ. (Lett.), **200**, L27.

41. Weiler, E.J.: 1978, Mon. Not. Roy. Astron. Soc., **182**, 77.

42. Weiler, E.J., Owen, F.N., Bopp, B.W., Schmitz, M., Hall, D.S., Fraquelli, D.A., Piirola, V., Ryle, M., and Gibson, D.M.: 1978, Astrophys. Journ., **225**, 919.

43. Baliunas, S.L., and Dupree, A.K.: 1980, in "Cool Stars, Stellar Systems and the Sun", (ed. A.K. Dupree), SAO Special Report No. 389, pp. 101-106.

44. Baliunas, S.L., Avrett, E.H., Hartmann, L., and Dupree, A.K.: 1979, Astrophys. Journ. (Lett.), **233**, L129.

45. Avrett, E.H.: 1981, this volume.

46. Baliunas, S., and Dupree, A.K.: 1979, Astrophys. Journ., **227**, 870.

47. Chaffee, F.H. Jr., Carbon, D.F., and Strom, S.E.: 1971, Astrophys. Journ., **166**, 593.

48. Lambert, D.L., Dominy, J.F., and Sivertsen, S.: 1980, Astrophys. Journ., **235**, 114.

49. Baliunas, S.L., Hartmann, L., and Dupree, A.K.: 1980, in "The Universe at Ultraviolet Wavelengths - the First Two Years of IUE", ed. R. Chapman, NASA Sp. Pub., in press.

50. Stern, R., Underwood, J.H., Zolcinski, M.C., and Antiochos, S.: 1980, in "Cool Stars, Stellar Systems and the Sun", (ed. A.K. Dupree), SAO Special Report No. 389, pp. 127-132.

51. Stern, R., Underwood, J.H., Zolcinski, M.C., and Antiochos, S.K.: 1981, this volume.

52. Faraggiana, R., and Selvelli, P.L.: 1979, Astron. Astrophys., 76, L18.

53. Hagen, W., Black, J.H., Dupree, A.K., and Holm, A.V.: 1980, Astrophys. Journ., 238, 203.

THE RS CANUM VENATICORUM BINARIES

Douglas S. Hall

Dyer Observatory, Vanderbilt University

Abstract. This review presents a brief up-to-date summary of observational properties of RS CVn binaries, with emphasis on the photometric. A table of the 69 systems known at this time gives the
orbital period, mean apparent V magnitude, spectral classification,
range of amplitude determinations for the photometric wave, and an
indication of known period variations and eclipses. Points considered are (1) the underlying cause of the extreme surface activity, (2) the evolutionary status, (3) the geometric configuration, (4) the temperature, size, shape, location, and lifetimes
of the dark spot regions, (5) synchronization of rotational and
orbital motion, (6) spectroscopic features in the optical and
far ultraviolet, (7) the wave migration, (8) the search for spot
cycles, and (9) the orbital period variations produced by the enhanced stellar wind.

1. INTRODUCTION

In just the last few years we have come to realize that the
active star in a typical RS CVn binary system displays virtually
all of the surface phenomena displayed by our sun. In the RS CVn
binaries, however, they are more pronounced and more dramatic by
many factors or by orders of magnitude. Useful background is
given by Hall (1972, 1976), Popper and Ulrich (1977), Eaton and
Hall (1979), Zeilik et al. (1979), Shore and Hall (1980), and Hall
and Kreiner (1980). Collections of papers concerning these binaries appeared in the December 1978 issue of the Astronomical
Journal (Hall 1978), were presented at the Joint Meeting of I.A.U.
Commissions 10, 27, 40, 42, and 48 (Larsson-Leander 1980), and of
course are being presented at this Institute.

The review paper by Hall (1976) treated not only the RS CVn

431

R. M. Bonnet and A. K. Dupree (eds.), Solar Phenomena in Stars and Stellar Systems, 431–447.
Copyright © 1981 by D. Reidel Publishing Company.

binaries but also binaries with similar properties: (a) the flare
stars = the UV Ceti stars = the dKe and dMe stars = the BY Dra
variables, (b) the W-type W UMa binaries, and (c) V471 Tau.
Since then we have found that the connecting thread is rapid rotation in a convective star (Bopp 1980) and have begun to build
models to explain the panolply of observed properties (Mullan
1979; Shore and Hall 1980). This short summary, however, will
concentrate on the RS CVn binaries themselves.

Hall (1976) suggested that a distinction be made tentatively
between short-period (P < 1 day), regular, and long-period (P >
2 weeks) RS CVn binaries. There does seem to be some relation
between observed properties and orbital period but it will be difficult to define specific domains of P(orb.) which segregrate certain sets of physical properties. This paper, therefore, will
consider a list of 69 RS CVn binaries known at this time which
range in P(orb.) from 0.479 days to 138.420 days. All are defi-
nately known or very probably are binaries. The one very long-
period binary β Sct, with P(orb.) = 834 days, has been excluded
but the short-period binary XY UMa has been included even though
no emission has never been observed (Geyer 1976).

For more specifics and greater detail the reader is referred
to the lengthier review paper of Hall (1981). That review will
treat the same 69 systems and for each will give the name and
alternate designations, orbital period, ephemeris for conjunction,
degree of period variability, apparent visual magnitude, spectral
type and luminosity class, eclipse durations, photometric wave
amplitude, migration period, references to Ca II H & K and H α
emission, reference to radio emission, reference to X-ray emission,
absolute dimensions, configuration (d, sd, or c), membership in
visual binary systems or clusters, and equatorial coordinates. In
addition it will include detailed listings of light curve determinations along with $\emptyset$(min.) and amplitude for the wave in each,
migration curves, and amplitude curves.

2. A LIST OF 69 RS CVn BINARIES
The short summary at this Institute can present only sum-
maries and generalizations of the many observed properties, with
emphasis on the photometric. Table 1 lists the 69 systems being
considered. The first column contains the name. The second column gives the orbital period. The third column gives an indication
of the extent to which the orbital period may be variable, taken
from Hall and Kreiner (1980), who fit quadratic ephemerides to
compilations of all known times of mid-eclipse. In general "yes"
indicates that the quadratic term was greater than thrice its error and "maybe" indicates it was between twice and thrice. The
fourth column gives the approximate mean apparent magnitude (V of
the UBV system), outside eclipse in the case of known eclipsing
variables. The fifth column contains the most informative spectral classification available, with the hotter star always given
first whether or not it is the relatively more luminous. The

TABLE 1

Known RS CVn Binaries

Name	P(orb.)	Var.P	V	Sp.Tp.	EB	Δ m
RT And	$0\overset{d}{.}629$	yes	$9\overset{m}{.}01$	F8V + G5-K0V	C	$0\overset{m}{.}02$-$0\overset{m}{.}06$
$\int$ And	17.769	–	4.06	K1 II	–	0.02
λ And	20.521	–	3.88	G8 IV-III	–	0.15-0.28
UX Ari	6.438	–	6.5	G5V + K0IV	–	0.02-0.15
CQ Aur	10.622	yes	9.0	G0 + ?	C	0.06-0.12
α Aur	104.023	–	0.06	G0III + G5III	–	0.15
SS Boo	7.606	–	10.3	G0V + K1IV	C	0.05-0.25
SS Cam	4.824	–	10.1	F5V-IV+K0IV-III	C	0.05-0.13
SV Cam	0.593	yes	8.40	G3V-IV + K3V	P	0.00-0.06
12 Cam	80.174	–	6.23	K0 III	–	0.13-0.14
54 Cam	11.076	–	6.40	F8 V	–	0.05
RU Cnc	10.173	maybe	10.1	dF9 + dG9	C	0.02-0.09
RZ Cnc	21.643	–	8.67	K1III + K3-4III	C	0.01-0.05
WY Cnc	0.829	–	9.51	G5 + M2	P	0.02
RS CVn	4.798	yes	7.93	F4V-IV + K0IV	C	0.05-0.22
AD Cap	2.96	yes	9.8	G5 + G5	–	–
42 Cap	13.174	–	5.18	G2 IV	–	–
13 Cet	2.082	–	5.20	F8 V	–	–
UX Com	3.643	maybe	10.0	G2 + K2IV	C	0.10
RT CrB	5.117	–	10.2	G0 + ?	C	0.04
σ^2 CrB	1.140	–	5.76	dF6-8	–	0.05
CG Cyg	0.631	yes	10.05	G9.5V + K3V	P	0.03-0.11
WW Dra	4.630	yes	8.22	sgG2 + sgK0	P	0.00-0.10
AS Dra	5.415	–	8.00	G3V + K0V	–	0.03-0.05
o Dra	138.420	–	4.67	K0 III-II	–	–
RZ Eri	39.282	–	7.69	A5-F5V + sgG8	C	–
σ Gem	19.603	–	4.25	K1 III	–	0.08-0.15
Z Her	3.993	–	7.25	F4V-IV + K0IV	P	0.02-0.03
AW Her	8.801	–	9.7	G2IV + sgK2	C	–

TABLE 1 (continued)

Name	P(orb.)	Var.P	V	Sp.Tp.	EB	Δ m
MM Her	$7.^{d}960$	–	$9.^{m}5$	G8 IV + ?	P	$0.^{m}06-0.^{m}12$
PW Her	2.881	–	9.9	G0 + ?	C	0.12
GK Hya	3.587	–	9.35	G0 + G7IV	C	yes
RT Lac	5.074	yes	8.84	G9IV + K1IV	C	0.02-0.17
AR Lac	1.983	yes	6.11	G2IV + K0IV	C	0.00-0.13
HK Lac	24.428	–	6.52	F IV + K0 III	–	0.10-0.22
V350 Lac	17.755	–	6.38	K2 IV-III	–	0.02
93 Leo	71.70	–	4.54	A + G5IV-III	–	0.01-0.03
RV Lib	10.722	–	9.0	G2-5 + K5	P	0.06
VV Mon	6.051	yes	9.4	G2V + K0III	C	0.01-0.09
AR Mon	21.209	–	8.63	G8III + K2-3III	C	–
II Peg	6.724	–	7.4	K2-3 V-IV	–	0.10-0.30
LX Per	8.038	–	8.20	G0V + K0IV	C	0.01-0.05
SZ Psc	3.965	yes	7.22	F8V + K1IV	P	0.02-0.15
UV Psc	0.861	yes	9.1	G2V-IV + K0IV	P	0.04-0.08
33 Psc	72.93	–	4.61	K1 III	–	–
TY Pyx	3.199	–	6.87	G5 + G5	P	0.04
V711 Tau	2.838	–	5.8	G5IV + K1IV	–	0.07-0.21
ς Tri	14.732	–	4.94	G5III + G5III	–	0.00
RW UMa	7.328	maybe	10.12	dF9 + K1IV	C	0.11
XY UMa	0.479	maybe	9.8	G2-5V + K5V	P	0.0 -0.2
ξ UMa(B)	3.980	–	4.87	G0 V	–	–
RS UMi	6.168	–	10.1	F8 + ?	P	–
ε UMi	39.481	–	4.23	dA8-F0 + G5III	P	0.00
ER Vul	0.698	maybe	7.27	G0V + G5V	P	0.00-0.04
HR 4665	64.44	–	5.42	K0III + K0III	–	0.08-0.18
HR 5110	2.613	maybe	4.97	F2 IV + K IV	–	0.00
HR 6469	?	–	5.60	dF8	–	–
HR 7275	28.59	–	5.81	K1 IV	–	0.15
HR 7428	108.571	–	6.36	A? + K2 III-II	–	–
HR 8703	24.649	–	5.65	K1 IV-III	–	0.16-0.23

TABLE 1 (concluded)

Name	P(orb.)	Var.P	V	Sp.Tp.	EB	Δm
HD 5303	$1\overset{d}{.}840$	–	$7\overset{m}{.}78$	FO + G2-5 V	–	$0\overset{m}{.}3$
HD 83442	?	..	9.0	K2 III p	–	0.3
HD 86590	1.070	–	7.8	KO V	–	0.12-0.15
HD 108102	0.962	–	8.16	F8 V	–	–
HD 155555	1.682	yes	6.67	G5IV + KOV-IV	–	0.16
HD 158393	30.9	–	8.7	G5 III	–	yes
HD 166181	1.810	–	7.66	G5V + dM	–	0.00-0.10
+61°1211	7.492	–	8.8	late G V-IV	–	0.32
W 92	?	–	11.69	KO:pIV + ?	–	0.04-0.10

sixth column indicates which systems are known to eclipse, with
"C" for complete (total + annular) eclipses and "P" for partial.
Other systems may, after more observation, prove to be eclipsing.
The last column contains the range of amplitudes (crest to trough)
which have been observed for the photometric distortion wave, for
most of them in V of the UBV system. Here again other systems
may, after more observation, prove to have measurable waves.

3. WHY THE EXTREME SURFACE ACTIVITY?
    As explained by Bopp (1980), rapid [V(equatorial)$>$5 km/sec]
rotation in a convective star appears to be the necessary and
sufficient condition for the development of extreme surface activity in BY Dra and RS CVn stars. Single BY Dra stars are young
and not yet rotationally braked; the binary objects are synchronously locked and thus have higher than normal rotational speeds.
This explanation easily explains also the similar extreme surface
activity in the K-type dwarf of V471 Tau and in the W-type W UMa
binaries (Hall 1976; Eaton, Wu, and Rucinski 1980).

    With the above rapid-rotation explanation in mind, you can
see that it is futile to attempt explaining the surface activity
phenomena as a direct consequence of a certain evolutionaly status. Popper and Ulrich (1977) argued that most of the regular
RS CVn binaries are post-main-sequence binaries of nearly unit
mass ratio in which one star, around 1.25 $M_\odot$, has evolved off
the main sequence, become a subgiant, but not yet filled up its
Roche lobe. Morgan and Eggleton (1979) demonstrated that even
more convincingly, adding that such objects will stand out as a
distinct class of binaries if criteria for membership in that
class include deep eclipses in SB2 binaries. Stars of $\sim$1.25 $M_\odot$
evolving off the main sequence will become convective and, as
subgiants constrained to rotate synchronously with the relatively

short orbital periods, will have relatively rapid equatorial velocities.

    Similarly, it is futile to attempt to explain the surface
activity phenomena as a direct consequence of a certain geometric
configuration (d, sd, or c). It is true that the overwhelming
majority of the RS CVn binaries (and the BY Dra binaries and V471
Tau) are detached (Hall 1976, 1981). This realization was important in making it clear that the surface activity phenomena seen
in the RS CVn binaries were fundamentally different from those
seen in the semidetached post-main-sequence mass-transferring
Algol-type binaries. Nevertheless, a few of the RS CVn binaries
are (or very nearly are) semidetached. There seem to be three
types. The first type includes RT Lac, RZ Cnc, AR Mon, and ϵ
UMi. In these systems the originally more massive component has
evolved off the main sequence, overflowed its Roche lobe, and
transferred enough mass to its companion to become the less massive component. The now-more-massive mass-gaining component happens to be late enough in spectral type to be convective and is
the active star which displays the RS CVn-type surface activity.
The second type includes at this point only HR 5110. Except for
the fact that it is noneclipsing, HR 5110 would appear to be a
typical semidetached Algol-type binary (Dorren and Guinan 1980a).
In addition, however, the late-type component which fills its
Roche lobe appears to be responsible for some surface activity
phenomena which mark HR 5110 as an RS CVn binary. Gibson (Feldman and Qwok 1979) has even suggested that the variable radio
emission in all Algol-type binaries is sufficiently similar to
that in RS CVn binaries to imply that the late-type component itself, not the many aspects of the ongoing mass transfer and assim-
ilation process, is responsible. The absence of obvious RS CVn
photometric features in the Algol-type binaries could easily be
explained by the fact that the active component in RS CVn binaries
is comparable to or greater in brightness than its companion
whereas the late-type component in Algol binaries is typically
only 10% as bright as its companion. The third type includes
SS Cam (Arnold et al. 1979) and SZ Psc (Jakate et al. 1976). These
two are not Algol-type binaries,because the more evolved (subgiant) component is still significantly more massive than its companion nearer the main sequence. On the other hand, the subgiant
is very very close to filling up its Roche lobe. Apparently then
such systems are poised and ready to begin the rapid phase of mass
transfer which will transform them quickly into Algol-type binaries.

4. THE PHOTOMETRIC WAVE
    There is little doubt now that the photometric variability in
the light outside eclipse (the wave) is produced by an area of low
surface brightness (a spot or group of spots) on one hemisphere of
one star which rotates very nearly in synchronism with the orbital
motion. See Eaton and Hall (1979) for a discussion of alternate
explanations, which are now generally discredited. Until recently

the case for starspots, though strong, was primarily circumstan-
tial. Now, however, direct evidence has been provided by several
different spectroscopic investigations (Vogt 1979; Ramsey and Nations 1980; and Fekel 1980).

Several investigations have resulted in quantitative determinations or estimates of the temperature (T) of the dark region
or the temperature difference (ΔT) between it and the surrounding unspotted photosphere (Eaton and Hall 1979; Bopp and Noah
1980; and Dorren and Guinan 1980b). Such determinations of ΔT
range from 800°K to 2000°K. It should be pointed out, however,
that the situation is complicated by realization that stars like
these could have hotter regions as well as cooler regions (Oswalt
1979).

But what about the size, shape, and location of the dark regions? For now we are stuck with a fundamental predicament in
the study of spotted stars: a given light curve cannot be solved
uniquely for the distribution of dark spots which produces it.
It has been proven that it is possible to generate light curves
of every conceivable shape by suitable distribution of dark spots
on the surface of a rotating star. It must be admitted that all
so-called solutions for sizes, shapes, locations, and orientations
of spots are quite model-dependent. Waves of a wide variety of
shapes have been observed in RS CVn binaries. By now we have a
real circus. The wave in 93 Leo (Hall et al. 1980) is sinusoidal,
although the scatter may be concealing departures from a perfect
sine curve. The wave in λ And in 1976-77 (Landis et al. 1978)
resembled the light curve of a W UMa-type eclipsing binary with
a period over 50 days long; in subsequent years its light curve
has resembled a damped harmonic oscillator. The wave in HR 4665
in 1978 (Hall et al. 1981) resembled the light curve of an RR Lyr-
type variable with a period over 60 days long. The wave in HR
7275 (Hall et al. 1981) resembled the light curve of a Cepheid
variable with a stillstand on its ascending branch. The wave in
RS CVn in 1976 (Ludington 1978) showed two minima, one not as
deep as the other. The wave in σ Gem in 1978 (Vaucher 1979) had
a broad maximum with a very shallow secondary minimum in the center of its plateau. And the wave in UX Com (Popper 1977) showed
an extended flat maximum with a depression occupying only 30% of
the cycle. Several of these light curves are illustrated in Figure 1. Although it would not be profitable to classify waves
morphologically, it can be concluded immediately that an asym-
metrical wave cannot be produced by one symmetrical spot situated
symmetrically anywhere on the surface of a uniformly rotating
spherical star. This fact has motivated several investigators to
consider various different spot models. Bopp and Noah (1980) and
Dorren and Guinan (1980b) modelled light curves with a two-spot
model. Ludington (1978) considered one spot shaped like a tad-
pole, with its head pointing in the direction of rotation. Eaton

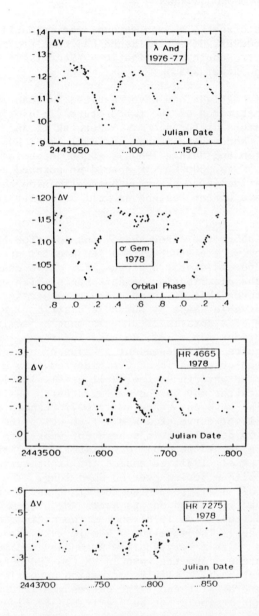

Figure 1.   Light curves of four RS CVn binaries, showing the photometric wave in each.   The ordinate is differential magnitude in
the sense variable minus comparison.   P(phtm.) ≈ P(orb.) in all
but λ And, for which P(orb.) = 20.521 days. These waves would be
poorly approximated by sine surves.   Note the sharp minimum in
λ And, the broad maximum in σ Gem, the rapid rises and slower
declines in HR 4665, and the stillstands in HR 7275.

and Hall (1979) used a model in which surface brightness varied
sinusoidally as a function of stellar longitude except for some
additional darkening over a certain longitude region.

In connection with spot shapes and sizes and the problem of
nonuniqueness, a few other considerations should be emphasized.
First, Rucinski (1979) has discussed the question of sizes of
spots and spot groups which might be expected from theoretical
consideration of the scale of granulation and supergranulation in
stars in various portions of the H-R diagram. Second, if more
progress is made in the development of theoretical starspot models,
it may be that a theoretical model can make predictions specific
enough to be tested explicitly by observation. Third, an obvious
but often overlooked point is that analysis of waves in eclipsing
RS CVn binaries will be much more fruitful than in those which do
not eclipse. This is because the inclination of the orbital (and
therefore also the rotational) axis is a known quantity and hence
is removed as an additional free parameter. Moreover, during
secondary eclipse, when the spotted star is transited by its
smaller companion, we have an entirely different handle on the
distribution of brightness across its surface, which must be consistent with whatever distribution is hypothesized to produce the
wave outside eclipse.

5. SYNCHRONIZATION
     One important generalization about the RS CVn binaries is
that the active star in virtually all systems is rotating in approximate synchronization with the orbital motion. In a light
curve with orbital phase as the abscissa, the wave moves (mig-
rates) very slowly, typically ~ 0.1 orbital cycles each year.
This wave migration indicates that the synchronization is not perfect but, because the migration periods are several years long,
the photometric period and the orbital period must differ by less
than 1%. Until recently it had been thought that rotation was not
synchronous in several RS CVn binaries: HK Lac, α Aur, HR 8703,
and λ And. Now it turns out that there is only one well established example of nonsynchronous rotation. According to Blanco
and Catalano (1970) P(phtm.) = 25.3 days for HK Lac, but according to Hall et al. (1981) 24.0026 days, which is much closer to
the orbital period, satisfies the observed times of wave minimum
better. According to Jackisch (1963) P(phtm.) = 366 days for
α Aur, but this result is not firmly established and needs to be
confirmed. According to Herbst (1973) his photometry of HR 8703
did not appear to be in phase with the orbital period, but the
newer photometry of Hall et al. (1981) is in phase. That leaves
λ And, with P(phtm.) = 54.20 days (Hall and Henry 1979) and
P(orb.) = 20.521 days, as the only well established example of
nonsynchronous rotation. It is interesting to compare this observational fact with the theoretical investigations of synchronization in binaries composed of convective stars (Zahn 1977).

6. SPECTROSCOPIC PROPERTIES

Let us look briefly at results provided by classical spectroscopic studies of RS CVn binaries. First, it is interesting that
radial velocities measured from the Ca II H & K emission peaks in
the active star seem to produce a radial velocity curve identical
to that produced by measures of that star's absorption lines.
This fact is consistent with the generally accepted picture of
the emission lines originating in that star's chromosphere, which
is very active but still reasonably close to and not systematically in motion with respect to the surface defined by the photosphere. Notice how very different the situation is in Algol-type
binaries with gas streams, circumstellar accretion disks, and expanding circumbinary shells. Second, rotational velocities determined from doppler halfwidths of absorption lines seem to be identical with those computed from the orbital period under the assumption of synchronous rotation. The same is true for the Ca II
H & K emission peaks also, but only in some RS CVn binaries, not
all. The H & K emission peaks in λ And, UX Ari, V711 Tau, and
HR 4665 seem to be broadened by some other mechanism besides rotation. Relevant references include Bopp and Fekel (1976), Naftilan and Drake (1977), Popper (1978), and Bopp et al. (1979).
Third, there are several well established examples of significant
orbital eccentricity, especially among the relatively long-period
RS CVn binaries. The largest eccentricity seems to be e = 0.37
in 12 Cam. This fact also is interesting to compare with theoretical investigations of orbit circularization in binaries composed of convective stars (Zahn 1977).

Even more interesting results have come from the newer far-
ultraviolet spectroscopy carried out from orbitting satellites
like Copernicus (OAO-3) and the International Ultraviolet Explorer (IUE). Most interest has centered on the strong Mg II h & k
and Lyman α emission. One of the most interesting findings is
the discovery of very complex emission line profiles in Mg II h
& k in UX Ari and V711 Tau, with satellite peaks displaced up to
250 km/sec from the central peak and sometimes stronger than the
central peak itself. Weiler et al. (1978) have interpreted these
as due to mass motions in the chromosphere, possibly a result of
gas trapped inside huge loop prominences. The amount of mass required to produce such satellite peaks is around 10(+19) gm (1000
times greater than the mass involved in active solar prominences)
and the lifetimes of the features are as short as a few hours.

Many spectroscopic observers have tried to look at variations
in emission strength which might be correlated with the phase of
the photometric wave. Simply stated, they expect that stronger
emission strength should be seen when the fainter = spotted =
active hemisphere is being observed. Some investigators have reported finding the expected correlation; others have not. Even
now the situation is not exactly clear and there are several problems inherent in the search. First, there might be different

behavior among the different emission lines: Ca II H & K, Hα,
Mg II h & k, and Lyman α . Second, any emission activity which
is related to flare activity will be sporadic and will not appear
strongly correlated with spot group orientation unless a statistically significant number of spectroscopic observations are analyzed. Relevant references are Weiler et al. (1978), Weiler
(1978), Ramsey and Nations (1980), and Smith (1980).

7. THE WAVE MIGRATION
       It should be extremely worthwhile to look in detail at the
migration of the wave, because that can tell us something about
the nature of the approximate synchronization, in particular about
the existence of differential rotation as a function of stellar
latitude. The original picture outlined by Hall (1972) was that
the enforced synchronization does not entirely suppress the active
star's tendency to have slightly different rotation rates at different latitudes, as the sun does. In this picture it would be
expected that spots at relatively low latitudes (near the equator)
would rotate slightly faster than synchronously, giving rise to
$P(phtm.) < P(orb.)$ and thus migration towards decreasing orbital
phase (retrograde migration), whereas spots at relatively high
latitudes (near the poles) would rotate slightly slower than synchronously, giving rise to $P(phtm.) > P(orb.)$ and thus migration
towards increasing orbital phase (direct migration). By now we
have observed retrograde migration in many RS CVn binaries; these
include UX Ari, CQ Aur, RU Cnc, WY Cnc, RS CVn, WW Dra, σ Gem,
Z Her, MM Her, RT Lac, HK Lac, VV Mon, II Peg, SZ Psc, and HR
4665. We may have observed direct migration in one RS CVn binary:
LX Per. And we know of three RS CVn binaries which have exhibited
at times both retrograde and direct migration: SS Boo, CG Cyg,
and V711 Tau. Figure 2 illustrates the wave migration in four of
these systems. In general the migration rate proves to be variable in most all systems, provided enough observational material
is available. An important goal will be to establish whether or
not such variable migration rates are a consequence of latitude
drift which occurs during the course of a spot cycle similar to
the sun's 11-year sunspot cycle.

       It is most fascinating but perplexing that the spots do not
appear distributed uniformly or randomly in stellar longitude but
rather they appear preferentially on one hemisphere of the active
star. The small amount of theoretical work on starspot formation
(Shore and Hall 1980) says little about this asymmetry. It is
ironic that this asymmetry, though not understood, is largely responsible for our knowing that starspots exist; starspots distributed uniformly or randomly in stellar longitude would produce no
photometric wave and would not call our attention to their existence. A fundamental question is how long these mysterious spot
groups exist. This question can be answered by looking at the
wave year after year and seeing how long it continues to reappear
at the same or only slightly different phase. Anytime the wave

is found far from where it was last observed, around a half orbi-
cycle away for example, then we cannot say with any assurance
that we are seeing the same spot group. By this criterion we
have found persistent spot groups which have moved slowly and
maintained their identity for several years: in λ And between
1976 and 1980, in SS Boo between 1967 and 1979, in RS CVn between
1956 and 1980, in CG Cyg between 1965 and 1976 and between 1977
and 1980, in σ Gem between 1977 and 1980, in MM Her between 1976
and 1979, in RT Lac between 1896 and 1932 and between 1946 and
1980, in LX Per between 1967 and 1979, in V711 Tau between 1976
and 1980, and in HR 4665 between 1977 and 1980. The reader is
referred again to the migration curves in Figure 2. In numerous
cases continuity in the migration curve has been lost. Examples
would be around 1979 in SS Boo, around 1895 and again around 1935
in RT Lac, and sometime in 1976 or 1977 in CG Cyg. Sometimes this
was because observation was interrupted for a couple of years or
more; other times the wave moved too far even in one year. In
these cases we simply do not know whether the spot group really
lost its identity and a new group appeared, or whether the same
spot group simply migrated rapidly. It seems to be true that
spot groups in short-period RS CVn binaries have such short lifetimes or have such rapid migration rates that the identity of
their spot groups can be traced for only a short time. The situation, however, is not clear, and observations which are more
careful and more extensive will be needed to answer this question
definitively.

8. SPOT CYCLES
     Of potential great significance is the search for spot cycles
in the RS CVn binaries and other spotted stars. Guided by our ex-
perience with the sun we might anticipate spot cycles a decade or
so in duration. The theoretical investigation of Shore and Hall
(1980), based on a generalization of the dynamo mechanism, suggests that such cycles should occur. Observationally, however,
the task is not easy. An obvious approach is to study variations
in the amplitude of the wave as a function of time, but proper interpretation of such amplitude changes is not trivial. First, if
ever a spot group grows enough to encompass more than 180° in
stellar longitude, the wave will diminish even though the area covered by spots may have grown. Second, if, as some investigators
have considered, there are two major spot groups rather than just
one, a simple drifting apart of these two groups could cause the
wave amplitude to diminish even though the area of those two
groups remained the same. Third, it might be that the formation
and dissolution of individual large spots occurring, as it does
on the sun, on timescales of months can influence the wave amplitude and obscure the search for long-term changes indicative of a
spot cycle. Figure 3 shows wave amplitude curves of four RS CVn
binaries. The first two, for MM Her and V711 Tau, show the wave
amplitude changing gradually and continuously over intervals of a
few years even though they do not clearly indicate progress through

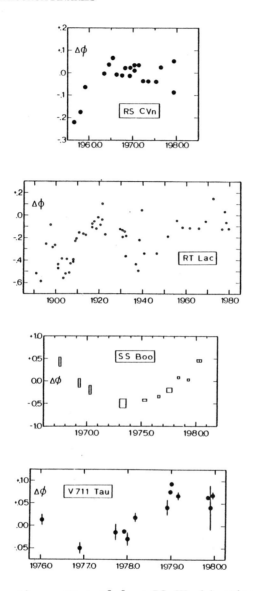

Figure 2. Migration curves of four RS CVn binaries. $\Delta\emptyset$ is the
difference between $\emptyset$(min.), the orbital phase of wave minimum as
observed, and $\emptyset'$(min.), the orbital phase of wave minimum computed with an assumed constant migration period. The value assumed for all four systems was 10.0 years, retrograde for RS CVn
and RT Lac and direct for SS Boo and V711 Tau. Wave migration
has always been retrograde in RS CVn and RT Lac but it has changed
from retrograde to direct in the other two: around 1977.0 in V711
Tau and sometime in the early 1970's in SS Boo.

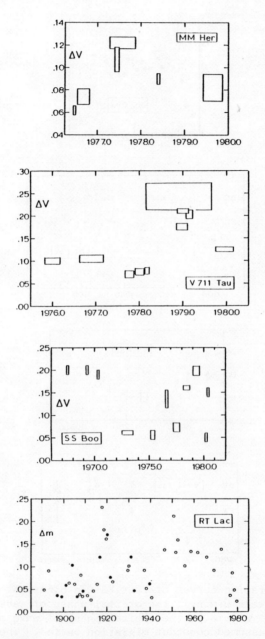

Figure 3. Wave amplitude curves for four RS CVn binaries. The ordinate is the wave amplitude, crest to trough, in magnitudes. The large wave amplitudes RT Lac had around 1918 and 1950 suggest a spot cycle 30 or 35 years in duration. If so, then RT Lac could be very near spot cycle minimum now, as of 1980.

a spot cycle. The third, for SS Boo, is the most confusing of all. The early transition from large wave amplitude ($\Delta V \sim 0.2$ magn.) to small wave amplitude ($\Delta V \sim 0.05$ magn.) later becomes confused as the wave amplitude changes back and forth from small to large amplitude within intervals of time as short as a couple of months. The last, for RT Lac, is the most impressive, with maxima in wave amplitude around 1918 and 1950 suggesting a spot cycle around 30 or 35 years in duration. Filled circles are means of two or three different light curves; the uncertainty of a typical point is $\pm$ 0.04 magn. The most recent values of Δm, with the last seven points coming from photoelectric light curves, suggest we may be at or very near a spot cycle minimum and that a relatively sudden rise to the next spot cycle maximum can begin to occur at any time. A further important question, not answered at this time, is if the identity of the dominant spot group ceases when the transition is made, at spot cycle minimum, from one cycle to the next.

Hartmann and Rosner (1979) clearly defined the so-called missing flux problem in spotted stars such as those among the BY Dra variables and the RS CVn binaries. One aspect of this problem is to learn how the mean light level changes with time as the stellar surface area covered with spots changes. At this point the situation is not at all clear. The mean light level in several RS CVn binaries definitely has been observed to change as the wave amplitude changes. A clear example is UX Ari (Hall 1977). One problem hindering further investigation into this matter is the lack of carefully standardized photometry and the use of different comparison stars with the same variable. Observers, although commended for obtaining and encouraged to continue their photometry of the RS CVn binaries and other spotted variables, are admonished to pay more attention to standardization of their differential magnitudes.

9. ORBITAL PERIOD CHANGES

This paper closes with a summary of the interesting question of orbital period changes. The most recent and comprehensive discussions are those of Hall, Kreiner, and Shore (1980) and Hall and Kreiner (1980). Of the known eclipsing binaries in Table 1, around 40% are known to have variable orbital periods. With additional observation presumably more variable periods will be discovered. Period variations among noneclipsing binaries will, of course, be far more difficult to detect. A nasty problem for some time was the fact that light curve asymmetry affecting the eclipse branches produces spurious O-C residuals which can mimic real changes in the orbital period. Even with this effect removed, however, it is clear now that there definitely are real long-term period changes. Hall and Kreiner (1980) interpreted these long-term period changes in terms of a model in which the active star loses mass in a convectively driven stellar wind which a magnetic field of roughly one kiloGauss constrains to co-

rotate with the binary system out to the Alfven radius. The resulting implication was that the active star is losing mass at a
rate of about 10(-9) $M_\odot$/yr on the average. Such a mass-loss rate
is consistent with the evolutionary picture of the active star being the first star, in a main-sequence binary originally of roughly unit mass ratio, to begin evolving off the main sequence into
the subgiant region of the H-R diagram. The overall situation
with respect to mass loss, orbital period changes, and synchronization is very complex (DeCampli and Baliunas 1979) and warrants
continued study both theoretical and observational.

This work was supported in part by N.A.S.A. Research Grant
NSG-7543, and funds for travel were provided by a Grant from the
Vanderbilt University Research Council.

REFERENCES

Arnold, C. N., Hall, D. S., Montle, R. E., Stuhlinger, T. W.:
    1979, Acta Astr. 29, 243.
Blanco, C. and Catalano, S.: 1970, Astron. Astrophys. 4, 482.
Bopp, B. W.: 1980, Highlights of Astronomy 5, 847.
Bopp, B. W. and Fekel, F.: 1976, A.J. 81, 771.
Bopp, B. W., Fekel, F., Griffin, R. F., Beavers, W. I.,
    Gunn, J. E., Edwards, D.: 1979, A.J. 84, 1763.
Bopp, B. W. and Noah, P. V.: 1980, P.A.S.P. in press.
DeCampli, W. M. and Baliunas, S. L.: 1979, Ap.J. 230, 815.
Dorren, J. D. and Guinan, E. F.: 1980a, A.J. 85, in press.
Dorren, J. D. and Guinan, E. F.: 1980b, B.A.A.S. 12, 452.
Eaton, J. A. and Hall, D. S.: 1979, Ap.J. 227, 907.
Eaton, J. A., Wu, C. C., Rucinski, S. M.: 1980, Ap.J., in press.
Fekel, F. C.: 1980, B.A.A.S. 12, 500.
Feldman, P. A. and Kwok, S.: 1979, J.R.A.S.C. 73, 271.
Geyer, E. H.: 1976, I.A.U. Symposium No. 73, 313.
Hall, D. S.: 1972, P.A.S.P. 84, 323.
Hall, D. S.: 1976, I.A.U. Colloquium No. 29, Part I, 287.
Hall, D. S.: 1977, Acta Astr. 27, 281.
Hall, D. S.: 1978, A.J. 83, 1469.
Hall, D. S.: 1981, Sp. Sci. Rev., in preperation.
Hall, D. S. and Henry, G. W.: 1979, B.A.A.S. 11, 630.
Hall, D. S., Henry, G. W., Vaucher, C. A., Landis, H. J.,
    Louth, H., Montle, R. E., Skillman, D. R.: 1980,
    I.B.V.S. No. 1798.
Hall, D. S. and Kreiner, J. M.: 1980, Acta Astr. 30, in press.
Hall, D. S., Kreiner, J. M., Shore, S. N.: 1980, I.A.U. Sym-
    posium No. 88, 383.
Hall, D. S. et al.: 1981, in preparation.
Hartmann, L. and Rosner, R.: 1979, Ap.J. 230, 802.
Herbst, W.: 1973, Astron. Astrophys. 26, 137.

Jackisch, G.: 1963, Veröff. Sternwarte Sonneberg 5, 231.
Jakate, S., Bakos, G. A., Fernie, J. D., Heard, J. F.: 1976, A.J. 81, 250.
Landis, H. J., Lovell, L. P., Hall, D. S., Henry, G. W., Renner, T. R.: 1978, A.J. 83, 176.
Larsson-Leander, G.: 1980, Highlights of Astronomy 5.
Ludington, E. W.: 1978, Ph.D. Thesis, University of Florida, Gainesville, Florida.
Morgan, J. G. and Eggleton, P. P.: 1979, M.N. 187, 661.
Mullan, D. J.: 1979, Ap.J. 231, 152.
Naftilan, S. A. and Drake, S. A.: 1977, Ap.J. 216, 508.
Oswalt, T. D.: 1979, P.A.S.P. 91, 222.
Popper, D. M.: 1977, Highlights of Astronomy, Part II, 397.
Popper, D. M.: 1978, A.J. 83, 1522.
Popper, D. M. and Ulrich, R. K.: 1977, Ap.J. 212, L131.
Ramsey, L. W. and Nations, H. L.: 1980, Ap.J.Letters, in press.
Rucinski, S. M.: 1979, Acta Astr. 29, 203.
Shore, S. N. and Hall, D. S.: 1980, I.A.U. Symposium No. 88, 389.
Smith, S.: 1980, Ph.D. Thesis, University of Toledo, Toledo, Ohio.
Vaucher, C. A.: 1979, M.A. Thesis, Vanderbilt University, Nashville, Tennessee.
Vogt, S. S.: 1979, P.A.S.P. 91, 616.
Weiler, E. J.: 1978, M.N. 182, 77.
Weiler, E. J., Owen, F. N., Bopp, B. W., Schmitz, M., Hall, D. S., Fraquelli, D. A., Piirola, V., Ryle, M., Gibson, D. M.: 1978, Ap.J. 225, 919.
Zahn, J. P.: 1977, Astron. Astrophys. 57, 383.
Zeilik, M., Hall, D. S., Feldman, P. A., Walters, F.: 1979, Sky and Telescope 57, 132.

STELLAR MAGNETIC STRUCTURE AND ACTIVITY (THEORY)

N. O. Weiss

Department of Applied Mathematics and
Theoretical Physics,
University of Cambridge, England.

Abstract. Both the overall behaviour of the solar cycle and the
underlying fine structure of magnetic fields in the sun have
been studied mathematically in some detail. These theories are
summarized and different phenomenological models of the solar
cycle are reviewed. In order to provide a description of the
magnetic fields in late-type stars it is necessary to extrapolate
boldly from what is known about the sun. In this way field
strengths and configurations can be estimated.

1. INTRODUCTION.

The observations of magnetically active stars that have been
described in the preceding lectures, by Dr. Hall, Dr. Hartmann
and Dr. Zwaan, leave the theoretician with a daunting and formidable task. Even for the sun, where there are perhaps too many
observational constraints, theoretical descriptions of either
large-scale or small-scale magnetic fields are woefully inadequate.
In magnetic stars, the relative paucity of detailed observations
makes it easy to construct theories that cannot readily be
falsified; but, for the same reason, it is hard to develop models
that have much credibility.

In this review, I shall mainly be concerned with problems
related to the origin of stellar magnetic fields: for, although
the effects of these fields may be observed in a stellar atmosphere, these effects are only symptoms of a disease that is
rooted deep in the convective zone. The treatment is not directed at experts and mathematical details will, wherever possible,
be avoided. My aim is to provide a brief and simple account of

R. M. Bonnet and A. K. Dupree (eds.), Solar Phenomena in Stars and Stellar Systems, 449–461.
Copyright © 1981 by D. Reidel Publishing Company.

the theory of hydromagnetic dynamos and turbulent magnetic
fields, distinguishing between what is known and what is sur-
mised and admitting where we are ignorant.

The next section contains a short account of the different
types of magnetic stars and their distinctive properties.
Dynamo theory is described in Section 3 and the intermittent
structure of turbulent magnetic fields is considered in Section
4. Different models of the solar dynamo, located at different
levels in the convective zone, are discussed in Section 5. In
conclusion, an attempt is made to describe the structure of
magnetic fields in rapidly rotating late-type stars.

2. STELLAR MAGNETISM

Stars are formed by the collapse and fragmentation of
diffuse gas clouds and some of the magnetic flux that threads
these clouds will be retained (e.g. Mestel 1974). During the
Hayashi phase, when protostars are wholly or partially convecting, dynamo action can generate magnetic fields in them. Thus
it is to be expected that any star will start its life on the
main sequence with a magnetic field. This field may, however,
be difficult to observe.

The magnetic properties of main sequence stars depend on
whether they have deep convective envelopes, and also on their
rotation rates. The so-called "magnetic stars", in which
magnetic fields have been measured by using the Zeeman effect,
are B and A type stars with spectroscopic anomalies. The fields
in these peculiar A stars show strictly periodic variations,
with the same periods as the anomalous abundances. Such
regular variability can only be explained as rotational modulation of a dipole-like field whose axis is inclined to the
axis of rotation. These stars have shallow convective zones
and, since the decay time for a primordial field is longer than
their lifetime on the main sequence, it is most likely that
they preserve, as fossils, the fields that they had when they
arrived there (Mestel 1976, 1981). There is no need, then, to
invoke fields generated by dynamos in their convective cores
(Krause 1981).

Why, then, do we not see magnetic fields in all early-type
stars? The answer must relate to the fact that the Ap and Am
stars all seem to be slow rotators. In a typical A star,
rotating near the breakup rate, meridional circulation may drag
the field lines below the surface, so that relatively little
magnetic flux emerges from the star (Mestel 1976); thus the
fields are concealed from observation unless the star happens
to be rotating slowly (and this occurs typically in binaries).

The X-ray observations show that early-type stars have coronae
and the recent observation of a red-shifted Zeeman pattern in
Hβ in Regulus, a rapidly rotating B7V star (Wolstencroft, Smith
and Clarke 1980) has been interpreted as showing the presence of
a localised magnetic field, corresponding to a starspot. Other
magnetic structures may yet be discovered.

Late-type main sequence stars, with deep convective envelopes, behave rather differently. In particular, magnetic activity becomes more vigorous as the rate of rotation is increased.
The sun's magnetic activity is comparatively feeble but it
provides a model for the magnetic cycles found by Wilson (1978)
in other late-type dwarfs. It is probable that the pattern of
magnetic behaviour varies continuously as the angular velocity
is increased. Nevertheless, it is schematically convenient to
distinguish stars with solar-type cycles from rapidly rotating
stars that exhibit the BY Dra syndrome. The characteristic
feature of the latter is the presence of starspots that recur as
the star rotates but vary irregularly over a longer timescale.
If the angular velocity is sufficiently high, concentrated
bundles of magnetic flux can fill a significant portion of the
surface area, though the net flux is hard to detect.

Finally, there are late-type evolved stars. The RS CVn
stars are subgiants with deep convective envelopes. All are
members of binary systems and therefore rotate rapidly. They
too have starspots which, unlike those in BY Dra stars, recur
persistently over many years. Indeed, the observations are
consistent with the presence of an almost steady field, maintained by dynamo action in their convective zones. Thus the task of
theoreticians is first to describe magnetic cycles in stars
like the sun, then to show how magnetic activity increases with
the angular velocity and, finally, to explain why magnetic
fields have a different structure in stars with more distended
envelopes. The first problem has proved difficult enough and we
can only guess at solutions to the others.

3. DYNAMO THEORY

The achievements and limitations of dynamo theory can only
be explained by introducing a short, formal treatment. Several
books on the subject, as well as a number of reviews, have
recently been published and further details can be found in them
(Weiss 1974; Cowling 1976; Stix 1976; Moffatt 1978; Parker 1979;
Krause and Rädler 1980; Gilman 1981). In a moving medium, Ohm's
law takes the form

$$\underline{j} = \sigma(\underline{E} + \underline{u} \times \underline{B}) \tag{1}$$

where σ is the electrical conductivity and other symbols have their usual menaings. The displacement current may be neglected, so that Maxwell's equations reduce to

$$\nabla \times \underline{E} = - \frac{\partial B}{\partial t} , \qquad\qquad \nabla \times \underline{B} = 4\pi\underline{j} , \qquad\qquad (2)$$

Combining (1) and (2), we obtain the induction equation

$$\frac{\partial B}{\partial t} = \nabla \times (\underline{u} \times \underline{B}) + \eta\nabla^2\underline{B} , \qquad\qquad (3)$$

where the magnetic diffusivity $\eta = (4\pi\sigma)^{-1}$ is assumed constant. The (linear) kinematic dynamo problem is that of prescribing a velocity $\underline{u}$ that will maintain the magnetic field, by induction, against the ohmic dissipation caused by imperfect conductivity.

The difficulty of solving the kinematic dynamo problem is shown by Cowling's theorem: a steady, axisymmetric magnetic field cannot be maintained. Suppose that such a field exists: then we can separate $\underline{B}$ into meridional and azimuthal (or poloidal and toroidal) components by writing

$$\underline{B} = \underline{B}_p + B_\phi\underline{e}_\phi \qquad\qquad (4)$$

where $\underline{B}_p.\underline{e}_\phi = 0$, referred to cylindrical polar co-ordinates (s,ϕ,z) with $\underline{e}_\phi$ a unit vector in the ϕ- direction. Now there must exist neutral points of the meridional field, where $\underline{B}_p = 0$ (for a dipole-like field these will lie in the equatorial plane) and the field is purely azimuthal. Let C be a circle passing through the neutral points. In a steady state, we have, on integrating (1) round the closed curve C,

$$\oint_C \underline{j} . d\underline{l} = \sigma\oint_C(\underline{E} + \underline{u} \times \underline{B}).d\underline{l} = \sigma\oint_C(\underline{u} \times \underline{B}).d\underline{l} = 0 \quad (5)$$

But the current $\underline{j}$ does not vanish at the neutral points and so the field cannot be maintained. It may also be shown that unsteady axisymmetric fields cannot be maintained in an incompressible fluid (e.g. Moffatt 1978) but no-one has yet proved that an oscillatory, axisymmetric field is impossible when the fluid is compressible. This last case is most relevant for stars.

To make the treatment more precise, we note that, since $\nabla.\underline{B}_p = 0$, we may introduce a flux function $\chi(s,z)$ such that

$$\underline{B}_p = \frac{1}{s}(- \frac{\partial\chi}{\partial z} , 0 , \frac{\partial\chi}{\partial s}) . \qquad\qquad (6)$$

Then, if we separate the velocity into meridional and azimuthal components by writing

$$\underline{u} = \underline{u}_p + s\omega\underline{e}_\phi \qquad (7)$$

so that $\underline{u}_p \cdot \underline{e}_\phi = 0$ and $\omega(s,z)$ is the angular velocity, we can express (3) in the form

$$\frac{\partial \chi}{\partial t} + \underline{u}_p \cdot \nabla \chi = \eta s D^2 \left(\frac{\chi}{s}\right) , \qquad (8)$$

$$\frac{\partial}{\partial t}\left(\frac{B}{s}\phi\right) + \nabla \cdot \left(\frac{B}{s}\phi \underline{u}_p\right) = \underline{B}_p \cdot \nabla\omega + \frac{\eta}{s}D^2 B_\phi \qquad (9)$$

where the Stokes operator $D^2 = \nabla^2 - 1/s^2$. In equation (8), the second term on the left describes the advection of flux by the moving fluid while the dissipative term is on the right. Similar terms appear in (9), together with a source term that describes the generation of toroidal flux from the poloidal field owing to differential rotation. There is no corresponding term in (8); indeed, it can be proved that the poloidal flux decays monotonically if $\nabla \cdot \underline{u} = 0$.

Of course, the sun's magnetic field is not axisymmetric and it is generated by turbulent motions in the convective zone. We can, however, discuss the behaviour of a suitably averaged field that is axisymmetric and equations (8) and (9) must then be modified to include the effect of small scale turbulent eddies. Parker (1955) pointed out that cyclonic eddies, with a systematic sense of rotation imposed Coriolis forces, could provide a regenerative term in (8). This regenerative mechanism is related to the presence of helicity, $\underline{u} \cdot \nabla \times \underline{u}$, that does not vanish when averaged. At the same time, turbulence produces an enhanced diffusivity $\tilde{\eta}$, so that (8) and (9) may be rewritten as

$$\frac{\partial \chi}{\partial t} + \underline{u}_p \cdot \nabla \chi = \alpha B_\phi + \tilde{\eta} s D^2 \left(\frac{\chi}{s}\right) , \qquad (10)$$

$$\frac{\partial}{\partial t}\left(\frac{B}{s}\phi\right) + \nabla \cdot \left(\frac{B}{s}\phi \underline{u}_p\right) = \underline{B}_p \nabla\omega + \frac{\tilde{\eta}}{s}D^2 B_\phi , \qquad (11)$$

with the regenerative process described by a co-efficient $\alpha(s,z)$.

The mean field dynamo equations (10) and (11) have been studied in considerable detail. The behaviour of solutions depends on the choice of the parameters α, ω and $\tilde{\eta}$. Typically, we might set

$$\alpha = \alpha_0(r)\cos\theta \qquad\qquad \omega = \omega(r,\theta) \qquad (12)$$

referred to spherical polar co-ordinates (r,θ,ϕ), and seek solutions that vary exponentially with time, so that $\underline{B} \propto \exp(pt)$, with $p = \Lambda + i\Omega$. For suitably chosen parameters there exist solutions with $\Lambda > 0$, $\Omega \neq 0$, corresponding to oscillatory dynamos. In particular, if the dynamo number

$$P = \frac{\alpha_0 \partial\omega/\partial r R^4}{\tilde{\eta}^2} \qquad (13)$$

is negative then fields migrate towards the equator as is observed on the sun. Many models of the solar cycle, based on solutions of the linear dynamo equations, have now appeared (see, for example, the review by Stix 1976) and detailed features of the butterfly diagram can be reproduced with remarkable fidelity by tuning the parameters in (12).

The success of these models shows, I believe, that the dynamo equations have captured the essential physics of the process that maintains the solar cycle. On the other hand, mean field electrodynamics has its limitations: the approximations used in deriving (10) and (11) cannot be justified in conditions that are relevant to the sun, nor is there any procedure that enables us to calculate α and $\tilde{\eta}$ from properties of the motion. What we have is a very plausible, parametrized theory which – like mixing length theory – must be used with caution.

So far we have considered only the linear problem, where $\underline{u}$ is prescribed and, if P is sufficiently large, the field grows without limit. In reality, dynamo action will be limited by the dynamical effect of the magnetic field. To solve the full non-linear problem it is necessary to obtain $\underline{u}$ from the equation of motion, including the Lorentz force. This requires a much more formidable calculation. Gilman and Miller (1981) have solved this problem numerically for a simplified model of the sun. They consider convection in a rotating, incompressible fluid contained between two spherical shells. The results are complicated and difficult to summarize: although dynamo action is found, there is no magnetic cycle. The fields are highly localized, or inter-mittent, and bear little resemblance to those adopted for mean field dynamo models. Moreover, attempts to calculate an effective α from the velocity $\underline{u}$ yield values that are much too large. The authors conclude that "we are much further away from a final solution to the solar dynamo problem than has been previously claimed to be the case."

Other studies have retained the formalism of mean field electrodynamics. In that case, the magnetic field may affect motions on the microscale by quenching the α-effect, or on the macroscale by driving a large scale flow: in either case growth of the magnetic field is limited. Finally, we should note that nonlinear problems do not necessarily have solutions that are

strictly periodic. There are many examples of systems of
nonlinear differential equations with aperiodic solutions, and
strange attractors have become a very fashionable topic for
research. Thus it is to be expected that in a nonlinear,
oscillatory dynamo solutions should vary irregularly, with
episodes of enhanced and reduced activity: the occurrence of
Maunder minima is compatible with theory.

4. INTERMITTENT MAGNETIC FIELDS

     In order to go beyond the parametrized theory contained in
the mean field dynamo equations we have to investigate the
structure of magnetic fields within turbulent convective zones.
Where we can observe such fields, at the surface of the sun,
they are found to be extremely intermittent. The field does
not vary smoothly: instead, magnetic flux is concentrated
into isolated tubes or ropes, with fields that are locally
intense. As Professor Parker has already emphasized, the
presence of such intense fields raises many problems for the
theoretician. To be sure, it has been known since the beginning
of the century that sunspots had strong magnetic fields, and
small scale intergranular fields had been predicted before they
were observed. Nevertheless, it came as a surprise when the
fields in photospheric flux tubes were found to be about 1200
gauss. For convection with a characteristic speed U in a gas
with density ϱ and pressure p we can define an equipartition
field B_e and a pressure-balancing field B_p, such that

$$B_e^2 = 4\pi\varrho U^2 \quad \text{and} \quad B_p^2 = 8\pi p \tag{14}$$

At the surface of the sun $B_e \sim 500$ gauss and $B_p \sim 1600$ gauss;
the observed field strengths approach B_p, which is of course
an upper limit (Parker 1979a).

     The presence of intense but intermittent fields raises
two problems. First, is there a mechanism that would convert
a weak uniform field into isolated flux tubes? Second, if
isolated flux tubes are injected into a turbulent convective
zone, will the field continue to be intermittent? It has been
known for some time that in a persistent cellular pattern of
motion, magnetic flux is expelled from the convective eddies
and concentrated into ropes between them, and that the ropes
are formed rapidly while flux expulsion takes much longer
(Weiss 1966; Parker 1979a; Galloway and Weiss 1981). Within
the flux tubes the field becomes strong enough to halt the
motion, so the tubes are almost stagnant (Galloway, Proctor and
Weiss 1981). In a turbulent region individual cells do not
survive for longer than the turnover time $\tau_0 \simeq L/U$, where L is
a characteristic length scale. It is not obvious, therefore,

that flux expulsion will occur. However, the topological
relationship between flux ropes and convection cells is not
affected by their ephemeral nature: strong fields will still
lie between convective eddies. Moreover, turbulent diffusion
can reduce the timescale for flux expulsion, $\tau_\eta = L^2/\tilde{\eta}$ so that
$\tau_\eta \simeq 5\tau_0$ (Galloway and Weiss 1981). So turbulent magnetic fields
are likely to be intermittent, as suggested by numerical experiments (e.g. Orszag and Tang 1979). In a tesselated pattern of
convection, like that seen in the solar photosphere, flux tubes
can form either on the central axes or at the vertices of cells.
So far kinematic calculations show a preference for central
flux ropes (Galloway and Proctor 1981) but this result is
probably dependent on the velocity that is chosen: in the sun,
downward velocities are higher and flux is preferentially confined to intergranular regions (Weiss 1978).

Thus it seems likely that a weak general field, if it
existed, would rapidly be converted into isolated tubes with
fields that were locally intense. In practice, the weak
field may never exist. (In the solar photosphere, new flux
emerges in isolated tubes and eventually ends up in the photospheric network but the evidence for weaker fields is still
controversial.) The mechanisms outlined above will suffice to
maintain flux ropes once they are formed. These ropes may be
shredded or amalgamated by convection, depending on the circumstances, and flux may pass from one rope to another but the
fields will still be intermittent. The best estimates available,
based on models of Boussinesq magnetoconvection, all suggest
that the fields produced will not be significantly greater than
B_e.

Compressibility introduces further complications. Except
near the surface of a star, $B_e \ll B_p$ and use of the Boussinesq
approximation can be justified. In the solar photosphere
$B_e/B_p \simeq 1/3$ and the effects of compressibility must be included.
As Dr. Spruit will explain, flux tubes become unstable to
adiabatic downflows and collapse, producing fields that are
significantly stronger and approach the limit set by B_p. The
intimate relationship between small-scale fields (e.g. filigree)
and granules at the solar surface shows that these intermittent
fields are strongly influenced by external convection but the
field strengths are determined by the gas within the flux tubes.

Magnetic buoyancy is another consequence of compressibility.
If an isolated flux tube is in thermal equilibrium with its
surroundings, static equilibrium cannot be maintained and it will
float upwards (Parker 1979a). Let p_i, ρ_i be the pressure and
density within the flux tube: then

$$p_i \; = \; p \; - \; \frac{B^2}{8\pi} \;\; , \;\;\;\; \rho_i \; \simeq \; \rho \; (1 \; - \; \frac{B^2}{8\pi p}) \hspace{3cm} (15)$$

and buoyancy makes the tube rise at a rate comparable to the Alfvén speed v_A. For $B \simeq B_e$ it follows that the upward velocity is of the same order as the convective velocity. Hence strong fields should escape from the convective zone of the sun in times that are short compared to the period of the solar cycle.

5.   SOLAR DYNAMO MODELS.

Any attempt to construct dynamo models for late-type stars must be based on some picture of the fields in their convective envelopes. In order to demonstrate both the range of possibil-ities that is available and the dangers of extrapolating from what is known about the sun, it is worth discussing some com-peting descriptions of the solar cycle. These phenomenological models are based on different assumptions about the distribution of magnetic fields in the sun's convective zone; each of them has its attractions and its weaknesses.

Weak Field Models.

The simplest possible picture is one in which there is a smoothly varying field that fills most of the convective zone. This configuration resembles that envisaged in mean field dynamo models. To accommodate the flux that emerges during the solar cycle the mean toroidal field must be about 100 gauss (Parker 1979a). Such weak fields have one great advantage: magnetic buoyancy is relatively ineffective and flux can be retained in the convective zone for several years. As we have seen, turbulent convection produces filamentary magnetic structures, with flux confined to isolated tubes or ropes and field strengths correspondingly enhanced. If these tubes are fairly small, with fluxes of order 10^{19} maxwells (Parker 1979b), the average field is unaffected though the regenerative process, described by α in the mean field dynamo equations, is reduced, since the tubes are partially isolated from the flow around them (Childress 1981). On the other hand, the total flux involved is large and relatively difficult to contain unless the field strengths are rather high. Moreover, it is hard to explain the systematic properties of large scale fields and active regions, especially their rotation rates, unless they are anchored deep in the convective zone.

Sea Serpent Models.

The appearance of active regions suggests that flux may be

confined to ropes containing $\sim 10^{22}$ maxwells throughout the sun's
convective zone. This leads to a picture, first put forward by
Babcock (1961), in which there are relatively few flux ropes,
originating deep in the convective zone, which emerge through
the photosphere to form active regions (Zwaan 1978). Although
flux may be exchanged between these ropes they remain essentially
isolated. The dynamo process depends on the distortion of
individual ropes, with strong magnetic fields ($\sim 10^4$ gauss), by
differential rotation and twisting caused by Coriolis forces
(Schmidt 1968). These flux rope dynamo models need to be
worked out in more detail but they seem capable of explaining
the observed phenomena. Their drawbacks are that the filling
factor is still rather high (about 20% of the lower half of the
convective zone) and that magnetic buoyancy should bring flux
rapidly to the photosphere, unless it is somehow held down by
the convective motions.

Shell Models.

    An alternative possibility is that the flux may be confined
to a relatively thin layer ($\sim$20000 Km thick) at the interface
between the convective and the radiative zones (e.g. Vainshtein
and Zel'dovich 1972). Within such a layer, instabilities driven
by magnetic buoyancy may lead to the formation of ropes which
emerge to form active regions. However, these instabilities
are partially stabilized owing to the effects of rotation,
giving a growth time

$$\tau \simeq \frac{\omega L^2}{V_A^2} \propto \frac{\omega}{B^2} \qquad (16)$$

(Acheson 1979). From the theoretician's point of view, a shell
dynamo is attractive because it seems more amenable to analytical
and numerical investigation. Some aspects have already been
studied (Spiegel and Weiss 1980; Galloway and Weiss 1981) though
it has not yet been established that dynamo action is possible
with realistic values of the relevant parameters. In my opinion,
these models offer the most promising hope of making progress.

Oscillator Theories.

    Most people working in this field accept that the solar
cycle is maintained by dynamo action, but there are still some
who regard the cycle as a dynamically driven oscillation
(Piddington 1978; Dicke 1978; Layzer, Rosner and Doyle 1979).
The principal argument against an oscillator theory is that it
requires a substantial modulation of angular velocity with a 22
year period. No such effect has been detected. The best available observations show much smaller changes, with a period of
11 years (Howard and LaBonte 1980), which is what one might
expect from a nonlinear dynamo.

6. STELLAR DYNAMOS.

Late-type main sequence stars should have magnetic fields
like that in the sun. In a general way, it seems inevitable
that increasing the angular velocity will increase the efficiency of dynamo action and so allow the production of stronger
fields. A greater rate of differential rotation speeds up the
production of toroidal flux in equation (11), while an enhancement of the Coriolis force should lead to higher values of α
and to stronger poloidal fields in equation (10). In shell
dynamo models, larger values of ω allow higher field strengths
before buoyancy driven instabilities develop, from (16)
(Rosner and Vaiana 1980). As yet, there is, however, no quantitative theory that predicts a net flux that is directly
proportional to ω, as suggested by the observations (Skumanich
1972).

For a given spectral type there is a wide range of magnetic
behaviour, shown by the variation in X-ray luminosities (Vaiana
et al. 1980). After allowing for magnetic cycles, this range
must depend solely on the angular velocity. Moving along the
main sequence, there are changes in the geometry of convection
that must also affect the generation of magnetic fields. In
particular, the convective zone becomes deeper as the stellar
mass decreases, until stars become fully convective around
spectral type M5. Dynamo action in a shell is plausible only if
the radius of the shell is significantly greater than the local
scale height. Otherwise a different model (for example, a flux
rope dynamo) is required. Moreover, the observations suggest
that as the convective zone extends deeper into the star the
dynamo becomes more effective.

Extremely rapid rotation leads to a different type of
behaviour with enormous flux concentrations that recur as
starspots. In main sequence stars the spots are part of the
BY Dra syndrome and they vary aperiodically. The surface
pressure is marginally greater than that on the sun, so fields
of 2000 gauss might be expected in these spots. The fact that
no such fields have been measured suggests that separate spots
of opposite polarity lie close together, in a restricted zone
of longitude. This zone drifts erratically and, as Dr.Hartmann
has explained, the activity varies on a longer time scale ($\sim$40
years). One is tempted to regard the BY Dra stars as extreme
cases of stars that exhibit magnetic cycles like the sun. If so,
the fields are more localized and the periods of the cycles are
considerably longer as well as being more irregular. Rotational
periods of 4 days rather than 25 days, increase the dynamo
number and may change the nature of the generated field. Indeed,
it is not obvious that the overall magnetic structure need
resemble any of the pictures outlined in section 5.

Finally, we have the RS CVn stars, subgiants with extended convecting atmospheres. The surface pressures are lower, corresponding to fields $B_p \simeq 500$ gauss. These stars also rotate rapidly, with periods of a few days, and all are members of binary systems. The starspots, unlike those in BY Dra stars, recur regularly. RS CVn itself has shown no sign of a spot cycle over the last 20 years. The observations that were described by Dr. Hall suggest the following speculative picture. Magnetic flux is generated by dynamo action in the convective zone, and tidal effects favour the production of a dipole-like field whose axis passes through the centres of the stars. On one side of the star this flux coagulates to form a single huge spot, which controls the large scale convective circulation over the entire star. This configuration leads to a quasi-permanent flow, which helps to maintain the spot. (Similarly, a moat cell is formed round a large sunspot, owing to the stabilization of supergranular convection by the central magnetic field.) The flux in the spot may gradually leak out but the timescale will be long, of the order of decades or more. Eventually the spot decays and the cycle is repeated. This conjectural model is fundamentally different from any of those that have been suggested for the sun.

REFERENCES

Acheson, D.J.: 1979, Solar Phys. 62, pp.23-50.
Babcock, H.W.: 1961, Astrophys.J. 133, pp.572-587.
Childress, S.: 1981, Geophys. Astrophys. Fluid Dyn., to be
            published.
Cowling, T.G.: 1976, 'Magnetohydrodynamics', Hilger, Bristol.
Dicke, R.H.: 1978, Nature 276, pp.676-680.
Galloway, D.J. and Proctor, M.R.E.: 1981, Geophys. Astrophys.
            Fluid Dyn., to be published.
Galloway, D.J., Proctor, M.R.E. and Weiss, N.O.: 1978, J. Fluid
            Mech. 87, pp.243-261.
Galloway, D.J. and Weiss, N.O.: 1981, Astrophys. J., in press.
Gilman, P.A.: 1981, in 'The Sun as a Star', ed. S. Jordan,
            NASA-CNRS.
Gilman, P.A. and Miller, J.: 1981 Astrophys. J. Supp., in press.
Howard, R. and LaBonte, B.J.: 1980, Astrophys. J., in press.
Krause, F.: 1981, Geophys. Astrophys. Fluid Dyn., to be
            published.
Krause, F. and Rädler, K.H.: 1980, 'Mean-field magnetohydro-
            dynamics and Dynamo Theory', Pergamon, Oxford.
Layzer, D., Rosner, R. and Doyle, H.T.: 1979, Astrophys. J. 229,
            pp.1126-1137.
Mestel, L.: 1974, in 'Magnetohydrodynamics', by L. Mestel and
            N.O. Weiss, pp.109-152, Swiss Soc. Astron. Astro-
            phys., Geneva.

Mestel, L.: 1976, in 'Physics of Ap Stars', ed. W.W. Weiss, H.
        Jenker and H.J. Wood, pp.1-23.
Mestel, L.: 1981, in 'Fundamental Problems of Stellar Evolution'
        (Proc. IAU Symposium No.93) pp. 129-130.
Moffatt, H.K.: 1978, 'Magnetic Field Generation in Electrically
        Conducting Fluids', Cambridge University Press.
Orszag, S.A. and Tang, C.M.: 1979, J. Fluid Mech. 90, pp.129-143.
Parker, E.N.: 1955, Astrophys. J. 122, pp.293-314.
Parker, E.N.: 1979a, 'Cosmical Magnetic Fields', Clarendon Press,
        Oxford.
Parker, E.N.: 1979b, Astrophys. J. 230, pp.905-913.
Piddington, J.H.: 1978, Astrophys. Space Sci. 55, pp.401-425.
Rosner, R. and Vaiana, G.S.: 1980, in 'High Energy Astronomy',
        ed. G. Setti and R. Giacconi, in press.
Schmidt, H.U.: 1968, in 'Structure and Development of Solar
        Active Regions', ed. K.O. Kiepenheuer, pp.95-107,
        Reidel, Dordrecht.
Skumanich, A.: 1972, Astrophys. J. 171, pp.565-567.
Spiegel, E.A. and Weiss, N.O.: 1980, Nature, in press.
Stix, M.: 1976, in 'Basic Mechanisms of Solar Activity' ed.
        V. Bumba and J. Kleczek, pp.367-388, Reidel,
        Dordrecht.
Vaiana, G.S. and 15 others: Astrophys. J., in press.
Vainshtein, S.I. and Zel'dovich, Y.B.: 1972, Usp. Fiz. Nauk 106,
        pp.431-457; Sov. Phys. Usp. 15, pp.159-172.
Weiss, N.O.: 1966, Proc. Roy. Soc. A. 293, pp.310-328.
Weiss, N.O.: 1974, in 'Magnetohydrodynamics', by L. Mestel and
        N.O. Weiss, pp.183-248, Swiss Soc. Astron.
        Astrophys., Geneva.
Weiss, N.O.: 1978, Mon. Not. R. Astron. Soc. 183, pp.63P-65P.
Weiss, N.O.: 1981, J. Fluid Mech., in press.
Wilson, O.C.: 1978, Astrophys. J. 226, pp.379-396.
Wolstencroft, R.D., Smith, R.J. and Clarke, D.: 1980, to be
        published.
Zwaan, C.: 1978, Solar Phys. 60, pp.213-249.

STELLAR MAGNETIC STRUCTURE AND ACTIVITY

Cornelis Zwaan

The Astronomical Institute, Utrecht, The Netherlands

1. INTRODUCTION:
SOLAR MAGNETIC STRUCTURE AND A FRAMEWORK FOR MAGNETIC STRUCTURE IN COOL STARS

The study of magnetic structure and activity in cool stars is largely inspired by the Sun. In the Sun the magnetic field has been concentrated in discrete elements which form a sequence, from the large dark sunspots down to the bright small elements in the magnetic network. The field strengths range from 3 kilogauss in sunspot umbrae to about 1.5 kilogauss in the network elements (for reviews, see Harvey, 1977 and Zwaan, 1980). This sequence may be explained by a set of magnetostatic flux tube models which depend on the magnetic flux through the tube (for a summary and references, see Zwaan, 1978). A basic assumption in the magnetostatic modeling is that the energy flux within the flux tube is no more than 20% of the energy flux in the nonmagnetic convection zone.

The magnetostatic models explain the magnetic elements as structures in equilibrium with the nonmagnetic surroundings, but how are these elements formed? Extensive numerical experiments by Weiss and coworkers (see Galloway et al., 1977; Peckover and Weiss, 1978) have indicated that any initially weak magnetic field in a convection zone will become locally concentrated at the boundaries of the velocity cells. In fact the magnetic field, when emerging in growing active regions, has already a field strength of several hundred gauss, which is close to the equipartition field strength in the top of the convection zone (Zwaan, 1978). Once the flux loops have arrived in the atmosphere the radiative cooling starts a downdraft which carries the cooling downwards. Spruit (1979) investigated the "convective collapse"

R. M. Bonnet and A. K. Dupree (eds.), Solar Phenomena in Stars and Stellar Systems, 463–486.
Copyright © 1981 by D. Reidel Publishing Company.

following an initial downdraft, which concentrates the field to
about 1.5 kilogauss.

The present understanding of the solar magnetic structure
strongly suggests that a magnetic field in any convective
envelope will be similarly concentrated in discrete features.
Although the dynamo theory for solar magnetism is still rather
sketchy, it seems plausible that similar dynamos operate in
other stars with convective envelopes. In any case, there are
many photospheric, chromospheric and coronal phenomena of cool
stars that resemble the solar phenomena of magnetic origin.
Stars of the same effective temperature and surface gravity
may show very different levels of magnetic activity, which
indicates that an extra parameter is needed for the description
of the structure of cool stars.

In order to give structure to the summary of observational
aspects of stellar magnetism, we first outline a framework
consisting of four hypotheses. These hypotheses are suggested by
solar studies and they can be applied to interpret data on cool
stars. Moreover, the hypotheses suggest observational tests,
which – if successfully passed – will put the framework on a
quantitative footing.
A.   *In stars with convective envelopes the magnetic field has
been concentrated in discrete elements to large strength, with
very low field strengths in between.* As in the Sun, the discrete
elements may be modeled as magnetostatic tubes. Large elements
would show up as dark ("starspots"), whereas the small elements
would appear as somewhat brighter in photosphere and low
chromosphere ("stellar faculae and network elements").
B.   *The chromospheric emission in the cores of the Ca II H and
K and the Mg II h and k lines, and the emission from chromosphere
and transition region in the UV lines (λ > 1000 Å) originate
predominantly from the magnetic flux tubes.* Close correlations
are expected between the fluxes in chromospheric and transition-
region lines because spectroheliograms in all these lines show
very similar spatial structure. Note, however, that these
correlations may depend on effective temperature and surface
gravity. The coronal emission (soft X rays, radio) will also
correlate with chromospheric emission but less closely: X-ray
heliograms show the same active regions, but the spatial structure
differs in detail from the structure in optical and UV spectroheliograms.
The statement, that the emission from the outer atmosphere
originates predominantly in magnetic structure, implies that the
heating of the outer atmosphere is correlated with magnetic
structure.

The number density of the magnetic flux tubes in the stellar
atmosphere depends on the efficiency of the dynamo operating in
the convective envelope. Since the dynamo depends on the

differential rotation brought about by convection and rotation,
it seems plausible that the measure of differential rotation
depends on the mean rotation rate.
C. *The efficiency of the dynamo decreases with decreasing
rotation rate.* This idea has been explored already by Skumanich
(1972). Note that the dynamo efficiency probably depends on the
structure of the convection zone as well (for instance, on its
depth). Hence a dependence on effective temperature and surface
gravity is to be expected.
D. *Stellar rotation is efficiently braked by the stellar wind
streaming out along the coronal and circumstellar magnetic fields*
(Schatzman, 1959, 1962, 1965). The loss of angular momentum per
unit mass in the stellar wind is quite large since the outflowing
matter is forced to corotate with the star to a large distance
because of the magnetic field. Note that here the term "stellar
wind" also includes transient mass ejections.

The hypotheses C and D bring in the time-dependent angular
momentum as an extra parameter determining stellar structure and
evolution (see Kippenhahn, 1973).

2. WAYS OF OBSERVING STELLAR MAGNETIC STRUCTURE AND ACTIVITY.

There is little hope of detecting magnetic fields in cool
stars by measuring circular polarization in the σ components
of magnetically sensitive lines, because the polarities cancel
in averaging. For instance, the field strength averaged over the
solar disk is of the order of 1 gauss. Possible ways of observing
magnetic structure and activity are listed below, in order of
decreasing directness.
(i) *Searching in spectral lines of large magnetic splitting
for σ components* (without circular polarization). These σ components, originating in the magnetic structure, are found on both
sides of the unshifted line from the nonmagnetic part of the
atmosphere and the π component formed in the magnetic structure.
The method requires large ratios: (magnetic splitting) to (line
width), signal-to-noise, and a fair fraction of the disk covered
with magnetic flux tubes. Even if it can only be applied to a
small sample of bright and very active stars, the method is of
great importance for the calibration of the less direct diagnostics mentioned below. It will yield quantitative data on the
magnetic structure: the typical field strength, and the fraction
of the disk covered with it. The first positive results have been
obtained by Robinson et al. (1980) for very active main-sequence
stars.
(ii) *Linear polarization* in narrow or broad spectral bands.
Leroy (1962) and Dollfuss (1965) noticed that magnetic areas not
far from the solar limb show a weak linear polarization (between
10^{-3} and 10^{-2}) perpendicular to the limb. Leroy (1962) explains

the polarization as the net effect of the π components of
strong spectral lines being more saturated than the σ components.
The polarization signal increases with the transverse component
of the magnetic flux and with the density of saturated magnetically sensitive lines in the passband.

Polarization surveys of stars in the solar neighborhood
suggest intrinsic linear polarization in many F, G, K and M stars.
Tinbergen and Zwaan (1981) suggest that this polarization is due
to magnetic regions in activity belts on both sides of the stellar
equator. They argue that a star may show detectable polarization
provided that it is more active than the Sun, and that it is not
observed nearly pole-on.

This method may be important in the study of active stars
(see Section 6.3), so a further exploration is needed.

Photometry provides indirect ways of searching for magnetic
structure.

(iii) *Broadband photometry* is used in searches for time
variations in the photospheric continuum which are attributed to
effects of starspots and stellar faculae not uniformly distributed
over the disk. Solar flux variations due to spots and faculae are
too small for detection with conventional stellar photometry.
However, flux variations due to starspots have been discovered
for very active cool stars (see Bopp and Evans, 1973, and the
contributions to this volume by Hall and by Hartmann).

(iv) *Spectral photometry and line photometry* can reveal hot
chromospheric structure by means of emission cores in strong
spectral lines in the optical spectrum. The classical example
is the Ca II H and K resonance doublet. The more sensitive Mg II
h and k doublet and many chromospheric emission lines further
down in the UV are made accessible for a large number of stars
by IUE.

(v) *Measurement of emissions from transition regions and
coronae.* The transition region from chromosphere to corona may
be investigated by means of the corresponding emission lines in
the UV (IUE). Coronae may be studied by means of soft-X-ray
emission (Einstein Observatory) and radio emission.

IUE spectra and Einstein data confirm that many cool stars
have hot outer atmospheres whose properties cannot be described
in terms of the two classical parameters: effective temperature
and surface gravity. Although the recent increase of interest
in cool stars has been provoked by the IUE and Einstein results,
this lecture concentrates on Ca II H and K observations since
these have been obtained for a large number of stars.

The Ca II H and K emission is measured or estimated in two
ways (Figure 1). In line photometry (Wilson, 1968; Vaughan et al.,
1978) the total fluxes in the line cores are measured and normalized by fluxes measured simultaneously in the neighboring
"continuum" background, yielding the Ca II H and K *line core
index* F_{H+K}/F_{R+V}, where R and V stand for the reference windows

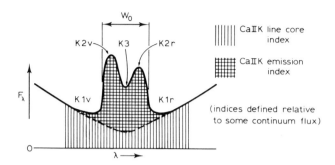

Figure 1: Core of the Ca II K line profile, with definitions

on the red and violet sides of the H and K doublet. For visual
estimates of the Ca II emission from spectrograms (e.g.,Wilson,
1976) the visibility of the K2 and H2 emission features is
estimated relative to the continuum background, which yields
a crude measure for the *emission index*. Note that an apparent
absence of emission features but a finite line core flux (dashed
in Figure 1) still requires a chromospheric temperature rise;
a radiative equilibrium model produces virtually black line
cores (see Linsky's contribution to this volume).

3.   CA II H AND K EMISSION

      Enhanced Ca II H and K line core fluxes are observed in
many main-sequence stars of spectral type later than F2 (see
below), and in giants later than about G0 (Wilson 1976). This
region in the Hertzsprung-Russell diagram (Figure 2) corresponds
to that of stars having well developed convective envelopes.
Hence the conditions for time-dependent discrete magnetic
structure (see Section 1.1) are satisfied.
      Note that in Figure 2 the magnetic Ap stars, with their
time-independent dipole-like magnetic structure, are located
in the area of stars without efficient convective envelopes.
Indeed the mechanisms for concentrating magnetic field in
discrete flux tubes (Section 1.) cannot work in stars without
convection zones (Zwaan, 1977).

3.1 Main-Sequence Stars

      Since O.C. Wilson's pionering work, Ca II H and K line core
indices have been measured at Mt. Wilson Observatory for many
stars of the same color very different H and K indices are found.
For single stars the strength of the H and K emission appears

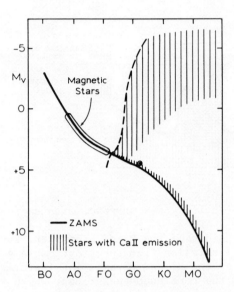

Figure 2: Schematic Hertzsprung-Russell diagram showing the domains of stars with Ca II H and K emission, and of magnetic Ap stars. The broken line indicates the boundary between stars with and without thick convective envelopes.

to decrease with age, as is demonstrated by the comparison of field stars and cluster stars (Wilson, 1963, 1970), stars of different Strömgren c_1 index (Wilson and Skumanich, 1964), and stellar groups of different spatial velocities (Wilson and Woolley, 1970; Vaughan and Preston, 1980).

Kraft's (1967) study of the rotational line broadening of cool main-sequence stars has shown that the rotation rate decreases with age as well. This suggests that the decrease of H and K emission with age is due to the rotation rate decreasing with time because of loss of angular momentum (this idea is incorporated in the framework in Section 1.).

Middelkoop (1981) has measured H and K indices for F-type stars showing large rotational line broadening. The results (Figure 4) indicate that the H and K emission decreases with decreasing rotation rate.

There is a remarkably steep rise of the H and K index with increasing b-y for b-y > 0.29 (Figures 3 and 4) in all groups of stars of similar rotation rate (or age), except for old, slowly rotating stars. This steep rise may be due to the dynamo efficiency increasing with the thickness of the convective envelope (Section 1., item C), which increases rapidly from spectral type F to G.

The short-period binaries shown in Figure 4 are discussed in Section 3.3 below.

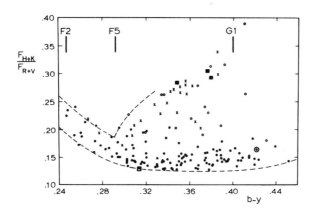

Figure 3: Ca II line core index F_{H+K}/F_{R+V} versus color (b-y) for field stars (Wilson, 1968) and Hyades (Wilson, 1970). Dots: field stars without emission visible on 10 Å/mm spectrograms; open circles: field stars with emission visible on spectrograms; crosses: Hyades; ⊙: quiet Sun (Wilson, 1978). Squares indicate stars observed with Einstein Observatory; filled squares: X-ray flux detected, open squares: upper limit (see Section 4.).

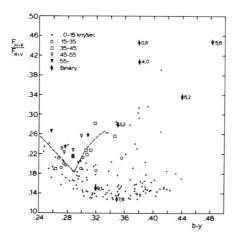

Figure 4: Ca II index F_{H+K}/F_{R+V} versus (b-y) for field stars measured by Wilson (1968) and Middelkoop (1981, for known vsini values indicated in the figure). For the binaries the periods are given in days. The broken line is the upper limit for Wilson's data, copied from Figure 3.

3.2 Giants

   For evolved stars quantitative measurements of H and K line
core indices have just been started. Wilson (1976) has published
eye estimates of the peak intensities of the emission features
relative to the neighboring continuum background from spectrograms.
He introduced six intensity classes, ranging from I_{H+K} = 0 (no
emission cores visible on 10 Å/mm spectrograms), to I_{H+K} = 5
(emission peaks about as high as, or higher than the
neighboring continuum). Note that I_{H+K} = 0 does not mean that the
line cores are black. Middelkoop and Zwaan (1981) have
reanalysed Wilson's data for G, K and M stars of luminosity
classes IV, III and II, which we call giants for short.

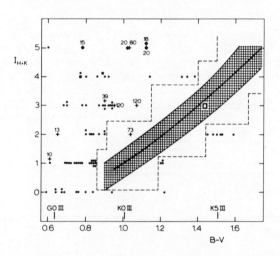

*Figure 5: Intensity class I_{H+K} versus color (B-V) for the giants
in Wilson's (1976) sample with M > -3.0. The heavy line
shows the main I_{H+K} (B-V) trend; the hatched area corresponds
to departures within 1 σ. Individual stars are plotted
if departing more than about 2 σ (broken line) from the I_{H+K} (B-V)
trend, and for I_{H+K} = 0. Binaries with periods P < 120 days
are indicated by crosses, with the periods in days. The squares
indicate stars observed with Einstein Observatory; filled: flux
detected, open: upper limit.*

   Figure 5 shows the intensity classes plotted against B-V.
Many late G-type giants, and the majority of the K and M-type
giants constitute a band of I_{H+K} increasing with B-V. We refer
to this band as the main I_{H+K} (B-V) trend. Wilson (1976) has
pointed out that this trend may be due to the continuum background
decreasing with increasing B-V.
   The G-type giants show a large spread in Ca II emission fluxes;

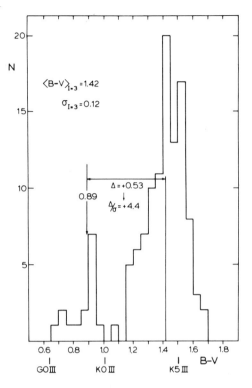

Figure 6: Histogram N(B-V) for stars of intensity class I_{H+K} = 3. The determination of the relative strength Δ/σ is explained for 57 Gem (HD 57727), G8 III, B-V = 0.89.

the corresponding region in the color-magnitude diagram seems to be bounded on the red side by B-V ≈ 0.95. The main trend I_{H+K} (B-V) among the M and K-type giants seems to continue as the lower envelope for the G-type giants.

For further analysis Middelkoop and Zwaan defined a crude measure for the strength of the Ca II H and K emission relative to the main trend I_{H+K} (B-V) among the K and M giants. For every intensity class I_{H+K} they determined the median value < B-V >$_I$ and the standard deviation σ_I (see Figure 6). For every giant star the relative strength Δ/σ is defined as

$$\frac{\Delta}{\sigma} \equiv \frac{< B-V >_I{}^* - (B-V)_*}{\sigma_I{}^*}$$

where $(B-V)_*$ is the color for the particular star, and < B-V >$_I{}^*$ and $\sigma_I{}^*$ are the median and the s.d. for the intensity class

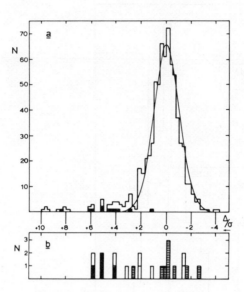

Figure 7a: Histogram N(Δ/σ) for all stars in Wilson's (1976) sample with M_V > -3.0. Binaries with periods P < 120 days are indicated in black.
Figure 7b: Histograms N(Δ/σ) for stars showing spectral-line broadening. Binaries are indicated in black for P < 120 days, and hatched for P > 120 days.

I^* of that star. Figure 7a is the histogram of the Δ/σ values for all stars of luminosity classes IV, III and II in Wilson (1976) sample. From the definition it follows that stars with Δ/σ > +3 exhibit strong Ca II H and K emission, whereas stars with $|\Delta/\sigma|$ < 1 do not differ significantly from the main trend. Figures 5, 6 and 7a show that the stars with enhanced emission are well separated from stars near the main I_{H+K} (B-V) trend.
      (A justification may be useful why   Middelkoop and Zwaan have deliberately chosen to measure the departure from the trend horizontally along the (B-V) direction. Clearly, in case of quantitative measurements of the H and K index the vertical departure from the lower envelope were the obvious measure of the relative H and K emission strength. However, for analysis of Wilson's I_{H+K} classes this does not seem the best procedure, because the   trend is not well defined in the range of the G stars, and because the stars are classified in discrete classes).

      We suggest that the large spread in Ca II emission among the G-type giants is due to a spread in the rotation rates. Wilson noticed indications for abnormal spectral-line broadening for several stars in his sample. For these stars the histogram of the

relative H and K emission strength Δ/σ is shown in Figure 7b. The
comparison with Figure 7a shows an excess of emission strengths
$\Delta/\sigma \gtrsim 2.0$, which is in harmony with the idea that a relatively
large rotation rate correlates with a large H and K emission.
(From Figure 7b it appears that some stars with line broadening
do not show a strong H and K emission: $\Delta/\sigma \lesssim +1.0$. There are two
possibilities: (i) the rotational broadening is so large that the
intensity of the Ca II emission peaks is reduced, and (ii) the
spectrum is contaminated by light from a star of earlier spectral
type which makes the spectral lines more shallow. The latter
explanation fits the large fraction of long-period spectroscopic
binaries among stars with $\Delta/\sigma \lesssim +1.0$ in Figure 7b).

3.3   Spectroscopic Binaries

It is well known that spectroscopic binaries of short period
exhibit strong Ca II H and K emission (see Hiltner, 1947; Gratton,
1950).

Two recent papers discuss the Ca II H and K emission from
spectroscopic binaries. Glebocki and Stawikowski (1977) plotted
the intensity class I_{H+K} against orbital period for binaries among
G- and early K-type giants. They conclude that the H and K
emission decreases monotonically with the period. Young and Koniges
(1977) found that many - but not all - binaries with periods less
than 100 days show enhanced H and K emission, and they also noticed
a tendency towards circular orbits among binaries with enhanced

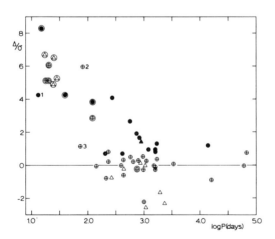

*Figure 8: H and K emission strength Δ/σ versus period P for
spectroscopic binaries with $+3.5 > M_V > -2.5$. Circles: from Wilson
(1976); triangles: from Glebocki and Stawikowski (1977). G-type
giants (B-V < 0.95) are indicated by filled symbols. Virtually
circular orbits (e < 0.05) are marked by circles around the symbols.*

emission. Both papers attempt to interpret the enhanced Ca II
H and K emission as a direct result of tidal interaction
affecting the chromospheric heating.

Middelkoop and Zwaan (1981) used the relative strength Δ/σ
of the H and K emission in their investigation of the spectroscopic
binaries among the giants in the samples of Wilson (1976) and
Glebocki and Stawikowski (1977). The enhancement of Ca II H and K
emission for short-period binaries among giants is apparent from
Figures 5 and 7. Figure 8 shows that the H and K emission is
enhanced for periods shorter than about 120 days; for P > 200 days
the emission is normal for the spectral type. Only the G-type
long-period binaries may show an enhanced emission, but in that
respect they do not differ from the single G-type giants. However,
short-period binaries with enhanced H and K emission are also found
for B-V > 0.95, outside the region where single G-type giants with
enhanced H and K emission are found.

It is striking that the majority of the binaries with enhanced
emission have virtually circular orbits (e $\lesssim$ 0.05), the transition
from nearly circular orbits to random eccentricities for a period
somewhere between 120 and 200 days (Figure 9). This critical period
is about equal to the period separating giants with enhanced H and
K emission from giants with normal emission.

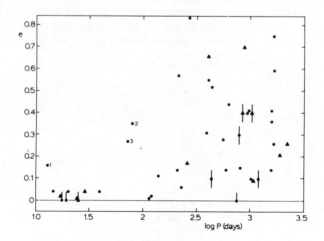

*Figure 9: Eccentricity e versus period P for binaries in the
samples of Wilson (1976, circles) and Glebocki and Stawikowski
(1977, triangles).*

Middelkoop and Zwaan suggest that the G- and K-type binaries
with periods less than about 120 days are in revolution-rotation
synchronization, which makes the primaries rotate faster than the
normal single giants of the same color. The H and K strength Δ/σ
increasing with decreasing period is in harmony with the

hypothesis that the dynamo efficiency increases with increasing rotation rate (see Section 1., hypotheses C and B). The circularity of the orbits supports the idea of synchronization: the theory of tidal friction in components with convective envelopes in close binaries predicts that the synchronization of rotation and orbital motion occurs before the orbits become circular (Zahn, 1977).

Among the *main-sequence stars* the binaries with enhanced H and K emission and circular orbits have periods below a critical period, which is much less than the critical period for giants. Middelkoop's (1981) measurements of F and early G-type dwarfs (Figure 4) indicate that binaries with periods less than about 8 days show an enhanced H and K emission. The same critical period of 8 days separates circular orbits from random eccentricities in the plot of the eccentricities against periods for main-sequence binaries (see Middelkoop, 1981). Here again it is suggested that binaries with periods shorter than the critical period are synchronized; the resulting relatively rapid rotation is the cause for the enhanced Ca II H and K emission.

(D.S.Hall has pointed out to me that similar ideas are advanced in a paper by Bopp and Fekel (1977). These authors suggest that rapid rotation is the ultimate cause for the BY Draconis phenomenon, that is, the photometric variation in very active late-type dwarfs which attributed to "starspots". Bopp and Fekel show that not all BY Dra variables are close binaries. They attribute the large fraction of binaries to the rapid rotation caused by synchronization).

3.4   Recent Measurements; Preliminary Conclusions

Middelkoop is surveying the giants using the Mt. Wilson H and K photometer. He measures stars from Wilson's (1976) sample; he adds some rapidly rotating single stars and spectroscopic binaries. The survey is not yet completed; some preliminary results will be published by Oranje and Middelkoop (1981). In this context it is of interest that the quantitatively measured Ca II H and K line core indices F_{H+K}/F_{R+V} substantiate the conclusions from Wilson's estimated I_{H+K} classes, mentioned in Sections 3.2 and 3.3.

The trend I_{H+K} (B-V) among the K- and M-type giants (Figure 5) appears as a narrow strip of F_{H+K}/F_{R+V} increasing with B-V, yet there seems to be a small intrinsic spread in F_{H+K}/F_{R+V}. There is a fairly dense extension of the strip in the late type G giants (B-V > 0.80). For 0.60 < B-V < 0.80 the trend extends as the lower envelope in the F_{H+K}/F_{R+V} versus B-V plot.

The wide spread in H and K indices among the G-type giants is confirmed. Furthermore, giants with large rotational line broadening vsini exhibit large Ca II H and K emission, which

confirms the indications from Wilson's (1976) data on rotational
broadening. The redward boundary to the region of active G-type
giants   near B-V = 0.95 needs some modification: Middelkoop found
some giants in the range 0.95 < B-V < 1.05 with a somewhat enhanced
H and K emission.

Figure 5 shows that short-period binaries with enhanced H and
K emission are also found beyond the redward boundary of the active
single G-type giants. The lack of active binaries for B-V > 1.15
is due to the scarcity of spectroscopic binaries of periods less
than about 200 days among K- and M-type giants; there is none in
Wilson's (1976) sample. Middelkoop included all known and accessible short-period binaries in his survey: all giant binaries with
periods less than about 150 days observed until present show
enhanced H and K emission; the reddest one HD 352 (B-V = 1,38,
P = 96 days, e = 0.12 ?) exhibits a strongly enhanced H and K
emission.

In this Section 3. we have compared Ca II H and K data with
the framework of hypotheses in Section 1. Below we summarize the
main conclusions, and we formulate some urgent questions.

(i) Relatively rapidly rotating single main-sequence stars
and many G-type single giants show enhanced Ca II H and K emission.
There are indications that relatively rapidly rotating G-type
giants show more Ca II emission than slowly rotating G- and K-type
giants. However, the quantitative dependence of the absolute Ca II
H and K flux on rotation rate, effective temperature and surface
gravity has not yet been determined.

(ii) The majority of short-period binaries with enhanced Ca II
H and K emission are probably synchronized, since the orbits are
virtually circular. The critical period separating these more
active binaries from the others depends strongly on the luminosity
class (about 120 days for G- and early K-type giants, and about
8 days for F- and early G-type main-sequence stars). The dependence
of the critical period on spectral type is not yet known. The
enhanced H and K emission is explained by the relatively rapid
rotation which is due to the synchronization of rotation and
orbital motion. However, it remains to be demonstrated that a
synchronized binary produces the same Ca II H and K flux as a
single star of the same rotation rate, surface gravity and effective temperature.

(iii) The important Wilson-Bappu relation, which is not
discussed in this paper, has defied a quantitative interpretation.
A difficult aspect is that the Wilson-Bappu width W_0 of the H and
K emission features is independent of the H and K line-core index.
Zwaan (1977) suggests the following explanation: the *width* W_0 is
an *average* property of *discrete* magnetic flux tubes (averaged over
the stellar disk). The *intensity* of the emission feature is
proportional to the *number of flux tubes* in sight. In this way
the width and the intensity of the emission features are uncoupled.

4. RELATIONS BETWEEN EMISSIONS FROM CHROMOSPHERES , TRANSITION REGIONS AND CORONAE

In Section 1., hypothesis B, correlations are suggested between the fluxes in the Ca II H and K lines cores, in the UV emission lines from chromospheres and transition regions, and in soft X-rays. This section summarizes some data bearing on the expected relations (a comprehensive review of recent IUE and Einstein data is given in Linsky's contribution to this volume).

Linsky and Haisch (1979) classify the IUE short-wavelength (1175 - 2000 Å) spectra of cool stars in two distinct groups: (i) "solar-type" stars showing all the chromospheric and trans- ition-region emission lines, and (ii) "non-solar-type" stars showing no transition-region lines, only chromospheric lines indicating temperatures up to 20 000 K. Main-sequence stars (F2 - M) and G-type giants fall in the group of solar-type stars, whereas K- and M-type giants are of non-solar type. Note that the domain of solar-type stars coincides approximately with the region where stars with enhanced Ca II emission are found, whereas the non-solar type stars correspond to giants in the I_{H+K} (B-V) trend.

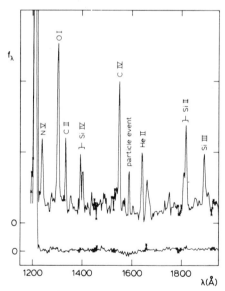

Figure 10: IUE short-wavelength spectrograms of the active binary HD 32357 (K0 III, B-V = 1.03) and the quiet single star HD 25604 (K0 III, B-V = 1.07). The fluxes have been reduced on the same relative scale by the factor $10^{-0.4m}$ allowing for differences in apparent brightness.

    Oranje obtained IUE short-wavelenth spectra of three G-type
single giants of high Ca II emission. These spectra do show all
the chromospheric and transition-region lines. In addition Oranje
obtained spectra of two stars just outside the region of active
G-type giants: HD 32357, a KO III spectroscopic binary (B-V = 1.03,
P = 80 days), with a very large Ca II H and K emission index, and
HD 25604, KO III (B-V = 1.07), with a very low H and K index (see
Figure 10). The spectrum of the active binary is similar to the
spectra of the three active G-type giants. However, the spectrum
of the quiet giant does not show any emission line, not even
chromospheric lines; its upper-limit fluxes are much less than for
the active stars (see Figure 10, and Oranje and Middelkoop, 1981).
    The IUE result suggests that the UV spectra correlate with
the Ca II H and K indices, but the correlation coefficients may
differ for different chromospheric and transition-region lines.
The correlations may depend on effective temperature and surface
gravity. We suspect that even in the domain of solar-type stars
(as indicated by Linsky and Haisch) the stars of low Ca II H and K
index will show UV spectra that are very weak in all spectroscopic
lines.

    Mewe and Zwaan selected cool stars for their guest-observer
programs with Einstein Observatory on the basis of the Ca II H and
K emission, which turns out to be a very useful criterion (Mewe
and Zwaan, 1980). All stars with large H and K indices are
efficient soft X-ray emitters. Moreover, in the first sample
(main-sequence stars F8 - K7; G-type giants) the X-ray and the
Ca II H and K fluxes correlate fairly well (Figure 11), bearing

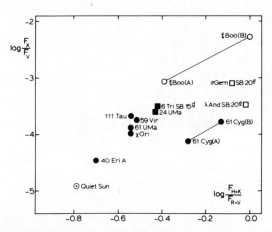

*Figure 11: X-ray flux normalized by the flux in the V band against
Ca II H and K index. The binaries ξ Boo and 61 Cyg are plotted
twice, assuming that all X-ray emission comes from component A or
from B. The periods of the spectroscopic binaries are given in days.
Circles:main-sequence stars; squares: giants; open symbols:HEAO-1;
filled symbols: HEAO-2.*

in mind that only a crude correlation is to be expected (Section
1., hypothesis B). Somewhat surprizing is that main-sequence stars
and giants seem to follow the same relation. Apparently there is
no systematic difference between short-period binaries and single
stars. (The correlation may be improved by normalizing both the
X-ray flux and the Ca II H and K flux by the bolometric flux).

Recently Mewe and Zwaan have obtained upper limits for the
X-ray fluxes F_x/F_v well below the value for the quiet Sun for two
stars of very small Ca II H and K index (these stars are indicated
in Figures 3 and 5).

Results obtained with IUE and Einstein Observatory up till now
indicate that the relations between chromospheric, transition-
region and coronal emissions are in agreement with the qualitative
predictions given in Section 1., hypothesis B. In particular, the
transition-region and the X-ray emissions are very small for stars
of low Ca II H and K index. Determinations of the quantitative
relationships are in progress.

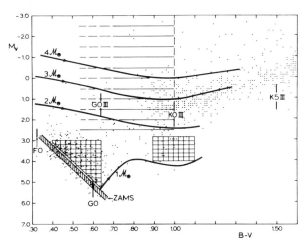

*Figure 12: Evolutionary tracks in the Hertzsprung-Russell diagram
(from Middelkoop and Zwaan, 1981). The region of G-type giants
of different levels of activity is indicated by a horizontal
hatching; the regions of the hot and cool components of RS CVn
binaries with periods P ≤ 10 days by cross hatchings (data from
Popper and Ulrich, 1977). The evolutionary tracks were kindly
supplied by Th.J. van der Linden, Amsterdam .*

5.   EVOLUTIONARY SCENARIO

The previous sections suggest that the chromospheric,
transition-region and coronal emissions are correlated manifestations of the stellar activity which depends on the rotation rate

of the star. In this section we complement the set of hypotheses
(Section 1.) with an evolutionary scenario, which attempts to
explain the variation of the stellar activity across the Hertzsprung-Russell diagram (Figure 12).

5.1 Single Stars

Stars with masses smaller than 1.5 solar masses. Near the
zero-age main-sequence (ZAMS) these stars have convective
envelopes, and they rotate relatively rapidly. As a consequence
the level of activity is high. However, the braking of the rotation reduces the dynamo efficiency, and hence the chromospheric
and coronal emissions decrease with age. These stars evolve
further as slowly rotating G to K subgiants, showing very little
Ca II H and K emission; they follow the I_{H+K} (B-V) trend (Figure
5).

Stars with masses between 1.5 and 4 solar masses. During the
main-sequence phase most of these stars are rapidly rotating.
Because there is no efficient convective envelope there is no
magnetic activity, and the star evolves without appreciable loss
of angular momentum until it enters the giant branch where a
convective envelope develops. Then the dynamo is turned on;
because of the still rapid rotation the level of activity is very
high. However, at the same time the star starts loosing angular
momentum, the dynamo efficiency drops, and so do the chromospheric
and coronal emissions. Eventually the rotation rate has dropped
to such a low value that the dynamo efficiency is small. From
there the K → M giant follows the I_{H+K} (B-V) trend (see Figure 5).
The observational data indicate that the active phase of
rapid rotation lasts for the short period of evolution through
spectral type G; if this picture is correct then the loss of
angular momentum should be very efficient.
      The spread in the Ca II H and K index for G-type giants is
due to a spread in rotation rate, which is probably largely due
to a spread in stellar mass. There is observational support for
this idea: from subgiants the Ca II emission is weak (I_{H+K} = 0 or
1), whereas giants show a large spread in I_{H+K}.
      Note that slowly rotating magnetic Ap stars are expected
to evolve into G-type giants that are *not* very active (although
transient active phenomena may happen when the onset of convection
destroys the original dipole-like magnetic structure).

5.2 Short-Period Binaries

      If the orbital period is below some critical period, then
the orbital motion and the rotation of the primary will become
synchronized some time after the primary has developed a convective envelope. After synchronization the primary rotates rapidly,
consequently the level of activity is high. For the loss of angular

momentum the primary can draw from the *total* angular momentum
in the binary system, which is many orders of magnitude larger
than the rotational momentum in the primary.

One important consequence is that for a primary in a
synchronized binary the *active* phase of the stellar evolution
lasts much longer than for a single star of the same mass. This
is in harmony with data discussed in Sections 3.2 and 3.3:whereas
for single giants the enhanced activity lasts only through spectral
type G (B-V $\lesssim$ 1.00), *all* primaries of binaries with periods less
than about 100 days show enhanced activity, even for B-V > 1.00.

The scenario predicts that all *single* subgiants show little
activity. However, subgiants in synchronized binaries are expected
to be very active, as is the case.

The so-called RS CVn variables, very active binaries of very
short period, discussed in Hall's contribution to this volume,
fit the scenario as extreme cases. The positions of the components
of short-period RS CVn variables are indicated in Figure 12.

6.   VARIABILITY OF MAGNETIC ACTIVITY

Repeated measurements of the Ca II H and K line core index
show variability for all cool stars, except those of a very low
H and K index. We restrict the discussion to the variability
of main-sequence stars (from F5 to M), for which there is a large
body of data. The *intrinsic scatter* in the H and K index shows
a wide range of time scales, from ten years (the length of longest
records),down to hours, and occasionally down to minutes.

6.1 Long-Period Variations, "Cycles"

O.C. Wilson has undertaken measurements of the Ca II H and K
line core indices of 91 main-sequence stars of spectral types F
to M over a period of 10 years. The results (Wilson, 1978) indicate that about one quarter of the sample shows waves in the H and
K emission that look similar to the solar cycle. The periods range
from 6 to more than 10 years. Of course the term "stellar cycle"
is somewhat premature until it has been demonstrated that the
wavelike variations continue for some periods (fortunately, O.C.
Wilson's work is being continued at Mt. Wilson by Vaughan and
Preston).

About one third of the stars investigated do not show cycle-
like variations but they vary on short time scales, and often the
variations seem stochastic.

Some patterns in the variability have been pointed out by
Vaughan (1980). In the diagram of Ca II H and K indices against
color (Figure 13) there are two branches separated by a gap. This
is not an artefact in Wilson's sample: the two branches separated
by a gap appear in the solar-neighborhood survey as well (Vaughan
and Preston, 1980). Note that the range of variation per star,

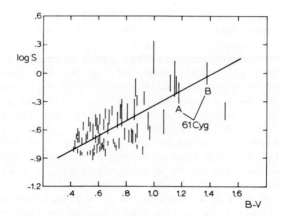

*Figure 13: A measure for the H and K index (S) against B-V for
the stars in Wilson's stellar-cycle program. The length of the
bars corresponds to the range of S due to the variations. The
straight line divides the stars in the two groups discussed in
the text (from Vaughan, 1980).*

though appreciable, is small as compared to the spread in the
mean H and K indices of the main-sequence stars; for one thing,
the time variations do not obliterate the gap between the branches.
    The young stars in the upper branch do not show cyclic
variations, the variations seem stochastic (with the exception of
the boundary cases 61 Cyg A and B). However, many of the stars
in the lower branch exhibit waves similar to the solar cycle; the
incidence of these "cycles" is largest among the G- and K-type
stars.
    It seems unlikely that the two branches are due to two waves
of star births: the stars, except the youngest, have revolved
several times about the galactic center, and consequently they
have been formed in very different parts of the galaxy. A coherent,
sudden change in the rate of star formation over such a large
volume is difficult to accept (Vaughan and Preston, 1980).

    The absence of cycles in the upper branch may be due to the
"noise" of chromospheric activity (e.g., flare-like brightenings)
obliterating a cyclic variation. The gap may indicate two dynamo
modes, which would follow the following rules. In young rapidly
rotating stars the dynamo operates in the first mode, which is
assumed to produce a dense magnetic structure in the atmosphere
(in this mode the dynamo may not be cyclic, or the cycles may be
lost in chromospheric noise). Due to the braking the rotation
rate drops. Once the rotation rate has dropped below some critical
value the dynamo is assumed to switch to a second mode, which
produces a less dense magnetic structure in the atmosphere.
In this second mode the dynamo is cyclic in many - or all - cases.

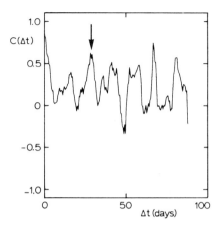

*Figure 14: An autocorrelation curve for the Ca II H and K index
of the Sun-as-a-star as a function of time. The arrow marks the
solar synodic rotation period (from Oranje, 1981).*

6.2   Short-Period Variations; Rotational Modulation

The study of the short-period variations in the Ca II H and
K line-core indices has just been started. A particularly interesting question is whether rotational modulation may be discovered.
The observational study of stellar dynamos would greatly profit
from the availability of accurate rotation *periods* for a wide
range of periods, and without the annoying inclination factor
sini.

Oranje (1981) is investigating time series of Ca II H and K
index of the Sun-as-a-star obtained with the adapted Utrecht
solar telescope and spectrograph. The autocorrelation curve
(Figure 14), based on about 100 days of observation out of a
period of 400 days, does show peaks at the solar rotation period
P and multiples of P, but other peaks as well. The comparison
between the Ca II H and K index of the full disk and Ca II
spectroheliograms is instructive. For instance, the peak in Figure
14 at about half the rotation period can be traced to a configuration of two large complexes of activity separated by about 180^{0}
in longitude, which lasted for several months. Moreover, short-
term variations in the H and K index can be attributed to active
regions coming and passing around the limb, and the emergence and
rapid growth of active regions. Flares will produce sharp upward
spikes in the H and K index. The solar sample suggests that the
rotation modulation may be detectable in the Ca II H and K index
from stars, but in many cases the assignment of unambiguous
rotation periods may not be simple.

Middelkoop et. al. (1980) investigated the feasibility of
finding rotation modulation by measuring the H and K indices
of 14 stars at least once per night during 10 nights. In 5 cases
they find waves in H and K emission that may be due to rotation;
in the case of the most active stars the H and K signal varies
strongly and irregularly. In nearly all cases there are short-
term variations with time scales down to a few hours, the
smallest interval in the measurements. (Most of the variations
can be interpreted in phenomena known on the solar disk. However,
occasionally deep drops in the signal have been observed, which
last less than a few hours. We see no explanation for this
phenomenon).

This summer (1980) observers from Harvard-Smithsonian Center
for Astrophysics, Sacramento Peak Observatory and Utrecht
Observatory participate with Mt. Wilson observers in a campaign
to measure some 50 stars for a period of three months.

6.3 Chromospheric Noise; Photospheric Measurements?

It appears that waves in the H and K index produced by
rotational modulation and by cyclic variation of the magnetic
activity may be masked by a large-amplitude, short-period noise
due to chromospheric activity.

The Sun indicates that the chromospheric emission varies more
strongly in time scales smaller than the rotation period than the
photospheric magnetic structure. So let us briefly reconsider
prospects of photospheric diagnostics for very active stars.

Broad-band photometry has been applied to the very active
RS CVn and BY Dra variables: the waves in the fluxes have been
attributed to rotation modulation produced by a star showing
large "starspots" (see Hall's and Hartmann's contributions to
this volume). Is broad-band photometry also capable of finding
long-period variations (cycles), and is it applicable to
moderately active stars?

Another possible photospheric diagnostic is the linear
polarization discussed in Section 2., item (ii). If the linear
polarization can be measured with sufficient accuracy for at
least the very active stars, then this method is very promising
for the discovery of rotation and cycle periods.

7.   CONCLUDING REMARKS

Recent developments in the study of solar and stellar
activity mark a vigorous start of the observational study of
stellar dynamos, that is, the interaction between convection and
rotation, producing magnetic structure in convective envelopes
and atmospheres. The dynamo and the heating of the outer
atmosphere are to be included in the theory of stellar structure;

the theory of stellar evolution should allow for losses of mass and angular momentum. As the observational counterpart both solar *and* stellar studies are required.

Solar studies are necessary for unraveling the physics of:
(i) the discrete magnetic structure, which mainly consists of flux tubes with diameters $\lesssim$ 200 km;
(ii) the heating of the outer atmosphere, which probably occurs by energy picked up in the convection zone and transferred by the individual flux tubes;
(iii) the magnetic patterns in space and time during the activity cycle.

The intrinsically small scales in the physics of the topics (i) and (ii) wait for the Solar Optical Telescope (SOT) to be brought in orbit by the end of 1985, in order to carry out solar investigations of great importance for stellar astrophysics.

Stellar studies are needed to determine general properties of activity, outer atmospheres and losses of mass and angular momentum in dependence on (i) mass (and chemical composition), (ii) age, and (iii) companionship in close binaries.

Acknowledgements

I wish to thank F. Middelkoop and B.J. Oranje for permission to quote their unpublished results, and for stimulating discussions. In the preparation of this paper I profited from thorough criticism by L.E. Cram, R.J. Rutten and O.C. Wilson on an earlier draft. I also wish to thank A.G. Hearn, F. Middelkoop and B.J. Oranje for comments on the draft, E. Landré for preparing the figures, and Mrs. J.G. Odijk who took care of the typescript.

References

Bopp, B.W., and Evans, D.S.: 1973, *Monthly Not. Roy. Astron. Soc. 164*, 343
Bopp, B.W., and Fekel, F.: 1977, *Astron. J. 82*, 490
Dollfuss, A.: 1965, in R. Lüst (ed.): "Stellar and Solar Magnetic Fields", *I.A.U. Symp.* No. 22, 176
Galloway, D.J., Proctor, M.R.E., and Weiss, N.O.: 1977, *Nature 266*, 686
Glebocki, R., and Stawikowski, A.: 1977, *Acta Astron. 27*, 225
Gratton, L.: 1950, *Astrophys. J. 111*, 31
Harvey, J.W.: 1977, in E.A. Müller (ed.): *Highlights of Astronomy 4*, Part II, 223
Hiltner, W.A.: 1947, *Astrophys. J. 106*, 481
Kippenhahn, R.: 1973, in S.D. Jordan and E.H. Avrett (eds.): "Stellar Chromospheres", NASA Sp. 137, p. 265
Kraft, R.P.: 1967, *Astrophys. J. 150*, 551
Leroy, J.-L.: 1962, *Ann. Astrophys. 25*, 127

Linsky, J.L., and Haisch, B.M.: 1979, *Astrophys. J.* _229_, L 27
Mewe, R., and Zwaan, C.: 1980, in A.K. Dupree (ed.): Proceedings
Workshop on Cool Stars, Stellar Systems and the Sun, Cambridge
(Mass.)
Middelkoop, F.: 1981, in preparation
Middelkoop, F., Vaughan, A.H., and Preston, G.W.: 1980, *Astron.
Astrophys.*, in press
Middelkoop, F., and Zwaan, C.: 1981, in preparation
Oranje, B.J.: 1981, in preparation
Oranje, B.J., and Middelkoop, F.: 1981, in preparation
Peckover, R.S., and Weiss, N.O.: 1978, *Monthly Not. Roy. Astron.
Soc.* _182_, 189
Popper, D.M., and Ulrich, R.K.: 1977, *Astrophys. J.* _212_, L 131
Robinson, R.D., Worden, S.P., and Harvey, J.W.: 1980, *Astrophys.
J.* _236_, L 155
Schatzman, E.: 1959, in J.L. Greenstein (ed.): "The Hertzsprung-
Russell Diagram", *I.A.U. Symp.* No. 10, 129
Schatzman, E.: 1962, *Ann. Astrophys.* _25_, 18
Schatzman, E.: 1965, in R. Lüst (ed.): "Stellar and Solar Magnetic
Fields", *I.A.U. Symp.* No. 22, 153
Skumanich, A.: 1972, *Astrophys. J.* _171_, 565
Spruit, H.C.: 1979, *Solar Phys.* _61_, 363
Tinbergen, J., and Zwaan, C.: 1981, in preparation
Vaughan, A.H.: 1980, *Publ. Astron. Soc. Pacific* _92_, 392
Vaughan, A.H., Preston, G.W., and Wilson, O.C.: 1978, *Publ.
Astron. Soc. Pacific* _90_, 267
Vaughan, A.H., and Preston, G.W.: 1980, *Publ. Astron. Soc.
Pacific* _92_, 385
Wilson, O.C.: 1963, *Astrophys. J.* _138_, 832
Wilson, O.C.: 1968, *Astrophys. J.* _153_, 221
Wilson, O.C.: 1970, *Astrophys. J.* _160_, 225
Wilson, O.C.: 1976, *Astrophys. J.* _205_, 823
Wilson, O.C.: 1978, *Astrophys. J.* _226_, 379
Wilson, O.C., and Skumanich, A.: 1964, *Astrophys. J.* _140_, 1401
Wilson, O.C., and Woolley, R.: 1970, *Monthly Not. Roy. Astron.
Soc.* _148_, 463
Young, A., and Koniges, A.: 1977, *Astrophys. J.* _211_, 836
Zahn, J.-P.: 1977, *Astron. Astrophys.* _57_, 383
Zwaan, C.: 1977, in B. Caccin and M. Rigutti (eds.): "The Sun, a
Tool for Stellar Physics", *Mem. Soc. Astron. Italiana* _48_, 525
Zwaan, C.: 1978, *Solar Phys.* _60_, 213
Zwaan, C.: 1980, in S.D. Jordan (ed.): "The Sun as a Star", NASA-
CNRS Publication

STELLAR SPOTS - PHYSICAL IMPLICATIONS

L. Hartmann

Harvard-Smithsonian Center for Astrophysics

ABSTRACT

Studies of certain late-type stars have revealed evidence for appreciable changes in their photospheric radiation. These stars are rotating much more rapidly than the Sun and/or are younger, and exhibit much enhanced examples of many kinds of solar activity, such as flares. The surface brightness distributions of these stellar photospheres are nonuniform; dark areas are called "spots" in solar analogy. Areas of up to 30% may be covered by such spots, again indicating activity on a far greater scale than typical of the Sun.

Despite the lack of definitive bolometric observations, the detection of visual light variations on the order of 30% may be translated into total luminosity variations on the order of 10% with reasonable confidence. Spots may appear or decay over timescales of days, but longer-term trends exist, with timescales of decades. No direct evidence for spot cycles on the order of 10 years in length has been found, although some stars exhibit quasi-sinusoidal variations in mean light with timescales $\sim$ 50 years. Other cool stars of generally greater age and slower rotation exhibit evidence for decade periods in Ca II emission.

I. INTRODUCTION

The study of spots on stars similar to the Sun is of course greatly hindered by the general impossibility of resolving stellar surfaces. Sunspots have very little effect on disk-averaged light; not surprisingly, no evidence for continuum light variations $\gtrsim 1\%$ has been found for solar-type stars (Jerzykiewicz and Serkowski 1967; Lockwood 1977, 1978, 1980). The only method presently able to track sunspot-type

R. M. Bonnet and A. K. Dupree (eds.), Solar Phenomena in Stars and Stellar Systems, 487–497.
Copyright © 1981 by D. Reidel Publishing Company.

activity is the measurement of Ca II emission, which on the Sun is enhanced in active regions near spots. This technique has been used with some success by Wilson (1978), who has found the first evidence for solar-type cycles on other stars with periods ∼ 10 years. The results clearly indicate that the Sun's magnetic activity pattern is not unique, so that the mechanisms underlying spot behavior on other stars of solar type by inference can be related to analyses of sunspot behavior.

While solar-type spot activity cannot be observed directly, there are two classes of cool stars, composed of dwarfs and subgiants, which manifest disk-averaged light variations that are hard to explain by any effect other than a nonuniform photospheric brightness. The observed changes in optical light can be enormous: up to 30% for most stars, and in one case ∼ 50%. Such light variations are obviously on a far greater scale than any conceivable solar variation. However, these stars also possess huge amounts of chromospheric, coronal, and flare activity, relative to the Sun. Therefore it is not impossible that the continuum light variations of these stars are due to an exaggerated form of solar activity.

Even if there is no directly analogous solar mechanism, the fact that these objects are late-type dwarfs and subgiants with thick convective envelopes and extreme variants of solar activity almost certainly ensures that continued study of this behavior will yield interesting implications for solar physics.

II. OBSERVATIONAL DATA

a) Evidence for Starspots

The two classes of variable stars producing such optical light changes are listed in Table 1. Both the BY Dra and RS CVn stars have extensive convective envelopes from which strong solar-type activity might arise. Although we know that stellar activity in general decays with age (Kraft 1967; Skumanich 1972), the two types of variables do not have youth in common. What is similar is that both types exhibit rapid rotation for their spectral types, and this is presumed to be the cause of the great activity (Hall 1972; Bopp and Espenak 1977). All of the RS CVn stars, and most of the BY Dra stars, are close binaries, so that tidal friction can keep the stars rotating rapidly even as they lose angular momentum due to the magnetic torque applied to their winds. (cf. deCampli and Baliunas 1980).

A typical light curve for one of these variables is shown in Fig. 1. The usual behavior involves a quasi-sinusoidal variation of amplitude up to 30%, with a period of a few days. The B-V color changes are small; and usually indicate redder continuum at minimum light (Bopp and Espenak 1977). In Fig. 2 we show a more unusual light curve for the dwarf variable CC Eri, which is "flat-topped."

Table 1

Spotted Stars

BY Dra (Dwarf)	RS CVn (Subgiant)
Strong chromospheric and coronal activity	
Typical: M $\sim$ 0.6M$_\odot$	Typical: M $\sim$ 1M$_\odot$
R $\sim$ 0.6R$_\odot$	R $\sim$ 2R$_\odot$
T$_{eff}$ $\sim$ 4000K	T$_{eff}$ $\sim$ 4500K
(Main Sequence Stars)	(Evolved Stars)
Rapid Rotators	
P$_{rot}$ $\lesssim$ 4 days	$1^d \lesssim P_{orb} \lesssim 15^d$
Not all binaries	Binaries
Generally young	

Many mechanisms to explain the observed light curves have been investigated. Generally speaking the periods and color changes are inappropriate for processes such as pulsation or orbiting circumstellar material. On the other hand, there is strong evidence that the observed periods are the same as the stellar rotational periods. In a few stars, such as YY Gem, V_{rot} can be measured directly, confirming this trend. In binaries, we normally assume tidal synchronization; the orbital period is generally close to the photometric period. One exception to this is BY Dra, which has a 4^d photometric period and a 6^d orbital period. The orbit of this system is very eccentric (e = 0.36), so that it is likely that BY Dra is so young that there has been time neither to circularize the orbit nor synchronize the rotation with the orbit. Some long-period RS CVn stars (with periods greater than the range given in Table 1) also do not show identical photometric and orbital periods.

The general property of a rotational period commensurate with the photometric period led to the notion of a nonuniform surface brightness, i.e. the presence of "starspots" (Kron 1952; Chugainov 1966; Krzeminski 1969). If this is correct, then in eclipsing systems one should be able to use the eclipses to demonstrate the nonuniformity of the surface. This has been done for YY Gem by Kron (1952) and for RS CVn by Eaton and Hall (1979). Fig. 3 shows the eclipse asymmetry

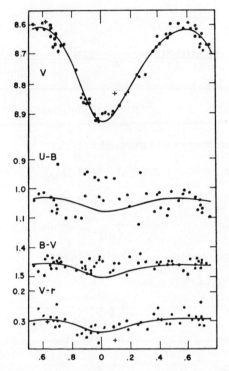

Fig. 1. V-light and color changes of AU Mic (Torres and Ferraz Mello
1973, Astron. Astrophys., 27, 231).

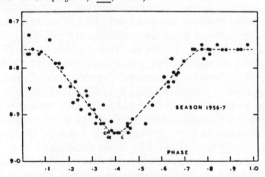

Fig. 2. Visual magnitude variation of CC Eri (Evans 1959, M.N.R.A.S.,
119, 526

for RS CVn. Eaton and Hall conclude that the observed asymmetry cannot be reproduced with a stellar surface that is smoothly and uniformly
varying in time. These analyses constitute the strongest evidence for
spotted areas.

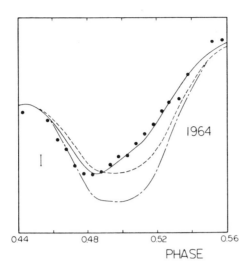

PHASE

Fig. 3. Light curve of RS CVn during secondary eclipse, along with model fits. Only an inhomogeneous surface brightness can reproduce the data (Eaton and Hall 1979, Astrophys. Journ. 227, 907).

If starspots are analogous to sunspots, they should be substantially cooler than the surrounding photosphere, and one might expect to see spectral variations. Ramsey and Nations (1980) have recently reported a strengthening of TiO bands upon the RS CVn variable HR 1099 as it gets fainter. A great deal more work remains to be done in this area.

The use of the term "spot" suggests a connection with sunspots. However, a direct identification is not possible at this time, for magnetic fields have not yet been measured (Vogt 1979). The sensitivity of such disk-averaged measurements does not yet appear to pose problems for the solar analogy; however, it is disappointing. New techniques may yield greater sensitivity (cf. Robinson, Worden, and Harvey 1980).

b) Time Behavior

Generally speaking, investigators have found that light curves remain stable for periods of months to years (Krzeminski 1969; Torres and Ferraz-Mello 1973; Bopp and Evans 1973; Vogt 1975; Hall 1972; Catalano and Rodono 1969). However, close watch has revealed amplitude and phase changes in BY Dra over timescales ∿ 1 day (Oskanyan et al. 1977).

Period changes have been reported for both BY Dra and RS CVn variables (Oskanyan et al. 1977; Vogt 1975; Hall 1972). In the case of the BY Dra variables, Hartmann and Rosner (1979) showed that it is not possible to separate the effects of differential rotation from patterns of spot growth and decay that seem plausible on the basis of the Oskanyan

et al. (1977) observations. The orbital period changes for RS CVn stars
were measured from eclipse timings, and were suggested to be related
to spot activity; this is a matter of controversy (Catalono and Rodono
1974; Hall, Kreiner, and Shore 1980).

Any evidence for solar-type cycles would of course enhance the
analogy of starspots with sunspots. Hall (1972) suggested that a 22-year
cycle in the light curve amplitude is present on RS CVn. However, the
data are so meager that the periodicity of the phenomenon has not been
demonstrated convincingly. The most complete and continuous series
of data on long-term spot activity is contained in the paper by workers
at Harvard using the archival plate collection (Phillips and Hartmann
1973; Hartmann, Londoño and Phillips 1979; Hartmann, Dussault, Bopp
and Dupree 1980). Although the accuracy of the photographic data
(~0.1 mg) is not great compared to the expected variations, the availability of a historical record dating back to the turn of the century is
invaluable.

Figs. 4-6 exhibit the long-term photometric behavior of several
spotted dwarfs. In the analysis we reduce a comparison star as if it
were a variable in order to get a better idea of the systematic errors
present. Thus the bars (standard deviation of the yearly mean) on the
photographic data are not true error flags, and the variation of the comparison star should be used as the error standard. The bars can be
misleadingly large if the star in question has a large (~0.3 mag) rotational variation, which gets averaged into the yearly mean.

In Fig. 4 BY Dra exhibits a 0.3 mag. variation, commensurate
with the range indicated by the modern photoelectric measurements, and
clearly real in comparison with the 0.05 mag. variation exhibited by
the standard star. Surprisingly, a long-term variation of timescale
~ 40-60 years is present.

The star HD 224085 presents an entirely different aspect in its
photometric behavior (Fig. 5). Within the accuracy of the photographic
magnitudes it appears that the star was constant, or at least relatively
quiescent, from about 1900 to 1940. However, subsequent to this period
a dramatic change in the level of activity occurred, with mean light
variations of up to 30% in the visual spectral region. This behavior is
confirmed by modern photoelectric data in 1974-76 (Krzeminski 1979)
which clearly confirms the general accuracy of the photographic measurements. Furthermore, in the modern period the standard deviation flags
for the variable are much larger than those for the comparison star,
consistent with the photoelectrically observed rotational modulation.
No difference in error flags between the variable and comparison star is
observed in 1900-1940. Hence neither the mean light changes nor the
rotational modulations were large for a period of ~40 years. One may
speculate that this behavior may be analogous to the Maunder minimum
(Eddy 1976); periods of relative spot quiescence would help explain why
all BY Dra stars are flare stars, but the converse is not true (Bopp and
Espernak 1977).

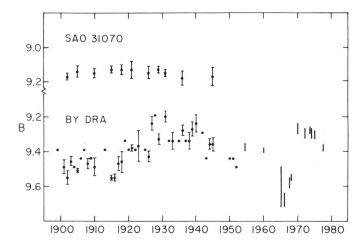

Fig. 4. Long-term variability of BY Dra. Dots are photographic data;
dots with bars, yearly means with associated standard deviations; lines,
photoelectric data. SAO 31070 is a constant comparison star (Phillips
and Hartmann 1978, Astrophys. Journ. 224, 182.)

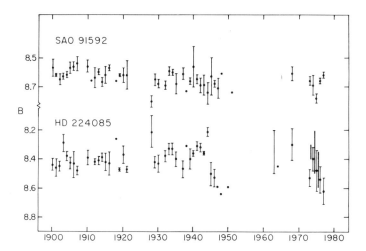

Fig. 5. Long-term variability of HD 224085. Prior to 1940 the star was
constant to ∿ 0. 1 magnitudes; in the 1940's, the variation was ∿ 0. 25 mag;
and in the 1970's, variations ∿ 0.3 mag are seen in both the photographic
and photoelectric data (Hartmann, Londoño, and Phillips 1979, Astrophys. Journ. 229, 183.)

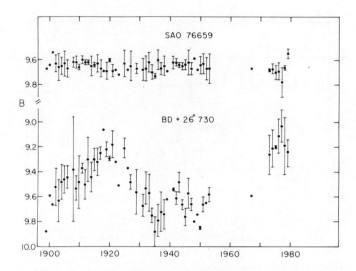

Fig. 6. Long-term variability of BD + 26°730, showing a huge amplitude of 0.5 mag and a suggestion of a period ~ 50 years (Hartmann,
Dussault, Bopp, and Dupree 1980, in preparation.)

The pièce de resistance is BD + 26°730 (Fig. 6). This K5 dwarf
emission-line star exhibits a huge variation ~ 0.5 mag in B. The observations clearly suggest a cycle ~ 40-50 years, very similar to BY Dra.
At the moment we do not know whether this star is a close binary and/or
rapid rotator, but it has been seen to exhibit short-term, BY Dra-type
light variations.

One might naively expect that rapid rotation should engender a
shorter dynamo period (cf. Stix 1976), but of course the pattern of differential rotation is unknown. The phenomenon of course may not have
a solar analogy. One confusing aspect is that pairs of cool stars with
similar spectral types, chromospheric, coronal, and flare activity, are
known such that one star has spot variations and the other does not
(e.g., AD Leo and EV Lac). Thus, the continuum light variations may
not be extremely closely connected with surface activity, responding
more to magnetic field variations that are deep within the convective zone.

III. PHYSICAL IMPLICATIONS

a) Spot Modelling

Several authors have derived areas and temperatures of starspots
on the basis of optical photoelectric data (Torres and Ferraz-Mello 1973,
Bopp and Evans 1973, Vogt 1975, Oskanyan et al. 1977, Eaton and Hall
1979). Typically, the results indicate spot temperatures cooler than
the photosphere by at least 300-400K, with areas of up to ~ 30% of a

hemisphere, although Oskanyan et al. (1977) invoked hot spots on BY Dra
at certain times. Vogt (1975) has emphasized the non-uniqueness of
temperature results based on BV data. Longer-wavelength data help to
pin down temperatures. With VR data, Bopp and Espenak (1977) found
$\Delta T \sim$ 200-500K indicated. However, as Hartmann and Rosner (1979)
have pointed out, there is still a large degree of non-uniqueness in such
modelling. The procedure assumes only one (or two spot area(s), with
the surrounding photosphere remaining constant. But large fractional
areas of the star covered by regions cooler by $\Delta T \gtrsim$ 300K imply significant changes in the bolometric output. In such a case there is really no
reason to assume a constant photosphere, until the "missing flux" is
shown not to be significantly redistributed from the spot to the surrounding
photosphere. Thus we can use the spot models to address the properties
of the spotted areas in a very rough way, but we cannot rely on the specific details of such models.

b) Energy Budget and Missing Flux

        The observations clearly show a persistence of the light variations
which are generally much longer than the convective overturning time
(Hartmann and Rosner 1979). If the photospheric surface brightness is
inhomogeneous, it is hard to imagine any mechanism other than the
modification of convective patterns (by magnetic fields) which can account
for such inhomogeneities. Hartmann and Rosner (1979) showed that the
responses of BY Dra-type stellar convective envelopes to ad hoc variations in the effeciency of convection are very similar to the luminosity
variations achieved in solar models (Dearborn and Newman 1978).
However, the large changes in optical light require truly global variations
in convective motions.

        Although complete wavelength coverage is not available for BY Dra
and RS CVn variables, it is almost certain that the optical light variations
represent true changes in the stellar bolometric luminosity of up to
5-10% (Hartmann and Rosner 1979). This is of some interest in connection
with the possibility that sunspots are associated with solar luminosity
changes (Eddy 1976). The standard sunspot model goes back to Biermann
(1941), who suggested that magnetic fields can inhibit convection and prevent normal energy transport from taking place in the spot. It is not
immediately clear what happens to the missing energy. Some must go
around the edges of the spot; whether this makes an observable brightening depends upon the size and depth of the magnetic field structure below
the photosphere (Cowling 1976).

        It is very difficult to see how all of the missing flux on BY Dra can
be spatially redistributed so that the overall luminosity remains constant
(Hartmann and Rosner 1979). Luminosity deficits $\sim$10% require photospheric
B-V changes on the order of 0.03 mag, even if the extra flux is uniformly
redistributed in the unspotted area. One would expect large, rotationally
modulated color changes, or that the star would be bluest when covered by
the most spotted areas; this has not been observed, although detectable.

It is true that rotationally modulated color changes have been observed on BY Dra in recent years, when spot amplitudes were generally small (Vogt 1975; Davidson and Neff 1977), and that surface flux redistribution may have occurred. Hot spots have also been suggested to account for some of the color variations observed, and this also may reflect redistribution (Vogt 1975; Oskanyan et al. 1977). However, the large V magnitude variations of 25% or larger occasionally seen do not seem to be accompanied by substantial rotationally modulated B-V changes (cf. Fig. 1). Long-term trends in B-V seem to have occurred (Oskanyan et al. 1977), which may indicate flux redistribution over timescales of years (Hartmann and Rosner 1979). Relatively little work has been done on the color changes of the spotted stars; long-term optical monitoring would be very helpful in elucidating the nature of the optical light variations and in defining the true bolometric changes.

REFERENCES

Biermann, L., 1941, Vierteljahresschr. Astron. Ges. 76, p. 194.
Bopp, B. W. and Espenak, F., 1977, Astron. Journ. 82, p. 916.
Bopp, B. W. and Evans, D. S., 1973, M.N.R.A.S. 164, p. 343.
Catalano, S., and Rodono, M., 1969, in "Nonperiodic Phenomena in Variable Stars", ed. L. Detre (Dordrecht: Reidel), p. 435.
_____, 1974, Pub. Astron. Soc. Pacific 86, p. 390.
Cowling, T. G. 1976, M.N.R.A.S. 177, p. 409.
Davidson, J. K., and Neff, J. S. 1977, Astrophys. Journ. 214, 140.
Dearborn, D. S. P., and Newman, M. J. 1978, Science 201, p. 150.
DeCampli, W. M., and Baliunas, S. L. 1979, Astrophys. Journ. 230, 815.
Eaton, J.A., and Hall, D.S. 1979, Astrophys. Journ. 227, p. 907.
Eddy, J.A. 1976, Science 192, p. 1189.
Evans, D. S. 1959, M.N.R.A.S. 119, p. 526.
Hall, D.S. 1972, Pub. Astron. Soc. Pacific 84, p. 323.
_____, 1976, in IAU Colloquium 29, "Multiple Periodic Variable Stars", Part I, ed. W.S. Fitch (Dordrecht: Reidel), p. 287.
Hall, D.S., Kreiner, J.M., and Shore, S.N., 1980, in "Close Binary Stars, IAU Symp. 73, ed. M.J. Plavec, D.M. Popper, and R.K. Ulrich (Dordrecht: Reidel), p. 383.
Hartmann, L., Dussault, M., Bopp, B.W., and Dupree, A.K. 1980, in preparation.
Hartmann, L., Londoño, C., and Phillips, M.J. 1979, Astrophys. Journ. 229, p. 183.
Hartmann, L., and Rosner, R. 1979, Astrophys. Journ. 230, p. 802.
Jerzykiewicz, M., and Serkowski, K., 1967, Lowell Obs. Bull. 6, p. 295.
Kraft, R.P. 1967, Astrophys. Journ. 160, p. 551.
Kron, G.E. 1952, Astrophys. Journ. 115, p. 301.
Krzeminski, W., 1969, in "Low Luminosity Stars", ed. S.S. Kumar (London: Gordon and Breach), p. 57.
Lockwood, G.W. 1977, Icarus 32, p. 413.
_____ 1978, Icarus 35, p. 79.
_____ 1980, private communication.
Oskanyan, V. S., Evans, D.S., Lacy, C., and McMillan, R.S., 1977,

Astrophys. Journ. 214, p. 430.
Phillips, M.J., and Hartmann, L. 1978, Astrophys. Journ. 224, p. 182.
Ramsey, L., and Nations, H. 1980, preprint.
Robinson, R.D., Worden, S.P., and Harvey, J.W. 1980, Astrophys. Journ. (Letters), 236, p. L155.
Skumanich, A. 1972, Astrophys. Journ. 171, p. 565.
Stix, M., 1976, in IAU Symposium 71, "Basic Mechanisms of Solar Activity", ed. K. Bumba and J. Kleczek (Dordrecht: Reidel), p. 367.
Torres, C.A.O., and Ferraz Mello, S., 1973, Astron. Astrophys. 27, p. 231.
Vogt, S.S. 1975, Astrophys. Journ. 199, p. 418.
_____, 1980, Astrophys. Journ., in press.
Wilson, O.C. 1978, Astrophys. Journ., 226, p. 379.

PHOTOSPHERIC MAGNETIC FIELD PATTERNS AND
THEIR CORONAL AND INTERPLANETARY CONSEQUENCES

Randolph H. Levine

Harvard-Smithsonian Center for Astrophysics
Cambridge, Massachusetts 02138 USA

Abstract. The purpose of this presentation is to discuss the
basic observational knowledge of the photospheric and
interplanetary magnetic fields, and to point out the minimal
theoretical considerations which are necessary to understanding
the connection between the photospheric and interplanetary
fields. In particular, I wish to point out that for most of the
solar cycle a consideration of only the dipolar component of the
photospheric field (including a "tilted dipole") does not lead to
an adequate explanation of even the most basic interplanetary
observations. In addition, I will emphasize the complementary
relationship between models of the coronal and interplanetary
magnetic field which are extrapolations of actual photospheric
data and models which are full MHD solutions for idealized cases.
Finally, an interesting complication to the interplanetary sector
pattern, in the form of a second pattern that exists primarily
near solar maximum and at higher latitudes, will be described.

1. BASIC OBSERVATIONS OF THE PHOTOSPHERIC MAGNETIC FIELD

The large-scale magnetic field in the solar photosphere has
been investigated in two ways: by spatially averaging high
resolution observations of the line-of-sight field (1-7) and by
measuring the line of sight field using integrated light from the
full solar disk (8-10). Results of the first of these procedures
are shown in Figure 1. These are synoptic charts of the measured
line of sight magnetic field for Carrington rotations 1601-1611.
Each panel represents the same data (from Kitt Peak National
Observatory) averaged over progressively coarser grids covering
the solar surface. As the "resolution" is thus decreased the

R. M. Bonnet and A. K. Dupree (eds.), Solar Phenomena in Stars and Stellar Systems, 499–508.
Copyright © 1981 by D. Reidel Publishing Company.

existence of large-scale patterns in the photospheric magnetic
field become apparent (7). At a scale of several tens of degrees
(heliographic) the individual regions of predominant polarity
often contain coronal holes, usually near their center, while the
reversals of polarity lie beneath either large arcades of closed
magnetic loops or prominent active regions. The polarity sectors
of the photosphere are long-lived and typically evolve slowly.
When significant changes in the pattern do occur, they often
happen over relatively short time scales (11), compared to the
interval between changes.

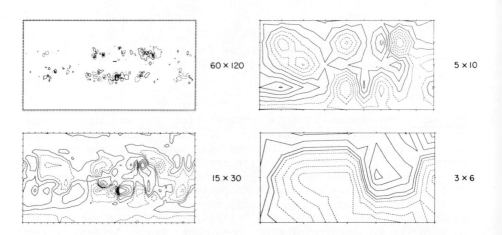

Figure 1. Contour plots of average magnetic field observed over
Carrington rotations 1601-1611. The four panels each represent
the same data averaged onto grids of different surface
resolution. The sizes of the grids used are indicated. Contour
levels (+ and -, in gauss) are (60 x 120): 5, 15, 25, 35; (15 x
30): 0, 1, 3, 5, 7; (5 x 10): 0, .4, .8, 1.2, 1.6; (3 x 6): 0,
.1, .3, .5, .7 (from (7)).

An important fact concerning the photospheric polarity
sectors is that their pattern is due to both strong average B and
weak average B. That is, regions where the magnetic flux within
the magnetograph aperture is large (e.g., greater than 30 gauss
averaged over 2.5 square arcseconds) show the same over-all
pattern when averaged into large-scale fields as do regions where
the magnetic flux in the magnetograph aperture is low (7). Thus
active regions, as well as quiet regions on the sun, contribute
to the over-all ordering of the large-scale magnetic field. Many
workers (2, 6, 8, 15, 16) have inferred or assumed that the
existence of large-scale photospheric patterns was due to a weak

background field with a strength of a few gauss. However, all measured fields contribute to the pattern and the visibility of the pattern is a question of spatial scale rather than of field strength. It is the imbalance of field strengths, rather than their absolute value, which establishes the systematic pattern.

This strongly suggests that the emergence of active regions occurs systematically in favor of contributing to the large-scale pattern. The same conclusion is found from studying the emergence of active regions and their contribution to the magnetic flux underlying coronal holes (12, 13). These and other features of the large-scale magnetic field have led to the suggestion that the large-scale photospheric field is a fundamental aspect of solar magnetism, and an important aspect of the solar dynamo (7, 11-14).

While the poles of the sun each exhibit a single (and opposite) predominant polarity at all longitudes, the pattern in the more equatorial regions, as deduced from either mean field or spatially resolved measurements, is typically divided into four polarity sectors. The combination of the patterns in the different latitude regions results in an over-all photospheric pattern of large-scale fields whose polarity boundaries typically can be thought of as resembling the seams on a baseball (14), or a tennis ball (for those not raised in North America).

2. BASIC OBSERVATIONS OF THE INTERPLANETARY MAGNETIC FIELD PATTERN

The fact that the interplanetary magnetic field near the ecliptic plane is divided into broad polarity sectors, much like the more equatorial regions in the photosphere, was deduced directly from spacecraft observations (15, 17). In addition, it is possible to infer the interplanetary polarity near the earth by observing geomagnetic variations near the earth's poles, and this method is reasonably accurate (18).

In Figure 2 the inferred interplanetary polarity at earth is plotted with each rotation shown twice. Negative (toward the sun) sectors are shaded (once as black and once as white areas). The cross-hatched regions which slant downward from left to right refer to another pattern to be discussed later. For now, it is important to note that the inferred interplanetary pattern almost always contains four sectors (i.e., the two negative sectors, each identified twice in Figure 2, and two positive sectors filling the regions between the negative sectors). A pattern with two sectors is present only for relatively brief periods, usually near sunspot minimum.

INFERRED SOLAR MAGNETIC SECTOR STRUCTURE DURING FIVE SUNSPOT CYCLES

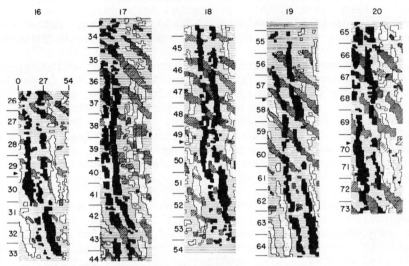

26.84 DAYS CALENDAR SYSTEM STARTING FEB 19, 1926

Figure 2. A plot of the inferred interplanetary magnetic structure during sunspot cycles 16–20. Two successive rotations are displayed horizontally to aid in pattern recognition. Sectors with field polarity toward the sun are shaded black if they are judged to be part of the four sector pattern, and have a dashed shading if they are judged to be part of the 28.5 day structure (from (19)).

The latitudinal extent of the interplanetary pattern is not known observationally in detail. From the angle of the ecliptic with respect to the heliographic equator and the existence of the sector pattern at all times an extent of more than 7 degrees on either side of the heliographic equator is necessary. Moreover, the Pioneer XI spacecraft observed the sector pattern at heliographic latitudes as high as 16 degrees (20).

3. THE "SIMPLEST" EXPLANATION

The photospheric and interplanetary sector patterns are reasonably (but not exactly) correlated, with a maximum correlation at a time lag corresponding to the approximate average transit time for the solar wind to carry the field from the sun to 1 AU. However, it is certainly not the entirety of the photospheric pattern that physically maps to the earth. Because most of the area of photospheric polarity regions is not likely to connect to interplanetary space (i.e., not all the field lines are open), the use of the extent of these regions as tracers of the extent of interplanetary magnetic field sectors is a fortuitous approximation. This procedure is successful only because open field lines concentrated in the photosphere diverge rapidly into the corona where they then tend to reflect the extent (and strength) of the region in which they are rooted. The longitudinal extent of this mapping is largely preserved at 1 A.U., although the latitudinal structure is substantially altered by electromagnetic and dynamical effects.

We now ask the question: what is the simplest explanation which connects the solar and interplanetary observations. By this I mean the specification of two things, (1) the physical processes and (2) the essential boundary conditions, which are necessary to understand the general character of both the photospheric and interplanetary magnetic fields. I put the issue this way not only because of its importance to the sun, but also to emphasize that progress in understanding the effects of extended stellar magnetic fields may require a similar exercise.

A. Dipole Fields

It is often argued that the major aspects of the interplanetary pattern of polarity can be understood in terms of the dipole component of the photospheric magnetic field (21). In order to produce a two-sector pattern the dipole discussed in this context is assumed to be tilted with respect to the solar rotation axis. This produces two polarity sectors at all latitudes less than the tilt of the dipole. Although this scenario works rather well when applied to specific intervals (e.g., the latter portion of the Skylab epoch), it falls short of providing an explanation of the interplanetary sector pattern in general. As noted above (Section 2) the interplanetary pattern seldom contains only two sectors; a four-sectored pattern is the norm. In addition, the dipole component of the photospheric field is a reasonable explanation for only the polar regions of the sun. Ignoring the influence of the lower latitude fields is a serious shortcoming because these fields are not randomly arranged with regard to polarity but form large sectors of predominant polarity in a systematic way. The general character of the interplanetary observations cannot be interpreted straightforwardly in terms of a dipole solar magnetic field.

B. Dipole and Quadrupole Fields

It is useful to continue this discussion in terms of multipole components of the magnetic field. After the dipole, which is not a sufficient description of either the photospheric or the interplanetary fields in general, the next simplest configuration is the superposition of a dipole and a quadrupole.

This describes the general character of the photospheric fields (mentioned above in terms of sporting equipment) reasonably well. That is, the poles can have opposite uniform polarities and the equatorial regions four sectors of alternating polarity (6, 14).

The inclusion of the quadrupole term is also advantageous in the description of the interplanetary field pattern (22). Figure 3 (from (23)) gives a quantitative illustration of this. Using a photospheric pattern which contains four polarity sectors (each of constant field strength) extending to 45 degrees north and south latitudes and uniform fields poleward of 45 degrees in each hemisphere, Figure 3 shows the polarity reversal lines in a simple model where the photospheric field is merely extrapolated outward with no electric currents. (Such a field has other multipoles than the dipole and quadrupole, but those are the most significant and dominate the structure of the pattern.) In addition, the field is required to be radial at some specified distance, in order to simulate the solar wind.

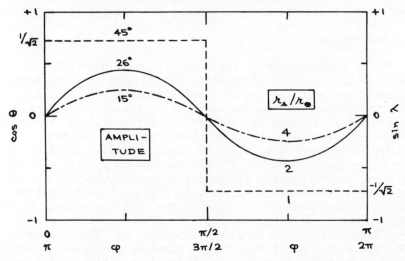

Figure 3. Plots of the zero contour of the radial magnetic field at the solar surface and at various values of a "source surface." The model maps the zero contour of the radial field at the source surface into an interplanetary current sheet having the same excursions in latitude.

This extremely simple model shows that the basic aspects of the interplanetary pattern, its division into four sectors and its confinement near the helioequator, can be understood as the straightforward extension of known properties of the photospheric field (at all latitudes) into the heliosphere.

There is cause for an objection at this point. The strength of the n-th order multipole of the field falls off as the n+2 power of the radius. Thus at a sufficient radius from the sun, the dipole will be by far the dominant term. Why is it not possible, when speaking of the field far from the sun, to use a description in terms of a dipole only? The answer is that the physics of the outer corona does not allow one to go to a "sufficient radius from the sun" and still retain the approximation of a multipole expansion. The MHD expansion of the coronal plasma and magnetic field causes field lines to become open at a radius of about 1.5 to 2.5 solar radii (12, 24, 25, 26). At this distance the strengths of equal dipole and quadrupole components of the field differ by only about a factor of two. Beyond this distance, the polarity pattern of the field, especially near the equator, is held fixed by the expansion of the solar wind, which overcomes the field. This is the rationale behind the calculation shown in Figure 3, and it explains why the quadrupole component of the photospheric field can be detected as readily as the dipole component several hundred radii from the sun.

C. N-tupole Fields

The logical extension of this exercise is to extend the expansion to higher order moments of the photospheric field. This is what is done in constructing potential (i.e., current-free) models of the coronal field with the observed line-of-sight photospheric field as a boundary condition (24, 25, 27), because the multipole components of the field define a harmonic expansion for the solution of Laplace's equation. An outer boundary condition in these models is specified as a simulation of the MHD expansion of the corona: a spherical "source surface" is defined where the field is required to be radial.

Such models are successful in showing the basic structure of most active regions, coronal holes, and other open field regions (12, 28, 29, 30), and they can give a good account of the polarity pattern (but not the strength) of the interplanetary field (31). They are not successful in reproducing the non-radial orientation of streamers or the boundaries of coronal holes in detail, and the method is not in agreement with a full MHD solution for a photospheric dipole (26). A recent improvement is the use of a non-spherical source surface (32, 22). This is capable of matching the orientation of streamers to within a few degrees, and it agrees in many important aspects with the MHD dipole. Thus it offers the most realistic hope of modelling interplanetary field strengths based on photospheric observations. However, the boundaries of coronal holes are not well represented, apparently because strong sheet currents exist at the interface between closed and open field lines.

D. Real MHD

An ideal procedure would be to establish a connection between each point in the heliosphere and a point on the solar surface using a full MHD model of a rotating, magnetized sun and its ionized corona. In practice this is well beyond present capability, for both physical and computational reasons (36). So while multipole expansions and potential field models have important shortcomings when examined in detail, MHD calculations cannot yet handle observed photospheric fields and are just beginning to include proper treatment of the energy equation. Thus both the approach of simplified MHD modelling and that of potential field models based on observed fields will, and should, be pursued for some time.

4. SOLAR CYCLE VARIATIONS

Beginning in 1977 the nominal four-sector pattern of the large-scale photospheric magnetic field was supplemented by a simultaneously existing two-sector pattern at higher latitudes. While the four-sector pattern at lower latitudes rotates with a period of approximately 27 days, the higher latitude pattern rotates more slowly, with a period of approximately 29 days (34). This is apparently the source of the slanting bands in Figure 2, which are evidence for an interplanetary sector pattern superimposed on the nominal four-sector pattern with a slower rotation period of about 28.5 days. The origin of this pattern was not clear at the time its significance was first suggested. We can now suggest that the higher latitude pattern in the photosphere near the maximum of the solar cycle is manifest in the interplanetary medium near the ecliptic.

A simplified schematic model of this process of two topologically distinct patterns rotating past one another leads to important implications for the corona and interplanetary medium during this period (35). Among these implications are 1) An interplanetary sector structure with a rotation period near 29 days and consisting of two sectors should exist, especially at higher latitudes, 2) The large-scale pattern of the coronal field should change significantly and systematically every 3 to 5 rotations, and 3) While a single interplanetary polarity reversal surface (or current sheet, or warped heliomagnetic equator) may be a basic feature of the solar minimum corona and interplanetary medium, the systematic patterns of the photosphere during the rising portions of the cycle may produce multiple polarity reversal surfaces in the corona and interplanetary medium which are topologically distinct.

Acknowledgements

I would like to thank Drs. J. Wilcox and M. Schulz for permission to use Figures 2 and 3, respectively. This work was supported by USAF contract F19628-80-C-0067 with Harvard University.

REFERENCES

(1) Babcock, H.W. and Babcock, H.D.: 1955, Astrophys. J., **121**, p. 349.
(2) Bumba, V. and Howard, R.: 1965, Astrophys. J., **141**, p. 1502.
(3) Bumba, V. and Howard, R.: 1966, Astrophys. J., **143**, p. 592.
(4) Wilcox, J.M. and Howard, R.: 1968, Solar Phys., **5**, p. 564.
(5) Stenflo, J.O.: 1972, Solar Phys., **23**, p. 301.
(6) Svalgaard, L., Wilcox, J.M., Scherrer, P.H. and Howard, R.: 1975, Solar Phys., **45**, p. 83.
(7) Levine, R.H.: 1979, Solar Phys., **62**, p. 277.
(8) Severny, A., Wilcox, J.M., Scherrer, P.H. and Colburn, D.S.: 1970, Solar Phys., **15**, p. 3.
(9) Scherrer, P.H., Wilcox, J.M., Kotov, V., Severny, A.B. and Howard, R.: 1977, Solar Phys., **52**, p. 3.
(10) Scherrer, P.H., Wilcox, J.M., Svalgaard, L., Duvall, T.L., Jr., Dittmer, P.H. and Gustafson, E.K.: 1977, Solar Phys., **54**, p. 355.
(11) Levine, R.H.: 1977, Solar Phys., **54**, p. 327.
(12) Levine, R.H.: 1977, Astrophys. J., **218**, p. 291.
(13) Levine, R.H.: 1977, in Zirker, J.B. (ed.), Coronal Holes and High Speed Solar Wind Streams, Boulder, Colo. Assoc. Univ. Press, ch. 4.
(14) Svalgaard, L., Wilcox, J.M. and Duvall, T.L.: 1974, Solar Phys., **37**, p. 157.
(15) Wilcox, J.M.: 1968, Space Sci. Revs., **8**, p. 258.
(16) Wilcox, J.M., Severny, A. and Colburn, D.S.: 1969, Nature, **224**, p. 353.
(17) Wilcox, J.M. and Ness, N.F.: 1965, J. Geophys. Res., **70**, p. 5793.
(18) Svalgaard, L.: 1973, J. Geophys. Res., **78**, p. 2064.
(19) Svalgaard, L. and Wilcox, J.M.: 1975, Solar Phys., **41**, p. 461.
(20) Smith, E.J., Tsurutani, B.T. and Rosenberg, R.: 1976, J. Geophys. Res., **83**, p. 717.
(21) Hundhausen, A.J.: 1977, in Zirker, J.B. (ed.), Coronal Holes and High Speed Solar Wind Streams, Boulder, Colo. Assoc. Univ. Press, ch. 7.
(22) Schulz, M.: 1973, Astrophys. Space Sci., **24**, p. 371.
(23) Schulz, M.: 1977, EOS, **58**, p. 770.
(24) Schatten, K.H., Wilcox, J.M. and Ness, N.F.: 1969, Solar Phys., **9**, p. 442.

(25) Altschuler, M.D. and Newkirk, G., Jr.: 1969, Solar Phys., 9, p. 131.
(26) Pneuman, G.W. and Kopp, R.A.: 1971, Solar Phys., 18, p. 258.
(27) Altschuler, M.D., Levine, R.H., Stix, M. and Harvey, J.W.: 1976, Solar Phys., 51, p. 345.
(28) Jackson, B.V. and Levine, R.H.: 1980, Solar Phys., in press.
(29) Burlaga, L.F., Behannon, K.H., Hansen, S.F. and Pneuman, G.W.: 1978, J. Geophys. Res., 83, p. 4177.
(30) Svestka, Z., Solodyna, C.V., Howard, R. and Levine, R.H.: 1977, Solar Phys., 55, p. 359.
(31) Wilcox, J.M., Scherrer, P.H. and Hoeksema, J.T.: 1980, Nature, in press.
(32) Schulz, M., Frazier, E.N. and Boucher, D.J., Jr.: 1978, Solar Phys., 60, p. 83.
(33) Levine, R.H., Schulz, M. and Frazier, E.N.: 1980, Solar Phys., submitted.
(34) Hoeksema, J.T., Scherrer, P.H. and Wilcox, J.M.: 1980, B.A.A.S., 12, p. 474.; Levine, R.H.: 1980, unpublished.
(35) Levine, R.H.: 1980, in preparation.
(36) Nerney, S.F. and Suess, S.T.: 1975, Astrophys. J., 196, p. 837.

SOLAR FLARES : MAGNETOHYDRODYNAMIC INSTABILITIES

A. W. Hood and E. R. Priest

Applied Mathematics Department
The University
St. Andrews
Scotland

Most flares may be regarded as either simple-loop (or compact) flares or two-ribbon flares. In the former, a single loop brightens and decays without moving, whereas the two-ribbon type involves the eruption of a magnetic arcade. The object of the present talk is simply to describe the basic MHD instabilities that may be responsible, but, first of all, a summary of some basic theoretical concepts is given. These include the tearing-mode instability, magnetic reconnection and the energy method.

A simple-loop flare may result when a new magnetic flux tube emerges from below the photosphere and interacts with the overlying magnetic field, as in the emerging flux model; triggering takes place when the current sheet separating the two flux systems reaches a critical height. Alternatively, it may be an example of thermal nonequilibrium, when a cool loop ceases to be in thermal equilibrium and heats up. A third possibility is that the loop becomes kink unstable (or resistive kink unstable) when it is twisted too much. Without the dominant stabilizing influence of line-tying, most loops would be completely unstable, but when its effect is included they become unstable only when the twist is too great.

A two-ribbon flare is caused by the instability of the magnetic configuration in which a filament is situated. The filament and magnetic field erupt outwards and then, during the main phase, the field reconnects back down to give rise to "post"-flare loops. The original eruptive instability is an MHD instability, which results when the arcade shear or height become too great. Possibilities for triggering the instability include emerging flux, thermal nonequilibrium or the tearing mode.

R. M. Bonnet and A. K. Dupree (eds.), Solar Phenomena in Stars and Stellar Systems, 509–531.
Copyright © 1981 by D. Reidel Publishing Company.

A large solar flare is an amazingly complex event that capti-
vates the onlooker with its beauty and stimulates the theorest to
try and understand it. Great strides have recently been made in
explaining the basic magnetohydrodynamic cause of a flare, and it
is these which we shall aim to summarize here. Several books on
the subject have appeared recently, namely Svestka (1976) mainly
on the observations, Sturrock (1980) on the results from Skylab,
and Priest (1980) on the MHD of the flare phenomenon. So we offer
here a personal view of our present understanding, rather than a
comprehensive review of all flare theories and secondary effects.

1. OBSERVED CHARACTERISTICS OF FLARES

First of all, we give a brief summary of flare observations.
Flares may be split into two classes, namely the large two-ribbon
flare and the simple-loop (or compact) flare. A typical schematic
profile for the intensity of a two-ribbon flare is shown in
Figure 1.

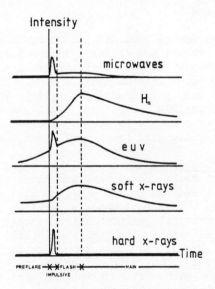

Figure 1. A schematic representation of the flare intensity in
several wavelengths. Typical time-scales for the various phases
are : preflare - ten minutes; impulsive - one minute; flash -
five minutes; and main phase - one hour to a day.

Four stages are apparent in the time-development of the
emission. The preflare phase lasts typically 10 minutes and is
characterised by a rise in the soft X-ray emission. This indicates that the region is being heated to a temperature of a few
million degrees, and, at the same time, there is evidence of
magnetic field activity. The flash phase normally lasts about

five minutes, during which the intensity and the flare area
increase rapidly. During the first minute of the flash phase,
many flares in magnetically active regions exhibit an <u>impulsive</u>
phase, which is characterised by a sharp increase in the intensity
of hard X-rays (> 30keV) and microwave emission. Following the
flash phase is the <u>main phase,</u> during which the intensity
decreases over a period of about an hour, although sometimes the
enhanced soft X-ray emission can continue for 1 or 2 days!

The total amount of energy released during a flare depends
on the size of the event. In a subflare, it is between 10^{21} and
5×10^{23}J (1J $= 10^7$ ergs). In an average flare the energy is
10^{23} to 10^{24}J, but in the largest events it may reach 3×10^{25}J.
Because other sources of energy seem inadequate for large flares,
it is usually assumed that their energy is provided by the magnetic field. For example, 3×10^{25}J would be released if the
whole of a 0.05 Tesla (500G) magnetic field in a cube of side
30Mm (1Mm $= 10^6$m) were annihilated. However, the difficult problem is to determine the mechanism for initiating such a release.

<u>Two-ribbon flares</u> are intimately related with active-region
filaments. During the preflare X-ray enhancement, the filament
is observed to become activated (Rust and Webb, 1977 and Kuperus
and Svestka, 1978).

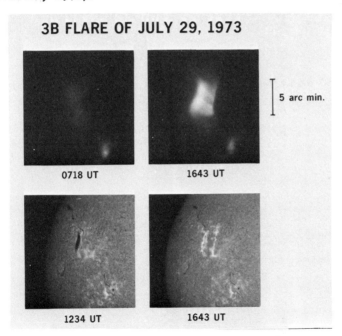

Figure 2. H_α and soft X-ray pictures before and after the start
of a two-ribbon flare (Courtesy A. S. Krieger, American Science
and Engineering).

Frequently, this takes the form of a slow rise of the filament before it disappears rapidly at the onset of the flare proper
(a "disparition brusque"). An excellent summary of events leading
up to a two-ribbon flare has been presented by the Preflare Group
during the Skylab workshop on Solar Flares (Chapter 2 of Sturrock,
1980). One important feature that has been stressed by, e.g.
Rust (1976) and Vorpahl (1973), is the emergence of new magnetic
flux from below the photosphere prior to most flares.

The simple-loop (or compact) flare is the category into which
most flares and subflares can be placed. Often a single magnetic
loop or flux tube brightens in X-rays, and, remains apparently
unchanged in shape and position throughout, while a brightening in
H_α shows up at the footpoints. Unfortunately, we are unable to
say, at present, whether the magnetic field in the loop changes
significantly or not. If such changes do occur, then a magnetic
instability seems the most plausible cause, but otherwise some
non-magnetic source must be sought.

2. BASIC THEORETICAL IDEAS

Before proceeding to a discussion of various flare models it
is perhaps appropriate to present a short summary of some basic
theoretical concepts that are involved in the relevant MHD instabilities.

2.1 Equations

The MHD equations that form the basis of our study may be
listed as follows.

Momentum
$$\rho \frac{D\underset{\sim}{v}}{Dt} = - \nabla p + \underset{\sim}{j} \times \underset{\sim}{B} + \rho \underset{\sim}{g}, \tag{1}$$

Continuity
$$\frac{\delta \rho}{\delta t} + \nabla \cdot (\rho \underset{\sim}{v}) = 0, \tag{2}$$

Gas law
$$p = 2n_e kT \tag{3}$$

Energy balance
$$\frac{\rho^\gamma}{\gamma - 1} \frac{D}{Dt} \left(\frac{P}{\rho^\gamma}\right) = \nabla \cdot (\underset{=}{\kappa} \nabla T) - n_e^2 Q(T) + H_w, \tag{4}$$

Induction
$$\frac{\delta \underset{\sim}{B}}{\delta t} = \nabla \times (\underset{\sim}{v} \times \underset{\sim}{B}) - \nabla \times (\eta \nabla \times \underset{\sim}{B}), \tag{5}$$

where
$$\nabla \cdot \underset{\sim}{B} = 0, \quad \rho = n_e m_H, \tag{6}$$

and $\underset{\sim}{j}$ and $\underset{\sim}{E}$ follow from
$$\underset{\sim}{j} = \nabla \times \underset{\sim}{B}/\mu, \quad \underset{\sim}{E} = -\underset{\sim}{v} \times \underset{\sim}{B} + \underset{\sim}{j}/\sigma \tag{7}$$

Here ρ is the density, $\underset{\sim}{v}$ the fluid velocity, $\underset{\sim}{B}$ the magnetic induction, $\underset{\sim}{j}$ the current density, $\underset{\sim}{g}$ the gravity vector acting towards the $\sim$ solar centre, T the temperature, m_H the proton mass, k the Boltzmann constant, n_e the electron number density, γ the ratio of specific heats, η ($= 1 / (\sigma\mu)$) the magnetic diffusivity, σ the electrical conductivity, μ the magnetic permeability (normally the vacuum value).

In the energy equation $\underset{\sim}{\nabla}.(\kappa\nabla T)$ represents thermal conduction, where κ is the tensor coefficient of thermal conductivity. Conduction may be a loss or a gain term. Energy loss due to optically thin radiation in a fully ionized hydrogen plasma is represented by $n_e^2 \, Q(T)$, where $Q(T)$, the radiated power loss function, has been calculated by several authors (Cox and Tucker, 1969; McWhirter et al, 1975; Raymond, Cox and Smith, 1976; Summers and McWhirter, 1979). It may be approximated by a piecewise-continuous function

$$Q(T) = \chi T^{\alpha} \tag{8}$$

where the constants χ and α depend on the temperature range involved (e.g. Hildner, 1974). H_w is a term that represents heating due to, for instance, waves or joule dissipation; it is not well-known and is normally parameterised and assumed proportional to $p^s T^d$, where s and d are constants.

In most cases the magnetic diffusivity is assumed uniform, so that the magnetic induction equation may be rewritten in its more normal form,

$$\frac{\delta \underset{\sim}{B}}{\delta t} = \underset{\sim}{\nabla}\times(\underset{\sim}{v} \times \underset{\sim}{B}) + \eta\nabla^2\underset{\sim}{B} \tag{9}$$

The importance of each of the two terms on the right-hand side depends on the size of the dimensionless parameter R_m ($= VL/\eta$), the magnetic Reynolds number, where V and L are typical velocities and lengths. If R_m is very large then the first term dominates and the magnetic field lines are "frozen" into the plasma, so that plasma and magnetic field move together. No reconnection is allowed in this ideal MHD limit. On the other hand, when R_m is very small, the diffusion term, $\eta\nabla^2\underset{\sim}{B}$, dominates. In this case, the magnetic field lines can slip through the plasma and they may break and reconnect to other lines (see Figure 3).

2.2 Ideal magnetic instabilities

To study ideal MHD instabilities, we usually neglect any thermal effects and take the adiabatic energy equation,

$$\frac{D}{Dt} (P/\rho^{\gamma}) = 0, \tag{10}$$

which is valid only when the time-scale is much smaller than the
time for radiation conduction and heating. In addition, the
plasma is assumed to be perfectly conducting, so that the induction equation reduces to

$$\frac{\delta \underset{\sim}{B}}{\delta t} = \underset{\sim}{\nabla} \times (\underset{\sim}{v} \times \underset{\sim}{B}).$$ (11)

The system of equations then possesses an energy integral
(Bernstein et al, 1958; Roberts, 1967; Bateman, 1978) such that

$$W = \int \{ \frac{B^2}{2\mu} + \frac{P}{\gamma-1} + \rho\phi + \frac{1}{2} \rho v^2 \}d\tau = \text{constant},$$ (12)

where ϕ is the gravitational potential and the integration is
over the whole domain of interest. It is intuitively obvious
that energy is conserved when there are no dissipation mechanisms
present such as magnetic diffusion or thermal conduction.

For a small perturbation to a static equilibrium the kinetic
energy only appears as a second order term, while the first-order
change in potential energy is zero, since it may be written as an
integral of the equilibrium equations. To second order it may
be shown that

$$\frac{1}{2} \int \rho_0 \left| \frac{\delta \underset{\sim}{\xi}}{\delta t} \right|^2 d\tau + \delta_2 W = 0,$$ (13)

where the perturbed velocity $\underset{\sim}{v}_1$ and displacement $\underset{\sim}{\xi}$ are related by

$$\frac{\delta \underset{\sim}{\xi}}{\delta t} = \underset{\sim}{v}_1,$$

and the perturbed potential energy $\delta_2 W$ may be written as

$$\delta_2 W = \frac{1}{2}\int \{\frac{1}{\mu}|\underset{\sim}{B}_1|^2 - \underset{\sim}{j}_0 \cdot (\underset{\sim}{B}_1 \times \underset{\sim}{\xi}) + \gamma p_0(\nabla.\underset{\sim}{\xi})^2 + (\nabla.\underset{\sim}{\xi})(\underset{\sim}{\xi}.\nabla p_0) - (\underset{\sim}{\xi}.\nabla\phi)(\nabla.\rho_0\underset{\sim}{\xi})\}d\tau,$$ (14)

in terms of the perturbed magnetic field $\underset{\sim}{B}_1 = \underset{\sim}{\nabla} \times (\underset{\sim}{\xi} \times \underset{\sim}{B}_0)$
The energy method, in this form, was first proposed by Bernstein
et al (1958). In essence, stability is tested by determining the
sign of $\delta_2 W$. If $\delta_2 W$ is positive for all possible trial displacements ξ then the kinetic energy must be negative! This just
means that the perturbations cannot grow in time and so the equilibrium is stable. However, if it is possible to find one $\underset{\sim}{\xi}$
such that $\delta_2 W$ is negative, then the system may feed potential
energy into motions and the equilibrium is unstable. Thus a
necessary and sufficient condition for the stability of a given
magnetic structure is that

$$\delta_2 W > 0 \qquad \text{for all perturbations } \underset{\sim}{\xi}, \qquad (15)$$

which was proved rigorously by Laval, Mercier and Pellat (1965).
In principle, one can Fourier analyse and consider the most
general perturbation, but in practice one often considers only a
certain class, in which case the results are limited : instability
may then occur at a lower threshold than the one discovered and
stability may not endure when other perturbations are included.

The most likely ideal MHD instability in the solar atmosphere
is the "kink instability" (§ 3.1), for which the equilibrium tries
to reduce its energy by relaxing into a corkscrew shape. As an
illustration, an elastic band may be twisted up until a critical
stage beyond which it tries to kink and bend back on itself. This
may model the effect of the magnetic tension, but not, of course,
of the magnetic pressure.

2.3 Resistive instability

In the ideal limit, field lines are not allowed to break and
reconnect. For instance, the effect of pressing together oppositely directed field lines of a <u>current sheet</u> (Figure 3) is just
to build up the magnetic field strength at the centre.

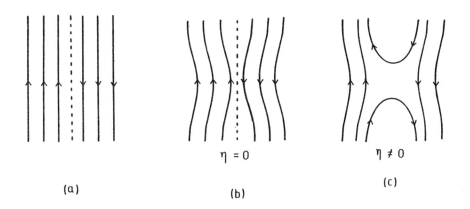

Figure 3. The effect of pressing together oppositely directed
field lines of a current sheet : (a) initial equilibrium current
sheet; (b) $\eta = 0$, magnetic pressure and tension oppose the displacement, with no reconnection; (c) $\eta \neq 0$, field lines may
diffuse and reconnect, to form magnetic islands.

However, if the electrical conductivity is allowed to be finite
($\eta \neq 0$), as in the full induction equation, then it is possible
for field lines to reconnect. New instabilities become possible,
such as the <u>tearing mode</u>, which can take place also in a sheared
magnetic structure. It is a long-wavelength instability, in the
sense that the ripples are longer than the sheet width, and it may
be understood qualitatively as follows. The time-scale for diffusion in a field that varies over a length-scale l is l^2 / η . Thus
when the field lines are squeezed together (Figure 3b), l becomes
smaller at the point of squeezing and so the field diffuses faster
there; this in turn implies that the field lines must reconnect,
and when the wavelength is long the restoring force due to magnetic tension is minimised. Alternatively, the current sheet may be
represented by current-carrying wires (Figure 4) which are equally
spaced in the equilibrium state.

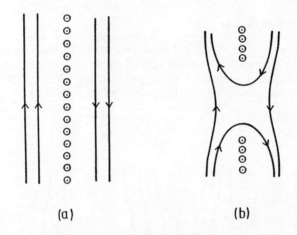

(a)   (b)

Figure 4. Qualitative representation of the tearing mode instability : (a) equilibrium state given by equally spaced current-
carrying wires; (b) small displacements result in bundles of wires
bunching together.

A small displacement that causes bundles of wires to bunch together will then continue due to the mutual attraction of the
wires. A comprehensive analysis of the tearing mode and other
resistive instabilities may be found in Furth <u>et al</u> (1963) and
Van Hoven (1980).

2.4 Magnetic reconnection

For most solar applications, R_m is the order of 10^7 or more
and the plasma is therefore, to a good approximation, a perfect
conductor. However, if the length-scales are small enough, as in
current sheets, the local value of R_m is much reduced and

reconnection may occur. This can develop in several ways:

(i)   it may be generated spontaneously by a linear resistive
       instability such as the tearing mode, either in a current
       sheet or throughout a sheared structure;

(ii)   it may be driven if the resistivity is locally enhanced;

(iii) it may be driven when two neighbouring flux systems are
       pushed together so that a thin reconnecting region forms at
       the boundary.

       In each of these cases the reconnection may develop into a
quasi-steady nonlinear state (Ugai and Tsuda, 1979a, b; Hayashi
and Sato, 1978), which has been studied in great detail. In case
(iii) the reconnection can take place at any rate, up to a maximum
value, and simply responds to the boundary conditions at large
distances. In cases (i) and (ii), where the boundary conditions
at large distances are free, reconnection tends to proceed at the
maximum rate.

       The only workable model for steady nonlinear reconnection,
in the solar atmosphere, is the Petschek mechanism (Petschek, 1964).
It incorporates a central diffusion region occupying a length L
which is very much smaller than the length under consideration.
From each corner of the diffusion region there propagates a slow
magnetohydrodynamic shock wave, which stands in the flow (Figure
5).

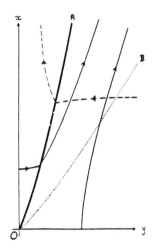

Figure 5. Magnetic field lines (——) and streamlines (----) for
the first quadrant of Petschek's mechanism. A finite Alfven wave
(or slow shock) is situated at OA. It is generated at the diffusion region, which is here regarded as a point O, and serves to
turn the flow and magnetic field through large angles.

The majority of the energy is released in the shock rather than
the diffusion region. The Petschek mechanism has been put on a
firm mathematical footing by Soward and Priest (1977); see also
Vasyliunas (1975), Priest and Soward (1976) and Priest (1981).
The maximum rate of reconnection (V_e) for it is typically (0.01 -
0.1) V_A and it depends weakly on R_m, where V_A (= B_e/ $(\mu\rho)^{1/2}$)
is the Alfven speed.

     The central diffusion region, although tiny, is an important
part of the model.

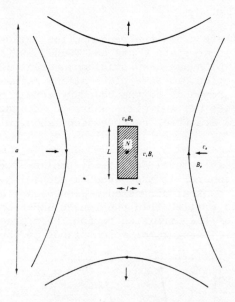

Figure 6. The configuration for steady magnetic reconnection.
Oppositely directed field lines of strength B_e , frozen to the
plasma, are carried towards one another at a speed v_e by converging flow. They enter a diffusion region (shaded) with dimensions
l and L, are reconnected at the neutral point N and are finally
ejected from the ends.

     The following order of magnitude relationships exist between
the input and output parameters. Since the diffusion region wants
to expand with a diffusion velocity η/ℓ , steady reconnection can
occur only if the inflow velocity provides a balance, so that

$$v_i = \eta/\ell. \tag{16}$$

Material is expelled, from the top and bottom (to prevent a pileup of mass) in the diffusion region at the hybrid Alfven speed

$$v_o = B_i / (\mu\rho_c)^{1/2}. \tag{17}$$

From mass continuity we must also have

$$\rho_i v_i L = \rho_c v_o \ell, \tag{18}$$

where ρ_i and ρ_c are the inflow and sheet densities respectively. In addition, from constancy of total pressure

or
$$P_c = p_i + B_i^2/2\mu \tag{19}$$

$$\frac{\rho_c}{\rho_i} = \frac{T_i}{T_c} (1 + \beta_i), \tag{20}$$

where the inflow beta is

$$\beta_i = p_i 2\mu/B_i^2. \tag{21}$$

The temperature in the current sheet, T_c , is determined from the energy equation, and then the values of l, v_o , L, ρ_c follow from (16) - (21) in terms of prescribed values of v_i , B_i , ρ_i , T_i (Tur and Priest, 1978). An approximate treatment of the flow equations within the diffusion region has been given by Milne and Priest (1981), who find the sheet-width may be 10 times thicker than previously predicted. Also, they have discovered a β - limitation, such that steady flow is impossible when β_i is too low.

3. SIMPLE-LOOP FLARE

How are all the above ideas of relevance to the solar flare? The majority of flares are small and appear to occur in a loop-type geometry. Three mechanisms now seem feasible for producing such a simple-loop flare.

(i) It has long been appreciated that flux tubes are subject to the kink instability, but the puzzle has been to understand how they can be stable before flaring. A recent breakthrough is the realisation of the importance of line-tying (Raadu, 1972; Hood and Priest, 1979), which stabilises a loop until it is twisted too much.

(ii) A more speculative mechanism is that a cool loop may be subject to thermal nonequilibrium, when it ceases to remain in thermal equilibrium and heats up. For this non-magnetic effect, the magnetic field is only important in channelling the heat along its length.

(iii) A third theory, the emerging flux model, has been inspired by the observations of new magnetic flux emerging from below the photosphere and presumably interacting with the over-lying magnetic field (Heyvaerts et al, 1977).

3.1 Kink instability

   One of the first MHD stability calculations for solar coronal
loops was performed by Anzer (1968), who proved that all coronal
loops should be unstable no matter what their twist is. Raadu
(1972), however, showed that loops may be stabilised by the effect
of the dense photosphere. This is modelled by assuming that all
perturbations, generated in the corona, must vanish at the photospheric footpoints. Physically, a sound wave or an Alfven wave
propagating down from the corona will be reflected when it strikes
the dense photosphere;   less than 0.4% of the wave energy is transmitted through the interface. Hood and Priest (1979, 1980)
repeated Raadu's analysis more precisely;   they also extended it
to include the effects of pressure gradients and derived both
necessary conditions and sufficient conditions for the stability
of coronal loops. (Giachetti et al (1977) also include pressure
gradients but model line-tying only crudely.) The effect of line-
tying is to make all cylindrical, force-free loops stable if they
are twisted by less than 2π : that is, if the angle

$$\Phi(r) \equiv \frac{2LB_\theta(r)}{rB_z(r)} < 2\pi,$$

where the cylindrical equilibrium field is assumed to be $B =$
$(0, B_\theta(r), B_z(r))$, r is the radial distance and 2L is the length.
As an example, Hood and Priest (1979, 1980) considered the force-
free field

$$\underset{\sim}{B} = \frac{B_o}{(1 + r^2/b^2)} (0, r/b, 1),\tag{22}$$

and showed that instability occurs when a critical twist, Φ_{crit}
is exceeded, lying in the range

$$2\pi < \Phi_{crit} < 3.3\pi.$$

The upper bound is derived by using a trial function, $\underset{\sim}{\xi}$, that
vanishes at the photosphere. It has the form

$$\underset{\sim}{\xi} = (\xi_r(r), -\frac{iB_z}{B} \xi_0(r), \frac{iB_\theta}{B} \xi_0(r))e^{i(m\theta+kz)} \sin\frac{n\pi z}{2L}\tag{23}$$

Notice that without the factor $\sin(n\pi z/2L)$, $\underset{\sim}{\xi}$ does not
vanish for all r at any location z along the loop, and therefore
one cannot simply model line-tying by putting $k = \pi/L$. By sub-
stituting (23) into $\delta 2W$ (14) and integrating over θ and z , ξ_0
may be eliminated algebraically, so that $\delta_2 W$ reduces to the
following form:

$$\delta_2 W = \int_0^\infty (F(r) \; (\frac{d\xi_r}{dr})^2 + G(r)\xi_r^2) \, dr, \qquad (24)$$

where F and G are complicated functions that depend on the basic state (Hood and Priest, 1979). Then a minimisation with respect to ξ_r gives the Euler-Lagrange equation

$$\frac{d}{dr} (F\frac{d\xi_r}{dr}) = G\xi_r. \qquad (25)$$

Newcomb (1960) has shown that the stability is governed by the number of zeros possessed by $\xi_r(r)$. For the kink instability (m = 1) the boundary conditions are

$$\frac{d\xi_r}{dr} = 0 \qquad\qquad \text{at } r = 0$$

$$\qquad\qquad\qquad\qquad\qquad\qquad\qquad\qquad (26)$$

$$\xi_r = 1 \qquad\qquad \text{at } r = 0$$

and (25) is integrated outwards from the origin. If $\xi_r > 0$ for all r then the equilibrium is stable to perturbations of the form (23), but if $\xi_r = 0$ at $r = r^*$, say, then the equilibrium is definitely unstable. Typical results are shown in Figures 7 and 8.

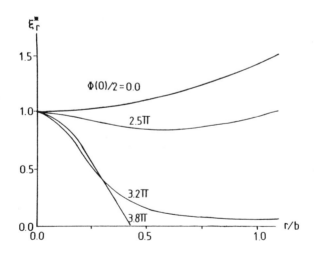

Figure 7. The effect on the radial perturbation, $\xi_r(r)$, of increasing the twist, Φ. The equilibrium remains stable until ξ_r first becomes negative, when the kink instability sets in.

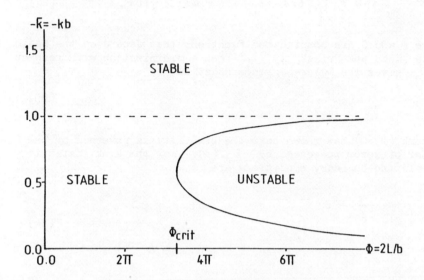

Figure 8. The stability diagram for the force-free, uniform-twist field. For a twist $\Phi > \Phi_{crit}$ the equilibrium is definitely unstable, whereas for $\Phi < 2\pi$ the equilibrium is definitely stable.

However, the problem still remained of calculating the actual critical twist, since the results so far left a "grey" region of uncertainty. Recently, therefore, Hood and Priest (1981b) solved the full normal mode equations numerically, for force-free fields,

$$- \omega^2 \underset{\sim}{\xi} = (\nabla \times \nabla \times \underset{\sim}{\xi} \times \underset{\sim}{B}_0) - (\nabla \times \underset{\sim}{\xi} \times \underset{\sim}{B}_0) \times (\nabla \times \underset{\sim}{B}_0) \tag{27}$$

in order to find the eigenvalue ω^2. The equilibrium is stable if $\omega^2 > 0$ and unstable if $\omega^2 < 0$. Without loss of generality, they write the perturbation as

$$\underset{\sim}{\xi} = (\xi_r(r,z), - \frac{B_z}{B} \xi_0(r,z), \frac{B_\theta}{B} \xi_0(r,z)) e^{i(m\theta+\omega t)}, \tag{28}$$

and solve the coupled partial differential equations (27), subject to the boundary conditions

$$\frac{\delta \underset{\sim}{\xi}}{\delta r} = 0 \quad \text{at } r = 0, \ (m = 1)$$

$$\underset{\sim}{\xi} = 0 \quad \text{at } z = 0, \ 2L, \tag{29}$$

$$\underset{\sim}{\xi} = 0 \quad \text{at } r = d$$

In theory the outer radius d should be taken at infinity, but in practice it was taken as far out as computational limitations would allow and the effect of varying it was considered. A reasonable value for d was found to be 10 times the typical width b for the equilibrium field (22). The critical twist was calculated, by putting $\omega^2 = 0$, as

$$\Phi_{crit} = 2.55\pi,$$

which of course lies between the previous bounds derived above.

The effect of non-zero resistivity on the kink instability is to allow the resistive kink or cylindrical tearing mode to occur (Spicer, 1977, 1980; Van Hoven, 1980). Spicer has stressed the importance of cylindrical tearing modes and has reviewed the work performed by others in connection with the laboratory; in particular, he points out that double tearing occurs considerably faster than the usual ordinary tearing and some nonlinear effects may increase dissipation. So far the effect of line-tying has not been incorporated mathematically, although Gibbons and Spicer (1980) and Van Hoven (1980) have argued, on intuitive grounds, that it is stabilizing.

Resistive effects for an infinite cyliner become important at a singular layer where

$$\underset{\sim}{k} . \underset{\sim}{B} = mB_\theta/r + kB_z = 0,$$

where m and k are the azimuthal and axial wavenumbers. This arises from a Fourier analysis in θ and z. However, when line-tying in a finite length is included, one cannot Fourier analyse in z, and the condition should be restated as

$$\left[\underset{\sim}{\nabla} \times (\underset{\sim}{v_1} \times \underset{\sim}{B_0}) \right]_r \equiv i\omega \left(\frac{B_\theta}{r} \frac{\delta\xi_r}{\delta\theta} + B_z \frac{\delta\xi_r}{\delta z} \right) = 0,$$

where ξ_r is the radial component of the perturbation. This condition can be evaluated from the numerical solution of Hood and Priest (1981b), and gives a curved surface where resistivity dominates.

3.2 Thermal nonequilibrium

Thermal nonequilibrium occurs because of the nonlinear nature of the energy equation, which may be written

$$\frac{d}{ds} \left(\kappa_\| \frac{dT}{ds} \right) = n_e^2 \chi T^\alpha - Hp^s T^d, \tag{30}$$

where

$$p = constant$$

and $n_e = p/ (2kT)$.

Here s is the distance along a magnetic field line, $\kappa_{||}$ ($=\kappa_0 T^{5/2}$)
is the coefficient of thermal conductivity along the field and
conduction across them is neglected ($\kappa_\perp / \kappa_{||} \simeq 2 \times 10^{-13}$)
Hood and Priest (1981a) used these equations to model approxi-
mately the cool core of an active-region loop. As the heating is
slowly increased, through the parameter H, so the summit temperature slowly rises as indicated in Figure 9.

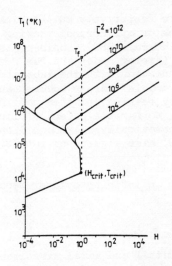

Figure 9. The equilibrium summit temperature, T_1, is shown as a
function of the heating parameter, H, for various values of the
dimensionless loop length $\overline{L}$. Critical conditions are denoted by
the point (H_{crit}, T_{crit}) and the flare temperature T_f is denoted
by a dot.

   However, once the critical state is reached, there is no
neighbouring equilibrium and the plasma must heat up rapidly in
its search for a new equilibrium. The multiplicity of the solutions results from the nonlinearity of the problem. The middle
solution is inaccessible, since it is thermally unstable, while
the lower branch represents a balance between radiation and heating, with conduction relatively unimportant. The critical point
exists because the radiative-loss function possesses a maximum
and, if the heating is increased beyond this value, radiation can
no longer provide a balance. The plasma therefore heats up until
thermal conduction becomes important at a few million degrees.

3.3 Emerging flux model

Heyvaerts et al (1977) proposed a model based on the fact that before most flares new magnetic flux is observed to emerge from below the photosphere. The type of flare that is produced depends on the magnetic environment in which the new flux finds itself. If the flux emerges near a unipolar sunspot or into a unipolar area near the edge of an active region, then a simple-loop flare may occur. (On the other hand, if the overlying field is the sheared field around an active-region filament, then a two-ribbon flare may occur (e.g. Neidig et al, 1978). In this case, the emerging flux triggers the release of the stored energy in the much larger field by providing a localised region of turbulent plasma, where the tearing mode can produce reconnection much faster than normal.)

In the emerging flux model, a current sheet forms between the old and new flux and the slow reconnection produces some preflare heating (Figure 10).

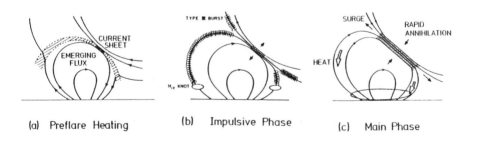

(a) Preflare Heating   (b)   Impulsive Phase   (c)   Main Phase

Figure 10. The emerging flux mechanism for a subflare or simple-loop flare (from Heyvaerts et al, 1977). (a) Preflare phase The emerging flux reconnects with the overlying field. Shock waves (dashed) radiate from a small current sheet and heat the plasma as it passes through them into the shaded region. (b) Impulsive phase The onset of turbulence in the current sheet (when it has reached a critical height) causes a rapid expansion. (c) Flash and main phases The current sheet reaches a new steady state, with reconnection based on a turbulent resistivity.

As more flux emerges, so the height of the current sheet rises, until a critical height is reached, at which its current density exceeds the threshold value for the onset of microturbulence. The flare is then triggered. These critical conditions may be written roughly in terms of the current-sheet temperature T_c as

$$T_c^2 > T_{turb}^2 \equiv 1.8 \times 10^6 B_i / v_i,$$

where the inflow field and speed are measured in Tesla and m/s respectively. The current-sheet temperature is determined from the energy balance inside the sheet, and it is often found that thermal nonequilibrium ensues if the height of the sheet is too great. The sheet heats up rapidly, T_{turb} is exceeded and the flare is triggered.

4. TWO-RIBBON FLARE

A two-ribbon flare is probably caused by the instability of the sheared magnetic configuration in which a filament is situated. The generally accepted scenario is that slow photospheric motions produce a sequence of largely force-free equilibria. However, when the shear is too strong, the equilibria becomes unstable and the field blows open. Subsequently, it closes back down, in the manner proposed by Kopp and Pneuman (1976) and Pneuman (1980), producing "post"-flare loops and H_α ribbons.

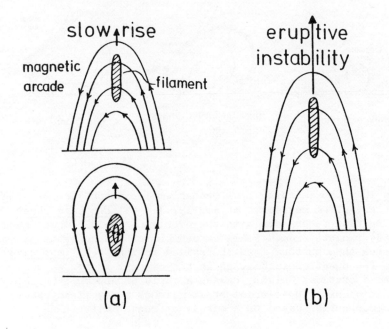

(a)   (b)

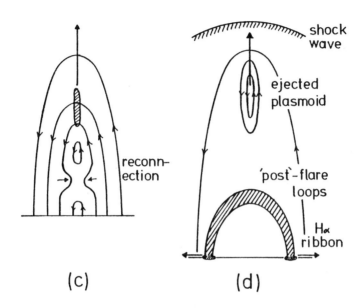

(c)   (d)

Figure 11.   The overall magnetic behaviour in a two-ribbon flare,
as seen in a section through the magnetic arcade. (a) Slow preflare rise of a filament, possibly due to thermal nonequilibrium,
emerging flux or the initial stages of an MHD instability. The
surrounding field may just be sheared (upper diagram), or it may
contain a magnetic island so that the filament lies along a flux
tube (lower diagram). (b) Eruptive instability of a magnetic
arcade and filament. (c) Field lines below the filament are
stretched out until reconnection can start. (d) As reconnection
proceeds, "post"-flare loops rise and Hα ribbons move apart.

However, until recently the cause of the instability has been
unsure; until, that is, Hood and Priest (1980) demonstrated
explicitly that the field can become MHD unstable, as outlined
below. The magnetic field structure near an active-region filament is not known in any detail, and so they considered two distinct types of configuration (Figure 11a):

(i)   a simple sheared arcade with the field lines crossing the
      filament;

(ii) a flux tube contained within a sheared arcade, so that the
      filament lies along the flux tube, which is consistent with
      the observations of flows along a filament.

Hood and Priest were unable to find any instability for (i), although a wide range of equilibria and perturbations were considered. This strongly suggests that the configuration is stable. However, when type (ii) is considered, instability is readily proved if the filament is twisted, by photospheric motions, beyond a critical amount. This is similar to the simple-loop flare but on a much larger scale. Alternatively, instability is produced if the filament rises above a critical height (d_c) so that the line-tying effect of the photosphere is minimised.

To study the stability in case (ii), the arcade is modelled by a cylindrical, force-free equilibrium, with its axis situated at a height d above the photosphere (in a similar manner to the lower diagram of Figure 11a). The line-tying condition* is just

$$\underset{\sim}{\xi} = 0 \qquad\qquad \text{at the photosphere} \qquad\qquad (31)$$

but the position of the photosphere is rather complicated.

The filament is assumed to have a finite length (2L) with its ends also line-tied. Thus, for a trial function sufficient to prove instability, one may consider

$$\underset{\sim}{\xi} = \begin{cases} (\xi_r(r), \; -\dfrac{iB_z}{B}\,\xi_0(r), \; \dfrac{iB_\theta}{B}\,\xi_0(r))e^{i(m\theta+kz)}\sin\dfrac{n\pi z}{2L}, & r \leqslant d \\[4mm] \underset{\sim}{0} & r > d \end{cases} \qquad (32)$$

As an example, the uniform-twist field

$$\underset{\sim}{B} = \frac{B_0}{(1 + r^2/b^2)}\,(0,\; r/b,\; 1), \qquad\qquad (33)$$

gives the necessary stability diagram shown in Figure 12.

---

* Alternative line-tying conditions to (31) and Section 3.2 have been suggested by Einaudi and Van Hoven (1981). They propose that only the perpendicular component of the displacement should vanish and that the parallel component is only restricted by the condition that the values at the two ends are equal. This is identical to (31) when the magnetic field is force-free. They also perform a general Fourier series expansion and minimise the potential energy to derive sufficient conditions for stability.

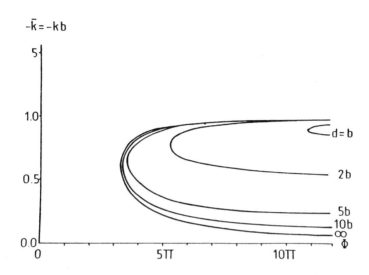

Figure 12. Sufficient conditions for the amount of twist, Φ , required to produce instability of a flux tube embedded in an arcade. The flux tube has a length 2L and its axis is situated at a height d above the photosphere. The equilibrium is definitely unstable to the right of each curve.

If the twist $\Phi = 5\pi$, for example, then the structure must become unstable if it rises beyond a height $d_{crit} = 2.16b$. Again if the filament height is $d = 5b$, instability has occurred if the filament is twisted beyond $\Phi_{crit} = 3.6\pi$.

The eruptive instability may occur spontaneously in the manner outlined above. Or else it may be produced by some other trigger. One possibility is the emergence of new magnetic flux near the active-region filament (Heyvaerts et al, 1977; Tur and Priest, 1978). Another is the occurrence of thermal nonequilibrium in the filament. It causes the preflare temperature-rise together with a fall in density and a draining of plasma along the filament legs. This in turn makes the filament crest rise slowly to maintain magnetohydrostatic equilibrium. When the critical height is reached the faster MHD instability described above will ensue (Figure 11).

In conclusion, it now appears that there are several viable mechanisms for a simple-loop flare. Perhaps, they each occur at different times, or maybe future observations of the magnetic field, of cool active-region loops, and of emerging flux will distinguish between them. Furthermore, the magnetic instability

or a coronal arcade to give a two-ribbon flare has at last been
demonstrated in a convincing manner.

ACKNOWLEDGMENT

The authors are grateful to the Science Research Council and
Royal Society for financial support.

REFERENCES

Anzer, U. : 1968, Solar Phys. 3, p.298.
Bateman, G. : 1978, MHD Instabilities, MIT Press.
Bernstein, I. B., Freeman, E. A., Kruskal, M. D. and Kulsrud, R.
    M. : 1958, Proc. Roy. Soc. (London) A244, p.17.
Brown, J. C., Smith, D. and Spicer, D. S. : 1980, "The Sun as a
    Star" to be published by NASA/CNRS.
Cheng, C. C. : 1977, Astrophys, J., 213, p.558.
Cox, D. P. and Tucker, W. H. : 1969, Astrophys. J., 157, p.1157.
Einaudi, G. and Van Hoven, G. : 1981, Phys. Fluids submitted.
Furth, H. P., Killeen, J. and Rosenbluth, M. N. : 1963, Phys.
    Fluids, 6, p.459.
Giachetti, R., Van Hoven, G. and Chiuderi, C. : 1977, Solar Phys.
    55, p.371.
Gibbons, M. and Spicer, D. S. : 1980, Proc. CECAM Workshop on
    Solar Flares, Paris (ed. Kuperus, M.).
Gibson, E. G. : 1977, Solar Phys. 53, p.123.
Hayashi, T. and Sato, T. : 1978, J. Geophys. Res. 83, p.217.
Heyvaerts, J., Priest, E. R. and Rust, D. M. : 1977, Astrophys.
    J., 216, p.123.
Hildner, E. : 1974, Solar Phys. 35, p.123.
Hood, A. W. and Priest, E. R. : 1979, Solar Phys. 64, p.303.
Hood, A. W. and Priest, E. R. : 1980, Solar Phys. 66, p.113.
Hood, A. W. and Priest, E. R. : 1981a, Solar Phys., in press.
Hood, A. W. and Priest, E. R. : 1981b, in preparation.
Kopp, R. and Pneuman, G. W. : 1976, Solar Phys. 50, p.85.
Kuperus, M. and Svestka, Z. : 1978, Proceedings of the 2nd Euro-
    pean Solar Meeting, CNRS, Toulouse.
Laval, G., Mercier, C. and Pellat, R. : 1965, Nuclear Fusion, 5,
    p.156.
McWhirter, R. W. P., Thoneman, P. C. and Wilson, R. : 1975,
    Astron. Astrophys. 40, p.63.
Milne, A. M. and Priest, E. R. : 1981, Solar Phys., in press.
Neidig, D. F., Demastus, H. L. and Wiborg, P. H. : 1978, Environ-
    mental Research Papers, No.637.
Parker, E. N. : 1963, Astrophys. J. Suppl. Ser. 77, 8, p.177.
Petschek, H. : 1964, AAS-NASA Symp. on Phys. of Solar Flares,
    Hess, W. N. (ed), NASA SP-50, p.425.
Pneuman, G. W. : 1980, in "Solar Flare Magnetohydrodynamics" (ed.
    Priest, E. R.) Gordon and Breach, London.

Priest, E. R. : 1976, "Physics of Solar Planetary Environment",
    Williams, D. J. (ed.), American Geophysical Union, Vol. 1,
    p.144.
Priest, E. R. : 1981, "Solar Magnetohydrodynamics", D. Reidel, in
    preparation.
Priest, E. R. and Soward, A. M. : 1976, IAU Symp. 71, p.353.
Raadu, M. : 1972, Solar Phys. $\underline{22}$, p.425.
Raymond, J., Cox, D. and Smith, J. : 1978, Astrophys. J.
Roberts, P. H. : 1967, "An Introduction to Magnetohydrodynamics",
    Longman.
Rust, D. M. : 1976, Phil. Trans. Roy. Soc. (London) $\underline{A281}$, p.427.
Rust, D. M. and Webb, D. F. : 1977, Solar Phys. $\underline{54}$, p.403.
Soward, A. M. and Priest, E. R. : 1977, Phil. Trans. Roy. Soc.
    $\underline{284}$, p.369.
Spicer, D. S. : 1977, Solar Phys. $\underline{53}$, p.305.
Spicer, D. S. : 1980, in chapter 3 of "Solar Flares" (ed.
    Sturrock, P. A.) Colo. Ass. Univ. Press.
Spicer, D. S. and Brown, J. C. : 1980, "The Sun as a Star" to be
    published by NASA/CNRS.
Sturrock, P. : 1980, "Solar Flares", Colo. Ass. Univ. Press.
Summers, H. and McWhirter, R. W. P. : 1979, J. Phys. $\underline{B\ 12}$, p.2387.
Svestka, Z. : 1976, "Solar Flares", D. Reidel.
Svestka, Z. : 1980, chapter 2 of "Solar Flare MHD" (ed. Priest,
    E. R.) Gordon and Breach, London.
Tur, T. J. and Priest, E. R. : 1978, Solar Phys. $\underline{58}$, p.181.
Ugai, M. and Tsuda, T. : 1979a, J. Plasma Phys. $\underline{21}$, p.459.
Ugai, M. and Tsuda, T. : 1979b, J. Plasma Phys. $\underline{22}$, p.1.
Van Hoven, G. : 1980, chapter 4 of "Solar Flare $\overline{MHD}$" (ed. Priest,
    E. R.) Gordon and Breach, London.
Vasyliunas, V. M. : 1975, Rev. Geophys. Space Phys. $\underline{4}$, p.207.
Vorpahl, J. A. : 1973, Solar Phys. $\underline{28}$, p.115.

DIFFUSE RECONNECTION : THE RESISTIVE TEARING-MODE

R. PELLAT

Centre de Physique Théorique
Ecole Polytechnique
91128 PALAISEAU, France.

Abstract. One gives the present status of our theoretical knowledge about the problem of diffusive reconnection which can be
considered as being related to the non-linear stage of the Tearing
mode instability.

1. INTRODUCTION

Magnetic field line reconnection in natural plasmas (solar
flares, magnetospheric substorms) or in laboratory (disruptions in
θ, Z, toroidal pinches[1]) has been a theoretical problem of great
interest for more than twenty years (1). There is now both experimental and theoretical evidence (some results have been obtained in
computer simulations) that reconnection can be achieved both in
collisional and collisionless plasmas ; reconnection may be sub-
sonic (Alfvenic) or supersonic as a function of boundary conditions
(open or closed system). Let us recall that in a perfect fluid the
Stokes theorem implies the conservation of the vorticity :
$\tilde{\Omega}$: $\frac{\delta}{\delta t} \tilde{\Omega} + \tilde{\nabla} \times (\tilde{\Omega} \times \tilde{v}) = 0$. In a magnetized ideal plasma, the
"perfect Ohm's Law" has the same meaning for the magnetic field :
$\frac{\delta \tilde{B}}{\delta t} + \tilde{\nabla} \times (\tilde{B} \times \tilde{v}) = 0$. In fluid mechanics it is well known that
viscosity or turbulence allows the flow to violate the Stokes
theorem. In a similar (but somewhat more complex) way electron-ion
collisions (i.e. finite resistivity) or Landau damping (finite
particle inertia in their motion parallel to the magnetic field
allows a plasma to violate the perfect Ohm's law and as a consequence reconnection is possible (2). Let us now precise what we call

R. M. Bonnet and A. K. Dupree (eds.), Solar Phenomena in Stars and Stellar Systems, 533–543.
Copyright © 1981 by D. Reidel Publishing Company.

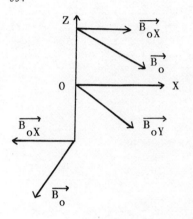
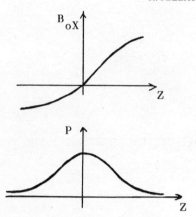

<u>Fig. 1</u>

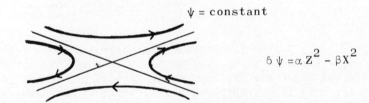

ψ_0 = constant

X

ψ_0 = constant

ψ = constant

$$\delta \psi = \alpha Z^2 - \beta X^2$$

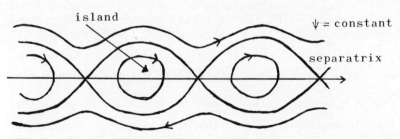

island

ψ = constant

separatrix

$$\psi = \alpha(t) \cos K X$$

<u>Fig. 2</u>

reconnection. The simplest mathematical model is the following
"slab" one. One starts with an initial magnetic field given by
(Figure 1) $B_o = \overset{\sim}{\nabla} \times (\Psi_o \overset{\sim}{e_Y}) + B_{oY} \overset{\sim}{e_Y}$ with for example

$\Psi_o = \lambda B_o \ln \text{Ch} (Z/\lambda)$ with $B_{o\perp}$, B_{oX} constant. In this model, the
plasma pressure P is confined in the Z direction and maximum in
the central plane (Z = 0). The pitch of the magnetic field lines
is a function of Z. The field lines projection in the (X, Z) plane
are the lines Z = constant. The reconnection in the plane Z = 0
appears as a modification of the potential Ψ which becomes
$\Psi = \Psi_o + \delta\Psi$ with $\delta\Psi = \alpha(t)Z^2 - \beta(t)X^2$ if there is formation of
one neutral point or $\delta\Psi = \alpha(t, Z) \cos KX$ if there is formation
of a periodic set of neutral points and magnetic island configuration (figure 2). The first case corresponds to what is called
reconnection in nature and the second to what is called reconnection
in laboratory. The reason is that in nature one is more interested
by the non-linear evolution of the system (and not by the initial
stage). In laboratory being interested by stable situation one
computes linearized situations (or non-linear but close to linear).
In that case the initial configuration being invariant in (Y, X)
one makes Fourier analysis in these directions with the previous
result. Apart from this difference one has still to precise the
flow properties in the vicinity of the neutral point(s). Two kinds
of "models" have been investigated :

   - a diffusive model (3) with subsonic plasma motion ; this
model seems to be relevant in laboratory configuration (such as
the Tokamak) in which neutral points (and "magnetic islands" formation) is an internal process which results of an instability of the
the initial toroidal axis symmetric plasma configuration. This
process may lead to non-linearly stable helicoidal configuration
or to a complete disruption of the plasma current. One does not
know yet if this disruption is due to the growth of the helicoidal
deformation up to the plasma boundary or to the superposition of
two or many helicoidal deformations which may finally break the
magnetic field topology by ergodisation of the magnetic field
lines and enhancement of the radial electron thermal transport.
Secondary micro-instabilities may provide an increase of resistivity which may plan an important role in this process. (We will
come back later to this problem). The diffusive model seems to be
relevant at the earth magnetopause when the interplanetary magnetic
field is southward ;

   - a strongly non-linear "super Alfvenic model" (4) in which
the plasma flow is driven from the "outside" ($|Z| \to \infty$) towards the
neutral point. This flow along $\overset{\sim}{e_Z}$ is turned in the $\overset{\sim}{e_X}$ direction in
the vicinity of the neutral point with a net acceleration which
leads to a super Alfvenic velocity. The main difficulty of this
model is related to the necessary formation of a pair of expansion
waves and a pair of shock waves in the vicinity of the neutral

point. Recent computer simulations (5) including anomalous resistivity provide an explicit example of this model. At the magnetospheric
nose the dimensions of the boundary layer are of order of thermal
proton Larmor radius and it is not very surprising that fluid
theory (based upon the opposite assumption) may not be relevant.
In fact two years satellite crossing of this boundary have observed one case which seems to be described in this model.

     In the following we will restrict our analysis to the results
which have been obtained in laboratory (theory - experiments -
computer simulations) for the slow reconnection process in a
"closed" system.

2. LINEARIZED RECONNECTION THEORY

     This is an old subject (2) but we shall briefly recall the
results (some of them have been obtained recently (2b)). In plane
geometry, one makes the simplifying (and justified) assumption of
plasma incompressibility and takes bidimensional magnetic and
velocity fields. With these assumptions one can take (for both the
initial stratified equilibrium and an "helicoidal" deformation of
one pitch) :

$$\tilde{B} = B_o \, e_Y^{\sim} + \nabla \times \left[\, (\Psi - \alpha Z B_o) \, e_Y^{\sim} \, \right]$$

$$\tilde{v} = \nabla \times (\Phi \, e_Y^{\sim})$$

where Ψ and Φ are function of $X - \alpha Y$ (if one makes a Fourier analysis
in X, Y, $K_Y = - \alpha K_X$). One obtains with a straightforward algebra
the following set of non-linear equations :

$$\frac{\delta \Psi}{\delta t} - \tilde{B} \cdot \nabla \tilde{\Phi} = \eta \Delta \Psi$$

$$\frac{\delta \Delta \Phi}{\delta t} + \tilde{\nabla} \cdot (\tilde{V} \Delta \Phi) = \tilde{B} \cdot \nabla \Delta \tilde{\Psi}$$

where η is the plasma resistivity, taken constant. The total current
is along $\vec{e}_Y$ and given by $J = \Delta \Psi$. One first linearize the equations
in the vicinity of the equilibrium ($\Psi = \Psi_o + \delta \Psi$). It appears that
the resistivity is negligible except in a boundary layer where
$B_{ox} - \alpha B_{oz}$ vanishes, i.e. where the pitch of the field lines is
the same as the pitch of the perturbation. In this layer the ideal
Ohm's law breaks (it would give an infinite velocity potential for
a finite $\delta \Psi$ necessary for reconnection). The finite resistivity
decouples the plasma and field line motions and allows the reconnection. One then matches the ideal MHD solution outside the layer to
the inside resistive solution (Figure 3). One gets two different
results for the growth rates of these so-called Tearing modes :

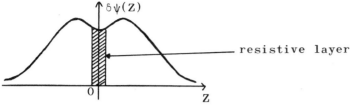

$\delta\psi$ (for $\alpha = 0$) as a solution of the MHD equation (with $\frac{\partial}{\partial t} = 0$):

$$\Delta\delta\psi = + \frac{\partial j_0}{\partial Z} \Big/ \frac{\partial \psi_0}{\partial Z} \delta\psi$$

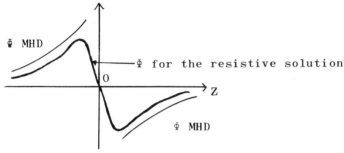

velocity potential in the resistive layer

Fig. 3

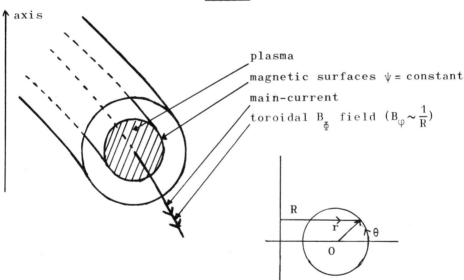

Tokamack configuration

Plasma poloidal cross section with "cylindrical coordinates"

Fig. 4

$\alpha K_X \sim \lambda\gamma$ scales as $\eta^{3/5}$ (2a) and for $\alpha K_X << \lambda,\gamma$ scales as $\eta^{1/3}$ (2b) ;
this result means that long wavelengths are more unstable than
short wavelengths (in the first case, $\delta\Psi$ is almost constant in the
boundary layer). One also finds instabilities related to the temperatures (or η) gradients which will not be covered in this talk (2a).
What are the modifications of this result in more realistic geometries (cylindrical or toroidal) ? The main modification is due
to the field lines curvature. An extensive litterature has been
devoted to this subject (6). Let us summarize the main findings.
In cylindrical geometry, let us first suppose that η vanishes : if
the pitch of the field lines is radially constant, the cylindrical
curvature is bad (if the plasma pressure is peaked on the axis) and
the local interchange modes are unstable. If the pitch of the field
lines varies then the exchange of two flux tubes increases the ma-
magnetic field energy and a plasma of sufficiently low pressure may
be stable with respect to local interchange (7).

     Let us come back to the effect of the resistivity when the
plasma is locally stable with respect to the interchange mode. In
this case, the boundary layer treatment shows that the stabilizing
effect of the variable pitch of the field lines disappear and the
bad cylindrical curvature drives locally "unstable resistive g
modes" (10) and increases the growth rate of the Tearing modes
(2a). In toroidal geometry the results are more complex. We have
first the fact that the curvature of the field lines is not constant along a magnetic field line : this follows from the fact that
now the main toroidal field has a 1/R dependence, if R is the
distance to the axis of the torus (Figure 4). This property has
two consequences : the local curvature is "bad" outside the torus
and "good" inside. The average curvature may be good or bad ; it
can be shown that it is good if a field line makes before closing,
more turns around the axis of the torus than around the quasi
cylindrical plasma poloidal cross section (and bad in the opposite
case). Now the finite plasma compressibility matters, because the
information along field lines propagates mainly with the ion sound
speed C_S ; as a consequence if R/C_S is large compared with a
growth time γ^{-1}, the plasma will be more unstable in a bad curvature
region. On the contrary if the information propagates faster
($R\gamma/C_S < 1$) the plasma "feels" the average curvature and the helicoidal perturbation may be less unstable or even stabilized (6,
11).

3. NON-LINEAR DIFFUSIVE RECONNECTION THEORY

     All these results are of course a consequence of a linearized
stability analysis. In fact they are expected to be completely
modified in their subsequent non-linear development. Let us comment
now the available non-linear theory of the Tearing modes. This theory

(12) follows a very simple approach. The magnetic island after the
initial stage of reconnection reaches a size larger than the boundary layer which appears in the linear treatment. To simplifly we
take $\alpha = 0$ and make the constant $\delta\Psi$ approximation. In the vicinity
of $Z = 0$ where the reconnection occurs the equation for the magnetic surface reads :

$$\Psi = B_o Z^2/2\lambda + \delta\Psi(t) \; \cos KX$$

the size of the periodic system of islands is simply $(2\lambda\delta\Psi/B_o)^{1/2}$
and increases in time. On such an increasing dimension one
expects the resistivity to drive a slow reconnection, i.e. a quasi
steady process (in the linearized theory the "fast" growth is
possible because the boundary layer has the dimensions of the
resistive skin depth). If one neglects the inertia, the perturbed
current δj becomes a function of Ψ and the Ohm's law gives its
value :

$$\delta j = \frac{1}{\eta} < \frac{\partial \delta\Psi}{\partial t} \; \cos KX >$$

where the average is taken along X at constant Ψ. One can then
match this solution inside the island to the MHD linear solution
outside :

$$\int_{island} j \, dZ = \left| \frac{\partial \delta\Psi}{\partial z} \right|_{-3}^{+3} \sim \frac{\delta\Psi(0)}{\lambda} \; .$$

By simple scaling arguments, the size of the island being proportion-
al to $(\delta\Psi)^{1/2}$ one finds that $\delta\Psi$ increases like ηt^2. This theory works
well if $\delta\Psi$ is almost constant in the island as demonstrated by
numerical simulations (13). It does not work in the long wavelength
case (14) and the unstable mode has still a fast, almost linear,
growth. It has been subsequently shown (15) that profile modifications following the island growth may non-linearly stabilize the
short wavelength mode . In cylindrical geometry, similar results
have been observed for respectively the short wavelength modes
$(m \geq 2)$ and the long wavelength mode $(m = 1)$.

An interesting situation happens when one has two helicoidal
modes (i.e. $m = 2$, $m = 3$) simultaneously unstable (Figure 5). When
these modes grow, even if the corresponding islands are initially
centered in two different plasma raddi (where $\vec{B}_o . \vec{\nabla}\Phi_{2,3}$ vanishes)
they can finally overlap ; in that case one can show that there
is no more possible coexistence of two different helical symmetries
and there is no more a nested set of magnetic surfaces. The magnetic
field lines wander from one island to the other one and the radial
thermal electron transport is expected to be very much enhanced
(16) by stochastic motion along field lines. In all the available
cylindrical numerical codes the two dimensional character of the
magnetic and velocity fields does not treat correctly the field
line curvature. This effect seems to matter only for the linearized

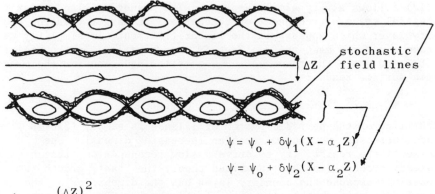

$$\psi = \psi_0 + \delta\psi_1(X - \alpha_1 Z)$$

$$\psi = \psi_0 + \delta\psi_2(X - \alpha_2 Z)$$

a) $\quad B_0 \dfrac{(\Delta Z)^2}{\lambda} \gg \delta\psi_1, \delta\psi_2$

b) $\quad B_0 \dfrac{(\Delta Z)^2}{\lambda} \lesssim \delta\psi_1, \delta\psi_2$

<u>Fig. 5</u>

a) 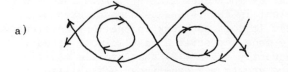 <u>initial</u> ψ

b) 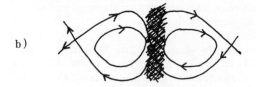 <u>after MHD instability</u>

c) 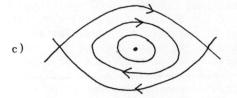 <u>after reconnection in</u> <u>shaded area</u>

<u>Fig. 6</u>

theory. It has been shown recently (17) that the non-linear development of the Tearing mode is <u>not</u> modified by cylindrical curvature
effets.

What are the results in toroidal geometry ? Apart from the
curvature effects already mentioned (which may well be unimportant
in a non-linear regime) the toroidal geometry brings a new effect
which we have not yet discussed. This effect is easy to understand
because it is purely geometrical. Let us expand the toroidal geometry as a function of (r/R). To the zero order, one recovers the
cylindrical approximation, but to the first order the corrections
will depend of r/R being a function of r and θ the cylindrical
coordinates (R ∿ R_o + r cosθ), it is now clear that a mode m will
be coupled to the modes m ± 1 by this geometrical effect.
follows that if the "island m" is unstable it will drive in toroidal geometry two islands m ± 1 with a size $(a/R)^{1/2}$. For Tokamak
experiments R/a ∿ 3,4 and this appears to be an important effect.
This has been verified in a recent numerical simulation (18).

An interesting possibility has been computed for a periodic
island configuration, which may also be relevant in some experiments.
In fact such a configuration, at least in plane geometry, is unstable with respect to the collapse of two adjacent pinches. This
is in the linear stage a pure MHD instability (19) without any
change of field topology. Nevertheless in the non-linear dissipative regime (20) it appears a new reconnection process which finally leads to one island (Figure 6). This process has not yet been
computed in toroidal geometry with a finite island number (m = 2,3)
and we do not know yet if it is implied in the disruption process.

The experimental observations in Tokomak's installations of
both saturation or explosive behavior of the reconnection process
have also suggested the preliminary following idea (21). The non-
linear solution for the algebraic growth ($\delta\Psi$ ∿ ηt^2) of a magnetic
island is not in fact very satisfactory. The inertia appears
globally negligible (the volume integrated kinetic energy decreases
in time) but locally divergent along the separatrix at the neutral
points. This divergence (and the possible related peaking of the
current density) may drive secundary instabilities. One interesting
candidate is the lower hybrid instability which will enhance the
resistivity (21a). This instability is directly related to density
and temperature gradients perpendicular to the magnetic fields which
could be created by this current peaking. An increase of resistivity as well as an increase of viscocity (21b) would accelerate the
island growth and could prevent its saturation to a finite size.

4. APPLICATION TO SOLAR FLARES

It seems to me very difficult (even if our theoretical know-

ledge has increased very much in the last years) to make any
definite prediction for a mechanism of Solar Flares from the
Tearing modes theory (whatever it would be a m = 1 or m = 2 modes).
The main reason, apart from the lack of a precise knowledge of the
magnetic field configuration, is in my opinion, related to the
sensitivity of the non-linear theory to the different plasma
parameters. Two examples will illustrate this opinion. The first
is related to the anomalous enhanced resistivity mentioned above.
In laboratory the plasma dimensions are "small" and the number
of ion Larmor radii in an equilibrium transverse scale length is
of order 10 in the largest installations. The lowest hybrid instability is important if the density or temperature gradients
are of the order of the ion Larmor radius which is possible for
an island. It seems to me that such a scaling is very unlikely in
a flare ! The second remark deals with the boundary conditions.
In a solar flux tube, the boundary conditions, along the magnetic
field, have to be taken in the Lithosphere. It is obvious from the
theory that if this plasma is hot, the short circuiting effect
along the field lines will supress the Tearing modes. If the plasma is cold and the resistivity sufficiently high the plasma may
be a priori stable or unstable but the theory remains to be done.

     In my opinion, up to now, there is no serious theoretical
understanding of the Solar Flares and I hope the SMM results will
improve our knowledge. If this is the case numerical codes in
progress may then become very useful.

REFERENCES

1.   Dungey, J.W. : 1958, Cosmic Electrodynamics, 48.
2a. Furth, H.P., Killeen, J., Rosenbluth, M.N. : 1963, Phys.
     Fluids, 6, 459.
2b. Coppi, B., Galvao, R., Pellat, R., Rosenbluth, M.N., Ruther-
     ford, P.H. : 1976, Fizika Plazmy, 961.
2c. Laval, G., Pellat, R., Vuillemin, M. : 1965, Conf. Proceed.
     Culham, 6-10 September, 1965, 259
3.   Parker, E.N., Sweet, P.A. : 1958, Electromagnetic Phenomena
     in Cosmical Phys., ed. by B. Lelmest, 123.
4.   Petschek, H.A. : 1964, NASA Spec. Publ. Sp. 50, 425.
5.   Sato, R. : 1979, J. Geophys. Res., 84, 7177.
6.   Glaser, A.H., Greene, J.M., Johnson, J.L. : 1975, Phys.
     Fluids., 18, 875 and references therein.
7.   Suydam, B.R. : 1958, Proceed. of the Genova Conference,
     31, 165.
8.   Shafranov, Y.D. : 1970, Zh. Tekh, Fiz., 40, 241.
9.   Edery, D., Laval, G., Pellat, R., Soulé, J.L. : 1976,
     The Physics of Fluids, Vol. 19, n° 2, 260.
10. Johnson, J.L., Greene, J.M. and Coppi, B. : 1963, The
     Physics of Fluids, Vol. 6, 1169.

11. Bussac, M.N., Edery, D., Pellat, R., Soulé, J.L. : 1977
     Plasma Phys. and Controlled Nuclear Fusion Research, Vol. 1,
     607.
12. Rutherford, P.H. : 1973, The Physics of Fluids, 16, 1903.
13. White, R.B., Monticello, D.A., Rosenbluth, M.N. : 1977,
     Phys. Rev. Lett., 39, 25, 1618.
14. Van Hoven, G., Cross, M.A. : 1973, Phys. Rev., A7, 1347.
15. White, R.B., Monticello, D.A., Rosenbluth, M.N., Waddell, B.V.
     1977, Phys. Fluids, 20, 800
16. Rechester, A.B., Rosenbluth, M.N., White, R.B. : 1980,
     Intrinsic Stochasticity in Plasmas, Ed. de Physique.
17. Pellat, R., to be published.
18. Edery, D., Pellat, R., Soulé, J.L., Frey, M., Somon, J.P. :
     Proceed. of the Bruxelles Conference on Plasmas Physics and
     Thermonuclear Fusion. IAEA CN 38/J.4.
19. Finn, J.M.., Kaw, P.K. : 1977, the Physics of Fluids, Vol. 20,
     72
20. Biskamp, D., Welter, H. : 1980, Phys. Rev. Lett., 44, 16, 1069.
21a. Sundasam, A.K., Sen, A. : 1980, Phys. Rev. Lett., 44, 5, 322.
21b. Kaw, P.K., Vales, E.J., Rutherford, P.H. : 1979, Phys. Rev.
     Lett., 43, 9, 1398.

NOTES

(1) One may find good introduction laboratory configurations
and stability theory in : Magnetohydrodynamic stability and thermo-
nuclear confinement (A. Jeffrey and T. Taniuti, Academic Press,
1966).

(2) When many helicoidal deformations with different pitch are
simultaneously present one cannot define the magnetic field by
a scalar function Ψ.

(3) Nevertheless the plasma remains unstable for example with
respect to non local long wavelength modes (m = 1) (8,9) ; m is
the azimutal wave number.

(4) There is still controverse about the relevance of the results
of numerical codes for the non-linear behavior of the long wavelength modes (the m = 1 mode in cylindrical geometry).

STELLAR ANALOGS OF SOLAR MICROWAVE PHENOMENA

David M. Gibson

Department of Physics, New Mexico Institute of Mining
and Technology, Soccoro, New Mexico, U.S.A.

I. INTRODUCTION

Because of the appearance of two complete (as of July 1979),
and readily available reviews dealing with solar-like microwave
emission from late-type stars, namely :

"Nonthermal Microwave Phenomena in Other Stars," by R.M.
Hjellming and D.M. Gibson (1980), and

"Binarity as a Factor in Stellar Radio Emission," by D.M.
Gibson (1980),

I think it unnecessary to present another similar review in this
NATO ASI Proceeding. Instead, let me first give a brief summary
of this important aspect of stellar activity in which I coalesce
the important findings of the above papers and references therein.
I will then follow with several addenda highlighting important new
observations which have been reported in the interim.

2. SUMMARY

The systematic study of microwave emission from late-type
stars has been in progress for ten years now. In a few cases it has
been possible to obtain observations of the temporal variability
of the radio flux, spectra, and both linear and circular polariz-
ations with sufficient signal-to-noise to determine the microphysical properties of the emission regions. In general, the emission
has been found to be flare-like, cyclotron or gyrosynchrotron in
character, with thermal gas acting as the principle opacity source.

545

R. M. Bonnet and A. K. Dupree (eds.), Solar Phenomena in Stars and Stellar Systems, 545–549.
Copyright © 1981 by D. Reidel Publishing Company.

Typical source conditions include mildly relativistic electrons
with $E \sim 5$ MeV, magnetic fields $B \sim 30$ Gauss, and ambient thermal
gas densities $n_e \sim 10^9$ cm^{-3}. The above parameters are also typical
of conditions found in solar microwave flares. Coupled with the
fact that microwave flaring has only been detected from stars with
convective envelopes, the most natural explanation for these
phenomena is that we are observing stellar analogs of solar microwave behavior. This conclusion is also consistent with the interpretation of phenomena observed at other bands.

However, the fact that we are able to observe microwave phenomena at all from objects with microphysical properties similar to
solar ones immediately suggests the typical size of a stellar
microwave emission region is much larger than its solar counterpart. This conclusion is supported by VLBI data. In addition,
stellar flares generally occur much more frequently than solar
flares and flare durations have been observed to be much longer.
Thus, while there is probably a close similarity in the microphysical processes associated with solar and stellar microwave
phenomena, the level of that activity —as measured my macrophysic-
al quantities such as source size, flare occurrence and duration,
microwave luminosities, and total energies— can be many orders of
magnitude greater in the stellar case.

Radio surveys of a wide variety of late-type stars have been,
and can be very helpful in determining the important factors in this
increased activity. The most active sources by far are classified
under the general heading of RS CVn binaries. Related binaries
appear to be somewhat less active. Very large flares have been
detected on a few single M-supergiants, but they are relatively
infrequent. Several M-dwarf flare stars have been recently detected
and they appear to exhibit activity intermediate between that of
the RS CVn's and that of the sun. By combining data —which are
summarized for the three stellar classes and the sun in Table I
hereafter— with basic stellar parameters, three factors appear
to have the largest effects on stellar microwave activity :
       Rotation : given that all other parameters are equal, stars
with short rotation periods, e.g. members of close binaries systems,
appear to be more active than those with longer rotation periods.
This factor probably affects both flare energy and flare frequency ;
       The gradient of the gravitational potential near the active
       stellar surface : stars with weak gradients, e.g. stars which
(nearly) fill their Roche lobes in close binary systems, or supergiants, appear to exhibit flares which are both stronger and of
longer duration than those in objects with strong gravitational
gradients, e.g. solar-type dwarfs. It seems likely that larger
emission regions could be formed in the weak gradient stars having
the combined effects of increasing the flare brightness and the
flare duration.

Evolutionary phase : stars in "short-lived" portions of the HR-diagram, e.g. subgiants, supergiants, and probably M-dwarf flare stars, appear to be more active than those in "long-lived" portions of the diagram. Possibly these rapidly evolving objects exhibit radially-dependent differential rotation which could also supply energy to power the flares.

Table I

A COMPARISON OF SOLAR AND STELLAR MICROWAVE BURST CHARACTERISTICS

TYPE	SUN	UV CETI's	RS CVn's	M I's
$L_{max}(\text{erg s}^{-1}\text{Hz}^{-1})$	3×10^{12}	2×10^{15}	2×10^{18}	6×10^{18}
DURATION	few secs - few hrs	few 10's secs - several hrs	40 mins - 8 days	<1 day
OCCURENCE	~3/month	~4/day	~4/day	~3/month
SOURCE LOCATION	top of loops	?	top of loops?	?
SOURCE DIAMETER (cm)	$<2 \times 10^{10}$	?	$\lesssim 10^{12}$	?
SOURCE MOVEMENT	some expansion	?	some expansion	?
BRIGHTNESS TEMPERATURE	$\lesssim 10^9 K$	?	$\sim 10^{10} K$	?
SPECTRAL PEAK	~10 GHz	~200 MHz?	~5 GHz	~10 GHz?
FREQUENCY DRIFT	slight to lower ν's	?	slight to lower ν's	?
CIRCULAR POLARIZATION	<50%	$\lesssim 100\%$	$\lesssim 70\%$	?
MODE	e	?	e and o?	?
LINEAR POLARIZATION	~0%	~0%	~0%	?
HIGHLY POLARIZED OSCILLATIONS	yes	?	yes	?
EMISSION MECHANISM	gyro-synchrotron	gyro-synchrotron	gyro-synchrotron	non-thermal
ASSOCIATED PHENOMENA	flares, X-ray bursts	flares	flares, X-ray bursts	?

3. ADDENDA

Since the aforementioned reviews were completed in July 1979
two new important observations have been reported and are reviewed
here for the sake of completeness.

Hurford et al. (1980) have made the first 2-dimensional
high-resolution observations of solar flares using the VLA. They
found that the initial phase of the flare began at a location
consistent with the top of a loop which joined two Hα footpoints.
Subsequently, the flare region appeared to exapand down the loop.
Since the size scale of loops is relatively small in terms of the
volume which relativistic electrons could fill in a short time,
the compact nature of the emission region must be a result of the
source conditions. Two possibilities suggest themselves :

3.1. Mirroring of the relativistic electrons takes place at the
edges of the emission region preventing them from reaching down
the loop. However, this conclusion must be weighed with evidence
which suggests the hard x-ray flare occurs near the footpoints and
is due to the collisional interaction of fast particles with the
chromosphere.

3.2. The breakdown of the magnetic field and the subsequent acceleration of and emission by fast electrons occurs where the B-field is
highest. The component of the magnetic field of interest here is
likely to be the toroidal component which may be significantly
larger than the poloidal field component in the loop due to the
twisting of field lines and the great expansion of the loop at
the top relative to its diameter at the bottom. The question as to
whether this is a viable possibility must await theoretical analyses
on the reconnection of toroidal field and the emission quasi-
relativistic electrons in regions with significant toroidal fields.

The implication of this observation for the geometry and
polarization of stellar sources is likely to be very important.

The second significant observation is that by Fisher and
Gibson (1980) of microwave emission from YZ CMi. This is the first
microwave interferometric detection of an M-dwarf flare star and it
is significantly below flux levels and of much longer duration
than flares reported by single dish observers (cf. Moffett et al.,
1978 ; Karpen et al. 1978). The flux density and timescale observed
suggest the emission is intermediate between solar flares and those
of the RS CVn binaries. The emission from YZ CMi had the very
interesting property of being 100% circularly polarized. This
could pose some difficulty in finding a suitable geometry for the
emission region unless radiative transfer effects are involved.

References

Fisher, P.L. and Gibson, D.M., 1980, Proc. Southwest Reg. Conf.
  Astron. Astrophys., vol. VI, in press.
Gibson, D.M., 1980, Close Binary Stars : Observations and Inter-
  pretations, (IAU Symp. 88), M.J. Plavec, D.M. Popper, and
  R.K. Ulrich (Eds.), (Reidel, Dordrecht), p. 31.
Hjellming, R.M., and Gibson, D.M., 1980, Radio Physics of the Sun
  (IAU Symp. 86), M.R. Kundu and T.E. Gegerly (Eds.), (Reidel,
  Dordrecht), p. 209.
Hurford, G.J., Marsh, K.A., and Zirin, H., 1980, Bull. A.A.S.,
  12, 478.
Karpen, J.T., Crannell, C.J., Hobbs, R.W., Maran, S.P., Moffett, T.J.
Bordas, D., Clark, G.W., Hearn, D.R., Li, F.K., Markert, T.H.,
McClintock, J.E., Primini, F.A., Richardson, J.A., Cristaldi, S.,
Rodono, M., Galasso, D.A., Magun, A., Nelson, G.J., Slee, O.B.,
Chugainov, P.F., Efimov, Y.S., Shakhovskoy, N.M., Viner, M.R.,
Venugopal, V.R., Spangler, S.R., Kundu, M.R., and Evans, D.S.,
  1977, Astrophys. J. 216, 779.
Moffett, T.J., Helmken, H.F., and Spangler, S.R., 1980, Publ.
  Astron. Soc. Pacific, 90, 93.

SUMMARY COMMENTARY

E.N. PARKER

Laboratory for Astrophysics and Space Research
University of Chicago

1. INTRODUCTION

A summary lecture is difficult to construct. We have all heard
the same review lectures and contributed papers. We may have
developed different impressions and convictions from those lectures,
but it is too soon after the first impressions to merit detailed
comparison. The Proceedings of this Advanced Study Institute
will be available by the time a review would be profitable. It
seems to me that the best course is to accept the responsibility
of a summary lecture in its simplest form, giving my own personal
views, biases, and bigotries of the science and how we might react
to it.

It is evident from the review lectures that we have accumulated
a considerable store of knowledge and understanding of the activity
of the sun and other stars. It is also evident that this accumulated
knowledge and understanding make up only a poor fragment of the
whole story of the active stars. There is a long road ahead of us,
but the unusual scenery along the way, and the attraction of an
eventual understanding of active stars, make the trip exceedingly
rewarding. I shall be curious to see how far over the next twenty
years we are able to advance our understanding of stellar activity.

We are participating in a sudden forward surge of an important part
of astronomy and physics, which we might call the physics of stellar
activity. The subject had a number of early antecedents. For
instance, the winking of Algol, the Demon Star, was known in ancient
times, with no way of understanding its real significance. A more
substantial line of inquiry began with the discovery and study of
sunspots, becoming a science following application of the telescope

551

R. M. Bonnet and A. K. Dupree (eds.), Solar Phenomena in Stars and Stellar Systems, 551–562.
Copyright © 1981 by D. Reidel Publishing Company.

to astronomical observation in 1610. The discovery of the H_α
flare, and the connection with auroral and geomagnetic
storms, were the first recognition of the existence of real activity
in the sun. The discovery by Edlen of the high temperature of the
solar corona was the next important step. The chromosphere, known
from naked eye observation of the total eclipse of the sun, is
another facet of solar activity, but it could not be interpreted
properly until the establishment of non-LTE. An awareness of the
activity in other stars became possible with the development of
large radio antennas and high gain-low noise receivers, making
possible the detection of the giant flares among the UV Ceti stars.
The systematic study of chromospheric lines of a number of young
stars by O.C. Wilson provided the first conclusive evidence for
cycles of activity in other stars, and showed that the level of
activity declines with increasing age. The development of comprehensive, high resolution optical observations of solar activity
have provided a steadily improving -and awesome- picture of the
vigor and complexity of the magnetic activity in the sun. The
observational studies of the sun provide the fundamental realization that most of stellar activity is really magnetic activity.
The chaotic superheated gases producing the observable effects are
created by the dynamical dissipation of magnetic fields.

These diverse lines of inquiry have been drawn together in the
last decade by the newly acquired ability to observe UV, EUV,
X-rays, plasmas, and fast particles (galactic and solar cosmic
rays) from spacecraft. The space observations show the supersonic
mass loss from the sun (the solar wind), the remarkable X-ray
features associated with active regions on the sun, and by contrast,
the coronal holes from which issue the fast streams of the solar
wind. Equally startling is the direct detection of EUV and X-rays
from the active atmospheres of so many other stars. Awesome as
the activity of the sun may be, the UV and X-ray observations show
that most other classes of stars are active too, with active regions,
extended coronas, and mass loss. Many of the other stars are
vastly more active than the sun, particularly in their youth.

So the inquiry into the physics of stellar activity has grown
from an extended infancy into a broad and coordinated investigation of the general phenomenon of suprathermal effects. The
basic problem is to understand the physics behind the fact that
the tranquil self-gravitating, self-luminous ball of gas, as
conceived by Emden, Eddington, Russell, Hertzsprung, et al, is
so tumultuous.

The situation challenges us to enlarge our horizons, toughen our
standards, and increase our efforts. To emphasize the scope of the
newly established field of stellar activity, remember that activity
begins with the physics of stellar interiors, which involves
fundamental questions in elementary particle physics, nuclear

physics, atomic physics, and the physics of electromagnetic
radiation, all mixed up together. The rotation rate, and a possible
slow circulation within stellar cores, is also an essential aspect
of the properties of the interior. Our knowledge of the stellar
interior is based entirely on theoretical inference from the
physics learned in the terrestrial laboratory. I would remind you
that the neutrino experiment conducted by Ray Davis in the Homestake
Mine is not proving out our expectations. So we all have a direct
concern with stellar interiors, and neutrino physics, etc...

Outside the radiative core of the star is the convecting envelope,
which brings in the physics of hydrodynamics and magnetohydrodynamics
in important ways. The best we have been able to manage is the mixing length representation of the convection, putting the effective
mixing length equal to some fixed number times the pressure scale
height. We really have no clear idea of the large-scale circulation
in the convective zone (and the degree to which the circulation
might extend into the radiative interior). All we know with certainty
about the circulation is that the sun, and presumably other stars,
possess a surface across which the angular velocity varies by some
50 percent. Studies in solar seismology promise to extend that
knowledge in coming years.

At the top of the convective envelope is the photosphere, followed
by the chromosphere, the transition region, and then the corona
which expands into the stellar wind. In some stars the stellar
wind is so massive as to accumulate a circumstellar cloud. The
magnetic fields of the active regions extend from deep in the
convective zone out through the photosphere into the corona and
stellar wind. The active regions produce swarms of fast particles
that escape outward from the star. The supersonic stellar wind
interacts with the magnetospheres of whatever planets there may be,
or with companion stars, producing further swarms of fast particles,
at the same time pushing back the galactic cosmic rays from a vast
region of circumstellar space. This we have learned from extensive
exploration and mapping of the high energy particles in the solar
system.

It is doubtful that any of us can maintain expertise over so fast
a domain of physics. However, the close connection between the
different facets of stellar activity demands that we be generally
informed on the state of knowledge and understanding in each. That
is the purpose of the Advanced Study Institute. It is apparent that
the participants have eagerly embraced the opportunity. We must
continue to maintain close communication in the future, if we are
to progress toward a goal of general understanding of stellar
activity. There will have to be meetings in the future for extended
periods of intensive review and discussion.

To look in another direction, it is obvious to all that the sun
-and in considerable degree the terrestrial magnetosphere- provides
the laboratory in which the direct quantitative study of activity
is possible. That laboratory is an essential part of the pursuit
of activity. We cannot understand distant active stars through
theoretical deductions and serendipity alone. For instance, who
could have anticipated such a thing as solar activity, or magnetospheric activity ? And what remarkably puzzling forms it takes !
Contemplate for a moment the complicated small-scale structure of
the umbral dots and light bridges, or the penumbral filaments of
a sunspot. We still stand in awe of the formation of sunspots.

Now reflect that some stars, too distant to be resolved, appear
to possess dark blotches on their sides that may be huge sunspots,
perhaps 3×10^5 km in diameter. Only the sun can provide a clue
as to what those "star blotches" might look like when viewed at
close range. Only the sun can suggest how the activity in the
magnetic fields around those huge star blotches might be understood. And only the other stars can suggest how remarkable may be
the relatives of the lowly spots on the sun.

It is my own personal view that our primary goal is new physics.
The sun and other stars are a fascinating means to that fundamental
end. They are the means to develop a new branch of physics -a new
chapter in the comprehensive elementary physics text. This new
branch involves such diverse effects as the hydrodynamics of the
rotating convective shell, the generation of magnetic fields by
fluid motions, magnetic buoyancy, magnetic-non equilibrium and
dissipation, the production of suprathermal gases and fast particles
the radiative transfer in an exploding flare, and the eruption of
coronal arches, to name but a few. For the most part these effects
are not, and cannot, be realized in the laboratory. They were not
and could not be anticipated by theoreticians. They were simply
thrust into our faces by observations of our local astronomical
neighborhood, establishing the boundless variation of nature,
far beyond what we can reliably produce by serendipity in our
own active imaginations. Humanity will know more physics than it
does now when we have progressed further with our work. And we will
know a lot more about the individual active objects with which we
share space and time.

This brings me to a point that I feel is not always fully understood. Progress into the field of stellar activity can be led
only by observation. As I have already indicated, theoretical
considerations of the chaotic magneto-hydrodynamical effects of
stellar activity can be deductive only in the most restrictive
cases. Theory did not and could not anticipate the solar flare, or
describe the variations of flares that arise in the sun and in
other more active stars (I was enormously impressed when one of
our reviewers pointed out that the active X-ray emission from some

O stars is equivalent to the entire surface of the star being
covered with solar flares). What theoreticians could have anticipated the fluctuations in solar luminosity reported from the
SMM detectors, etc., etc. The theoreticians can work out a firm
explanation for a phenomenon only when it is adequately defined
by observations. The more complex the phenomenon, the more completely
it must be described before the theoretician can spell out a
unique "deductive" explanation.

There are, of course, many theoretical problems that can be
reduced to unique solutions. The reduction of the six equations
for the steady flow of a highly conducting fluid in a magnetic
field (with one ignorable coordinate) to a single quasi-linear
second order elliptic equation permits a complex statement of the
boundary conditions necessary for uniqueness of the solution. But
when we deal with time dependent three dimension fluids and fields,
in vigorous motion, there is no way to solve the equations directly,
even with the fastest computers that we can reasonably imagine.
Nor do we have available any general mathematical methods for
abstracting the average behavior of turbulent fields. Yet it is
just the violent turbulent agitation of fluids and fields that makes
up the visible stages of stellar activity. It is evident, therefore,
that there are limits (at least at present) to our development of
the physics of stellar activity. Theory can, in most cases,
progress only as far as observation can describe the way.

Pellat gave a review lecture on some special aspects of plasma
dynamics, as it applies to magnetically confined columns of plasma
in the laboratory. He showed how much work goes into making even
modest progress in the non linear instabilities that develop. The
work is a model of what we must do in the future to understand the
well defined problems of stellar activity. It is evident that our
quarry will have to be stalked with guile and ingenuity by both
the observer and the theoretician. There are new instruments that
will have to be invented to get at crucial observations. There are
new dynamical effects that will have to be invented to explain
existing, as well as future, observations. We can expect that
sooner or later we will pass from the present era of rapid
discovery to an epoch of hard analytical observation and theory,
to get at the basic physics that now is so vague. Solar physics
has been in the analytical stage for some time. A state of gradually diminishing returns will arise, broken from time to time by
a new discovery or a bold observational or theoretical stroke.
The siren song of new exotic objects, discovered by the Space
Telescope perhaps, will lure some of us into other inquiries,
such as the related subject of active galaxies.

II. CONSPICUOUS PROBLEMS

This is perhaps an appropriate point to write down a partial list

of obvious problems. This list is not complete, but its length
alone is impressive and serves to illustrate the breadth and
diversity of the physics of stellar activity mentioned earlier.
In particular the list shows how broadly we should be interested
in the various scientific activities carried out around the world.

First on the list again is the solar interior with the fundamental
question of energy generation, atomic abundances, angular velocity,
and neutrino production. In view of the apparent discrepancy
between the measured and expected neutrino fluxes, and the recent
preliminary laboratory suggestions of a finite neutrino rest mass
(of the order of 10-100eV) we are directly concerned with elementary particle physics. The laboratory determination of the neutrino
rest mass, or a small upper bound if no mass is detected, is a
question of vital concern, which, hopefully, through the efforts
of a number of physicists, will progress substantially in the next
couple of years. The new gallium experiment, to detect the neutrinos
from the basic proton reaction, is getting underway. It should
give significant results in five years. We all will be watching it
closely.

Questions of global mixing throughout the interior of the sun,
and particularly of the rotation of the core, may prove to have
direct bearing on the activity of the sun. The Star Probe Mission
(perhaps 10 years from now) proposes to measure the quadrupole
moment J_2 with considerable precision to determine the internal
state of rotation. It is of direct concern to all of us.

Turning to theoretical problems, the hydrodynamics of the convective envelope is a fundamental and very difficult theoretical
problem, from the turbulent transfer of heat to the global circulation, all presently little known. The high Reynolds number and
the intense density stratification are the main obstacles to
theoretical progress. I am hopeful that the new field of solar
seismology can help us along substantially.

Then there is the magnetohydrodynamic problem of the convective
zone in which we recognize the dynamical role of the magnetic field
created by the hydrodynamic convection and circulation. A variety of
dynamo models have been considered, and more are being worked on,
but they are plagued by the problem of magnetic buoyancy and none
have fully addressed the dynamics. Most models specify the fluid
motions and then consider what fields they generate.

There is the problem of relating the magnetic fields in the dynamo,
in the lower convective zone, with the fields that are observed to
burst through the surface of the sun. Are the fields that reach
the surface a direct indication, and a major portion, of the
fields generated at depth ? Or is nature hiding important parts
of the field from us, or deceiving us with peculiar time delays

between generation and appearance at the surface.

At the surface of the sun we are confronted with the real flowering
of magnetic activity, beginning with the concentration of field to
1500 gauss, the clustering of small tubes to form sunspots, and the
extraordinary dissipation of the fields above the visible surface
to form the plage, the flare, the active and quiet coronal regions,
and the coronal holes and fast streams in the wind. Yet we are
still not sure as to how it is all done, and we cannot write down
any unique form for the heat function. There has been real progress
in the theory of efficient particle acceleration, in solar flares
and active planetary magnetospheres, but the precise location and
nature of the acceleration in any particular flare is still a matter
of guess work. The 10 hour relativistic electron beacon at Jupiter,
synchronized with the rotation of the Jovian magnetosphere in the
solar wind, is a fascinating problemn that has as yet only the
vaguest solution. It would be interesting at some future meeting of
us "stellar activists" to have a review of the fast particle and
cosmic ray phenomena observed in the solar system. It would serve
to acquaint us with that facet of solar activity and it would
provide us with some idea of what may be unseen in the space around
distant active stars. Indeed the fast particle populations around
some of the very active stars, may, by extrapolation from solar
particle emission, reach intellectually fascinating intensities.

We should not forget the theoretical dynamical problems posed by
coronal transients and coronal loops, and by the general
theoretical problems of stellar winds driven by steady or
transient, thermal, wave and radiation effects. Nor should we
underestimate either the importance or the difficulty of the
radiative transfer problems involved in the nonsteady chromosphere,
the spicules, the coronal loops, and the coronal transients. The
satisfactory solution of any of these problems involves a combination of dynamics and radiative transfer, and a reliable magnetic
dissipation or heating function. The Solar Maximum Mission is doing
a beautiful job in showing examples of the transient dynamical
aspects of solar activity.

In this connection, gamma rays will begin making their contribution to the quantitative study of solar activity before long, with
the advent of the Gamma Ray Observatory. It is a space mission of
vital concern to us all.

The long and short term variations of the bolometric luminosity of
the sun and other stars are a fundamental field of inquiry that has
only begun to be explored. I understand that there are some precision studies of the relative magnitudes of a number of stars getting
underway. We are looking for variations perhaps as small as one
part in a thousand or two. The variations may occur over a week, a
year, or a decade or more, probably over all time periods. The

studies will require ingeniuty and patience and high scientific
standards. Everything will have to be checked and double checked.
The variations of the solar luminosity of $\pm$ 0.02 percent over a
day and $\pm$ 0.05 percent over a month, reported from the Solar
Maximum Mission, provide a fascinating insight into the problem of
the variation of instruments and stellar luminosities. Similar
variations should be looked for in other stars, as a check on the
results of the Solar Maximum Mission, and as a source of new
information in its own right.

A couple of days ago several of us amused ourselves by standing in
front of the black board and sketching how the magnetic field
configuration might look inside a dwarf star sporting a magnetic
star spot with a diameter equal to the radius of the star. The
geometrical proportions of the field cause it to penetrate to the
core of the star. It is an interesting question as to where the
field goes after reaching the core. It is also an interesting
question whether the blotch on the surface is a single huge mono-
polar magnetic region or a compact herd of bipolar spots. May I
suggest that some of you let your imaginations direct your pens
in making similar sketches, to see what you can turn up. It is
apparent that observation must somehow settle the basic questions.
The theory of sunspots must in some way be capable of extension,
or diversification, to cover the enormous single star spot or a
large number of bipolar spots clustered closely together over one
face of the star.

This brings us to the problem of describing and understanding the
remarkable behavior of the emerging flux tubes in active regions
on the sun. The ground based studies of the evolution of flux
tubes in active regions are making giant studies. The recent work
of Harvey and Zwaan spells out the fascinating, and so far in-
explicable, combinations of concentrated fields and downdrafts
that make up the developing active region. Eventually, the studies
will have to be carried to resolutions well below 10^2 km to be
sure that we are seeing all of the action. On theoretical grounds
alone (based on minimum electrical conductivities of 10^{11} esu in
the photosphere and characteristic granule periods of 500 sec)
there may be dynamical scales as small as 15 km in the magnetic
field. We must have a comprehensive understanding of the behavior
of convection and magnetic fields in the sun before we can go very
far in discussing the causes of the magnetic activity in, say, the
RS CVn stars. Hence the high resolution ground based observations,
the Solar Maximum Mission, and then the Solar Optical Telescope
(designed to resolve down to 100 km) and eventually the Star Probe
(designed to resolve down to 10 km if the thermal problems with
the optical telescope can be overcome) are all fundamental steps
along the way to the physics of stellar activity. The Solar Polar
Mission, to map out the plasma, fields, and fast particles over the
poles of the sun, is another fundamental step in which we all have

a direct concern. It is our one opportunity to compare and contrast the polar and equatorial activity in the space around a star. Finally I would remind you again of the importance of the Gamma Ray Observatory, the gallium neutrino experiment, the determination of J_2 with Star Probe, the Space Telescope, as well as the UV Explorer. We shall be busy indeed following the results of these observations in the years ahead. I am curious to see how far image reconstruction of fields of low contrast can be pushed for ground based observations of the sun. Can ground based telescopes reach 0.1" (80 km) resolution ?

As a final problem on the list, let me commend to your attention the long term variability of the activity of the sun, presumably from a tumultuous youth to a more sedentary middle age in which centuries of activity are interspersed with an occasional century of inactivity and an occasional century of enhanced activity (perhaps a dim recollection of early aeons). The present long term variability is well established from the Maunder Minimum and the presently available 7000 years of C^{14} production data. We are without an explanation for this remarkable behavior of the sun. We may infer that other stars exhibit similar behavior, equally unexplainable, perhaps on a grander scale.

There is a growing body of data that suggests a connection between the mean terrestrial climate and the long term level of solar activity. A temperature correlation appears in connection with the century long variations, with the mean annual temperature in the northern temperature zone running a degree lower during the periods of solar inactivity. The Little Ice Age of the 15th-17th centuries is the most recent cold period that occurred in association with low solar activity. The economic and political consequences of the Little Ice Age are clearly recorded in both the East and the West. More recently there is evidence that drought in the arid western half the United States is influenced by the 22 year magnetic cycle of the sun. The physical mechanisms are not known for either the temperature or the dought effects, but the data are sufficiently extensive that a priori rejection of a connection represents a point of view that is hard to defend.

III. A QUESTION OF STYLE

In this final section it may be interesting to have a critical look at our methods —our "style". It is often pointed out that the observations represent the reality toward which the theorician is striving. It should also be pointed out that the activity observed with high resolution in the sun represents the reality that is not apparent when we observe the distant unresolved objects. We should all, theoreticians and observers, pause occasionally to contemplate the realities.

Now we are aware of the enormous diversity of the physics of stellar activity. There is another kind of diversity that is important in formulating our methods and standards of procedure. That is the diversity in the degree of definition and solution of various problems. For instance, the theory of stellar interiors is precisely formulated for most stars because it is time independent and a function of a single variable. We think that we know all of the relevant physics (although a new contribution to the opacity occasional rears its ugly head).

The major uncertainties are the abundances of helium and the metals, and perhaps the rotation and general slow circulation in the core. The uncertainties introduced by application of the mixing length theory to the convective envelope seem not to be serious. The whole model can be checked, in principle, by the neutrino experiments. The picture is clearly defined, in spite of what surprises may be in store for us in the future. In contrast to the stellar interior there are a number of ill-defined problems, such as the detailed state of the convection and circulation in the lower convective zone, which is essential for deducing the form of the magnetic field in the convective zone. The escape of the magnetic field from the dynamo region through the surface of the star, to produce sunspots, flares, and plages is a poorly defined problem, with only fragmentary understanding.

My point is that we have to deal with all of these problems –the quantitative and the qualitative, the precise and the vague. We have to distinguish hard facts from intuitive (and often personal) inferences. We have to distinguish unambiguous theoretical conclusions from "hand waving" ideas. The problem is made difficult by the deeply ingrained psychological tendency to believe what we hear repeated. "There, I have told you thrice. What I tell you three times is true". Our vocabulary is not well suited to making the necessary distinctions.

I suggest that it might be useful to institute the "conjecture" in the sense that the mathematician uses the term. The conjecture is a well considered, extensively tested idea which, however, has not been established by direct "proof" (of one sort or another). The word strongly conveys the preliminary nature and cries out for a more definite foundation. Mathematical conjectures have been extremely useful in the development of mathematics. Even those that are eventually proved wrong in some way are invaluable in getting the ideas pulled together in coherent form to stimulate new work. I have discovered that the word sometimes arouses strong negative passions, because of the human tendency to take one's own ideas more seriously than the facts sometimes warrant. But I think the conjecture has a useful role to play in our work and I strongly urge its general application. To make a first move, I state that the ideas that I presented in my earlier

review lecture, on the fundamental role of a converging flow and
down draft in the separated flux tubes beneath a sunspot, are a
conjecture. It may prove to be wholly or partly correct or incorrect
in the end. But I think it is a useful conjecture. I recall that
Bullard's conjecture, of nearly thirty years ago, that there is
a "super Cowling" antidynamo theorem, was extremely valuable,
leading eventually to formulation of the $\alpha\omega$ dynamo, even though
Herzenberg and others later provided counter examples that proved
the conjecture wrong. The important fact was that the conjecture
stated a problem in clear and concise terms and suggested a fruitful new line of inquiry. So it contributed to progress.

Our own work on stellar activity involves a variety of conjectures
today. I suggest that it would be useful to organize our thoughts
more closely on the question of fact, conjecture, and speculation.

Turning to another point, I have noticed over the years that some
workers in astronomy are inclined to draw a straight line through
a swarm of data points, stating that data confirm their theory to
within the usual sloppy standards of astronomy. I can only presume
that they speak for their own standards, because they do not speak
for mine, or for most others active in the field. I see astronomy
as one of the more fascinating branches of physics. I presume that
one puts together the best data that are available. Then either the
data substantiates the theoretical result, or the data is inadequate
to substantiate or refute the theory, or the data refutes the
theory. We can use only the usual standards of physics to judge the
validity of the result. None others have any meaning that is
apparent to me. We astronomers should strongly reject the insinuat-
ion that we use other standards.

Finally, it is my impression that our ultimate goal in astronomy
in general, and the physics of stellar activity in particular, is
to discover and understand the general physical principles. We
discover many effects that have broad application and we discover
effects often times that have little or no application in the final
analysis, but the physics is interesting anyway. The effects
tucked away in our general store of knowledge —our comprehensive
elementary physics text. Now politicians scorn such pure science,
of course. If you cannot eat it or sell it, what good is it, they
say. They can be excused as being outsiders who do not wish to see
the inside of pure science and who do not wish to see the long
road by which their sacred applications have arisen from the earlier,
and quickly ignored, knowledge of pure science. But what excuses
can I make for an astronomer who rejects a newly discovered
theoretical effect with the statement that the effect does solve
his problem with the solar flare, or the time dependent transition
region over an active region, etc. We who work in pure science
must not forget that we are interested in discovery, both observ-

ational and theoretical, irrespective of immediate applications
to any specific problem, commercial or astronomical.

Well, I think you have heard enough of this summary. Let us look
forward to our research over the next few years, and let us look
forward to the time when we get together again to share our progress
in the physics of stellar activity. This has been about the most
rewarding meeting I have ever had the pleasure of attending.

INDEX of NAMES

INDEX of SUBJECTS